CAMBRIDGE LIBRARY COLLECTION

Books of enduring scholarly value

Technology

The focus of this series is engineering, broadly construed. It covers technological innovation from a range of periods and cultures, but centres on the technological achievements of the industrial era in the West, particularly in the nineteenth century, as understood by their contemporaries. Infrastructure is one major focus, covering the building of railways and canals, bridges and tunnels, land drainage, the laying of submarine cables, and the construction of docks and lighthouses. Other key topics include developments in industrial and manufacturing fields such as mining technology, the production of iron and steel, the use of steam power, and chemical processes such as photography and textile dyes.

Electromagnetic Theory

Oliver Heaviside (1850–1925) was a scientific maverick and a gifted self-taught electrical engineer, physicist and mathematician. He patented the co-axial cable, pioneered the use of complex numbers for circuit analysis, and reworked Maxwell's field equations into the more concise format we use today. In 1891 the Royal Society made him a Fellow for his mathematical descriptions of electromagnetic phenomena. Along with Arthur Kennelly, he also predicted the existence of the ionosphere. Often dismissed by his contemporaries, his work achieved wider recognition when he received the inaugural Faraday Medal in 1922. Published in 1899, the second of three volumes of Heaviside's collected work argues that physical problems (such as the age of the Earth) drive mathematical ideas, and then goes on to compare the propagation of electromagnetic waves with physical analogues.

Electromagnetic Theory

VOLUME 2

OLIVER HEAVISIDE

CAMBRIDGE
UNIVERSITY PRESS

CAMBRIDGE UNIVERSITY PRESS

Cambridge, New York, Melbourne, Madrid, Cape Town,
Singapore, São Paolo, Delhi, Tokyo, Mexico City

Published in the United States of America by Cambridge University Press, New York

www.cambridge.org
Information on this title: www.cambridge.org/9781108032162

This edition first published 1899
This digitally printed version 2011

ISBN 978-1-108-03216-2 Paperback

ELECTROMAGNETIC THEORY.

BY

OLIVER HEAVISIDE.

VOLUME II.

LONDON:
"THE ELECTRICIAN" PRINTING AND PUBLISHING COMPANY'
LIMITED,
SALISBURY COURT, FLEET STREET, E.C.

Printed and Published by
"THE ELECTRICIAN" PRINTING AND PUBLISHING CO., LIMITED,
1, 2 and 3, Salisbury Court, Fleet Street,
London, E.C.

PREFACE TO VOL. II.

From one point of view the present volume consists essentially of a detailed development of the mathematical theory of the propagation of plane electromagnetic waves in conducting dielectrics, according to Maxwell's theory, somewhat extended. From another point of view, it is the development of the theory of the propagation of waves along wires. The connection of the two subjects was thoroughly explained in Chapter IV. of Volume I., which should be understood. But on account of the important applications, ranging from Atlantic telegraphy, through ordinary telegraphy and telephony, to Hertzian waves along wires, I have usually preferred to express results in terms of the concrete voltage and current, rather than the specific electric and magnetic forces belonging to a single tube of flux of energy. The translation from one form to the other is quite easy, when understood. As far as space would permit, I have tried to develop the theory as thoroughly as possible, considering every kind of wave, and including the calculation of the waves produced by multiple reflections. Even the theory of the latest kind of so-called wireless telegraphy (Lodge, Marconi, etc.) has been somewhat anticipated, since the waves sent up the vertical wire are hemispherical, with their equatorial bases on the ground or sea, which they run along in expanding. (*See* §60, Vol. I.; *also* §393 in this volume.) The investigations are based upon those in my "Electrical Papers," with considerable extensions. My old predictions relating to skin conduction, and to the possibilities of long-distance telephony have been abundantly verified in advancing practice; and my old predictions relating to the behaviour of approximately distortionless circuits have also received fair support in the quantitative observation of Hertzian waves along wires. The reader need not therefore fear that he may be muddling himself over fantastic theories void of practical significance, whatever the scienticulist may say.

The mathematical methods employed are those which have proved themselves to me by practice to be those best suited to obtaining new results and advancing natural knowledge. The general idea is to make the differential equations themselves perfectly definite, so that the differential equation of a problem is actually its full solution, the operational or differential solution, though it may not be in obvious quantitative form. The process of algebrisation, or conversion from differential to algebraical form admitting of numerical treatment is, of course, very important. Though it may be easy when the proper way of treatment has been found, yet there has been a good deal of exploring work which makes no appearance.

In Chapter VIII. I have given a condensed account of my researches on generalised differentiation and series, a subject that grows naturally out of the operational way of working. Although I think this subject has a large future, yet I must warn the reader that there is no pretence of logical rigour, and that much of the matter was rejected some years ago by persons who ought to be good judges.

The several appendices relate to electromagnetic waves in general, save the one on rational units. There is some progress to report. Of the three stages to Salvation, two have been safely passed through, namely the Awakening and the Repentance. I am not alone in thinking that the third stage, the Reformation, is bound to come.

I have good reason to be satisfied with the reception given to the first volume of this work. Nearly all parts of it, the outline of general theory, the nomenclature, the rational units, the vector analysis, and the waves and their application, have been approved in this and other countries. But I regret that I have been able to make so little impression upon British official science as expressed by its late leader. It is true that the "K.R. law," which set such unnecessary and unwarrantable restrictions upon telephony, is not much heard of now. With advancing practice it became so ridiculously wrong (say 1,000 per cent.) that it was impossible to save appearances by any manipulation of figures. But a dangerous and alarming official error has been pressed forward, even to the extent of experimentation with the public funds. I refer to Mr. Preece's proposal to increase the capacity of telephone cables, with a view to Atlantic telephony, by bringing the twin conductors as close together as possible. It is, indeed, very true that by Mr. Preece's ingenious plan of flattening the wires on one side, and bringing the flat sides closely together, the capacity may be considerably, and even greatly increased. But it is not the working capacity that is increased, but the electrostatic capacity! Faraday knew that much.

And this blundering is so unnecessary. For if it be beneath the dignity of one who sat at the feet of Faraday and afterwards rose to be the leading authority on electrical matters (according to *Answers*), to consult the works of an insignificant person, still there are other ways. Why not ask someone else? It may be too late to consult the family doctor; but there are many young gentlemen going about who have been to technical colleges and are quite competent to give information concerning the capacity of condensers.

It is to be hoped and expected that the late important removals in the British Telegraph Department will lead to much improvement in the quality of official science. The above two examples show how much improvement has been needed. Others could be given. This volume may help.

April 10, 1899.

CONTENTS OF VOLUME II.

[The dates within brackets are the dates of first publication.]

CHAPTER V.

MATHEMATICS AND THE AGE OF THE EARTH.

(Pages 1 to 29.)

SECTION		PAGE
223	[*Nov. 23, 1894.*] Mathematics is an Experimental Science ...	1
224	Rigorous Mathematics is Narrow, Physical Mathematics Bold and Broad	4
225	Physical Problems lead to Improved Mathematical Methods ...	8
226	[*Dec. 14, 1894.*] "Mathematics—and Mathematics." Remarkable Phenomenon	10
227	The Age of the Earth. Kelvin's Problem	12
228	Perry's Modification. Remarkable Result	14
229	Cooling of an Infinite Block composed of Two Materials ...	16
230	Large Correction for Sphericity in Perry's Problem	18
231	Remarks on the Age of the Earth	19
232	[*Dec. 21, 1894.*] Peculiar Nature of the Problem of the Cooling of a Homogeneous Sphere with a Resisting Skin	20
233	Cooling of a Body of Variable Conductivity and Capacity but with their Product Constant	22
234	Magnitude of the Correction for Sphericity in Various Cases ...	22
235	Explanation of the Last	24
236	Investigation by the Wave Method of the Cooling of a Homogeneous Sphere with a Resisting Skin. Effect of Varying the Constants	25
237	Importance of the Operational Method	28

CHAPTER VI.

PURE DIFFUSION OF ELECTRIC DISPLACEMENT.

(Pages 30 to 274.)

SECTION		PAGE
238	[*Jan. 11, 1895.*] Analogy between the Diffusion of Heat in a Rod and the Diffusion of Charge in a Cable	30
239	The Operational Method assists Fourier	32

SECTION PAGE

240 The Characteristic Equation and Solution in Terms of Time-Functions 33

241 Steady impressed Force at Beginning of Cable. Fractional Differentiation. Simply Periodic Force 35

242 Effect of a Terminal Arrangement. Two Solutions in the Case of a Resistance 37

243 [*Jan. 25, 1895.*] Theory of a Terminal Condenser 40

244 Theory of a Terminal Inductance 42

245 The General Nature of Electrical Operators 44

246 [*March 15, 1895.*] The Simple Waves of Potential and Current 49

247 The Error Function. Short Table 50

248 The Way the Charge and Current Spread 52

249 Theory of an Impulsive Current produced by a Continued Impressed Force 54

250 Diffusion of a Charge initially at One Point. Arbitrary Source of Electrification 57

251 [*March 29, 1895.*] The Inversion of Operators. Simple Examples 58

252 The Effect of a Steady Current impressed at the Origin ... 60

253 Nature and Effect of Multiple Impulses 63

254 [*April 12, 1895.*] Convenient Way of denoting Diffusion Formulæ 66

255 Reflected Waves. Cases of Simple Reflection 67

256 An Infinite Series of Reflected Waves. Line Earthed at Both Ends 69

257 The Method of Images. The Waves are really Successive ... 72

258 [*April 26, 1895.*] Reflection at an Insulated Terminal 74

259 General Case. Effect of an Impressed Force at an Intermediate Point, with any Terminal Conditions 75

260 Four Cases of Elementary Waves 77

261 The Reflection Coefficients in terms of the Terminal Resistance Operators 79

262 Cases of Vanishing or Constancy of the Reflection Coefficients ... 80

263 General Case of an Intermediate Source of Electrification subject to any Terminal Conditions 81

264 [*May 10, 1895.*] The Two Ways of expressing Propagational Results, in terms of Waves, or of Vibrations 83

265 Conversion of Operational Solutions to Fourier Series by Special Ways. (1) Effect due due to *e* at A, when earthed at A and B 88

266 (2) Modified Way of doing the Last Case 90

267 [*June 7, 1895.*] (3) Earth at Both Terminals. Initial Charge at a Point. Arbitrary Initial State 91

268 (4) Line Cut at Both Terminals. Effect of Impressed Force ... 95

269 (5) Line Cut at Terminals. Effect of Initial Charge 96

270 (6) Earth at A and Cut at B. Effect of Initial Charge... ... 96

271 Periodic Expression of Impulsive Functions. Fourier's Theorem 98

SECTION PAGE
272 [*June 21, 1895.*] Fourier Series in General. Various Sorts needed even when the Function is Cyclic and Continuous. Expansions of Zero 100
273 Special Forms of Fourier Integrals. Interchangeable Property. Use in Transforming Definite Integrals 104
274 Continuous Passage from Wave Series to Fourier Series, and its Reversion 107
275 How to Find the Meaning of a Fourier Series Operationally ... 108
276 [*July 12, 1895.*] Taylor's Theorem in its Essentials Operationally Considered 110
277 A Fourier Series involving a Parabola interpreted by Taylor's Theorem 114
278 Representation of a Row of Impulses by Taylor's Theorem, leading to Fourier's 116
279 [*Aug. 9, 1895.*] Laurent's Theorem and Fourier's 119
280 On Operational Solutions and their Interpretation 122
281 Sketch of Way of Extending Fourier's Method to Fourier Series in General 124
282 The Expansion Theorem. Operational Way of getting Expansions in Normal Functions 126
283 [*Aug. 30, 1895.*] Examples of the Use of the Expansion Theorem :—(1) Inductance Coil and Condenser separately... 129
284 The Treatment of Simply Periodic Cases 131
285 (2) Coil and Condenser in Sequence. Also in Parallel ... 134
286 (3) Two Coils under Mutual Influence 137
287 [*Sept. 20, 1895.*] (4) Cable Earthed at A and B. Impressed Force at A 138
288 (5) Cable Earthed at A and Cut at B. Impressed Force at A 140
289 (6) Earth at A, Cut at B. Impressed Force at y 140
290 (7) Earth at A and B. Impressed Force at y 142
291 (8) General Terminal Conditions. Impressed Force at A ... 143
292 [*Oct. 4, 1895.*] (9) General Case of an Intermediate Impressed Force 148
293 The Determinantal Equation 150
293A Subsidence of Special Initial States 152
294 (10) General Case of an Arbitrary Initial State in the Cable... 153
295 (11) Auxiliary Expansions due to the Terminal Energy. Case of a Condenser 156
296 Points of Infinite Condensation. Exceptions to Fourier's Theorem 158
297 [*Oct. 18, 1895.*] Abnormal Fourier Series 159
298 Origin of Two Principal Abnormal Cases 160
299 First General Case : $Z_0 = -Z_1$ 162
300 Equal Positive and Negative Terminal Resistances 163
301 Equal Positive and Negative Terminal Permittances 165
302 Physical Interpretation of the Abnormal Case of § 300... ... 167

SECTION		PAGE
303	[*Nov. 8, 1895.*] Positive Terminal Resistance and Negative Terminal Permittance	170
304	Equal Positive and Negative Terminal Inductances	171
305	Impressed Force in the Case $Z_0 = -Z_1$	172
306	Singular Extreme Case of Z_0 or $-Z_1$ being a Cable equivalent to the Main One	173
307	More General Case to Elucidate the Last. Terminal Cable Z_0 not equivalent to Main One	176
308	[*Nov. 22, 1895.*] Arbitrary Initial State when Z_0 or $-Z_1$ is a Cable. Singular Case	179
309	Real Terminal Conditions. Terminal Arbitraries. Case of a Coil. Two Ways of Treatment	181
310	Terminal Coil and Condenser in Sequence	184
311	Coil and Condenser in Parallel	185
312	Two Coils in Sequence or in Parallel	186
313	[*Dec. 6, 1895.*] A Closed Cable with a Leak. Split into Two Simpler Cases	187
314	Closed Cable and Leak. Another Way	190
315	Closed Cable with Intermedite Insertion. Split into Two Simpler Cases	192
316	Same as last without Initial Splitting	194
317	Closed Cable with Discontinuous Potential and Current ...	196
318	[*Dec. 27, 1895.*] A Cable in Closed Circuit without Constraint	196
319	Theory of a Leak. Normal Systems	198
320	Theory of a Leak. Operational Solution	201
321	Evaluation of Energy in Normal Systems	203
322	[*Jan. 10, 1896.*] Initial States in Combinations	207
323	Two Cables with different Constants in Sequence, with an Insertion	208
324	Cable involving the Zeroth Bessel function	211
325	A Fourier and a Bessel Cable in Sequence	213
326	Construction of a Normal System in General	215
327	[*Jan. 24, 1896.*] Construction of Operational Solutions in a Connected System	217
328	Remarks on Operational and Normal Solutions. Connection with the Simply Periodic	220
329	A Bessel Cable with one Terminal Condition. Two Bessel Cables in Sequence	222
330	General Solutions for Sources in a Bessel Cable with Two Terminal Conditions	225
331	[*Feb. 28, 1896.*] Conjugate Property of Voltage and Gaussage Solutions	228
332	Operational Solutions for Sources in the General Case ...	229
333	Conversion of Wire Waves to Cylindrical Waves. Two Ways	231
334	Special Cases of Zeroth Bessel Solutions	234
335	Numerical Interpretation of Formulæ. The Divergent Series	236

SECTION PAGE
336 [*March 27, 1896.*] The Divergent Formulæ are Fundamental. Generation of Waves in a Medium whose Constants vary as the nth Power of the Distance 238
337 Construction of General Solution by Waves 241
338 Construction of General Solution by the Convergent Formulæ. 244
339 Comparison of Wave and Vibrational Solutions to deduce Relation of Divergent to Convergent Bessel Functions ... 246
340 Nature of Algebraical Transformation from Divergent to Convergent Formulæ... 248
341 [*April 24, 1896.*] Rationality in p of Operational Solutions with two Boundaries. Solutions in terms of I_m and K_m ... 250
342 The Convergent Oscillating Bessel Functions, and Operational Solutions in terms thereof 252
343 The Divergent Oscillating Bessel Functions 255
344 Physical Reason of the Unlikeness of the Two Divergent Functions H_m, K_m 256
345 Electrical Argument showing the Impotency of Restraints at the Origin unless $1 > n > -1$ 259
346 [*May 29, 1896.*] Reduced Formulæ when one Boundary is at the Origin 260
347 The Expansion Theorem and Bessel Series. The Potential due to Initial Charge 262
348 Time Function when Self-Induction is allowed for 265
349 The Potential due to Initial Current 266
349A [*July 17, 1896.*] Uniform Subsidence of Induction and Displacement in Combinations of Coils and Condensers ... 268
349B Uniform Subsidence of Mean Voltage in a Bessel Circuit ... 270
349C Uniform Subsidence of Mean Current in a Bessel Circuit ... 272

APPENDIX C.

RATIONAL UNITS 275

CHAPTER VII.

ELECTROMAGNETIC WAVES AND GENERALISED DIFFERENTIATION.

(Pages 286 to 433.)

SECTION PAGE
350 [*June 26, 1896.*] Determination of the Value of $p^{\frac{1}{2}}1$ by a Diffusion Problem 286
351 Elementary Generalised Differentiation. Value of p^m1 when m is Integral or Midway between 288
352 Cable Problem :—$C=(K+Sp)^{\frac{1}{2}}(R+Lp)^{-\frac{1}{2}}e$. (1.) Elementary Cases by Inspection 290
353 (2.) Algebrisation when c is constant and K zero Two ways. Convergent and Divergent Results 291

SECTION PAGE
354 (3.) Third Way. Change of Operand 293
355 [*Aug. 28, 1896.*] (4.) V due to steady C when R=0. Instantaneous Impedance and Admittance 295
356 (5.) C due to steady V when S=0. The Error Function again 296
357 (6.) V due to steady C when L=0. Two ways... 297
358 (7.) V due to steady C when S is zero. Two ways 298
359 (8.) All Constants Finite. C due to V varying as ϵ^{-pt} ... 298
360 (9.) C due to V varying as $\epsilon^{-Kt/S}$. Discharge of a Charged into an Empty Cable 299
361 (10.) C due to steady V, and V due to steady C 300
362 (11.) C due to impulsive V, and V due to impulsive C... ... 300
363 Cubic under Radical. Reversibility of Operations. Distribution of Operators 301
364 [*Sept. 18, 1896.*] (12.) Development of Equation (60). C due to steady V 303
365 (13.) Another Development of Equation (60) 304
366 (14.) Third Development of Equation (60). Integration by Parts 305
367 (15.) Generalisation. The Complete Wave of C due to V at the Origin varying as $\epsilon^{-Kt/S}$ 306
368 (16.) Summary of Work showing the Wave of C due to V at Origin varying as $\epsilon^{-Kt/S}$... 308
369 (17.) Derivation of the Wave of V from the Wave of C 309
370 (18.) The Wave of V Independently Developed... 310
371 (19.) The Waves of V and C due to Initial Charge or Momentum at a Single Place 311
372 [*Oct. 9, 1896.*] The Waves of V and C due to a Steady Voltage Impressed at the Origin... 312
373 Analysis of Transmission Operator to Show the Deformation, Progression and Attenuation of Waves 315
374 Three Examples. The Wave of V due to impressed Voltage varying as ϵ^{-pt}, $\epsilon^{-Kt/S}$, or steady 317
375 [*Nov. 6, 1896.*] Expansion of Distortion Operators in Powers of p^{-1} 319
376 Example. The Current Wave due to Impressed Voltage $e_0\epsilon^{-Kt/S}$ at the Origin 322
377 Value of $(p^2-\sigma^2)^n I_0(\sigma t)$ when n is a Positive or Negative Integer. Structure of the Convergent Bessel Functions 324
378 Analysis Founded upon the Division of the Instantaneous State into Positive and Negative Pure Waves 326
379 [*Nov. 27, 1896.*] Simplest Solutions. Waves of Infinite Length, and of Length $2\pi v/\sigma$ 329
380 Development of General Solution in $u^m P_m$ and $w^m P_m$ Functions 330
381 Derivation of C Wave from V Wave, and Conversely, with Examples. Condition at a Moving Boundary. Expansion of $\epsilon^{\sigma t}$ 331

SECTION PAGE

382 Deduction of the V and C Waves when V_0 is Constant from the Case $V_0 = e\epsilon^{-Kt/S}$. Expansion of ϵ^{pt} in I Functions. Construction of the Wave of V due to any Impressed Voltage ... 335

383 [*Dec. 18, 1896.*] Identical Expansions of Functions in I_n Functions. Formula for $(\frac{1}{2}x)^n$. Electromagnetic Applications ... 337

384 Expansion of any Power Series in I_n Functions. Examples ... 340

385 Expansion of a Power Series in J_n Functions. Examples ... 344

386 [*Jan. 15, 1897.*] The Waves of V and C due to any V_0 developed in $w^m P_m(z)$ Functions from the Operational Form of $V_0\epsilon^{pt}$ ... 346

387 Remarkable Formula for the Expansion of a Function in I_n Functions, and Examples. Modification of §386, and Example. ... 348

388 Impulsive Impressed Voltages, and the Impulsive Waves and their Tails Generated ... 350

389 Tendency of Distortion to vanish in rapid Fluctuations. Effect of increased Resistance in Rounding off Corners and Distorting 353

390 [*Feb. 26, 1897.*] Wave due to Impressed Voltage varying as $\epsilon^{-pt}t^{-1}$ ... 357

391 Wave due to V_0 varying as $\epsilon^{-pt}\log t$ ... 361

392 [*April 9, 1897.*] Effect of a Terminal Resistance as expected in 1887 and as found in 1896 ... 363

393 Reflection at the Free Ends of a Wire. A Series of Spherical Waves ... 367

394 Reflection at the End of a Circuit terminated by a Plane Resisting Sheet ... 369

395 Long Wave Formulæ for Terminal Reflection ... 373

396 [*May 21, 1897.*] Reflection of Long Waves in General ... 375

397 Terminal Reflection without Loss. Wave Solutions ... 378

398 Comparison with Fourier Series. Solution of Definite Integrals 380

399 [*June 18, 1897.*] General Way of Finding Second and Following Waves due to Terminal Reflection ... 384

400 Application to Terminal Resistances. Full Solutions with the Critical Resistances. Second Wave with any Resistance ... 387

401 Inversion of Operations. Derivation of First Wave from the Second ... 390

402 Derivation of Third and Later Waves from the Second ... 391

403 Summarised Complete Solutions ... 392

404 [*Aug. 6, 1897.*] Reflection by a Condenser ... 393

405 Reflection by an Inductance Coil ... 395

406 Initial States. Expression of Results by Definite Integrals ... 396

407 The Special Initial States $J_0(\sigma x/v)$ and $J_n(\sigma x/v)$ ... 398

408 The States of V and C resulting from any Initial States, V_0 and C_0, expansible in J_n functions ... 400

SECTION PAGE
409 Some Fundamental Examples 402
410 [*Dec. 17, 1897.*] The General Solution for any Initial State, and some Simple Examples 403
411 Conversion to Definite Integrals. Short Cut to Fourier's Theorem 406
412 The Space Integrals of V and C, due to Elements at the Origin 408
413 The Time Integral of C, due to Elements at the Origin ... 409
414 Evaluation of the Fundamental Integral... 410
415 Generalisation of the Integral. Both kinds of Bessel Functions 412
416 [*Jan. 14, 1898.*] The C due to intial V_0. Operational Method, and Modification 413
417 The V due to initial V_0. 415
418 [*Feb. 18, 1898.*] Final Investigation of the V and C due to initial V_0 and C_0 417
419 Undistorted Waves without and with Attenuation 422
420 [*March 25, 1898.*] Effects of Resistance and Leakance on an Initial State of Constant V_0 on one side of the Origin ... 424
421 Division of Charge Initially at the Origin into Two Waves with Positive or Negative Charge between them 426
422 The Current due to Initial Charge on one side of the Origin ... 428
423 [*April 29, 1898.*] The After-effects of an Initially Pure Wave. Positive and Negative Tails 429
424 Figs. 1 to 13 described in terms of Electromagnetic Waves in a doubly Conducting Medium 431

CHAPTER VIII.

GENERALISED DIFFERENTIATION AND DIVERGENT SERIES.

(Pages 434 to 492.)

SECTION PAGE
425 A Formula for $\lfloor n$ obtained by Harmonic Analysis 434
426 [*May 27, 1898.*] Algebraical Construction of $g(n)$. Value of $g(n)g(-n)$... 436
427 Generalisation of Exponential Function 439
428 Application of the Generalised Exponential to a Bessel Function, to the Binomial Theorem and to Taylor's Theorem 441
429 Algebraical Connection of the Convergent and Divergent Series for the Zeroth Bessel Function... 443
430 Limiting Form of Generalised Binomial Expansion when Index is -1 445
431 [*July 1, 1898.*] Remarks on the Operator $(1+\Delta^{-1})^n$ 445
432 Remarks on the Use of Divergent Series 447
433 Logarithmic Formula derived from Binomial 450
434 Logarithmic Formulæ derived from Generalised Exponential ... 451

SECTION PAGE

435 [*August 19, 1898.*] Connections of the Zeroth Bessel Functions 453
436 Operational Properties of the Zeroth Bessel Functions ... 456
437 Remarks on Common and Generalised Mathematics 457
438 [*Oct. 7, 1898.*] The Generalised Zeroth Bessel Function Analysed 463
439 Expression of the Divergent $H_0(x)$ and $K_0(x)$ in Terms of Two Generalised Bessel Functions. Generalisation of ϵ^{-x} .. 465
440 The Divergent $H_n(x)$ and $K_n(x)$ in Terms of Two Special Generalised Bessel Functions 467
441 The Divergent $H_n(x)$ and $K_n(x)$ in Terms of any Generalised Bessel Function of the Same Order 469
442 [*Dec. 2, 1898.*] Product of the Series for ϵ^x and $\epsilon^{-x}\cos r\pi$. Possible Transition from ϵ^{-x} to ϵ^x 472
443 Power Series for log x 474
444 Examination of some Apparent Equivalences, and Rectification 476
445 Determination of the meaning of a Generalised Bessel Function in Terms of H_0 and K_0 478
446 Some Apparent Equivalences 481
447 [*Jan. 13, 1898.*] Cotangent Formula and Derived Formula for Logarithm. Various Properties of these and other Divergent Series 482
448 Three Electrical Examples of Equivalent Convergent and Divergent Series 487
449 Sketch of Theory of Algebrisation of $(1-bp^r)^{-1}1$ 490

APPENDIX D.

[*Nov. 12, 1897.*] On Compressional Electric or Magnetic Waves 493

APPENDIX E.

[*Aug. 7, 1896.*] Dispersion 507

APPENDIX F.

[*Dec. 20, 1893.*] On the Transformation of Optical Wave-Surfaces by Homogeneous Strain 518
(1) Simplex Eolotropy 518
(2) Properties Connected with Duplex Eolotropy 519
(3) Effects of Straining a Duplex Wave Surface 520
(4) Forms of the Index- and Wave-Surface Equations, and the Properties of Inversion and Interchangeability of Operators 521
(5) General Transformation of Wave-Surface by Homogeneous Strain 525
(6) Special Cases of Reduction to a Simplex Wave-Surface ... 527
(7) Transformation from Duplex Wave- to Index-Surface by a Pure Strain 528

PAGE

(8) Substitution of Three Successive Pure Strains for One. Two Ways 529
(9) Transformation of Characteristic Equation by Strain ... 530
(10) Derivation of Index Equation from Characteristic 531

APPENDIX G.

[*Jan. 14, 1898.*] NOTE ON THE MOTION OF A CHARGED BODY AT A SPEED EQUAL TO OR GREATER THAN THAT OF LIGHT 533

APPENDIX H.

[*June 11, 1897.*] NOTE ON ELECTRICAL WAVES IN SEA WATER ... 536

APPENDIX I.

[*Nov. 11, 1898.*] NOTE ON THE ATTENUATION OF HERTZIAN WAVES ALONG WIRES... 538

ERRATA.

Page 42, equation (30) : For $(h/\pi t)$ read $(h/\pi t)^{\frac{1}{2}}$. Equation (31): For $h^{\frac{1}{2}}/\pi t$ read $(h/\pi t)^{\frac{1}{2}}$.

Page 81, line 17 : For " conection " read " connection."

Page 97, equation (45) : For $= sl$ read $= -sl$. Equation (46) : For $=($ read $=-($; two lines after, for sl read $-sl$.

Page 115, equation (91) : For 2Σ read Σ.

Page 160, equation (100) : For $+\mathrm{R}s$ read $-\mathrm{R}s$.

Page 163, line 8 from end : For " thus " read " then."

Page 165, equation (126) : For $\epsilon^{h\ x}$ read $\epsilon^{h(x}$.

Page 210, line 12 : For Z read z.

Page 229, line 5 : For " reciprocal " read " conjugate."

Page 245, equation (29) : For I read I_m.

Page 246, 11th line from end : For " is ' read " is a."

Page 248 : For § 339 read § 340.

Page 255, 8th line from end : For " when " read " where."

Page 274, 3rd line from end : For " notation " read " standardization."

CHAPTER V.

MATHEMATICS AND THE AGE OF THE EARTH.

Mathematics is an Experimental Science.

§ 223. That the study of the theory of a physical science should be preceded by some general experimental acquaintance therewith, in order to secure the inimitable advantage of a personal acquaintance with something real and living, will probably be agreed with by most persons. After, however, the general experimental knowledge has been acquired, accompanied with just a sufficient amount of theory to connect it together and render its acquisition easier and more interesting, it becomes possible to consider the theory by itself, as theory. The experimental facts then go out of sight, in a great measure, not because they are unimportant, but because they become subordinate to the theory in a certain sense, and, we might also add, because they are fundamental, and the foundations are always hidden from view in well-constructed buildings. So it comes about that a great theoretical work like Maxwell's treatise on Electricity and Magnetism contains so little explicit information regarding the experimental facts of the science. Theory always tends to become abstract as it emerges successfully from the chaos of facts by the processes of differentiation and elimination, whereby the essentials and their connections become recognised, whilst minor effects are seen to be secondary or unessential, and are ignored temporarily, to be explained by additional means.

There is the same tendency in the most abstract and logical of all sciences—pure mathematics. Geometry, for instance, has most certainly an experimental foundation, like any other science. We make our geometry, in the first place, to suit the state of things in which we are born and live. We all make our acquaintance with geometry first through our senses, and become saturated, so to speak, with the essence of the geometry of nature, even though this be unaccompanied by any intelligent comprehension and expression. Now take a hint from nature, and we can see plainly how much it is to the advantage of the learner that he should continue, in the first place, to acquire geometrical ideas through his senses—though now, of course, with his attention specially directed to the subject—freely assisted by models, solid and skeleton, before being set to work upon the more intellectual theory on a formal basis. For, disguise it as we may, no strictly formal basis, apart from experience, can ever be possible. There is always something very considerable to start with of an experiential nature in the background, besides the formal axioms and definitions and postulates which would be unintelligible without it. A straight line can never be intelligibly defined *per se*. One must actually know the practical straight line before any definition of the abstract straight line can be understood. Then our understanding and acceptance of the definition is a recognition that it states what we knew already, in accumulated experience, though we may have never openly thought about it. As regards an axiom, such as the one that asserts that two straight lines cannot enclose a space, its acceptance involves the existence of a very extensive experience of the geometry of nature as it is found to be. How impossible, then, must it be to prove anything rigorously as absolute truth, independent of nature. We come down to axioms, definitions, and postulates at last, and these are only understandable by experience.

There is also no self-contained theory possible, even of geometry considered merely as a logical science, apart from practical meaning. For a language is used in its enunciation, which implies that developed ideas and complicated processes of thought are already in existence, besides the general experience associated therewith. We define a thing in a phrase, using words. These words have to be explained in other words, and so on, for ever, in a complicated maze. There is no bottom to anything. We are all antipodeans and upside down.

It is by the gradual fitting together of the parts of a distinctly connected theory that we get to understand it, and by the revelation of its consistency. We may begin anywhere, and go over the ground in any way. Some ways will be preferable to others, of course, since they may be easier, or give broader and clearer views, but no strict course is necessary. It may not even be desirable. It may be more interesting and instructive not to go by the shortest logical course from one point to another. It may be better to wander about, and be guided by circumstances in the choice of paths, and keep our eyes open to the side prospects, and vary the route later to obtain different views of the same country. Now it is plain enough when the question is that of guiding another over a well-known country, already well explored, that certain distinct routes may be followed with advantage. But it is somewhat different when it is a case of exploring a comparatively unknown region, containing trackless jungles, mountains and precipices. To attempt to follow a logical course from one point to another would then perhaps be absurd. You should keep your eyes and your mind open, and be guided by circumstances. You have first to find out what there is to find out. How you do it is quite a secondary consideration. Later on, no doubt, much easier and perhaps better ways will be found, so that a crowd can push along. It is obvious, I think, that complaints of the want of perfection of the ways and manners of work of explorers on the part of men who are accustomed to more rigorous methods have a considerable element of the ludicrous in them. However harmless in intention, they may operate unfairly in effect, if they lead, as sometimes happens (of which a case was quite lately brought to my notice), to the rejection of honest work which failed to be appreciated by the judges, who had no doubt different ways of working and thinking, and different experiences. When this result arises, it has the effect of putting a learned society in the unfortunate position of *appearing* to exist not merely for the encouragement of research along established lines, but also for the active discouragement of work of a less conventional character. That this is the case in reality it is impossible to believe. But then, papers are so cheap, and one more or less does not matter. Again, the probable fact that the judges were animated by benevolent motives, and only desired to turn the misguided author from the error of his ways into more rigorous paths approved by themselves, does not make the matter any the better for the author. He has his own ways, and must follow them, even though he be told (virtually) that his work is valueless, and not worth printing, and of course, by inference, that he must not continue to send more of it.

Rigorous Mathematics is Narrow, Physical Mathematics Bold and Broad.

§ 224. Now, mathematics being fundamentally an experimental science, like any other, it is clear that the Science of Nature might be studied as a whole, the properties of space along with the properties of the matter found moving about therein. This would be very comprehensive, but I do not suppose that it would be generally practicable, though possibly the best course for a large-minded man. Nevertheless, it is greatly to the advantage of a student of physics that he should pick up his mathematics along with his physics, if he can. For then the one will fit the other. This is the natural way, pursued by the creators of analysis. If the student does not pick up so much logical mathematics of a formal kind (common-sense logic is inherited and experiential, as the mind and its ways have grown to harmonise with external Nature), he will, at any rate, get on in a manner suitable for progress in his physical studies. To have to stop to formulate rigorous demonstrations would put a stop to most physico-mathematical inquiries. There is no end to the subtleties involved in rigorous demonstrations, especially, of course, when you go off the beaten track. And the most rigorous demonstration may be found later to contain some flaw, so that exceptions and reservations have to be added. Now, in working out physical problems there should be, in the first place, no pretence of rigorous formalism. The physics will guide the physicist along somehow to useful and important results, by the constant union of physical and geometrical or analytical ideas. The practice of eliminating the physics by reducing a problem to a purely mathematical exercise should be avoided as much as possible. The physics should be carried on right through, to give life and reality to the problem, and to obtain the great assistance which the physics gives to the mathematics. This cannot always be done, especially in details involving much calculation, but the general principle should be carried out as much as possible, with particular attention to dynamical ideas. No mathematical purist could ever do the work involved in Maxwell's treatise. He might have all the mathematics, and much more, but it would be to no purpose, as he could not put it together without the physical guidance. This is in no way to his discredit, but only illustrates different ways of thought. There have been enormous advances made in pure mathematics in the last half century, as is right and proper to match the advance in physical science. But along with this has come a tendency for purists to object to the introduction of physical ideas in mathematics, with a possible lack of rigour as result. It may be that there is no lack of rigour sometimes, but an increased generality and simplified treatment. Maxwell was severely taken to task by a distinguished purist for his use of Green's Theorem in Spherical Harmonics, a method which is excellently to the purpose, and which commends itself to the electrician, and it is probably quite rigorous. But no doubt there is frequently a lack of rigour. I have seen with much pleasure some remarks on this point in the Preface to the recently-published second edition of Lord Rayleigh's treatise on Sound, which I cannot do better than reproduce :—

> In the mathematical investigations I have usually employed such methods as present themselves naturally to a physicist. The pure mathematician will complain, and (it must be confessed) sometimes with justice, of deficient rigour. But to this question there are two sides. For, however important it may be to maintain a uniformly high standard in pure mathematics, the physicist may occasionally do well to rest content with arguments which are fairly satisfactory and conclusive from his point of view. To his mind, exercised in a different order of ideas, the more severe procedure of the pure mathematician may appear not more but less demonstrative. And further, in many cases of difficulty to insist upon the highest standard would mean the exclusion of the subject altogether in view of the space that would be required.

Particularly notice the words "not more but less demonstrative." This is exceedingly true, especially in the subject of the expansion of functions in series of other functions, which occupies so large a part of the treatise in question. And I would add that if the physicist does sometimes get carried too far, the proper time to find out the reservations and corrections is later on, in order not to interrupt his work. But this purification is more especially suitable for the purist to undertake. If one sort of work is as necessary as the other, it is certain that the physicist would get very little work done by trying to do both, having the fear of the rigourists before him. What is more hateful than a physical work done in propositions and corollaries? It is bad enough in pure mathematics, and I wish purists would take a lesson from Fourier, Thomson and Tait, Maxwell, or Rayleigh, and tell their tale differently and make it interesting by letting in a little imagination. I have had occasion to go through a considerable part of a very big Theory of Functions in search for what I did not find. The work is most admirably painstaking and severely rigorous, but how different from physical mathematics, how hard to read from its severe formalism, and how narrow it seems from the want of physical illustration. Perhaps the subject might be greatly lightened by having a physical theory to rest upon or to illustrate.

When mathematics is cleared away from physics it becomes set in logical form. But it is to be remembered that the men who have in the past initiated great advances in mathematics have usually been men who were employed in working out physical questions. They supplied the purists with raw material to be made coherent and elaborated. The expansion of functions in series, already mentioned, arose physically. It is an enormous and endless subject, and there is a striking difference in the ways in which it is regarded by the physicist and the rigourist, with the peculiarity that the former is far in advance of the latter. To understand this, consider that a physicist can have practical certainty that a certain expansion is possible, because his physical problem tells him so, when he seeks and finds the solution, even though he has not investigated the properties of the functions used in the expansion. He does not arrange long and severely disagreeable demonstrations to prove what he *knows*; although it be a fallible kind of knowledge, it only differs in degree from most mathematics in this respect. But eliminate the physics, and put it simply as a question of functions. Given certain functions, can an arbitrary function be expressed in terms of them? What a pretty piece of work there must be to answer this question! This is true even in the case of circular functions in the particular case rigorously treated by purists. But by changes in the conditions the physicist gets an endless number of expansions of one and the same arbitrary function in circular functions, and what becomes of the rigorous demonstration then? Then there is an endless number of other sorts of functions, every one of which can represent an arbitrary function in an endless number of ways, as is perfectly clear from physical considerations. It is evident that for comprehensive rigorous demonstrations (not special) we need enlarged ideas about functions, and perhaps the purist would obtain the necessary broadness of view by a study of the physics in which such comprehensiveness of expression is found. Certainly the purist is bound to complete the logical treatment some day, but the hard-bound rules of the purist make it difficult. Thus at present, although the purist carries his mathematical developments so far in some directions as to be far beyond physics, out of sight in fact, yet in other respects the purist lags behind. And this is true in other matters than that mentioned. For a physicist may use daily with success, and as a matter of course, methods which he knows work usefully to his assistance, but which are logically unintelligible to a purist, and which have to wait for a proper development.

The best result of mathematics is to be able to do without it. To show the truth of this paradox by an example, I would remark that nothing is more satisfactory to a physicist than to get rid of a formal demonstration of an analytical theorem and to substitute a quasi-physical one, or a geometrical one freed from co-ordinate symbols, which will enable him to *see* the necessary truth of the theorem, and make it be practically axiomatic. Contrast the purely analytical proof of the Theorem of Version well known to electrical theorists, with the common-sense method of proof by means of the addition of circuitations. The first is very tedious, and quite devoid of luminousness. The latter makes the theorem be obviously true, and in any kind of co-ordinates. When seen to be true, symbols may be dispensed with, and the truth becomes an integral part of one's mental constitution, like the persistence of energy.

Physical Problems lead to Improved Mathematical Methods.

§ 225. There is a curious analogy to be found between extensions of mathematical ideas and extensions of electrical theory. Take, for instance, Maxwell's theory in the form presented in Vol. I. of this work, Chapter II. If we do not inquire too minutely into the consequences, we may easily be temporarily under the impression that the two circuital equations and their accessories express the dynamics of a quite self-contained system. But when we go to the very verge of the system, and find that mechanical forces of the ordinary kind are involved when electromagnetic disturbances pass through the suppositional ether which supports them, we come to a stop. Now, there is nothing impossible or incredible in the result. It is simply unintelligible at present. The plain meaning to be given to it without introducing additional data is that the ether, as regards electromagnetic disturbances traversing it, should be regarded like any other dielectric; that is, as having a *substantial* existence. To complete the matter evidently requires a theory of the ether itself, and the suggestion is that it should not be regarded as an elastic solid (generalised) in which the actual displacements in bulk represent the electromagnetic disturbances. But, however this may be (and the matter is difficult and speculative), the point here is that on the outskirts of the theory we come to matters needing interpretation and a larger theory.

So it is in mathematics. The fundamental notions are so simple that one might expect that unlimited developments could be made without ever coming to anything unintelligible. But we do, and in various directions. To say nothing of the interpretation of negative quantity (which is a sort of imaginary), there is *the* imaginary, which has only become understood and its properties developed comparatively recently. But, besides these, there are much more obscure and ill-understood questions, such as the meaning and true manipulation of divergent series, and of fractional differentiations or integrations, and connected matters. It is customary to keep to convergent series and whole differentiations and regard divergent series and fractional differentiations as meaningless and practically useless, or even to ignore them altogether, as if they did not exist. The latter is the usual attitude of moderate and practical mathematicians, for obvious reasons. If they can be ignored, why trouble about them at all?

But when these things turn up in the mathematics of physics the physicist is bound to consider them, and make the best use of them that he can. I am thinking more particularly here of the solution of the differential equations to which physicists are led by quasi-algebraical processes. The reader will see to what I refer by reference to § 203, Vol. I., when I allude to the definiteness of meaning of the operator R'' in the equation $E = R''C$, R'' being a complicated function of a differentiator. See also § 221 later on. When C is explicitly given as a function of the time we have to find E to match through the operator R'', and when it is E that is given, then C has to be found by the operation on E of the inverse of R''. It is in the carrying out of these processes in the investigation of various electromagnetic problems that we are obliged to regard certain kinds of divergent series as representing fully significant functions, and the execution of the processes involved in R'', which assumes various algebraical forms, as being legitimate and feasible, however ill-understood may be the theory involved therein.

Nor is the matter an unpractical one. I suppose all workers in mathematical physics have noticed how the mathematics seems made for the physics, the latter suggesting the former, and that practical ways of working arise naturally. This is really the case with resistance operators. It is a fact that their use frequently effects great simplifications, and the avoidance of complicated evaluations of definite integrals. But then the rigorous logic of the matter is not plain! Well, what of that? Shall I refuse my dinner because I do not fully understand the process of digestion? No, not if I am satisfied with the result. Now a physicist may in like manner employ unrigorous processes with satisfaction and usefulness if he, by the application of tests, satisfies himself of the accuracy of his results. At the same time he may be fully aware of his want of infallibility, and that his investigations are largely of an experimental character, and may be repellent to unsympathetically constituted mathematicians accustomed to a different kind of work.

There is another point of view. Convergent mathematics is often excessively unpractical in the labour it involves in the numerical interpretation of results. I have noticed this particularly in spherical harmonic expansions, where the labour is sometimes prohibitive. Under these circumstances the substitution of equivalent divergent series which may be readily calculated becomes a matter of great practical importance. If the example fails, others may readily be given. But what about the theory of functions which deliberately ignores the treatment of divergent series? Can it really be *the* theory of functions? Is not a more comprehensive theory needed, including both convergent and divergent functions in a harmonious whole?

"Mathematics—and Mathematics." Remarkable Phenomenon.

§ 226. If it should ever come to pass that there prevailed in this world a so-called religion in which the minor virtues of mercy, charity, meekness, resignation, and so forth, were unduly exalted at the expense of the supreme virtue of justice, and if this religion were carried into practice generally, it would be a very bad thing for the world. For such a religion would be a snivelling religion, only fit for the weakminded of both sexes; and if the strong-minded and the just allowed it to prevail, then would the liars, rogues, hypocrites, slanderers, and other wicked people have it all their own way, and the just would be crushed along with the meek and lowly. But just men are better than their religions, and in self-defence would assert the paramount importance of justice, in tacit defiance of the nominal religion in which it was made a secondary virtue. Let us above all things try to be just. Even Cambridge mathematicians deserve justice. I cannot join in any general attack upon them.* I regret exceedingly not to have had a Cambridge education myself, instead of wasting several years of my life in mere drudgery, or little more. It is to Cambridge mathematicians that we are indebted for most of the mathematico-physical work done in this country. Do not most mathematical physicists hail from Cambridge? Are not

* *See* article in *The Electrician*, November 23, 1894, p. 100.

Thomson and Tait, Maxwell and Rayleigh Cambridge mathematicians, to say nothing of the large crowd of other and mostly younger men whose names will suggest themselves? We must take the good with the bad, in this as in other matters; and though legitimate and serious objection may be raised to the distressing and soul-destroying style in which some Cambridge mathematicians do their work, and to the unpractical conservative tendency that exists (conserving the bad as well as the good, and resisting innovations), we should also bear in mind the great volume and value of the work done, and not unduly depreciate or make invidious comparisons. As regards their want of sympathy with less conventional men, it is not sympathy that is particularly wanted—perhaps it would be unreasonable to expect any at all. What one has a right to expect, however, is a fair field, and that the want of sympathy should be kept in a neutral state, so as not to lead to unnecessary obstruction. For even men who are not Cambridge mathematicians deserve justice, which I very much fear they do not always get, especially the meek and lowly, and those who long suffer under slight.

On this question of "Mathematics—and Mathematics," I may mention a somewhat remarkable phenomenon which has lately occurred. Orthodox mathematicians, when they cannot find the solution of a problem in a plain algebraical form, are apt to take refuge in a definite integral, and call that the solution. It is certainly one form of the solution. But it may be just as hard, or harder, to interpret than the differential equation of the problem in question, from the difficulty of evaluating the integral, and so finding out what the solution means. In such cases we might as well keep to the differential equation, and be just as wise. Now, it has come to my knowledge that a man who is not a Cambridge mathematician, and who does not pretend to be much of any sort of mathematician, but who is a practical physicist, capable of discussing with proper judgment such a question as the age of the earth, a higher limit to which he finds (and with good reason) to be very likely hundreds of times greater than the most probable previous estimate (which conclusion has obviously important interest to geologists and astronomers), recently made the discovery that a certain unconventional mode of treating the mathematics of the question (explained to him by myself) conducted him immediately to the exact solution of the problem he had in hand, in a few lines in fact; whereas by the methods generally employed he might have spent days over it, without any final success beyond obtaining a definite integral of too complicated a nature to be practically discussed or obviously evaluated.

It has naturally given me much pleasure to find that the method in question, which professes to obtain the solution in plain language directly from the differential operator, and, so to speak, to evaluate the definite integral without the trouble of finding it, should receive such ready appreciation from a practical physicist. Of course, he has no prejudices of the rigorous kind; but makes use of what he finds useful, as soon as he has got to know how to go to work. It is the fact that he is a practical physicist, without mathematical pretensions, that constitutes the importance of the phenomenon. For this reason, I shall have no further hesitation in making use of the method in question occasionally in the course of the rest of this work, at least in such simple cases as the above experience shows are fairly and without much trouble within the reach of practical physicists and electricians; not mathematicians of the Cambridge or conservatory kind, who look the gift-horse in the mouth and shake their heads with solemn smile, or go from Dan to Beersheba and say that all is barren; but of the common field variety, who take the seasons as they come and go, with grateful appreciation. It is really not a question of high mathematics at all, in these diffusion problems at least, but of the substitution of simpler and more direct processes for the indirect and complicated processes of the highly cultivated mathematician with too rigorous proclivities.

The Age of the Earth. Kelvin's Problem.

§ **227.** Now Prof. John Perry has suggested to me that I should write something on the subject. Therefore, as he has made the matter of the age of the earth interesting at the present time, I give some particulars regarding simple solutions. They are not so much out of place as may appear at first, for they all represent electrical problems of interest, as we shall see later.

The main problem is: Given the earth initially at a constant temperature V_0 throughout, find its way of cooling, and, in particular, find the gradient of temperature in the earth's crust, for that is the observed datum. It can then be deduced how long it takes to arrive at its present state. In accordance with general practice, much simpler problems are substituted, and approximations made with various data. Lord Kelvin,* who started this branch of inquiry in his celebrated paper on the Age of the Earth, substituted for the earth an infinite body of uniform capacity and conductivity, the same as those in the crust, with a plane boundary kept at zero temperature.

Now consider first the converse problem. The earth is initially at zero temperature, and by means of surface sources its skin is thereafter maintained at constant temperature V_0. Find the temperature gradient in the skin which results. Let V be the temperature at distance x from the plane face; the well-known characteristic of V is

$$\frac{d^2V}{dx^2} = \left(\frac{c}{k}p\right) V = q^2V, \text{ say,} \tag{1}$$

where c is the capacity per unit volume, k the conductivity, and p means the time-differentiator. Therefore,

$$V = \epsilon^{-qx}V_0 \tag{2}$$

gives V in terms of V_0. This is easily integrated, but we only want the surface gradient, say g. Thus,

$$g = -\frac{dV}{dx} = q\epsilon^{-qx}V_0. \tag{3}$$

So, between (2) and (3) we get

$$g = qV_0, \tag{4}$$

which is the solution.

To turn it to algebraical form, we have

$$q = \left(\frac{c}{k}p\right)^{\frac{1}{2}}, \quad \text{and} \quad p^{\frac{1}{2}}1 = (\pi t)^{-\frac{1}{2}}, \tag{5}$$

so that (4) is the same as

$$g = V_0\left(\frac{c}{\pi kt}\right)^{\frac{1}{2}}. \tag{6}$$

Since the final state due to our source is V_0 everywhere in the earth, it follows that in the subsidence problem, starting

* "On the Secular Cooling of the Earth," *Trans.* R. S. Edin., 1862; or App. D., Thomson and Tait's "Natural Philosophy."

from a uniform temperature V_0, the gradient has the same value, only reversed in direction, so that (6) is still the required solution. This is the formula, due to Fourier, used by Lord Kelvin. Taking

$$V_0=4{,}000^\circ\text{C.},\quad g=\frac{1}{2{,}743},\quad \frac{k}{c}=0\cdot 0118,$$

which are data used by Perry, we find that t is 103 million years. That is, it takes about 100 million years for the gradient in the skin to fall to its present value, under the assumed circumstances. The correction necessary for the finite size of the earth, other data being the same, is not large. I find that the time requires to be reduced by $\frac{1}{29}$ part.*

Perry's Modification. Remarkable Result.

§ 228. The next step, due to Perry,† is to assume that the capacity and conductivity are higher in the earth than in its skin. That this will prolong the time of subsidence may not be difficult to understand in the case of an infinite block with a plane face; but the result is most distinctly not an obvious one in the case of a sphere, for it may be readily shown that the time of cooling may be either less or greater, according to circumstances. But we want numerical results. So take a distinct case. Let the skin be so thin that its capacity may be ignored, whilst its conductance per unit of surface, which is k_1/l, where k_1 is the conductivity, and l the depth, is finite. Then the current of heat inwards through the skin equals that entering the inner earth. This gives the condition that at $x=0$ we have

$$\frac{V_0-V_1}{R}=-k\frac{dV_1}{dx}=kqV_1, \tag{7}$$

if V_0 is the temperature outside the skin, V_1 that just inside. So

$$V_1=\frac{V_0}{1+Rkq} \tag{8}$$

* This may be proved by formula (39) below, or by the same investigation simplified to meet the present case.

† John Perry, "On the Age of the Earth," privately circulated in MS. October 14, 1894; further circulated with additions in pamphlet form in November, and published in revised form with other matter in *Nature*, January 3, 1895, p. 224.

is the solution for V_1 in terms of V_0 impressed. R is the resistance of unit area of the skin.

There are two ways of converting (8) to algebraical form, one convergent, the other divergent. The latter is most useful. Thus, by division,

$$V_1 = [1 - Rkq + (Rkq)^2 - (Rkq)^3 + \ldots] V_0, \tag{9}$$

which, by the use of (5), is turned to

$$V_1 = V_0 \left[1 - \left(\frac{a}{\pi t}\right)^{\frac{1}{2}} \left\{1 - \frac{a}{2t} + 1.3\left(\frac{a}{2t}\right)^2 - \ldots\right\}\right], \tag{10}$$

from which the subsidence solution, due to V_0 constant all over initially, is*

$$V_1 = V_0 \left(\frac{a}{\pi t}\right)^{\frac{1}{2}} \left\{1 - \frac{a}{2t} + 1.3\left(\frac{a}{2t}\right)^2 - 1.3.5\left(\frac{a}{2t}\right)^3 + \ldots\right\}. \tag{11}$$

Here $a = ckR^2$. The gradient is got by dividing by l, so that the first term, which is sufficient when t is big enough, is

$$g = \frac{V_0}{k_1}\left(\frac{ck}{\pi t}\right)^{\frac{1}{2}}. \tag{12}$$

So, with the same k_1 in the skin, the value of t varies as the value of ck in the earth when t is big enough, which is an important conclusion.

Now (11) is unsuitable when t is small, on account of the divergency. Then apply an alternative method to (8), viz.,

$$V_1 = \frac{V_0}{Rkq\left(1 + \frac{1}{Rkq}\right)} = \left\{\frac{1}{Rkq} - \frac{1}{(Rkq)^2} + \frac{1}{(Rkq)^3} - \ldots\right\} V_0, \tag{13}$$

which by means of $p^{-n}1 = t^n / \lfloor n$, is turned to the algebraical form

$$V_1 = 2V_0\left(\frac{t}{a\pi}\right)^{\frac{1}{2}}\left(1 + \frac{2t}{3a} + \frac{1}{3.5}\left(\frac{2t}{a}\right)^2 + \frac{1}{3.5.7}\left(\frac{2t}{a}\right)^3 + \ldots\right)$$
$$+ V_0(1 - \epsilon^{t/a}). \tag{14}$$

* I am informed that this is Riemann's solution for the surface temperature of an infinite block with a plane face, cooling from an initially uniform temperature; the boundary condition being constancy of ratio of the rate of loss of heat to the temperature. The problem is formally the same as Perry's problem of a resisting skin, because the boundary condition on the inside of the skin is formally the same as in Riemann's case. Perfectly free escape of heat means R = 0.

So the subsidence solution due to initial V_0 is

$$V_1 = V_0 \left\{ \epsilon^{t/a} - 2\left(\frac{t}{a\pi}\right)^{\frac{1}{2}}\left(1 + \frac{2t}{3a} + \frac{1}{3.5}\left(\frac{2t}{a}\right)^2 + \ldots\right) \right\}. \quad (15)$$

The equivalence of (11) and (15) is merely an example of the generalised exponential, but it is interesting to see how it arises.

The solution for the temperature at any distance x may be similarly obtained and without difficulty, but we need not consider that here.

Cooling of an Infinite Block composed of Two Materials.

§ 229. Going a step further, let us examine the influence of the capacity of the skin itself. This was done by Perry immediately on receipt of the above, so far as the beginning and most important part of the solution was concerned. The skin may now be of any depth, and is treated in the same way as the inner earth, but with different constants. Let c, k, V belong to the earth, and c_1, k_1, v to the skin. Measure x in the earth from the inside of the skin, and z in the skin also from its inside. Then we have

$$V = \epsilon^{-qx} V_1, \quad (16)$$

$$v = \epsilon^{-q_1 z} C + \epsilon^{q_1 z} D, \quad (17)$$

where V_1 is the temperature at $x=0$ and C, D are unknown, and to be eliminated. The boundary conditions are $v = V_0$ at $z = l$, and $v = V_1$ at $z = 0$, besides continuity of current at the interface of skin and earth, or

$$k_1 \frac{dv}{dz} = -k \frac{dV}{dx} \quad (18)$$

at $z=0$, $x=0$. These find C and D, and lead us by ordinary algebraical work, which need not be given, to the following expression for the gradient of temperature at the outside of the skin :—

$$g = \frac{1+sy}{1-sy} q_1 V_0, \quad (19)$$

where $$s = \frac{1-(c_1k_1/ck)^{\frac{1}{2}}}{1+(c_1k_1/ck)^{\frac{1}{2}}}, \qquad y = \epsilon^{-2q_1 l}. \quad (20)$$

Notice that $s=0$ when $ck = c_1k_1$. That is, the flow of heat in the skin due to applied V_0 on its outside is the same for any material inside, provided $ck = c_1k_1$, which reduces (19) to the elementary form (4) first considered.

The approximate solution involving the first power of l may be got at once from (19), and it leads to the former results when l is small. To get the complete solution, expand the fraction in (19) by division. Thus,

$$g = (1 + 2sy + 2s^2y^2 + 2s^3y^3 + \ldots)q_1\mathrm{V}_0. \tag{21}$$

Only one term needs transformation, since they are formally alike. We have

$$y^n q_1 \mathrm{V}_0 = \left(q_1 - 2nq_1^2 l + \frac{4n^2 q_1^3 l^2}{\underline{|2}} - \ldots\right)\mathrm{V}_0, \tag{22}$$

by (20), using the ordinary exponential expansion. Integrating by (5) we obtain a series which also belongs to the exponential kind, being

$$\mathrm{V}_0\left(\frac{c_1}{\pi k_1 t}\right)^{\frac{1}{2}}\epsilon^{-c_1 n^2 l^2/k_1 t}. \tag{23}$$

Here n has to be 0, 1, 2, 3, &c., in the successive terms of (21), so we obtain

$$g = \left(\frac{c_1}{\pi k_1 t}\right)^{\frac{1}{2}}\{1 + 2s\mathrm{A} + 2s^2\mathrm{A}^4 + 2s^3\mathrm{A}^9 + \ldots\}, \tag{24}$$

where $\mathrm{A} = \epsilon^{-c_1 l^2/k_1 t}$.

In fact (21) expresses the initial effect of V_0 in the first term on the right, and of an infinite series of images due to the change of medium.

Although (24) is a very neat form of solution, we may want a solution arranged in inverse powers of $t^{\frac{1}{2}}$, as in (11) for example. We may get this by picking out the coefficients properly from the expansions of A and its powers in (24). Or, from the operational equation (21) itself, which becomes

$$\begin{aligned} g = \Big\{1 + \frac{2s}{1-s} &- 2q_1 l \,.\, 2s\,(1 + 2s + 3s^2 + 4s^3 + \ldots) \\ &+ \frac{(2q_1 l)^2}{\underline{|2}} \,.\, 2s\,(1 + 2^2 s + 3^2 s^2 + 4^2 s^3 + \ldots) \\ &- \frac{(2q_1 l)^3}{\underline{|3}} \,.\, 2s\,(1 + 2^3 s + 3^3 s^2 + 4^3 s^3 + \ldots) + \&c.\Big\} q_1 \mathrm{V}_0, \end{aligned} \tag{25}$$

when arranged in powers of q_1. Here the even powers of q_1 are ignorable, as they involve whole differentiations. So the

q_1 term gives the factor of $t^{-\frac{1}{2}}$, the q_1^3 term that of $t^{-\frac{3}{2}}$, and so on. It is easily seen by inspection that the functions of s that occur in (25) are derivable in succession from one another by the operator $s(d/ds)$, the first one being $2s/(1-s)$. So they may be written finitely if we like. Or,

$$g = V_0\left(\frac{c_1}{\pi k_1 t}\right)^{\frac{1}{2}}\left\{1 + \left(1 - \frac{c_1 l^2}{k_1 t}(s\Delta)^2 + \left(\frac{c_1 l^2}{k_1 t}\right)^2 \frac{(s\Delta)^4}{\underline{|2}} - \ldots\right)\frac{2s}{1-s}\right\} \quad (26)$$

where Δ is short for d/ds, is the complete solution, of which the first part is the same as

$$g = \frac{V_0}{k_1}\left(\frac{ck}{\pi t}\right)^{\frac{1}{2}}\left\{1 - \left(1 - \frac{c_1 k_1}{ck}\right)\frac{ckR^2}{2t} + \ldots\right\}, \quad (27)$$

where $R = l/k_1$ as before. Comparing with (11), we can see the effect of the capacity of the skin.

Large Correction for Sphericity in Perry's Problem.

§ 230. The next step would be to go on to consider the case of a spherical earth, first with no surface resistance to the escape of heat, next with a resisting skin, and, thirdly, with a skin of any depth, the capacity being allowed for. But these are too complicated for the present purpose, and would frighten timid readers, and perhaps some Cambridge mathematicians as well. At the same time, I may remark that the solutions can be got through the operators in the form of Fourier series with much less work than by Fourier's way.

Prof. Perry has examined Fourier's solution for a homogeneous sphere with constant surface emissivity (equivalent to a resisting skin) and a very interesting result comes out, viz., that the plane solution gives much too big a result with the same data as regards conductivity and capacity, internal and external.

Thus, according to Perry, take the radius of the earth at 6380^6, and the depth of the skin at 40^5 centim., or 4 kilom., the initial temperature as 4000°C., the present surface gradient 1° in 2743 centim., and $c = 2{\cdot}86$, $k = {\cdot}47$, $c_1 = {\cdot}507$, $k_1 = {\cdot}00595$, so that k in the earth is 79 times that of surface rock, and k/c in the earth is 14 times that of surface rock. Then the time of subsidence to the present state is 960^8 years, or 96 times the estimate of 10^8 years given with the same data except that the earth has the same c and k throughout as in the skin, when calculated by the plane theory, or the linear diffusion of heat.

But, according to the plane theory adapted to meet the case of a plane slab of skin instead of a spherical shell, the time of subsidence would, by (12) above, be ck/c_1k_1 times as great as the estimate without the increased c and k inside, or about 4630^8 years. So the effect of having an infinite conductor with a plane face to represent the earth is to increase the time of subsidence from 960^8 to 4630^8 years. Perhaps it does not matter in very rough estimates; but it is interesting to note the difference made by the finite size of the earth, and that it lessens the time of subsidence in the ratio 463 : 96, or 4·7 to 1.

In order to ascertain whether the objection made to Perry's neglect of the capacity of the skin had any serious basis, I have worked out the corresponding formula with the capacity allowed for. There is very little difference numerically. Thus, Perry found two terms in the Fourier expansion to be necessary. But the first term is a large multiple of the second, so we may take it alone for our comparison. By the first term of Perry's formula, the time of subsidence comes to 9020^7 years. Allowing for the capacity of the skin, I find it comes to 9030^7 years. But I have not taken special pains to get the third figures right.

Remarks on the Age of the Earth.

§ 231. Now a few remarks (which I make with much diffidence) on the practical outcome of Prof. Perry's investigations. It is known that geologists demand long ages of time for the earth's evolution in its geological aspect. Physicists, on the other hand, have offered them what geologists consider the miserable allowance of from 20 to 400 million years. Prof. Tait, I believe, offers them only 20. They want more.*

There are two evident ways of getting more. The first is by not requiring so much, by allowing that the earth's evolution went on at a far more rapid pace in former times than at present, wholly apart from catastrophes. This seems to

* It is to be remarked, however, that geologists have come down in their demands remarkably, and probably in consequence of Lord Kelvin's work.

be a reasonable view. The other is Prof. Perry's way. He advances arguments to show that it is reasonable to suppose that the earth's effective capacity and conductivity, especially the latter, may be even at present much greater in the earth's interior than in its skin, so that the time of subsidence to the present skin gradient is prolonged. But it is not a mere question of the present state of the earth's interior. The state during the whole period of geological evolution has to be considered, and Prof. Perry's argument seems to me to apply with greater and greater force the further we go back in time. I presume that the earth's birth, for geological purposes, should be reckoned from the time when it became encrusted, or when the crust attained some notable thickness to give some sort of stability. Is it necessary to solidify the earth all through before beginning its life? If we allow it to solidify gradually by the natural increase of depth of its solid crust, consequent upon its cooling, it is evident that the age of the earth may possibly be much extended.

As for the origin of life upon this planet, the only reasonable view seems to me to be Topsy's theory. She was a true philosopher, and "she spekt she growed." Any other theory is of the elephant and tortoise kind, a sort of evasion, which explains nothing, whilst it increases one's difficulties. Prof. Tyndall was of Topsy's persuasion. So am I, as I firmly believe (subject to correction) in the truth of his view as to the "promise and potency" of life in so-called dead matter under the influence of the forces of nature.

Peculiar Nature of the Problem of the Cooling of a Homogeneous Sphere with a Resisting Skin.

§ 232. In connection with the above problems in cooling by diffusion and escape of heat at the surface of bodies there are a number of incidental matters of great interest, some of the most noteworthy of which may be briefly noticed. In the first place, as was remarked in § 228, the prolongation of the time of cooling of a sphere from a given uniform initial temperature until a given gradient of temperature is reached at the surface of escape, produced by augmentation of the conductivity and capacity of the inner portion only, is not by any means an obvious result, though not difficult to understand in the case of an infinite block with a plane boundary when there is a similar augmentation of conductivity and capacity within its skin. In fact, we can easily make it be either a retardation or an acceleration at pleasure, when it is a sphere that is in question. To show this, let the conductivity be made infinitely great as an extreme case, except in the skin, where it remains finite, without changing the capacity either in the skin or body of the sphere. The theory of the cooling of the sphere is then like that of the discharge of a condenser through a resistance. The resistance here is the resistance of the skin, and the capacity is that of the inner body. Now by reducing the depth of the skin, and therefore the resistance, we may accelerate the discharge as much as we please. Thus, with the skin conductivity as in § 230, or 0·006, and the internal capacity per unit volume also the skin value, or 0·5, the time taken to fall from an initial uniform temperature of 4000°C. until the present gradient of temperature is reached in the skin is 160^8 years when the skin's depth is 10 kilometres, but only 330^7 years when it is 1 kilometre. If it is 4 kilometres, as in Prof. Perry's example, the result is 920^7 years, which is only 9 times the standard result of 10^8 years found by Lord Kelvin. It is raised to 960^8 years, as Perry has shown, by reducing the internal conductivity from infinity to 79 times that of the skin, whilst at the same time increasing the internal capacity to 5·7 times that of the skin. To obtain the standard result, 10^8 years, with infinite internal conductivity and with internal capacity as in the skin, requires the skin to be only $\frac{1}{4}$ kilometre in depth, or a little less. When made thinner still, the time required falls off to any extent. These examples will show the danger of over-hasty generalisations regarding the effect of varying the internal conductivity and capacity. It is a general principle that increasing the conductivity accelerates the subsidence of a normal system, or a distribution of temperature which will subside according to the condenser law, and Prof. Perry's case is no exception. But there are other considerations, and the case is considerably mixed. If, in Prof. Perry's 960^8 years problem, we raise the internal conductivity to infinity, making no other change, we *reduce* the time to 520^8 years. Here the internal capacity is still 5·7 times that in the skin. Now reduce it to the same value as in the skin, and the time falls to 920^7 years, as just mentioned. Lastly, reduce the internal conductivity to the skin value, and it falls to 10^8 years, being now Lord Kelvin's case.

Cooling of a Body of Variable Conductivity and Capacity but with their Product Constant.

§ 233. After these illustrations of the curious nature of the problem, consider another matter. It was mentioned in § 229 that the flux of heat into an infinite homogeneous block due to sources maintaining its plane face at the constant temperature V_0 was not altered by changing the material under the skin to another having the same value of the product *ck*. Or thus, by equations (2) and (3), the flux of heat, say C, per unit area, is

$$C = kg = (ckp)^{\frac{1}{2}} V_0. \qquad (28)$$

This is unaltered by a change of material not altering the value of *ck*.

The result may be extended to include any number of slabs of different materials put together to make a block, provided *ck* is the same for all; or, in the limit, to a continuously heterogeneous material in which *ck* is constant, with, however, homogeneity in every slice parallel to the plane face.

Conversely, since the final temperature due to the impressed V_0 is a state of uniform temperature V_0 everywhere, if we start with V_0 everywhere constant, and let it subside by internal diffusion and free escape at the surface, the flux of heat at the surface will be unaltered by any change of material, provided every plane stratum is homogeneous in itself, and *ck* is the same for all.

A similar result applies to a sphere, with concentric shells instead of plane slabs, provided the correction for sphericity, due to the finite size of the sphere, be insensible: or if it be sensible, then we may have approximately the same result.

Magnitude of the Correction for Sphericity in Various Cases.

§ 234. Another interesting point is the magnitude of the correction for sphericity. As mentioned in § 230, this is very large in Perry's 9,600 millions problem of a shell of depth 4 kilom., surrounding a homogeneous sphere of greater capacity and very much greater conductivity. The time for the corresponding infinite block is then 4·7 times that for the sphere. This was so remarkable that I suspected and suggested to Perry an error in his calculation of Fourier's formula. But I confirmed the result, and also obtained very nearly the same result from an entirely different formula which allowed for the capacity of the skin.

On the other hand, when the surface values of c and k extend all through the earth, as in Lord Kelvin's problem, the correction for sphericity is quite small, as mentioned in § 227. It only reduces the time of cooling by $\frac{1}{29}$th part of the 10^8 years which belongs to the infinite block.

Now in Perry's case we have a large increase in c and a very large increase in k beginning at a moderate depth. But if we increase them gradually, so as only to become very big near the centre, we do not get the Perry effect. To illustrate this I have calculated a few cases of continuously heterogeneous material.

When c and k both vary inversely as the distance from the centre of the earth, with the same values at the surface as before, I find that the correction is reduced to $\frac{1}{58}$ part. That is, the 10^8 years of the infinite homogeneous block is reduced by $\frac{1}{58}$ part to represent the new case of variable c and k in the earth, instead of by $\frac{1}{29}$ part, as when they are constant.

Also, to accentuate this effect, let the c and k in the earth vary inversely as the square of the distance from the centre. Then I find that the correction vanishes. That is, the time of subsidence from the initial state of 4,000°C. to the given gradient of temperature at the surface is 10^8 years, the same as for the infinite homogeneous block.

Similar results occur in other cases of gradual variation, with, it may be, very large changes in c and k near the centre, but very little near the surface, or between the surface and half-way down; and clearly great latitude in the law of variation is permissible, provided we do not introduce great changes in c and k near the surface.

Explanation of the last.

§ 235. We may get an insight into the meaning of these corrections by dividing the sphere into a series of shells of unit depth. Since the flow of heat is radial, it is like the diffusion of heat through a series of flat plates of variable conductivity and capacity. Now the total conductance of a shell is proportional to its area, and so is its total capacity. Therefore, if the c and k in the sphere be constant, the conductance and capacity of a shell vary as the square of the distance from the centre, being zero at the centre, very small round it, and greatest and increasing most rapidly at the surface. So we see that large variations in c and k near the centre may make only trifling differences in the state of things at the surface, as the conductance and capacity of the innermost shells are naturally low. Besides that, they are so far away from the surface where the escape takes place that their c and k may become of little moment in the practical problem concerning the gradient of temperature in the skin. These considerations may help one to understand why, when c and k vary as the first or as the second inverse power of the distance from the centre, so little difference is made in the time of cooling to the present gradient.

Also, by substituting a block for the sphere, say a block of length equal to the radius of the sphere, and of cross-section equal to the surface of the sphere, this block to be insulated at its sides and open at its ends, we see that to represent the sphere of uniform c and k, the c and k in the block must vary directly as the square of the distance from the far end (corresponding to the centre of the sphere) where they are zero (equivalent to insulation). On the other hand, when c and k in the sphere vary inversely as the distance from the centre, c and k in the block must vary directly as the distance from the far end. Finally, when c and k in the sphere vary inversely as the square of the distance from the centre, c and k in the block must be *uniform*. That is, we have a homogeneous block, only it is of finite depth instead of infinite.

We now see why the correction in the last case disappears, the time for the sphere being the same as for the block. It is not asserted that the complete solution of the problem is the same in both cases, but that the problem is reduced to one of linear diffusion in a homogeneous medium, and that under the circumstances the finiteness of depth of the block does not influence the result sensibly, the secondary waves to and fro along the block due to its finite depth being of insensible effect because the depth is so great.

Investigation by the Wave Method of the Cooling of a Homogeneous Sphere with a Resisting Skin. Effect of Varying the Constants.

§ 236. In contrast with the above results with continuously varying c and k, Prof. Perry's case involves so large a correction for sphericity as to deserve an independent confirmation by a method not requiring the use of the Fourier expansion. For it is by the consistency of results obtained in different ways that a conviction of the accuracy of the results of complicated processes may best be obtained. It is very easy to make mistakes in calculating Fourier series of complicated forms. Fortunately, in this case, I find that my operational method leads straight to the solution by a simple process.

We found in § 228, equation (8), that

$$V_1 = \frac{V_0}{1 + Rkq} \tag{29}$$

expresses the temperature V_1 just inside the skin due to V_0 impressed on its outside, when R is the resistance of unit area of the skin, k the internal conductivity, and $q = (cp/k)^{\frac{1}{2}}$. This is in the plane problem.

Now get the corresponding solution for the sphere. Let V be the temperature at distance r from the centre of a sphere of radius a and of uniform c and k due to V_0 impressed on the outside of an enveloping skin. Then V is given by

$$V = \frac{\dfrac{a}{r}\dfrac{\text{shin}\, qr}{\text{shin}\, qa}V_0}{1 - \dfrac{Rk}{a} + Rkq \coth qa}. \tag{30}$$

To prove this, note that in the first place V satisfies the spherical characteristic

$$\frac{1}{r^2}\frac{d}{dr}r^2\frac{dV}{dr} = q^2V; \tag{31}$$

next, that it is finite at the centre; and, lastly, that at the inside of the skin, where $r=a$, it satisfies the condition of continuity of the flux of heat there, or

$$\frac{V_0 - V_1}{R} = k\frac{dV}{dr}. \tag{32}$$

This is complete. But what we want is V_1. So put $r=a$ in (30). This makes, if $s=Rk/a$,

$$V_1 = \frac{V_0}{1 - s + Rkq \coth qa}, \tag{33}$$

which gives V_1 in terms of V_0. Comparing with (29) we see that 1 becomes $1-s$ (which is a trifle less) and Rkq receives the factor $\coth qa$, which brings in an infinite series of secondary diffusive waves between the centre and boundary.

To show them explicitly, we may develop (33) by long division to the form

$$V_1 = (a_0 + a_1 y + a_2 y^2 + \dots)V_0, \tag{34}$$

where

$$y = \epsilon^{-2qa}. \tag{35}$$

Here $a_0 V_0$ is the result of the primary wave from the source V_0 outside the skin, as modified by sphericity; the second term is the result of the first wave reflected from the centre, the third term results from the weaker second reflected wave, and so on.

But all these secondary waves are of insensible effect in our problem, as we know by the solutions previously given when the proper numbers for c and k, &c., are put in. The significant solution is merely the first part independent of y. This amounts to the same as making a infinite in the coth function, when it reduces to 1. So, by (33),

$$V_1 = \frac{V_0}{1 - s + Rkq}, \tag{36}$$

is the practical solution in operational form. We see that it is equivalent to (29), only with a changed constant.

We also know, by (11) and (12), § 228, that only the first power of q is significant in the earth problem. This makes (36) become

$$V_1 = \frac{1}{1-s}\left(1 - \frac{Rkq}{1-s}\right)V_0, \tag{37}$$

which, by (5), is converted to the algebraical form

$$V_1 = \frac{V_0}{1-s}\left\{1 - \frac{R}{1-s}\left(\frac{ck}{\pi t}\right)^{\frac{1}{2}}\right\}, \tag{38}$$

the required result. The subsidence solution is got by subtracting the right member from V_0. This makes

$$g = -\frac{V_0 s}{(1-s)l} + \frac{V_0}{k_1(1-s)^2}\left(\frac{ck}{\pi t}\right)^{\frac{1}{2}}, \tag{39}$$

if $V_1 = gl$, where g is the gradient of temperature in the skin of depth l. We may write (39) thus,

$$g' = \frac{V_0}{k_1}\left(\frac{ck}{\pi t}\right)^{\frac{1}{2}}, \tag{40}$$

if

$$\frac{g'}{g} = (1-s)^2\left(1 + \frac{V_0 s}{(1-s)lg}\right). \tag{41}$$

Now (40) is of the same form exactly as (12), with a changed value of the gradient. The effect of the sphericity is, therefore, the same as changing the gradient in the plane problem from g to g'.

Now put in the numerical values as in § 230. That of s is 0·0495, l is 40^5, and V_0 is 4000. So

$$g' = 2{\cdot}1819\, g, \tag{42}$$

which is to be used in (40). This increased value of g requires t to be reduced as its square. So the time required to make g be $\frac{1}{2743}$ is $(2{\cdot}1819)^2$, or 4·76 times as long for the infinite block as for the sphere. Q.E.D.

If desired, the full expression for the secondary waves can be developed from (34), but all we wanted was a direct corroboration of the result got from the Fourier expansion. The method followed is an example of the theory of § 12 of my paper "On Operators in Physical Mathematics," *Proc. R. S.*, Vol. LII., 1893, which is of very general application.

The formula (39) allows us to see readily the effect of varying the constants. The time of cooling varies directly as c and as the square of V_0, so these may be dismissed at once. There are left l and k. Varying k only, I find that t has a maximum and a minimum when l is under 7 kilometres. The minimum is of no consequence. With $l = 4$ kilom., $c = 2{\cdot}86$, the maximum occurs when $k/k_1 = 73$, and is $t = 95{\cdot}5$, the unit being 10^8 years. That Perry should have spotted the maximum so closely is (unless he is a witch) one of the most remarkable coincidences in ancient or modern history. But the hump is so flat-topped that much smaller values of k will do for big t. Thus $k/k_1 = 30$ makes t be about 90.

Another (and perhaps physically better way than increasing k) of getting big t is to increase the depth of the crust. Then a smaller k will do. Thus with $l = 20$ kilometres we get $t = 68{\cdot}5$ when $k/k_1 = 15{\cdot}95$ only. And $l = 30$ kilometres allows us to have $t = 53{\cdot}4$ when k/k_1 is only $10{\cdot}65$, and $t = 95$ when k/k_1 is $21{\cdot}3$. These results altogether favour Perry's view, and are better than his own example.

Importance of the Operational Method.

§ 237. We now leave heat problems, and pass to the theory of electrical matters involving diffusion. Pure diffusion, as of heat, comes in principally in two different ways. There is, first, Lord Kelvin's electrostatic diffusion in a submarine cable when perfectly insulated and free from self-induction. Secondly, there is Maxwell's diffusion of magnetic induction in electrical conductors. There are also two comparatively unimportant cases, viz., diffusion in a cable or other circuit, when it is the self-induction and the leakage that control matters, and a kind of diffusion in a magnetic conductor. Of these, the electrostatic diffusion involves the simplest fundamental ideas, and will therefore occupy us first. After that, diffusion in electrical conductors will naturally follow.

How these diffusive propagations arise from the general theory of electromagnetic waves has been explained in Chapter IV. in considerable detail, including the more difficult case of elastic diffusion. What we have now to do is to consolidate the knowledge by actual exemplification. We shall then be able to explain the meaning of the operational mathematics above employed, as it turns up naturally. The physics itself will serve to guide us along to useful methods and results. At present the above illustrations from the theory of heat diffusion will serve a double purpose. First, to illustrate Lord Kelvin's theory of the age of the earth and its recent extension by Prof. Perry, the practical import of which, however, remains to be discovered, as very uncertain and speculative data are involved. Next, to show that my operational method of dealing with these and similar more advanced problems is of importance. I assert that by its means problems can be attacked and successfully solved with greater power than by any other known method. Furthermore, that it is essentially simple in operation; so that, although it goes deeper, yet it requires less work and less mathematics of the complicated kind. And, finally, that it is for the above reasons and others quite practical. It is rather disagreeable to have to be self-assertive and dogmatic (especially when one thinks of the always possible risk of error); but there may be times when it becomes a duty—*e.g.*, when mathematical rigourists are obstructive.*

* [March 21, 1895.] After writing the above, Prof. Perry wrote asking about the case of capacity and conductivity functions of the temperature, saying, "X. says he can't do it, doesn't know anyone who can, and is sure *you* can't." The general case is perhaps hard (I did not try it), but I found at once that when c and k vary together, according to any power, integral or fractional, of the temperature, the solution was quite easy, the characteristic becoming linear. This is obvious when done. I sent Perry some solutions of this kind. He then himself extended the matter by taking c and k to be any similar functions of the temperature. This is also obvious when done (Perry, *Nature*, Feb. 7, 1895, p. 341). He finds that if c and k increase s per cent. per 100deg., then Lord Kelvin's age is multiplied $(1+s/5)^2$ times; *e.g.*, by 121 if $s=50$. His data were due to Dr. R. Weber, and indicated a large increase in c and k with temperature. If correct, Prof. Perry would be fully justified, though to an uncertain extent. But Dr. R. Weber has supplied fresh data which do not show any notable increase in k, whilst that in c is much less than Perry assumed. So Lord Kelvin (*Nature*, Mar. 7, 1895, p. 438) concludes that Perry is wrong. He is also inclined to reduce the initial temperature, and so bring down the age even to 10 million years; or, allowing for other things, to about 24 millions, in agreement with Mr. Clarence King's conclusion in comparing the calculations of Helmholtz, Newcomb, and Kelvin on the age of the sun. It will be interesting to see whether the geologists will continue their downward course to 24 or 10 millions (Sir A. Geikie, *Nature*, Feb. 14, 1895, p. 367, is quite satisfied with only 100), or whether mathematical physicists will, by fresh data, be obliged to go up to meet them. Prof. Perry said (*Nature*, Jan. 3. 1895, p. 224) that his conclusions were independent of the correctness of R. Weber's results (the old ones). "Lord Kelvin has to prove the impossibility of the rocks inside the earth being better conductors (including convective conduction in case of liquid rock in crevices) than the surface rocks." "The rocks at 20 miles deep are not merely at a high temperature, but also under great pressure." In any case, however, it must be difficult to come to a reliable estimate as to how far Prof. Perry's important principle is really operative.

CHAPTER VI.

PURE DIFFUSION OF ELECTRIC DISPLACEMENT.

Analogy between the Diffusion of Heat in a Rod and the Diffusion of Charge in a Cable.

§ 238. In order that the problem of the propagation of electrical disturbances along a telegraphic or telephonic circuit shall reduce practically to that of the diffusion of the electric displacement after the manner of heat in the celebrated theory of Fourier, it is necessary for the self-induction to be ignorable, and that the external disturbances to which circuits are liable should be removed. It would not be at all desirable to bring a practical telegraph circuit to such a state closely, because it is a state of relative inefficiency, accompanied by the greatest possible distortion in transit, and is therefore a state to be avoided by, as before explained,* making self-induction be of importance, if efficient rapid signalling with little distortion be required. The nearest approach to the theory of diffusion being in slow signalling through a long cable, we make believe now that this case is truly represented by the reduced forms of the more general equations appropriate to elastic diffusion.

On this understanding the two circuital equations, when suitably transformed as explained in §§ 200-202, reduce to

$$-\frac{dC}{dx}=SpV, \qquad -\frac{dV}{dx}=RC, \tag{1}$$

* §§ 215-218, Chap. IV., Vol. I.

where R and S are the resistance and permittance per unit length of line, whilst V and C are the transverse voltage and the circuital gaussage of the more general theory, but which may now be called the potential difference of the wires and the current in them, if there be a pair of wires. Or, if we have a cable in question, using only one wire, then we may call V simply the potential and C the current.

From (1) we derive the characteristic

$$\frac{d^2\mathrm{V}}{dx^2} = \mathrm{RS}p\mathrm{V} = q^2\mathrm{V}, \text{ say.} \qquad (2)$$

In order to translate to heat problems, perhaps the easiest way is to consider the longitudinal conduction of heat in a rod. Then V is the temperature and C the flux of heat, whilst R^{-1} and S are the conductance and capacity for heat per unit length of rod. But the rod should be insulated laterally. It is easy to insulate a rod electrically; but it is much harder, if not impossible, to insulate it thermally to an equivalent extent. So, if the flow of heat in a real rod be rejected for want of a sufficiently close similarity to the electrical problem, we may imagine an infinite number of rods fitted together in contact side by side. Just as jerry-built houses in a street mutually support one another, and prevent the collapse that would occur were they separated, so will the rods prevent the lateral escape of heat from their neighbours, so that a longitudinal flux of heat is possible in the same way as in a perfectly insulated rod. This is the case of the linear flow of heat in an infinite homogeneous conductor. These remarks are to enable the reader to translate from electrical to heat problems readily. On the whole, the cable is preferable in the study of diffusion, on account of the facility with which terminal and other auxiliary arrangements can be imagined, and, if need be, practically realised. The heat problems are not so convenient in this respect. On the other hand, there is no doubt greater scientific interest in heat problems when they concern such stupendous questions as the age of our common mother earth; but since this is primarily an electrical work, I cannot go on further with that question, but leave it to David and Goliath.

The Operational Method assists Fourier.

§ 239. We have now to consider a number of problems which can be solved at once without going to the elaborate theory of Fourier series and integrals. In doing this, we shall have, preliminarily, to work by instinct, not by rigorous rules. We have to find out first *how* things go in the mathematics as well as in the physics. When we have learnt the go of it we may be able to see our way to an understanding of the meaning of the processes, and bring them into alignment with other processes. And I must here write a caution. I may have to point out sometimes that my method leads to solutions much more simply than Fourier's method. I may, therefore, appear to be disparaging and endeavouring to supersede his work. But it is nothing of the sort. In a complete treatise on diffusion Fourier's and other methods would come in side by side—not as antagonists, but as mutual friends helping one another. The limitations of space forbid this, and I must necessarily keep Fourier series and integrals rather in the background. But this is not to be misunderstood in the sense suggested. No one admires Fourier more than I do. It is the only entertaining mathematical work I ever saw. Its lucidity has always been admired. But it was more than lucid. It was luminous. Its light showed a crowd of followers the way to a heap of new physical problems.

The reader who may think that mathematics is all found out, and can be put in a cut-and-dried form like Euclid, in propositions and corollaries, is very much mistaken; and if he expects a similar systematic exposition here he will be disappointed. The virtues of the academical system of rigorous mathematical training are well known. But it has its faults. A very serious one (perhaps a necessary one) is that it checks instead of stimulating any originality the student may possess, by keeping him in regular grooves. Outsiders may find that there are other grooves just as good, and perhaps a great deal better, for their purposes. Now, as my grooves are not the conventional ones, there is no need for any formal treatment. Such would be quite improper for our purpose, and would not be favourable to rapid acquisition and comprehension. For it is in mathematics just as in the real world; you must observe and experiment to find out the go of it. All experimentation is deductive work in a sense, only it is done by trial and error, followed by new deductions and changes of direction to suit circumstances. Only afterwards, when the go of it is known, is any formal exposition possible. Nothing could be more fatal to progress than to make fixed rules and conventions at the beginning, and then go on by mere deduction. You would be fettered by your own conventions, and be in the same fix as the House of Commons with respect to the despatch of business, stopped by its own rules.

But the reader may object, Surely the author has got to know the go of it already, and can therefore eliminate the preliminary irregularity and make it logical, not experimental? So he has in a great measure, but he knows better. It is not the proper way under the circumstances, being an unnatural way. It is ever so much easier to the reader to find the go of it first, and it is the natural way. The reader may then be able a little later to see the inner meaning of it himself, with a little assistance. To this extent, however, the historical method can be departed from to the reader's profit. There is no occasion whatever (nor would there be space) to describe the failures which make up the bulk of experimental work. He can be led into successful grooves at once. Of course, I do not write for rigourists (although their attention would be delightful) but for a wider circle of readers who have fewer prejudices, although their mathematical knowledge may be to that of the rigourists as a straw to a haystack. It is possible to carry waggon-loads of mathematics under your hat, and yet know nothing whatever about the operational solution of physical differential equations.

The Characteristic Equation and Solution in terms of Time-Functions.

§ **240.** Now, consider the characteristic equation (2) above. If q were a constant, its solution would obviously be

$$V=\epsilon^{qx}A+\epsilon^{-qx}B, \tag{3}$$

where A and B are any constants. That is, there are two independent functions of x which satisfy (2). The constancy of A and B means independence of x.

It is equally true that (3) is the solution in the same sense when q^2 has the operational meaning RSp, because the formal satisfaction by test is the same. The constants are still constants with respect to x, but they are now any functions of t. That is, (3) is a form of the general solution of the characteristic. To go further, we have to find A and B to suit special cases, and then by the execution of the processes implied by the exponential operators convert the solutions from operational to algebraical form. There is a lot of assumption here; for example, that the operations can be effected, as they involve preliminarily unintelligible ideas. The best proof is to go and do it.

The easiest solutions are those relating to the effects produced at a given spot by causes acting there. Those produced at a distance can be easily deduced later. So, now we take some special cases to begin the treatment. Let an infinitely long cable be laid in any depth of water. It need not be laid straight, so by winding it about, even the finite size of the seas of the earth might be sufficient to contain a sufficient length for our purpose, which is, that the near end of the cable is to be freely at our disposal to operate on, whilst the far end is so very far off that it cannot react sensibly on the near end, and to a great distance therefrom, in a large interval of time.

Let the cable be initially free from charge, and be then operated upon by a battery of voltage e and no resistance at its beginning, where $x=0$. That is, one end of the battery is put to line, and the other to earth, the absence of resistance being merely a practical simplification. The impressed voltage e may be regarded as any function of the time t (real, of course, but not necessarily continuous). The effect is to raise the potential at the beginning to $V_0=e$, and V_0 may be regarded as the sole cause of disturbance in the cable itself further away. Then $A=0$ and $B=V_0$ in (3), making

$$V=\epsilon^{-qx}V_0, \tag{4}$$

for the characteristic is satisfied, and $V = V_0$ at $x = 0$, and there is no other imposed condition. Other cases, in which A is not zero, will come later.

By (4), and the second of (1),

$$C = \frac{q}{R} V = \frac{q}{R} \epsilon^{-qx} V_0 ; \tag{5}$$

the first equation giving C in terms of V at the place, the second in terms of V_0, We also have

$$C = \epsilon^{-qx} C_0, \tag{6}$$

if C_0 is the current at $x = 0$; and also

$$C_0 = \frac{q}{R} V_0 = \left(\frac{Sp}{R}\right)^{\frac{1}{2}} V_0. \tag{7}$$

The last equation is the simplest in general, because V_0 can be made simple. The operator q/R turns potential to current. It is, therefore, the conductance operator of the cable. Similarly,

$$V_0 = \frac{R}{q} C_0 = \left(\frac{R}{Sp}\right)^{\frac{1}{2}} C_0. \tag{8}$$

So the resistance operator is $(R/Sp)^{\frac{1}{2}}$.

We see at once from (7) that the current entering the cable depends only upon the *ratio* of S to R. Its propagation in the cable itself depends on their product.

Steady Impressed Force at Beginning of Cable. Fractional Differentiation. Simply Periodic Force.

§ **241.** Now let V_0 be such a function of the time as to be zero before and constant after $t = 0$. What is C_0? To find it, we require to know the meaning of $p^{\frac{1}{2}} V_0$. Now the problem stated is a well-known one in Fourier's theory of heat conduction; and when by Fourier's methods we develop the solution we find that it is

$$C_0 = \left(\frac{S}{R\pi t}\right)^{\frac{1}{2}} V_0. \tag{9}$$

Comparing with (7), we see that we require to have, on removal of unnecessary constants,

$$p^{\frac{1}{2}} 1 = (\pi t)^{-\frac{1}{2}}, \tag{A}$$

a fundamental formula. The 1 means that function of the time which is zero before and unity after $t=0$. We are only concerned with positive values of the time. This way of finding the meaning of a fractional differentiation of a given function is purely experimental. Any problem involving $p^{\frac{1}{2}}1$ in the operational form of solution will do for the determination of its value, by comparison with the solution by Fourier's method. On the other hand, the result is a simple fundamental one in fractional differentiation, and does not need Fourier. But the reader presumably cannot take in the idea of a fractional differentiation yet. So, for the present, let it be taken as a fact that the value of $p^{\frac{1}{2}}1$ *is* $(\pi t)^{-\frac{1}{2}}$. We can make use of this fact extensively in Fourier mathematics with much advantage, without necessarily going a step further in the direction of fractional differentiation.

By (9) we see that the current entering the line is infinite at the first moment (because of the absence of self induction), and then falls, according to the inverse square root of the time, to zero. At first, the slope of V in the line is infinite at its beginning, and so is C_0. But as the cable gets charged the slope gets smaller. Finally, the potential is V_0 everywhere and the current is zero. Or we may say that the final current is zero, because the resistance is infinite. There can only be current when the charge is increasing. It really never stops increasing, but the potential near the beginning gets to be so nearly V_0, as to prevent the very distant parts of the cable receiving their charge except at an insensible rate. Mathematically speaking, we say that $V=V_0$ everywhere, when $t=\infty$.

The final states of V and C may also be seen from (4) and (5). Put $p=0$ in them and they reduce to $V=V_0$, and $C=0$. This process is general. Putting $p=0$ in an operator destroys time-variation, and gives the ultimate steady form, when there can be a steady state.

Another way of looking at the matter is to consider how we get the simply periodic solution out of an operational solution, when the impressed force is simply periodic. If the frequency is $n/2\pi$, we put $p=ni$ in the operator. Now $p=0$ is equivalent to $n=0$, or an infinitely prolonged period, which means a steady state.

In the present case, the simply periodic solution is, by (7),

$$C_0 = \left(\frac{Sni}{R}\right)^{\frac{1}{2}} V_0 = \left(\frac{Sn}{2R}\right)^{\frac{1}{2}} (1+i) V_0. \qquad (10)$$

Here i means p/n; that is, differentiation with respect to nt; so (10) is complete when V_0 is given in amplitude and phase. Say it is $e \sin nt$, then

$$C_0 = \left(\frac{Sn}{2R}\right)^{\frac{1}{2}} (\sin nt + \cos nt)e. \qquad (11)$$

It should be understood, though, that time must be allowed to enable this state to be arrived at.

Effect of a Terminal Arrangement. Two Solutions in the Case of a Resistance.

§ 242. Still keeping to the beginning of the cable, let us examine the effect of a terminal arrangement. Let $V = ZC$ be its equation *per se*, so that Z is its resistance operator. Now, that of the cable is $(R/Sp)^{\frac{1}{2}}$, as before seen; so if Z is put between the cable and earth with the impressed voltage acting, we have

$$C_0 = \frac{e}{Z + (R/Sp)^{\frac{1}{2}}} \qquad (12)$$

to express the current through Z and entering the cable. This is because the operators are additive like resistances. Also, we have $V_0 = (R/Sp)^{\frac{1}{2}} C_0$ as before; consequently by (12)

$$V_0 = \frac{e}{1 + Z\left(\frac{Sp}{R}\right)^{\frac{1}{2}}}. \qquad (13)$$

This finds V_0, the potential at the beginning of the cable, in terms of e.

The operational solution (13) may be readily algebrized (or converted to algebraical form) in various cases of Z, practical and unpractical. One case will do to begin with to illustrate the conversion.

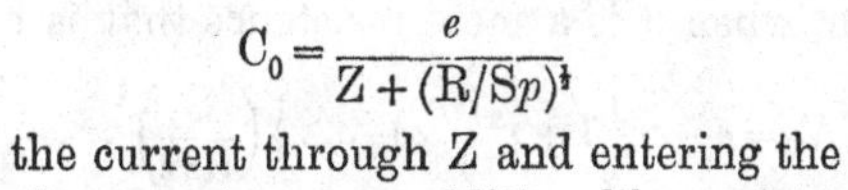

Let Z be the resistance operator of a coil, say

$$Z = r + lp, \qquad (14)$$

where r is its resistance and l its inductance. Then

$$V_0 = \frac{e}{1 + (r + lp)\left(\frac{Sp}{R}\right)^{\frac{1}{2}}} \tag{15}$$

is the operational solution giving V_0 in terms of e.

This may be algebrized as follows. By division,

$$V = \left\{1 - Z\left(\frac{Sp}{R}\right)^{\frac{1}{2}} + Z^2\frac{Sp}{R} - Z^3\left(\frac{Sp}{R}\right)^{\frac{3}{2}} + \dots\right\}e. \tag{16}$$

Here in Z we have only complete differentiations, therefore in union with the even powers of $(Sp/R)^{\frac{1}{2}}$ we still have complete differentiations. All these terms may be ignored when e is, as we shall suppose, constant after $t=0$, having previously been zero, and Z is a mere resistance. The cases of a permittance and an inductance will follow. So (16) reduces to

$$V_0 = \left\{1 - Z\left(1 + \frac{Z^2Sp}{R} + \frac{Z^4S^2p^2}{R^2} + \dots\right)\left(\frac{Sp}{R}\right)^{\frac{1}{2}}\right\}e. \tag{17}$$

We know $p^{\frac{1}{2}}1$ already, so the solution is found by complete differentiations performed upon it. Thus, in the case of no self-induction, when it is a mere resistance that is concerned,

$$V_0 = e - er\left\{1 + r^2\frac{Sp}{R} + \dots\right\}\left(\frac{S}{R\pi t}\right)^{\frac{1}{2}}. \tag{18}$$

This makes a series in descending powers of $t^{\frac{1}{2}}$. Thus,

$$V_0 = e - er\left(\frac{S}{R\pi t}\right)^{\frac{1}{2}}\left\{1 - \frac{r^2S}{2Rt} + 1.3\left(\frac{r^2S}{2Rt}\right)^2 - \dots\right\}. \tag{19}$$

When t is big enough, the only significant term is e, the final value. When t is smaller, the next becomes significant. When smaller still another term requires to be counted, and so on. But we must never pass beyond the smallest term in the series. As t decreases, the smallest term moves to the left. As it comes near the beginning of the series, the accuracy of calculation becomes somewhat impaired. When it reaches the first t term, so that the initial convergency has wholly disappeared, then we can only roughly guess the value of the series. So (19) is unsuitable when t is small enough to make the initial convergency be insufficient.

It is said that every bane has its antidote, and some amateur botanists have declared that the antidote is to be found near the bane. We have an example here. The antidote is got by algebrizing (15) in a different way. Keeping for the present to the simple case of a resistance only—$l=0$ in (15)—we may write

$$V_0 = \frac{\left(\frac{R}{r^2Sp}\right)^{\frac{1}{2}} e}{1+\left(\frac{R}{r^2Sp}\right)^{\frac{1}{2}}}; \tag{20}$$

or, by division,

$$V_0 = \left\{1 - \left(\frac{R}{r^2Sp}\right)^{\frac{1}{2}} + \left(\frac{R}{r^2Sp}\right) - \left(\frac{R}{r^2Sp}\right)^{\frac{3}{2}} + \dots\right\}\left(\frac{R}{r^2Sp}\right)^{\frac{1}{2}} e. \tag{21}$$

This we may split into two series, viz. :—

$$V_0 = \left\{\frac{R}{r^2Sp} + \left(\frac{R}{r^2Sp}\right)^2 + \dots\right\}\left(\frac{r^2Sp}{R}\right)^{\frac{1}{2}} e$$
$$-\left\{\frac{R}{r^2Sp} + \left(\frac{R}{r^2Sp}\right)^2 + \dots\right\} e. \tag{22}$$

In the second line we have complete integrations to perform on e. This is done by

$$p^{-n}1 = \frac{t^n}{\lfloor n}, \tag{B}$$

which is obvious enough, when, as at present, n is integral; viz., 1, 2, 3, &c. In the first line we have to make the same complete integrations upon the function $p^{\frac{1}{2}}1$. This is also done at sight by (B), when the matter of fractional differentiation is understood. But at present we can do without it, and integrate directly thus :—

$$p^{-1}t^{-\frac{1}{2}} = \frac{t^{\frac{1}{2}}}{\frac{1}{2}},$$
$$p^{-2}t^{-\frac{1}{2}} = \frac{t^{\frac{3}{2}}}{\frac{1}{2}\cdot\frac{3}{2}},$$

and so on, which is easy enough. So, by using (A) in the first line of (22) we obtain the complete solution in the form

$$V_0 = 2e\left(\frac{Rt}{r^2S\pi}\right)^{\frac{1}{2}}\left\{1 + \frac{2Rt}{3r^2S} + \frac{1}{3.5}\left(\frac{2Rt}{r^2S}\right)^2 + \dots\right\} - e(\epsilon^{Rt/r^2S} - 1). \tag{23}$$

We see now that we can calculate V_0 conveniently when t is small. But (23) is bad when t is big. Then we may consider (23) the bane, and (19) the antidote. They are complementary, though not mutually destructive.

Theory of a Terminal Condenser.

§ **243.** There is another simple case in which substantially the same process obtains as in the last example. Superficially considered, the problem of the effect of a terminal condenser in modifying the action of an impressed force on a cable is entirely different from that of the effect of a terminal resistance. Yet there is a very close analogy. Thus, let the terminal arrangement be a condenser of permittance s. Its equation is $C = spV$, so its resistance operator is $(sp)^{-1}$. Put this for Z in equation (13). Then

$$V_0 = \frac{e}{1 + \frac{1}{sp}\left(\frac{Sp}{R}\right)^{\frac{1}{2}}} = \frac{e}{1 + \frac{1}{aq}}, \tag{24}$$

if $a = s/S$, or that length of cable whose permittance equals that of the condenser, and q is as before.

Now, to show the analogy with the effect of a terminal resistance, put $Z = r$ in (12), making

$$C_0 = \frac{e}{r + \frac{R}{q}}, \quad \text{or} \quad rC_0 = \frac{e}{1 + \frac{1}{bq}}, \tag{24A}$$

if $b = r/R$, or that length of cable whose resistance is the same as the terminal resistance in the changed problem.

Comparing (24) with (24A), we see that the operational solutions are of the same form, only differing in the changed constant, a becoming b. So, if they are equal, we see that the potential at the beginning of the cable due to the impressed force runs through the same course when the condenser is interposed as the current (multiplied by r) does when a resistance is interposed.

To obtain the effect at a distance requires in both cases the introduction of the same operator ϵ^{-qx}. Consequently, we know that the course of the potential throughout the whole cable in the condenser problem is the same as that of the current in the resistance problem, due to the same impressed force, which may be any function of the time. And, we do not need to algebrize the solutions in order to predict this result.

Perhaps some people will say (as usual) that they do not like "algebrize"; that it is un-English, &c., &c. People are always saying something. What is more important is that a word to express the idea of conversion from operational to algebraical form is much wanted, and that "algebrize" seems to answer the purpose very well. Similarly we might say that we logarize a number when we take its logarithm, and delogarize it when we find the number whose logarithm it is; and so on.

When e is a steady force, beginning when $t=0$, we may algebrize (24) in two ways as before, and I will do it rather fully now, merely remarking that the work can be done at sight after a little practice by using equations (A) and (B), extended in the latter case to fractional degrees, a matter to be considered later. Thus, to obtain a convergent solution, expand the operator in descending powers of aq by division, making

$$V_0=\left(1-\frac{1}{aq}+\frac{1}{a^2q^2}-\frac{1}{a^3q^3}+\ldots\right)e. \tag{25}$$

Here the even powers of q involve complete integrations, to be done by (B) at sight. The odd powers involve complete integrations performed upon $p^{-\frac{1}{2}}1$, with limits 0 and t. Thus,

$$V_0=\left(1+\frac{1}{hp}+\frac{1}{h^2p^2}+\frac{1}{h^3p^3}+\ldots\right)e$$

$$-\left(1+\frac{1}{hp}+\frac{1}{h^2p^2}+\ldots\right)\frac{e}{(hp)^{\frac{1}{2}}}, \tag{26}$$

where $h=\text{RS}a^2$, which is a time constant. Also, we know already that $p^{-\frac{1}{2}}1=2(t/\pi)^{\frac{1}{2}}$, so (26) is converted to

$$V_0=e\,\epsilon^{t/h}-2e\left(\frac{t}{h\pi}\right)^{\frac{1}{2}}\left\{1+\frac{1}{3}\left(\frac{2t}{h}\right)+\frac{1}{3.5}\left(\frac{2t}{h}\right)^2+\ldots\right\}. \tag{27}$$

This is complete, and answers well, except when t is big, so that many terms have to be used.

To get the alternative solution, expand the operator in (24) in rising powers of qa. Thus,

$$V_0=\frac{qa}{1+qa}e=(1-qa+q^2a^2-q^3a^3+\ldots)\,qae. \tag{28}$$

Here the even powers of q contribute nothing (which is not so simple a matter as it looks), so

$$V_0 = (1 + q^2a^2 + q^4a^4 + q^6a^6 + \dots)\,qae. \tag{29}$$

Here we have to find $qa\,1$, which is known, and then execute complete differentiations upon it. Thus,

$$V_0 = e\,(1 + hp + h^2p^2 + h^3p^3 + \dots)\left(\frac{h}{\pi t}\right) \tag{30}$$

$$= \left(\frac{h}{\pi t}\right)^{\frac{1}{2}} e\left\{1 - \frac{h}{2t} + 1.3\left(\frac{h}{2t}\right)^2 - \dots\right\}. \tag{31}$$

This formula answers well for big t, and also when t is not so small as to render the initial convergency insufficient.

The condenser acts like a short-circuit at the first moment, so that the potential at the beginning of the cable acquires the full value e instantly. It then falls to zero as the condenser gets charged, in accordance with (27) and (31). Of course, the cable receives the same charge as the condenser; that is, the current is continuous through the condenser into the cable, according to Maxwell's now orthodox theory. But as the charge spreads itself over a condenser of infinitely great permittance, its density attenuates to nothing, so that $V = 0$ is the final state of the cable, although the total charge is finite. That the final V is zero is also to be seen by the operational solution (24), when we put $p = 0$ in it.

Theory of a Terminal Inductance.

§ **244.** As a third example, let the terminal arrangement be an inductance coil. For simplicity, let its resistance be zero. If really small the resistance may be merged in that of the cable itself without much error, and this is allowable when we desire to exhibit the effect of the inductance alone, which is materially different from that of a resistance or a permittance.

The terminal Z is now lp, so that, by (13), we have

$$V_0 = \frac{e}{1 + lp\left(\dfrac{Sp}{R}\right)^{\frac{1}{2}}} = \frac{e}{1 + (fq)^3}, \tag{32}$$

if $f^3 = l/R^2S$, a constant. It is quite easy to obtain the convergent algebraical solution. Expand in rising powers of fq, thus

$$V_0 = \frac{(fq)^{-3}e}{1+(fq)^{-3}} = \{(fq)^{-3} - (fq)^{-6} + (fq)^{-9} - \dots\}e. \tag{33}$$

Here the even powers of q involve complete integrations on 1, and the odd powers complete integrations performed upon $(fq)^{-1}$, so there is no new difficulty. To ease matters, put g for RSf^3. It is a time constant. Thus (33) is the same as

$$V_0 = \left(\frac{1}{gp} + \frac{1}{(gp)^4} + \frac{1}{(gp)^7} + \dots\right)\frac{e}{(gp)^{\frac{1}{2}}}$$

$$-\left(\frac{1}{(gp)^3} + \frac{1}{(gp)^6} + \frac{1}{(gp)^9} + \dots\right)e. \tag{34}$$

So, using (B) in the second line, and the known value of $p^{-\frac{1}{2}}1$ in the first line, we obtain

$$V_0 = \pi\frac{4e}{3}\left(\frac{t}{g\pi}\right)^{\frac{3}{2}}\left\{1 + \frac{(2t/g)^3}{5.7.9} + \frac{(2t/g)^6}{5.7.9.11.13.15} + \dots\right\}$$

$$-e\left\{\frac{(t/g)^3}{\lfloor 3} + \frac{(t/g)^6}{\lfloor 6} + \frac{(t/g)^9}{\lfloor 9} + \dots\right\}. \tag{35}$$

It is not laborious to calculate the curve of V_0 from this formula, at least up to $t = 5$ or 6 times g. I get the results in the following table :—

t/g	$\frac{1}{2}$	1	2	3	4	5	6
V_0/e	0·25	0·603	1·152	1·297	1·208	1·07	0·97

The inductance stops the current completely at first, so that V_0 is then zero. But, later on, the inertia of the current in the coil causes V_0 to rise above its final permanent value, which is e, and oscillate above and below it. An analogy in heat diffusion would need something far-fetched to illustrate the terminal condition and the inertia it brings in. A mechanical analogy is plainer, as in § 215. Have a long flexible elastic string of insensible mass suspended from fixed supports in a viscous medium which resists the transverse motion of the string with a force varying as its velocity. Let the string be first in equilibrium. Then apply a force e close to the fixed end. The string will at once be transversely displaced to a distance V_0, say, proportional to e, and the rest of the string will follow suit in time, but without any vibration, owing to the absence of inertia. This illustrates the case of no terminal inertia in the cable problem, V_0 becoming e immediately, and V becoming e everywhere in the cable later.

But next attach a mass to the string at the place of application of the force, close to the fixed end. When the force is applied it will now take time to fully displace the mass, which will then swing past its equilibrium position and oscillate about it. The attached mass corresponds to the coil in the cable problem. The table on the preceding page shows the initial rise of V_0 and its passage beyond the value e to its first maximum, and back again to a little below the equilibrium position.

The alternative formula is more difficult to obtain, and as its derivation from the operational solution involves more advanced ideas than have yet presented themselves, I will merely give the result here. It is

$$V_0 = e - \frac{4e}{3}\epsilon^{-t/2g}\cos\frac{t\sqrt{3}}{2g}$$

$$+\frac{e\pi}{2}\left(\frac{g}{\pi t}\right)^{\frac{3}{2}}\left\{1 - 3.5.7\left(\frac{g}{2t}\right)^3 + 3.5.7.9.11.13\left(\frac{g}{2t}\right)^6 - \dots\right\}, \quad (36)$$

which is useful in the later oscillatory part of the phenomenon. The period is $4\pi g/\sqrt{3}$. The descending series must be counted up to the smallest term; but, of course, when it is close to the beginning of the series, and the accuracy of calculation becomes impaired to the possible extent of the size of the smallest term, or, more likely, to the extent of half its size, the previous convergent series should be employed. The oscillatory function in (36) arises from the infinite series of even powers of q in the operator when expanded in rising powers of q, a matter to be returned to.

The General Nature of Electrical Operators.

§ **245.** I have worked out the above examples (except the end of the last one) in a manner suited to one who has not done any work of the kind before, with a considerable amount of detail in the transformations. But when the go of it is perceived, the transforming work may be simplified by attention to certain rules which are obeyed. So now, before passing to problems of an elementary kind, concerning the propagation of effects to a distance, I interpolate some explanatory remarks about operators in general. Later on, we may be concerned with the theory of fractional differentiation.

Observe, in the above, that we first obtain the operational solution, and that this is usually easily got and is of simple form—at least in the examples used, which admit of generalisation. Now, the operational solution is got by algebraical processes, of the same nature as if we were dealing with merely conductive circuits, only replacing the resistances concerned by the appropriate resistance operators, though treating them as if they were still resistances; that is, constants. Thus, in getting (24) for example, if the condenser were a a resistance, say Z, and the cable also a resistance, say z, then the current due to e would be

$$C_0 = \frac{e}{Z+z}, \tag{37}$$

obviously, and the potential on the right side of Z would be

$$V_0 = e - ZC_0 = \frac{e}{1 + Z/z}. \tag{38}$$

Now, in the real problem, we work in the same way, with different meanings attached to Z and z. They become the resistance operators. They are the functions of p, the time-differentiator, which take the place of resistance in the equation $V = RC$; viz., Ohm's law applied to a simple conductor, which connects the V and C thereof, V being the fall of potential through R in the direction of C, the current. If this becomes $V = ZC$ when there is stored electric and magnetic energy concerned, we call Z the resistance operator, because it replaces resistance, and reduces to resistance in steady states. That the Z's may be treated as resistances may be seen by considering the nature of the well-known problem of a conductive net of wires. We have an equation $V = RC$ for every branch, or, more generally,

$$e + V = RC, \tag{39}$$

if e is an additional impressed force therein. If, then, we sum up along any path in the net, we get

$$\Sigma e + \Sigma V = \Sigma RC. \tag{40}$$

But ΣV is zero in any circuit, so we have

$$\Sigma e = \Sigma RC \tag{41}$$

for every circuit in the net. This being the case, or, more generally, (40) being true for any particular path in the net, there is only one thing more required to determine C in all the branches due to all the e's, and that is the circuital nature of the current itself, which connects together the values of all the C's meeting at a junction, and makes $\Sigma C = 0$ there. The problem is now determinate, and the algebra of simple equations enables us to write down the expression for the current in any branch due to the impressed force in the same or in any other branch. When it is a very complicated net determinants are useful; but in most practical problems they are a useless complication, and the work is easier without them, and is more instructive from the physical point of view.

Now, instead of the branches of the net being simple conductors following Ohm's law in the above way, let them be arrangements storing electric and magnetic energy—that is, arrangements of condensers and coils; but still such that the current in any branch is the same at both ends, and such that there is no mutual action between one branch and another, though there may be mutual action between the constituents of a branch. Clearly, then, the currents, though now variable with the time when the forces are steady, are subject to identically the same conditions of continuity. But the equations of voltage are changed. We now have

$$e + V = ZC \tag{42}$$

in any branch, where Z has to be found from its detailed structure. Also

$$\Sigma e + \Sigma V = \Sigma ZC \tag{43}$$

along any path in the net, and, ΣV being zero in a circuit,

$$\Sigma e = \Sigma ZC \tag{44}$$

in any circuit in the net. There is, therefore, a complete formal similarity between the problem of merely conductive circuits and the general one involving stored energy. Every R becomes a Z. The equations which find the C's in terms of the e's are, therefore, identically the same, only with the R's replaced by Z's. These equations are the operational solutions. So the rule is, work out the given problem as if the independent branches were mere resistances; then give to the Z's their actual functional expressions in terms of p and constants; the result is the operational solution. It follows that anybody can work out electrical problems of an advanced nature so far as the operational solutions are concerned, by common algebra, assisted by electrical ideas. Nor need he stop there, for the very important case of simply periodic variations can be fully investigated by a continuation of the algebra from the operational solution to the algebraical. For, when a single simply periodic impressed e acts with frequency $n/2\pi$, the *power* of p^2 in the operators is $-n^2$; so, by putting $p=ni$, we obtain an algebraical solution which may be reduced to the simple form $(a+bi)\,e$, where the i signifies p/n or $d/d(nt)$. It is then fully realised.

Geometrical methods are sometimes used, involving the rotation of vectors in a plane. Their value seems to me to be principally illustrative. Their drawback is the great complication of the diagrams that arise when we depart from very simple problems, and the hard thinking and labour required to work out results. The algebraical method, on the other hand, works with admirable simplicity, even in complicated problems. It is, however, only a special case of resistance operators, in the general use of which we are not confined to simply periodic variations, the e's being any functions of the time in the operational equations. The application of these operators is not confined to condensers and coils, but extends to electromagnetism in general, with waves in conductors and dielectrics, and dissipation in space, the ultimate reason being the linear nature of the equations. Nor is it confined to electrical problems, but applies generally to the mathematical sciences involving linear equations, and can be used with advantage therein.

Returning to the network before considered, if a branch is itself complex, its Z must be got by properly eliminating all the internal V's and C's, so as to lead to a resultant equation

$V=ZC$, where V is the voltage on and C the current in the branch as a whole—that is, at the terminals. But should there be mutual influence between a branch and some other one, a further generalisation is required, which presents no difficulty save in the extra work involved, which, however, is still of the same nature in treating p and functions of p as constants for the time.

A remark should be made here about the figure in § 242. The impressed force is put between the Z and earth. It is therefore necessary that the current should be the same at both ends of Z. But if we put the e between Z and the cable, which will make no difference in the state of the cable in the examples above, we can attack more general problems. For Z may now have many branches; for example, a complicated arrangement of condensers and resistances like the cable itself. Thus we shall have

$$C_0 = \frac{e}{\left(\frac{R_1}{S_1 p}\right)^{\frac{1}{2}} + \left(\frac{R}{Sp}\right)^{\frac{1}{2}}} \tag{45}$$

when e is put between two cables, R_1 and S_1 being the constants of the new one, on the left side. If $R_1/S_1 = R/S$, the current is halved by the substitution of the second cable for direct earth. To find V_0, multiply by the resistance operator of the first cable, viz., by $(R/Sp)^{\frac{1}{2}}$. To find V_1, the potential of the beginning of the second cable, multiply by the negative of $(R_1/S_1 p)^{\frac{1}{2}}$, the resistance operator of the second cable. The changed sign is necessary on account of the current being *from* one cable *to* the other. If $R_1/S_1 = R/S$, the potentials are $\frac{1}{2}e$ and $-\frac{1}{2}e$. But in general they will not have the same numerical value, though $V_0 - V_1 = e$ always.

It should be understood that these potentials have nothing indeterminate about them, like the electrostatic potential, for they are really transverse voltages in the dielectric of the cables. They are proportional to the displacement, and to the charge, so that the diffusion of V in the cables is representative of the diffusion of the charge on the wires. No constant can be added to this kind of V, of course, in our problems, as it would introduce extra energy, having no connection with our impressed force.

The Simple Waves of Potential and Current.

§ 246. Let us now pass to some simple cases concerning effects produced at a distance. I remarked before that the solutions concerning the effects produced on the spot by an impressed force were the easiest to investigate. This is true when the constraint on the spot (or terminal condition) is not too complicated. But some cases of the effects produced at a distance are quite easily examined operationally, provided the terminal conditions are of the simplest kind.

Go back to § 240. To find V at x due to V_0 at $x=0$, we have the operational solution

$$V=\epsilon^{-qx}V_0. \qquad (1)$$

This expands to

$$V=\left(1-qx+\frac{(qx)^2}{\underline{|2}}-\frac{(qx)^3}{\underline{|3}}+\dots\right)V_0$$
$$=(\cosh qx-\text{shin}\, qx)\,V_0. \qquad (2)$$

Here we have even and odd powers of q, so there is nothing new in the way of operations. Taking $V_0=e$, constant, beginning when $t=0$, we may discard the even powers of q, and write

$$V=(1-\text{shin}\, qx)e$$
$$=e-\left(1+\frac{(qx)^2}{\underline{|3}}+\frac{(qx)^4}{\underline{|5}}+\dots\right)qxe, \qquad (3)$$

and since this involves complete differentiations performed upon $q1$, which is known, the full algebraical result follows at once:—

$$\frac{V}{e}=1-\left(\frac{RSx^2}{\pi t}\right)^{\frac{1}{2}}\left\{1-\frac{1}{3}\left(\frac{RSx^2}{4t}\right)+\frac{1}{5\underline{|2}}\left(\frac{RSx^2}{4t}\right)^2-\dots\right\}. \qquad (4)$$

This is an exceedingly important formula in diffusion, both in itself and as the basis of other formulæ, so we may as well give some details about it.

If we differentiate to x we shall obtain the formula for the current. Thus

$$\frac{RC}{e}=\left(\frac{RS}{\pi t}\right)^{\frac{1}{2}}\left\{1-\left(\frac{RSx^2}{4t}\right)+\frac{1}{\underline{|2}}\left(\frac{RSx^2}{4t}\right)^2-\dots\right\}. \qquad (5)$$

Here we recognise the exponential formula, so we may write it finitely, thus,

$$\frac{\mathrm{RC}}{e} = \left(\frac{\mathrm{RS}}{\pi t}\right)^{\frac{1}{2}} \epsilon^{-\mathrm{RS}x^2/4t}, \tag{6}$$

which is another important formula in diffusion.

We got (5) by differentiating (4) to x. But if we please we may get it in the same way from any of the previous operational forms. For example, from (3) we obtain

$$\mathrm{RC} = \left(1 + \frac{(qx)^2}{\underline{|2}} + \frac{(qx)^4}{\underline{|4}} + \ldots\right) qe, \tag{7}$$

which gives rise to (5) or (6) on development.

Or we may start from the initial operational solution for C, viz.,

$$\mathrm{RC} = \epsilon^{-qx}\mathrm{RC}_0 = \epsilon^{-qx} qe. \tag{8}$$

On expansion, this makes

$$\mathrm{RC} = (\cosh qx - \mathrm{shin}\ qx) qe. \tag{9}$$

In developing V we rejected the cosh function, excepting the constant term 1. But now we must reject the shin function, because (on account of the q factor) the even powers of q go with it. So we get

$$\mathrm{RC} = \cosh qx \,.\, qe, \tag{9A}$$

which is equivalent to (7). I give these variations to let the reader see that the solutions do not arise by fortuitous accident, but that there is a consistent fitting together.

That these solutions for V and C are *the* solutions may be tested by their satisfying the necessary conditions: (1), the characteristic; (2), the terminal condition; (3), the time condition, that V and C are zero initially everywhere except at the origin. The last, however, is troublesome numerically, on account of the very slow convergence when t is small. But the functions are well known, so there is no need to be frightened.

The Error Function. Short Table.

§ 247. The V/e formula (4), observe, is a function of $\mathrm{RS}x^2/4t = y^2$, say. So we may write it

$$\frac{\mathrm{V}}{e} = 1 - \frac{2}{\pi^{\frac{1}{2}}} \left\{ y - \frac{y^3}{3} + \frac{y^5}{5\underline{|2}} - \frac{y^7}{7\underline{|3}} + \ldots \right\}, \tag{10}$$

which is the same as

$$\frac{V}{e} = 1 - \frac{2}{\pi^{\frac{1}{2}}}\int_0^y \epsilon^{-y^2}\,dy \tag{11}$$

$$= 1 - \operatorname{erf} y. \tag{12}$$

A comparison of (11) and (12) defines the function which is sometimes called the error function, and denoted by erf y. A pretty full table was given by De Morgan in the *Ency. Met.*, "Theory of Probabilities," going by steps of 0·01 from $y=0$ to 2. This table is reproduced by Lord Kelvin in his "Physical Papers," Vol. III., p. 434. But a much briefer table is all that is needed for general purposes, and for curve tracing, say with step 0·05. Perhaps even step 0·1 would be enough.

y.	erf y.	Δ.	y.	erf y.	Δ.
0·05	05637	5637	1·05	86243	1973
0·10	11246	5609	1·10	88020	1777
0·15	16799	5553	1·15	89612	1592
0·20	22270	5471	1·20	91031	1419
0·25	27632	5362	1·25	92290	1259
0·30	32862	5230	1·30	93400	1110
0·35	37938	5076	1·35	94376	976
0·40	42839	4901	1·40	95228	852
0·45	47548	4709	1·45	95969	741
0·50	52049	4501	1·50	96610	641
0·55	56332	4283	1·55	97162	552
0·60	60385	4053	1·60	97634	472
0·65	64202	3817	1·65	98037	403
0·70	67780	3578	1·70	98379	342
0·75	71115	3335	1·75	98667	288
0·80	74210	3095	1·80	98909	242
0·85	77066	2856	1·85	99111	202
0·90	79690	2624	1·90	99279	168
0·95	82089	2399	1·95	99417	138
1·00	84270	2181	2·00	99532	115

We see that the error function rises from 0 to 1, as y goes from 0 to ∞; but as it reaches 0·995 when $y=2$ only, the subsequent rise to the full 1 is very slow work. The values are very easily calculated by formula (10), and without great labour; but the table is useful for reference.* The Δ column shows the differences or steps in erf y, corresponding to the

* The figures in the erf columns are decimals.

equal steps 0·05 in y itself. They therefore serve to plot the curve of the derivative, that is, the curve of current, represented by formula (6), which may also be used directly.

The Way the Charge and Current Spread.

§ 248. In Fig. 1 is represented the way the potential (or the charge) extends itself into the cable when the potential at its beginning is raised to and maintained at a steady value. The abscissa is length from the beginning, and the ordinate is the potential. The curve 1 is got by making $RS = 4t$, and varying x. Or if $t' = 4t/RS$, then curve 1 represents the state of things when $t' = 1$. We may use any unit of length we please,

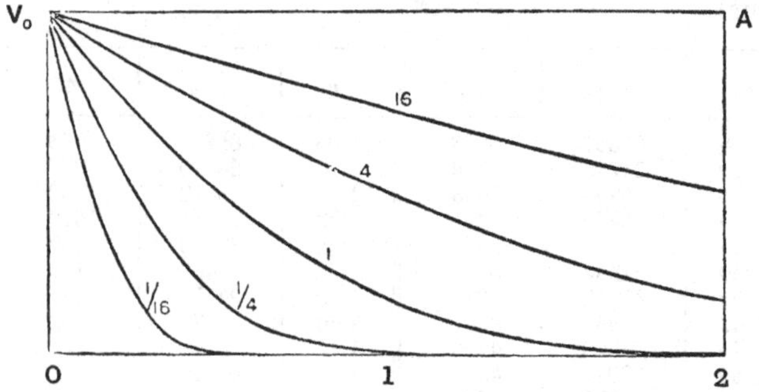

FIG. 1.

so that the base line may be 2 centim. or 2 kilom., or 2,000 kilom. if we like.

Now V/e is unchanged by altering x^2 in the same ratio as t'. So when $t' = 4$ we have the same values as before with doubled values of x. This is shown by curve 4, representing the potential distribution when $t' = 4$. Doubling x again for the same values of V/e, we get the curve 16, showing the potential when $t' = 16$. Similarly, by halving x in the curve 1, we obtain the curve $\frac{1}{4}$, showing the potential when $t' = \frac{1}{4}$; and halving x again brings us to the curve $\frac{1}{16}$, showing the potential when $t' = \frac{1}{16}$. The initial potential curve is simply the vertical line OV_0 (up and down) and the base line 012. The final potential curve is the vertical OV_0 and the horizontal line V_0A.

We may imagine the curves to represent the shapes assumed by a rope resting initially on the base, when the end of it is lifted from 0 to V_0, there being enough friction or as little inertia as will stop oscillations. But the exact mechanical analogy before employed requires an elastic massless string, moving in a medium which resists its transverse motion with a force varying as its velocity. The string should be fixed at O, and be initially along 012. Then a force applied close to

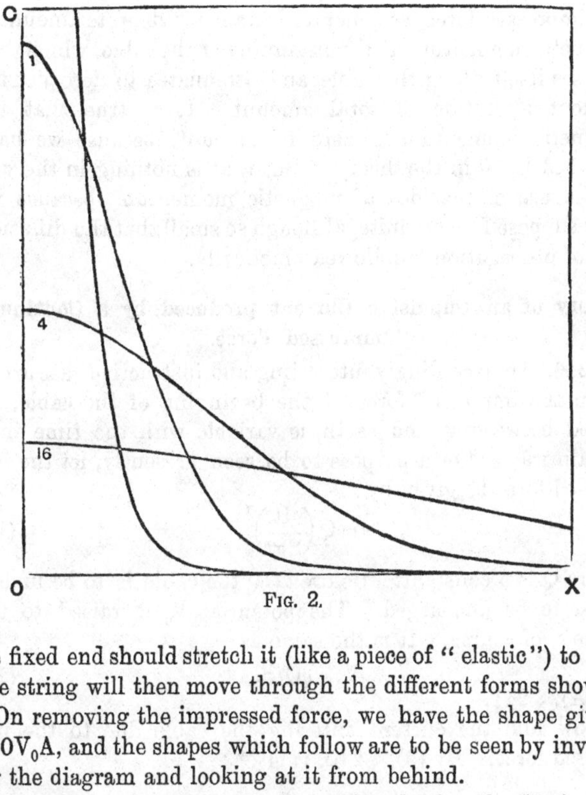

FIG. 2.

the fixed end should stretch it (like a piece of "elastic") to V_0. The string will then move through the different forms shown.

On removing the impressed force, we have the shape given by OV_0A, and the shapes which follow are to be seen by inverting the diagram and looking at it from behind.

The curves of current may be got by plotting the Δ's in the table. When $t'=1$, 4, and 16, we have the curves marked with these numbers in Fig. 2. It will be seen by (6) that by doubling x and quadrupling t we get a halved C. So curve 4 is got from curve 1 by halving the ordinates and doubling the abscissæ. Similarly as regards the other curves. The unnumbered one, with the top cut off, belongs to $t'=\frac{1}{4}$. These curves all bound the same area, that is, the area bounded by the curve and the vertical and horizontal axes is constant. In another form, the line-integral of the current is constant, the reason being that

$$V_0 - V = \int_0^x RC\,dx \tag{13}$$

by Ohm's law, and $V=0$ at $x=\infty$. That is, when we apply the impressed force we generate instantly a definite amount of magnetic momentum at the beginning of the cable, which then diffuses itself along the cable and attenuates to zero density without alteration of total amount. It is true that the magnetic momentum is zero in amount, because we have assumed $L=0$ in the theory. But that is nothing in the way of the use of the idea of magnetic momentum, because we may suppose L to be finite, although so small that the diffusion law of propagation is followed practically.

Theory of an Impulsive Current produced by a Continued Impressed Force.

§ 249. An exceedingly interesting and instructive case arises when the impressed force at the beginning of the cable, inserted between it and earth, is variable with the time in a certain way. For a purpose to be seen presently, let the impressed force be given by

$$e = Q\left(\frac{R}{S\pi t}\right)^{\frac{1}{2}}, \tag{14}$$

where Q is a constant. Before $t=0$ the cable is to be understood to be uncharged. The potential V_0 is raised to the value e, of course. It is the same as

$$V_0 = qQ/S, \tag{15}$$

by (A), § 241.

Now find the current entering the cable due to the impressed force. By (5), § 240, it is

$$C_0 = \frac{q}{R}V_0 = \frac{q^2}{RS}Q = pQ, \tag{16}$$

where the second equation arises by (15), and the third by the definition of q^2. Since Q is constant for any finite value of time, the result is zero. That is, there is no current entering the cable under the action of the continuously-present impressed force at any finite value of the time.

Is this nonsense? Is it an absurd result indicating the untrustworthy nature of the operational mathematics, or at least indicative of some modification of treatment being desirable? Not at all. It is the fact under the circumstances stated, and the principal remarkability is the instantaneous arrival at the result. For the above details concerning (14), (15), (16) are not wanted by an experienced worker; the operational solution $RC_0 = qV_0$ of § 240 being sufficient to show that if V_0 is $q1$, then C_0 vanishes.

We have to note that if Q is any function of the time, then pQ is its rate of increase. If, then, as in the present case, Q is zero before and constant after $t=0$, pQ is zero except when $t=0$. It is then infinite. But its total amount is Q. That is to say, $p1$ means a function of t which is wholly concentrated at the moment $t=0$, of total amount 1. It is an impulsive function, so to speak. The idea of an impulse is well known in mechanics, and it is essentially the same here. Unlike the function $p^{\frac{1}{2}}1$, the function $p1$ does not involve appeal either to experiment or to generalised differentiation, but involves only the ordinary ideas of differentiation and integration pushed to their limit. Our result $C_0 = pQ$ therefore means that an impulsive current, that is a charge, is generated by the impressed force at the first moment of its application; that the amount of the charge is Q, and that there is no subsequent current. It is the same as saying that the charge Q is instantaneously given to the cable at its beginning, which charge then spreads itself without loss anywhere.

Next work out the solutions for V and C anywhere in the cable. We have, by (15),

$$V = \epsilon^{-qx} V_0 = \epsilon^{-qx} q(Q/S). \tag{17}$$

This was algebrized before, equations (8) and (6), so the result is

$$V = Q\left(\frac{R}{S\pi t}\right)^{\frac{1}{2}} \epsilon^{-RSx^2/4t} \tag{18}$$

By differentiation to x, the current is

$$C = -\frac{1}{R}\frac{dV}{dx} = \frac{Qx}{2t}\left(\frac{RS}{\pi t}\right)^{\frac{1}{2}} \epsilon^{-RSx^2/4t}. \tag{19}$$

It will also be useful to obtain the current formula directly. We have, by (16),

$$C=\epsilon^{-qx}C_0=\epsilon^{-qx}pQ. \tag{20}$$

Here reject the even powers of q, and we get

$$C=-p\,\text{shin}\,qxQ, \tag{21}$$

which expands to

$$C=-p\left(1+\frac{(qx)^2}{\lfloor 3}+\frac{(qx)^4}{\lfloor 5}+\dots\right)qxQ, \tag{22}$$

involving only complete differentiations upon $q1$. Thus

$$C=-xQ\left(p+\frac{RSx^2p^2}{\lfloor 3}+\frac{(RSx^2)^2p^3}{\lfloor 5}+\dots\right)\left(\frac{RS}{\pi t}\right)^{\frac{1}{2}},$$

which leads directly to the formula (19) above or, rather, its full expansion.

We may also note that (20) or (21) gives C by a time differentiation upon the function $\epsilon^{-qx}1$, already obtained, (1) and (4) above. This again leads to the result (19).

That the previous special results for V_0 and C_0 are contained in the general formulæ for V and C anywhere is clear enough—that is, they make $V_0=e$ and $C_0=0$ at $x=0$. We have now only to explain why there is no current after the initial charge. It is for the same reason as why there is no current in the galvanometer in Poggendorff's way of comparing two battery voltages. If there be current in a network of conductors, and two points A and B thereof be joined through an external conductor, there will be a derived current in the new wire usually—that is, if there be any voltage between A and B due to the original arrangement. But if we introduce in the new wire an impressed voltage equal to and acting against the former voltage, there will be no current, and everything will be the same as if A and B remained insulated.

Applying this to the cable, we see that the impressed force in the above case is so artfully graduated in its strength as to be exactly equal to the potential at the beginning of the cable due to the charge Q when redistributing itself without terminal loss—that is to say, if we remove e and insulate the beginning of the cable, everything will go on as before. Of course, the removal must be done after its initial effect in charging the cable.

So, dismissing the impressed force altogether, we see that if we suddenly communicate a charge Q to the beginning of a cable, and then immediately insulate it, the potential and current which result will be given by (18) and (19) above. The potential (or the charge) redistributes itself in the same way as the current in the previous problem of a steady impressed force, that is, according to the curves in Fig. 2, p. 53. It has been mentioned already that areas are conserved in that diagram. In the present case this means that the electrification is conserved.

Diffusion of a Charge initially at One Point. Arbitrary Source of Electrification.

§ 250. If the cable be infinitely long both way, and have a charge 2Q suddenly introduced at $x=0$, the resulting V and C will still be given by (18) and (19) on the right side of the origin, where x is positive, because the charge 2Q will split into two equal charges, one of which will go to the right and the other to the left, and obviously in a symmetrical manner, so that there is no current at the origin. On the left side, V will be the same, and C the same negatived as at the corresponding points on the right side. We shall also obtain these results in another way, without the use of e. Thus, let a doubly infinite cable have an auxiliary wire attached at the point $x=0$, through which current is artificially sent into the cable. Let $2h$ be this current. It is equivalent to a source of electrification of strength $2h$; that is, electrification is generated at $x=0$ at the rate $2h$ per second. This splits equally right and left, so that $C_0=h$ may be taken to be the terminal datum for the positive half of the cable. Consequently

$$C=\epsilon^{-qx}h \tag{23}$$

gives the current at distance x, and since $RC=qV$,

$$V=\frac{R}{q}\epsilon^{-qx}h \tag{24}$$

is the corresponding V. Here h may be any function of the time.

Now, if h is impulsive, acting only at the moment $t=0$, we shall have

$$h=pQ, \tag{25}$$

where Q is the time-integral of h. Putting this in (23) and (24) we obtain the expressions (17) and (20) for V and C. This proves formally the previous statements about the behaviour of e, so far as the general electrical property of balancing voltages entered into the explanation.

The Inversion of Operators. Simple Examples.

§ 251. We have hitherto usually supposed that the potential at the beginning of the cable is produced by an impressed force there, and the potential and current anywhere have been derived from it. We may, however, regard the matter differently; as, given V_0, find C_0 (and V and C); or, given C_0, find V_0 (and V and C); in all cases on the understanding that the state of the cable depends entirely upon the state at the origin due to a cause acting there. This dependence upon the state at the origin implies that V and C are initially zero except at the origin, and that the cable is not subjected to impressed force or allowed to receive electrification anywhere else. The state at the origin need not, however, be due to an impressed force there, but may result (for instance) from its connection with a continuation of the cable on the other side of the origin, this continuation being initially charged. Thus, the case of a steady impressed voltage e between line and earth at the origin may be imitated by having the whole of the imagined continuation on the negative side initially charged to the uniform potential $2e$, and allowing it to discharge freely into the cable on the positive side. For the assumed distribution is equivalent to the combination of two distributions, one expressed by $V = e$ constant all over, and therefore not subject to change, whilst the other consists of $V = e$ on the negative and $V = -e$ on the positive side of the origin, with a node between them at the origin itself. So the potential at the origin will be made permanently equal to e, and the current is uncontrolled save by its natural connection with the potential, therefore the result on the positive side must be the same as that due to e considered as an impressed force acting at the origin between the line and earth.

But suppose it is the current at the origin that is controlled and led through a sequence of values. Then we may find the potential that results there and elsewhere. Thus, we have in general

$$V_0 = (R/q)C_0 \tag{26}$$

at the origin. So, by taking C_0 conveniently, we can immediately find V_0. If, for example, C_0 is proportional to $q1$, that is, to $t^{-\frac{1}{2}}$, the q in the denominator of (26) is cancelled, and V_0 comes out constant. We found before that V_0 constant made C_0 vary as $t^{-\frac{1}{2}}$. So now we solve the inverse problem; that is, if C_0 is zero before $t=0$, and is then made to vary as $t^{-\frac{1}{2}}$, the potential V_0 will be suddenly raised to and be maintained at a constant value. Furthermore, since C and V are derived from C_0 and V_0 by the operator ϵ^{-qx}, we see that the state of things everywhere is determined as well by the condition $C \propto t^{-\frac{1}{2}}$ as by the other condition $V_0 = \text{constant}$. They are equivalent under the circumstances mentioned.

Again, suppose C_0 varies as $q^2 1$. Then C_0 is impulsive, as before seen, so that we may call it pQ, where Q is constant, being the charge in the impulse. The V_0 that results is therefore

$$V_0 = \frac{R}{q} pQ = \left(\frac{Rp}{S}\right)^{\frac{1}{2}} Q = \left(\frac{R}{S\pi t}\right)^{\frac{1}{2}} Q. \tag{27}$$

This is the inversion of the problem beginning § 249, equations (14) to (16), which was, given V_0 varying as $t^{-\frac{1}{2}}$, to find C_0. The result was that C_0 was initially impulsive, and zero later.

Next, suppose that V_0 is impulsive, say

$$V_0 = pf, \tag{28}$$

where f (when regarded merely as a constant) is the measure of the impulse, that is, the time-integral of V_0. The result is

$$C_0 = \left(\frac{Sp}{R}\right)^{\frac{1}{2}} pf = fp\left(\frac{S}{R\pi t}\right)^{\frac{1}{2}} = -\frac{f}{2t}\left(\frac{S}{R\pi t}\right)^{\frac{1}{2}}, \tag{29}$$

which indicates the current coming out of the cable after its initial charging.

The reader may have noticed in the above, and perhaps previously, that we change the order of operations at convenience, as in $f(p)\phi(p)1 = \phi(p)f(p)1$, and that it goes. But I do not assert the universal validity of this obviously suggested transformation. It has, however, a very wide application, and transforms functions in a remarkable manner.

Reservations should be learnt by experience. The present example is, of course, elementary.

Observe that the current due to impressed e when it is made impulsive is got by writing pf for e, f being the time-total of e. This process is general. If we have worked out the solution arising from a steady impressed force, or a steady source, we obtain that for a momentary force or source by differentiating the former solution to the time, and substituting the strength of the impulse for the intensity of the former force or source. Or, in another form, if $C = Ye$ is the primary operational solution, giving C in terms of any e through the operator Y, then $C = Ypf$ is the operational solution when e is impulsive, of total f, the moment of time of the impulse being $t = 0$.

It is generally better to work out a solution due to e constant than that due to an impulse, because the former leads, as above, to the latter by a simple differentiation, whereas there might be some trouble in rising from the developed impulsive solution by integration to that for a steady force. But, knowing the developed impulsive solution, either directly, or by derivation from the other, the solution for a continued force varying anyhow with the time is at once expressible by a definite integral, because the continued force may be regarded as consisting of an infinite series of successive infinitesimal impulses. The definite integral, however, is of little use unless the integration can be readily effected in the case of the special function of the time that e is chosen to be. Moreover, it is not uncommon for the result of the integration to be obtainable more easily directly from the operational solution itself. The simply periodic solution is an obvious example, and others may be given.

The Effect of a Steady Current impressed at the Origin.

§ 252. Returning to the mutual dependence of C and V and the inversion of operators, let C_0 be made constant. That is, there is to be no current before, and a steady current after $t = 0$, at the origin. Clearly this would be impossible were it not for the permittance of the cable, which allows of any amount of electrification being sent in, practical limitations arising from the finite strength of the dielectric being nowhere in a theoretical discussion. Of course the potential rises infinitely, if sufficient time be allowed. The potential at the origin is

$$V_0 = \left(\frac{R}{Sp}\right)^{\frac{1}{2}} C_0 = 2C_0\left(\frac{Rt}{S\pi}\right)^{\frac{1}{2}}, \tag{30}$$

by using the known value of $p^{-\frac{1}{2}}1$.

Inversely, we may put it thus:—Given that there is an impressed force at the origin between line and earth varying as $t^{\frac{1}{2}}$, what is the resulting current? The answer is, the current is steady. This contrasts well with the case of a steady e, which leads to zero final current, though it is initially infinite, varying intermediately as $t^{-\frac{1}{2}}$. For in our present case we get a steady current from the first moment by an e rising from zero to infinity according to $t^{\frac{1}{2}}$.

As a typical example, it is worth while algebrising fully the solutions when the current at the origin is steady. From

$$C = \epsilon^{-qx} C_0, \tag{31}$$

we see that C follows the same law as was worked out for V due to steady e, as was exemplified in Fig. 1. Those curves now show how C distributes itself in the cable. From there being initially no current anywhere save at the origin, where it is C_0, the final state that is tended to is a constant current everywhere. The formula being the same, it is unnecessary to repeat it. The operational form (31) is fully explicit and understandable.

As regards V, we have to develop

$$V = \frac{R}{q} \epsilon^{-qx} C_0. \tag{32}$$

Here reject, as in previous cases, the even powers of q, and we get

$$V = -RxC_0 + \frac{R}{q} \cosh qx \,.\, C_0, \tag{33}$$

where the term involving the first power of x comes from the term in the previous equation which is independent of q; that is, it would partly express the final steady state were such a state possible, which is obviously not the case in the present problem, since the next term is $(R/q)C_0$, which tends to infinity. There is nothing new in the way of operations in equation (33). Written fully, it is the same as

$$V = -RxC_0 + \left(1 + \frac{RSx^2p}{\lfloor 2} + \frac{(RSx^2p)^2}{\lfloor 4} + \dots\right)\left(\frac{Rt}{S\pi}\right)^{\frac{1}{2}} 2C_0, \quad (34)$$

which only requires the complete differentiations to be effected to give the full development, namely,

$$\frac{V}{C_0} = -Rx + 2\left(\frac{Rt}{S\pi}\right)^{\frac{1}{2}}\left\{1 + y^2 - \frac{y^4}{3\lfloor 2} + \frac{y^6}{5\lfloor 3} - \frac{y^8}{7\lfloor 4} + \dots\right\}, \quad (35)$$

where $y^2 = RSx^2/4t$, as before. This formula shows the infinitely great ultimate increase of V, and how it spreads into the cable. As a check, obtain the current by differentiation to x. We get

$$\frac{C}{C_0} = 1 - \frac{2}{\pi^{\frac{1}{2}}}\left\{y - \frac{y^3}{3} + \frac{y^5}{5\lfloor 2} - \frac{y^7}{7\lfloor 3} + \dots\right\}. \quad (36)$$

Comparing with (10), we confirm the previous statement about the law followed by C now being the same as that followed by V when e is steady.

If, on the other hand, we should start with the last formula for C, and attempt to obtain that for V by integration with respect to x, there might be some difficulty initially. The indefinite integral is the formula (35) without the term independent of x. Now

$$V = \int_x^\infty RC\,dx, \quad (37)$$

so to get the missing term (that involving $t^{\frac{1}{2}}$) we require to evaluate a series for an infinite value of the argument, which is inconvenient. But instead of that we may note that

$$V_0 - V = \int_0^x RC\,dx, \quad (38)$$

so by this formula we get what we want from the value of V at the origin, which is already known. That is, it is known by the operational method; otherwise it would have to be specially obtained. The operational method usually avoids auxiliary evaluations of this kind, as we may see by the way (35) was got. The work is done automatically, as it were, and V is made to vanish at $x = \infty$ for any t, and also when $t = 0$ at any x.

Fig. 1 serving to show how the current spreads, the new formula (35) requires calculating to show how the charge spreads. Without making a fresh table fully, like that for the "error function," the following special results which I have calculated may be quite sufficient to allow the reader to draw the curves roughly, should he care to do so. Write (35) in the form

$$\frac{V}{C_0} = \left(\frac{2Rt}{S}\right)^{\frac{1}{2}} z, \tag{38A}$$

where

$$z = -y\sqrt{2} + \left(\frac{2}{\pi}\right)^{\frac{1}{2}}\left\{1 + y^2 - \frac{y^4}{3\underline{|2}} + \ldots\right\}. \tag{38B}$$

Then z is a function of the former $y^2 = RSx^2/4t$, and the values of z corresponding to some of those of y^2 are as follows :—

y^2	0	$\frac{1}{16}$	$\frac{1}{8}$	$\frac{1}{4}$	$\frac{1}{2}$	1	$1\frac{1}{2}$
z	0·798	0·508	0·395	0·282	0·167	0·071	0·034

Since y varies with x as well as with t, the last equation and the numbers in the table will allow the form of the V curve to be traced for a series of values of the time. But this case of steady impressed current is by no means so important as that of steady impressed potential, although interesting and instructive enough in its way.

Nature and Effect of Multiple Impulses.

§ 253. We have already considered the cases of an impulsive V_0 and an impulsive C_0. Those impulses were simple. But we may also have multiple impulses, to understand which it is best to take a special case. We know that

$$C = \epsilon^{-qx} C_0 \tag{39}$$

expresses the current due to C_0 arbitrary at the origin, or to a source of strength $2C_0$ at the origin when it is free to spread both ways equally, as if an auxiliary wire supplied electrification to the middle point of a doubly infinite cable without earth connection. Also, if C_0 is impulsive, say pQ, then

$$C = \epsilon^{-qx} pQ, \tag{40}$$

as before examined, § 250, equations (23), (25). Now, if this impulse be followed at time Δt later, by the impulse $-pQ$, the negative of the former, it will nearly undo the effect of the first one. If Q be finite, the differential effect will decrease with Δt, and become zero when it does. But if, whilst decreasing Δt, we simultaneously increase Q in the same ratio, the differential effect is maintained of finite size, and in the limit, where Δt vanishes and Q is infinite, whilst their product $Q\Delta t$ is finite, the result takes a special simplified finite form.

It is like the old way of making the magnetic force of a magnetised molecule out of the combined magnetic forces of two equal poles of positive and negative magnetism. If the poles are kept of finite strength, their joint effect vanishes when they are brought to coincidence. But if the product of strength of pole into distance apart, or the magnetic moment, be maintained finite, the resultant magnetic force is finite, and takes a special form in the limit. This is the way to form a solid harmonic distribution of potential, and Maxwell showed how to carry on the process with multiple poles, so as to generate solid harmonics of any order.

In our present case, however, it is not a space distribution of poles that we are concerned with, but a time distribution. The "moment" of the differential impulse is $Q\Delta t$, where $\pm Q$ is the strength of the poles (or impulses), and Δt is their distance apart in time. The finally resulting C is

$$C = \epsilon^{-qx} p^2 (Q\Delta t), \tag{41}$$

where Δt is infinitesimal. Or, if $Q_1 = Q\Delta t$,

$$C = \epsilon^{-qx} p^2 Q_1 \tag{42}$$

is the current due to a differential impulse of the first degree. It is clear that we may extend the above to impulses of any degree, say

$$C_0 = pQ_0, \text{ or } p^2 Q_1, \text{ or } p^3 Q_2, \text{ \&c.},$$

but at present let us keep to the middle one.

The potential corresponding to (42) is obtained in the usual way by multiplying by the operator R/q. This makes

$$SV = \epsilon^{-qx} pq Q_1, \tag{43}$$

from which we see that the impressed force at the origin between line and earth required to produce the same effect is given by

$$Se = pqQ_1. \tag{44}$$

This, however, needs close and literal interpretation. Let $Q_1/S = 1$ for simplicity, then, by algebrising $pq1$ we get

$$e = pq1 = p\left(\frac{RS}{\pi t}\right)^{\frac{1}{2}} = -\frac{1}{2t}\left(\frac{RS}{\pi t}\right)^{\frac{1}{2}}. \tag{45}$$

Now this merely represents the dregs of e, acting when t is finite, and, as before explained with respect to the impressed force $e = q1$, serving (in Poggendorff's way) to prevent current at the origin. But we must understand $pq1$ more literally. Thus, $q1$ is the function of the time which is zero before $t = 0$ and is $(RS/\pi t)^{\frac{1}{2}}$ later. So $pq1$, which is its rate of increase, is zero before $t = 0$, then jumps to ∞, then jumps through zero to $-\infty$, and lastly rises to zero again gradually. Of course it is only the last part that is explicitly represented by the developed time-function in (45). On the other hand, $pq1$ represents it all. Similarly, p^2q1 is to be interpreted as the rate of increase of $pq1$ just considered; and so on. And, going back to the impulses without any residual effect, pQ_0 has been already interpreted as a simple impulsive current; next, p^2Q_1 we see is the time-rate of increase of pQ_1, and is therefore a double impulse, first positive and then negative; and so on.

As regards the charge in the cable produced by the impulse p^2Q_1, just after the first moment it is confined practically to the region close to the origin, and consists of a positive wave followed by a negative one. These tend to neutralise and do neutralise one another by mutual diffusion to a large extent. But since at the same time the positive charge in front diffuses itself forward into the cable, there cannot be a complete neutralisation. The result is that the region occupied by the charge is continuously enlarged, the node between the positive and negative charges advancing along the cable, all to the right thereof being positive and to the left negative, whilst the density rapidly attenuates to zero. Since the formulæ (43) and (42) for V and C are derivable from previous formulæ by simple time differentiation, viz. (43) from (8) or (6), and (42) from (20) or (19), it is unnecessary to write the full developments.

Moreover, since the operator RSp is equivalent to a double differentiation to x, as is indicated by the characteristic of the potential and current, the new curves of V and C may be found by making their ordinates be proportional to the curvatures of the old ones. Similar remarks apply to multiple impulses of higher degrees, as regards their development by differentiation; so, as we do not want them at present, it is sufficient merely to point out, as above, how they may be got.

Convenient Way of denoting Diffusion Formulæ.

§ 254. Our results, so far, depend upon the fundamental function $q1$, where q is the square root of a differentiator. From it follow various other results, notably the functions $\epsilon^{-qx}1$ and $q\epsilon^{-qx}1$. Since they represent the elementary waves of potential and current due to a steady impressed force, they are of particular importance; and since they form the elements in more advanced problems that the preceding, their meanings should be carefully noted. Thus, collecting some formulæ for reference, let

$$q = (RSp)^{\frac{1}{2}}, \qquad y = (RSx^2/4t)^{\frac{1}{2}},$$

for shortness. Then we have

$$\epsilon^{-qx}1 = 1 - \text{erf}\, y$$

$$= 1 - \frac{2}{\pi^{\frac{1}{2}}}\left\{y - \frac{y^3}{3} + \frac{y^5}{5\lfloor 2} - \frac{y^7}{7\lfloor 3} + \dots\right\}, \qquad (1)$$

which gives the V/V_0 curves, and

$$\epsilon^{-qx}q1 = \left(\frac{RS}{\pi t}\right)^{\frac{1}{2}}\epsilon^{-y^2}, \qquad (2)$$

which gives the C/V_0 curves, when V_0 is steady, beginning at the moment $t=0$. Equation (1) belongs to Fig. 1, and equation (2) to Fig. 2, § 248.

Most solutions of problems in mathematical physics are in the form of infinite series. Finite solutions are quite exceptional. When of fundamental importance and of a relatively simple nature, the series functions receive special names and have conventional short expressions. Mathematicians get so accustomed to working with the short expressions that when they get solutions in terms of one or a finite number of such functions they sometimes say that they have got the solution in a finite form. But if the functions were such as had not received special names, the case would be substantially the same. The difference is merely conventional. Compare (1) and (2) above, for example. In (2) we have a case of the familiar exponential function; in (1) a relatively unfamiliar function which seems a great deal more transcendental than the other, which is its slope. The name erf y for the deficit of $\epsilon^{-qx}1$ from unity was invented by Glaisher, and by using erf we have a condensed empirical form for the functions we are concerned with.

Now, granting the desirability of having special short ways of representing important functions, we may remark first that $\epsilon^{-qx}1$ and $\epsilon^{-qx}q1$ may be themselves regarded as the special short ways, shorter than the other ways, in fact. This would be an undesirable deviation from common practice were no advantage gained. But in the present case the forms in question are actually indicative of the functions themselves in their structural meaning, through the operators generating them. Moreover, they are the forms which present themselves naturally in the mathematics. Furthermore, they are the proper forms for the easy and immediate performance of operations on the functions far more easily than upon either the full series or the exp and erf forms. Lastly, when we pass to more complicated cases, we shall see that they are made up of the $\epsilon^{-qx}1$ and $\epsilon^{-qx}q1$ functions, occurring in these particular forms. Considering all these things, we see that there may be great advantages in using these forms in their naked simplicity, serving not merely as empirical abbreviations, but as structural formulæ. Practice confirms this conclusion, as will be evident in the following.

Reflected Waves. Cases of Simple Reflection.

§ 255. When we employ an infinitely long line we do away with reflected waves, and exhibit the essentials in the simplest manner feasible. The conditions prevailing are the same all along, so no change occurs in the behaviour of the V and C waves generated at a source. Now, passing to more practical cases, the easiest are those in which there is only one reflected wave. Let, for example, the impressed force e be situated at the point $x=a$, and the beginning $x=0$ be earthed, whilst the end of the line is at an infinite distance.

The potential V generated by e is

$$V_1 = +\tfrac{1}{2}\epsilon^{-q(x-a)}e, \qquad \text{on the right,} \tag{3}$$

$$V_2 = -\tfrac{1}{2}\epsilon^{-q(a-x)}e, \qquad \text{on the left,} \tag{4}$$

because there is a rise of potential to the amount e at the place of e, and the two potentials $+\frac{1}{2}e$ and $-\frac{1}{2}e$ are the sources of elementary V waves, V_1 to the right, and V_2 to the left. The currents, however, are the same at equal distances from e on either side, viz.,

$$RC_1 = \tfrac{1}{2}q\epsilon^{-q(x-a)}e, \qquad RC_2 = \tfrac{1}{2}q\epsilon^{-q(a-x)}e, \tag{5}$$

which follow from (3) and (4) by space-differentiation, or by using the operator q on the right side of e, and $-q$ on the left, because C is reckoned + from the source e on the right side and to the source on the left side.

The positive wave V_1 to the right suffers no change, except what is involved in its known expression. But with the negative wave V_2 it is different. When it reaches the origin it has gone through the length a, and is, therefore, attenuated to $-\frac{1}{2}\epsilon^{-qa}e$. But the potential is constrained to be zero at the origin. This requires $+\frac{1}{2}\epsilon^{-qa}e$ to be superposed on the negative wave at the origin. Taken by itself, this would mean that the line is raised to potential $+\frac{1}{2}\epsilon^{-qa}e$ at the origin, which, by the previous, means a wave

$$V_3 = \epsilon^{-qx} \times \tfrac{1}{2}\epsilon^{-qa}e. \tag{6}$$

This is the reflected wave, and is positive, or from left to right. Since it suffers no further reflection, the complete potential is

$$V = V_1 + V_3 = \tfrac{1}{2}\epsilon^{-q(x-a)}e + \tfrac{1}{2}\epsilon^{-q(a+x)}e \tag{7}$$

on the right side of the source, and

$$V = V_2 + V_3 = -\tfrac{1}{2}\epsilon^{-q(a-x)}e + \tfrac{1}{2}\epsilon^{-q(a+x)}e \tag{8}$$

on the left side. These waves are of the former kind precisely, only the constants being different. The e may be any function of the time. If e is steady, we may employ the developed formula (1) above. The final result in this case (by $q=0$ in (7) and (8)) is $V=e$ on the right side, and $V=0$ on the left side.

Note that when e is shifted up to the origin (by making $a=0$), the two positive waves are made equal and coincident, whilst the negative vanishes. That is, the case of e at the origin (earthed) previously treated may be regarded as the case of instantaneous generation of a reflected wave of identically the same nature as the primary wave to the right, their co-existence making a single wave of doubled size.

Another easy case is when the line is cut at the origin. This makes $C=0$ there. The initial waves V_1 and V_2 from e at $x=a$ are the same as before, but the reflected wave differs. It is the negative of the former reflected wave, because it has to cancel the current due to the primary wave V_2. So

$$V = +\tfrac{1}{2}\epsilon^{-q(x-a)}e - \tfrac{1}{2}\epsilon^{-q(a+x)}e, \qquad \text{(right)} \qquad (9)$$

$$V = -\tfrac{1}{2}\epsilon^{-q(a-x)}e - \tfrac{1}{2}\epsilon^{-q(a+x)}e, \qquad \text{(left)} \qquad (10)$$

are the potentials when the line is cut at the origin. The final state when e is steady is $V=-e$ on the left and $V=0$ on the right side of the source. The final current is zero. These steady results are obvious.

An Infinite Series of Reflected Waves. Line Earthed at Both Ends.

§ 256. After the above easy cases in which only one reflected wave is generated, pass to a case more nearly allied with practice, in which we have an infinite series of reflected waves all of the same type. Let the line be of length l, and be earthed at both beginning $x=0$ and at the end $x=l$. Call these A and B for descriptive convenience. Have e on at A. We will first build up the result in a physical manner. The initial wave from A is

$$V_1 = \epsilon^{-qx}e, \qquad (11)$$

and there would be no other if the line were infinitely long. But when V_1 reaches the end B, and has attenuated to $\epsilon^{-ql}e$, it would raise the potential there to that value, were it not for the constraint forcing V to be zero. So $V_2 = -\epsilon^{-ql}e$ has to be superposed to cancel the effect of the first wave. This is the value at B of the reflected wave starting from B, and going towards A. At distance x from A it has traversed the distance $l-x$, and therefore has attenuated to $-\epsilon^{-q(l-x)}\times\epsilon^{-ql}e$; so the second wave is

$$V_2 = -\epsilon^{-q(2l-x)}e. \qquad (12)$$

When this wave reaches the origin A, it has attenuated to $-\epsilon^{-2ql}e$. It would lower the potential there to that extent. But the potential is kept constant. So we have the same kind of reflection as at B, viz., with reversal of potential. The third wave is therefore $+\epsilon^{-2ql}e$ at A, so that, generally,

$$V_3 = \epsilon^{-q(2l+x)}e \qquad (13)$$

expresses the third wave. It is positive, like V_1. When it reaches B, it becomes $\epsilon^{-3ql}e$, and (in the same way as the first wave) generates a fourth wave with the potential reversed. The fourth wave is therefore $-\epsilon^{-3ql}e$ at B, and becomes

$$V_4 = -\epsilon^{-q(4l-x)}e \qquad (14)$$

at x. It is unnecessary to elaborate further, because the process is the same for all the succeeding waves. The complete potential is therefore

$$\begin{aligned} V &= V_1+V_2+V_3+V_4+\dots \\ &= (\epsilon^{-qx}-\epsilon^{-q(2l-x)}+\epsilon^{-q(2l+x)}-\epsilon^{-q(4l-x)}+\epsilon^{-q(4l+x)} \\ &\qquad -\epsilon^{-q(6l-x)}+\epsilon^{-q(6l+x)}-\epsilon^{-q(8l-x)}+\epsilon^{-q(8l+x)}-\dots)e. \qquad (15)\end{aligned}$$

The current to correspond is given by

$$RC = (q\epsilon^{-qx}+q\epsilon^{-q(2l-x)}+q\epsilon^{-q(2l+x)}+\dots)e, \qquad (16)$$

where it should be noted that all the signs of the waves are +. That is, if the initial wave produces positive V_1, the reflected waves are alternately − and + as regards potential, but the currents to correspond are all the same way.

If e is steady, (15) gives V in a series of elementary waves of the kind (1) above, and (16) gives C in terms of the functions (2) above. The successive waves are smaller and smaller, of course. But the above reasoning is the same whether e be steady or be any function of the time, so the above results are fully expressive of the solutions in general, understanding that the waves will have different meanings quantitatively in different cases.

We may readily transform the above solution to a compact form. The odd terms in (15) make one geometrical series, and the even terms make another. So, summing up, we get

$$V = \frac{\epsilon^{-qx} - \epsilon^{-q(2l-x)}}{1 - \epsilon^{-2ql}} e = \frac{\text{shin}\, q(l-x)}{\text{shin}\, ql} e. \tag{17}$$

This is the condensed and most convenient form of the operational solution.

Putting $q=0$ in (17), we see that a steady e gives the final steady state

$$V = e\left(1 - \frac{x}{l}\right), \qquad C = \frac{e}{Rl}, \tag{18}$$

as is obvious by Ohm's law when the cable has become permanently charged. The potential is kept down by the reflections, whilst at the same time they allow the current to rise to a steady value.

The above way is instructive, and should certainly be followed sometimes. But when understood, much of it may be taken for granted, because the operational method gives the waves automatically and easily. Thus, to obtain (17), we have the general solution

$$V = \epsilon^{qx} F + \epsilon^{-qx} G, \tag{19}$$

where F and G are time functions (constants as regards x) to be determined. We should not call them "indeterminate," as is sometimes done in similar cases, because they *are* determinate, or determinable, and in fact have to be determined. They are determined by the terminal conditions $V=e$ at $x=0$, and $V=0$ at $x=l$, which give

$$e = F + G, \qquad 0 = \epsilon^{ql} F + \epsilon^{-ql} G. \tag{20}$$

Finding F and G from these, and inserting them in (19), we obtain the solution (17) at once.

Expand it by division, and the developed wave solution (15) results. This is the practical way to work in more complicated cases. We arrive easily and speedily at the condensed form, and may then develop it if we like. That the development of the operator in (17) into elementary wave operators by division is legitimate, is obvious from the above.

If the simply periodic solution be got from (15) by the property $p^2 = -n^2$, or $p = ni$, we obtain an infinite series of simply periodic trains of waves. It may be that only the first, or the first two or three, are wanted. If so, this way may be useful. We have

$$\epsilon^{-qx} \sin nt = \epsilon^{-\mathrm{P}x} \sin (nt - \mathrm{P}x) \tag{21}$$

when $p = ni$, the value of P being

$$\mathrm{P} = (\tfrac{1}{2}\mathrm{RS}n)^{\frac{1}{2}}, \text{ by using } i^{\frac{1}{2}} = \frac{1+i}{2^{\frac{1}{2}}}.$$

Equation (21) shows the wave train in an infinitely long cable. When of finite length, it is the type of the individual members of the infinite series of wave trains, x having to receive the series of values indicated in (15).

We may sum up this infinite series if we please, and obtain a resultant formula in a complicated way. But if this resultant be wanted, it can be got much more easily by making $p = ni$ in the condensed operational solution, because the summation has been already effected in it.

The Method of Images. The Waves are really Successive.

§ 257. There is another way of regarding the matter. It is usual in heat problems involving reflections, to consider the extra terms to be due to images. In our present problem, it would work out thus. Let the straight line represent part of an infinitely long cable without any external connections. Let the dots divide it into equal lengths l. At every second point

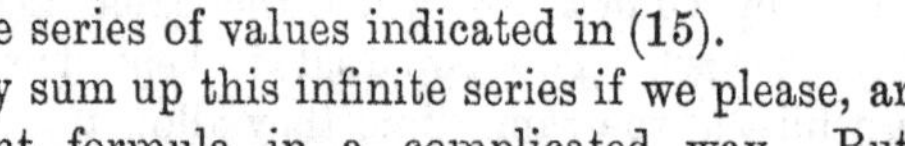

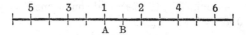

let impressed forces, each of strength $2e$, act simultaneously and in the same direction, say from left to right. Then the potential at A, due to the pair of forces at 2 and 3, is zero. So is that due to the pair 4 and 5, and so on. The potential at A is therefore that due to the $2e$ there only; that is, e on the right and $-e$ on the left of A. Now consider B. The forces at 1 and 2 produce no potential at B. Nor do the forces at 3 and 4, and so on. The result is that between A and B, the potential is the same as if the line were AB only, but earthed at both A and B, and with an impressed force e inserted at A. This is a mathematical equivalence.

But the way previously followed is, I think, preferable. It is as easy to follow, if not easier. But more importantly, it represents the true physics of the matter. We take the line as it is, and do not leave it, but introduce the reflected waves just as they arise at the terminals. The order of the waves is 1, 2, 3, 4, 5, &c. In the figure these mark the places of the source and its imagined images. But the real sources are at A and B.

It will also be noted that I described these waves as if they came into existence one after the other. The formula (15) or (17), on the other hand, says that they are contemporaneous. This is right enough for the formula, but is only a mathematical fiction in reality. The waves are successive, in the way described. The speed of propagation of disturbances is $v = (\mu c)^{-\frac{1}{2}}$ or $(LS)^{-\frac{1}{2}}$ approximately, where L is the inductance per centimetre. Our ignoration of L makes v be infinite. This accounts for the contemporaneity of the waves in the diffusion formula.

But give L a finite value, no matter how small, and v is finite, and the waves are successive. By taking L small enough, they will differ as little as we please from the above waves in type, whilst being successive. So we are justified in using the above natural way of description.

All diffusion formulæ (as in heat conduction) show instantaneous action to an infinite distance of a source, though only to an infinitesimal extent. It is a general mathematical property; but should be taken with salt in making applications to real physics. To make the theory of heat diffusion be rational as well as practical, some modification of the equations is needed to remove the instantaneity, however little difference it may make quantitatively in general.

Of course, to rationalise the theory in our immediate problem, we have merely to take L into account. We then change the type of the waves as well. The change may be little or great. It is very great in some telephone circuits, and, of course, with Hertzian waves. In the more advanced treatment of the subject, including L (and K the leakance as well), we have just the same kind of wave analysis as above, though q has a different meaning, and the waves are of another kind.

There is only one exception to the rule that an infinite series of waves results from terminal electrical arrangements made up of a finite number of parts, and that is in the theory of the distortionless circuit. It is possible to completely absorb an arbitrary wave by means of a suitable terminal resistance. Then there are no reflected waves.

Reflection at an Insulated Terminal.

§ 258. If the line is insulated at the far end B, instead of being earthed, as in § 256, other things remaining the same, namely, earth at the beginning A, where the impressed force e is situated, the change made in the waves is very easily settled.

The initial wave (11) from A is the same as before, but it must now be reflected positively at B, or without reversal of the sign of the potential, in order that the current may be maintained zero. So the second wave V_2 differs from the old one of equation (12) in sign only, being now positive. This V_2 is reflected negatively at A. Therefore V_3 is negative, and since it is reflected positively at B, V_4 must be negative too. This makes a cycle of signs. So we now have the following series:—

$$V = + + - - + + - - + + - - \&c.,$$

writing down only the signs of the waves, which are otherwise identically the same as in equation (15), where they make the arrangement $+ - + -$ &c.

Summing up as before, by the law of geometric series, we obtain

$$V = \frac{\epsilon^{-qx} + \epsilon^{-q(2l-x)}}{1 + \epsilon^{-2ql}} e = \frac{\cosh q(l-x)}{\cosh ql} e, \qquad (22)$$

which is the condensed form of the solution. The final V is e all over, and the final C is zero, as is obvious. To test (22) generally, observe that it makes $V = e$ at $x = 0$, and $C = 0$ at $x = l$, which are the two terminal conditions.

General Case. Effect of an Impressed Force at an Intermediate Point, with any Terminal Conditions.

§ 259. When the impressed force is not at the terminal A, but is at some intermediate point, say at $x = a$, the case is a little more complicated, because there are two series of waves. The primary wave is double, going both ways, and each of its members suffers reflection at A and B. At an earth, the reflection of V is positive; at a disconnection it is negative. By means of these considerations it is easy to write down without further calculation the full solution in terms of the waves for any case in which the terminal conditions are earth or insulation, as two earths, or two insulations, or one earth and one insulation (two cases); and these solutions are fully realised algebraically by the functions $\epsilon^{-qx}1$ and $\epsilon^{-qx}q1$ when the impressed force acts steadily. As there is no difficulty in this process, it will be as well to take a more comprehensive case for illustration.

Let the terminal conditions be unstated except that it is given that if v is a wave incident at B, then βv is the reflected wave; and that if v is a wave incident at A, then αv is the reflected wave. Here α and β are the terminal operators defining the nature of the reflection, the coefficients ot reflection, so to say.

Now let there be an impressed force e at $x = a$, and let us find its effect at a point x on the right side,

or between the impressed force and B. The wave from e going to the right is

$$v_1 = \epsilon^{-q(x-a)}\tfrac{1}{2}e. \qquad (23)$$

In this put $x = l$ to obtain its value at B. Then multiply by β to obtain the value of the reflected wave v_2 at B. Finally, multiply by $\epsilon^{-q(l-x)}$ to obtain the value of the last at distance $l - x$ from B, that is, at x from A. The result is

$$v_2 = \epsilon^{-q(l-x)}\beta\epsilon^{-q(l-a)}\tfrac{1}{2}e = \beta\epsilon^{-q(2l-a-x)}\tfrac{1}{2}e, \qquad (24)$$

which shows the second wave, going towards A.

Put $x=0$ to get the value of v_2 at A. Then multiply by α to produce the value of the reflected wave v_3 at A. Finally, multiply by ϵ^{-qx} to get its value at x. The result is

$$v_3 = \epsilon^{-qx}\alpha\beta\epsilon^{-q(2l-a)}\tfrac{1}{2}e = \alpha\beta\epsilon^{-q(2l-a+x)}\tfrac{1}{2}e, \tag{25}$$

showing the third wave.

Put $x=l$ to get its value at B. Multiply by β to get value of the reflected wave v_4 at B. Then multiply by $\epsilon^{-q(l-x)}$ to get its value at x. The fourth wave is therefore

$$v_4 = \epsilon^{-q(l-x)}\beta\alpha\beta\epsilon^{-q(3l-a)}\tfrac{1}{2}e = \alpha\beta^2\epsilon^{-q(4l-a-x)}\tfrac{1}{2}e. \tag{26}$$

After this, it is the same over and over again. So we have the series

$$\left.\begin{aligned} v_1 &= \epsilon^{-q(x-a)}\tfrac{1}{2}e, & v_2 &= \beta\epsilon^{-2q(l-x)}v_1, \\ v_3 &= \alpha\epsilon^{-2qx}v_2, & v_4 &= \beta\epsilon^{-2q(l-x)}v_3, \\ v_5 &= \alpha\epsilon^{-2qx}v_4, & v_6 &= \beta\epsilon^{-2q(l-x)}v_5, \end{aligned}\right\} \tag{27}$$

and so on. The first wave is given explicitly, and the rest are obtained in succession by multiplying by one or the other of two factors in turn. So, by the law of geometric series, the sum of the v waves is

$$\Sigma v = \frac{v_1+v_2}{1-\alpha\beta\epsilon^{-2ql}} = \frac{\epsilon^{-q(x-a)}(1+\beta\epsilon^{-2q(l-x)})}{1-\alpha\beta\epsilon^{-2ql}}\tfrac{1}{2}e, \tag{28}$$

which represents the potential at x so far as it arises from the initial wave to the right.

But there is also the initial wave to the left to be considered. It is $-\epsilon^{-q(a-x)}\frac{1}{2}e$ between a and A; becomes $-\epsilon^{-qa}\frac{1}{2}e$ at A; generates the new wave $-\alpha\epsilon^{-qa}\frac{1}{2}e$ by reflection, which becomes $-\alpha\epsilon^{-q(a+x)}\frac{1}{2}e$ when it reaches x. After that, we have a succession of reflections at B and A precisely in the former manner. So, if this new set of waves be called w, we have

$$\left.\begin{aligned} w_1 &= -\alpha\epsilon^{-q(a+x)}\tfrac{1}{2}e, & w_2 &= \beta\epsilon^{-2q(l-x)}w_1, \\ w_3 &= \alpha\epsilon^{-2qx}w_2, & w_4 &= \beta\epsilon^{-2q(l-x)}w_3, \\ w_5 &= \alpha\epsilon^{-2qx}w_4, & w_6 &= \beta\epsilon^{-2q(l-x)}w_5, \end{aligned}\right\} \tag{29}$$

and so on. The total is

$$\Sigma w = -\frac{\alpha\epsilon^{-q(a+x)}(1+\beta\epsilon^{-2q(l-x)})}{1-\alpha\beta\epsilon^{-2ql}}\tfrac{1}{2}e, \tag{30}$$

showing the part of the potential at x due to the initial wave going to the left. The real potential is, therefore, the sum of the v's and w's.

But when the point x lies between A and the impressed force, the case is somewhat different. If we consider the

effect at x of the initial wave to the right, making allowance for the fact that it only gets to x by producing a reflected wave at B, we have the series of waves

$$\left.\begin{array}{ll} v_1 = \beta\epsilon^{-q(2l-a-x)}\tfrac{1}{2}e, & v_2 = \alpha\epsilon^{-2qx}v_1, \\ v_3 = \beta\epsilon^{-2q(l-x)}v_2, & v_4 = \alpha\epsilon^{-2qx}v_3, \end{array}\right\} \qquad (31)$$

and so on, whose sum is

$$\Sigma\, v = \frac{\beta\epsilon^{-q(2l-a-x)}(1+\alpha\epsilon^{-2qx})}{1-\alpha\beta\epsilon^{-2ql}}\,\tfrac{1}{2}e. \qquad (32)$$

Similarly, the initial wave to the left and its consequences are represented by the series

$$\left.\begin{array}{ll} w_1 = -\,\epsilon^{-q(a-x)}\tfrac{1}{2}e, & w_2 = \alpha\epsilon^{-2qx}w_1, \\ w_3 = \beta\epsilon^{-2q(l-x)}w_2, & w_4 = \alpha\epsilon^{-2qx}w_3, \end{array}\right\} \qquad (33)$$

and so on, whose sum is

$$\Sigma\, w = -\,\frac{\epsilon^{-q(a-x)}(1+\alpha\epsilon^{-2qx})}{1-\alpha\beta\epsilon^{-2ql}}\,\tfrac{1}{2}e. \qquad (34)$$

Finally, denote by V_1 the potential on the right side of e, and by V_2 the potential on the left side. The former is the sum of (28) and (30), and the latter the sum of (32) and (34). That is to say,

$$V_1 = \frac{\epsilon^{qa}-\alpha\epsilon^{-qa}}{\epsilon^{ql}-\alpha\beta\epsilon^{-ql}}\,(\epsilon^{q(l-x)}+\beta\epsilon^{-q(l-x)})\tfrac{1}{2}e, \qquad (35)$$

$$V_2 = -\,\frac{\epsilon^{qx}+\alpha\epsilon^{-qx}}{\epsilon^{ql}-\alpha\beta\epsilon^{-ql}}\,(\epsilon^{q(l-a)}-\beta\epsilon^{-q(l-a)})\tfrac{1}{2}e, \qquad (36)$$

represent the complete solution in compact form, equivalent to the previous development in terms of waves.

Four Cases of Elementary Waves.

§ **260.** There are several cases in which these solutions are reducible to the elementary waves of the kind before considered. These occur when α and β are $+1$ or -1, or zero, or constants lying between the limits -1 and $+1$. Under these circumstances the reflected waves are copies of the incident either full-sized or reduced, with or without reversal of sign.

Thus, α or β being ± 1 makes four important cases.

(1.) Both ends earthed. The V waves are reflected negatively, so $\alpha = -1 = \beta$, and

$$V_1 = +\frac{\cosh qa}{\operatorname{shin} ql} \operatorname{shin} q(l-x).e, \tag{37}$$

$$V_2 = -\frac{\operatorname{shin} qx}{\operatorname{shin} ql} \cosh q(l-a).e. \tag{38}$$

(2.) Both ends cut. The V waves are reflected positively, so $\alpha = 1 = \beta$, and

$$V_1 = +\frac{\operatorname{shin} qa}{\operatorname{shin} ql} \cosh q(l-x).e, \tag{39}$$

$$V_2 = -\frac{\cosh qx}{\operatorname{shin} ql} \operatorname{shin} q(l-a).e. \tag{40}$$

(3.) Earth at A, and cut at B. Then $\alpha = -1$, $\beta = 1$, and

$$V_1 = +\frac{\cosh qa}{\cosh ql} \cosh q(l-x).e, \tag{41}$$

$$V_2 = -\frac{\operatorname{shin} qx}{\cosh ql} \operatorname{shin} q(l-a).e. \tag{42}$$

(4.) Earth at B, and cut at A. Here $\alpha = 1$, $\beta = -1$, and

$$V_1 = +\frac{\operatorname{shin} qa}{\cosh ql} \operatorname{shin} q(l-x).e, \tag{43}$$

$$V_2 = -\frac{\cosh qx}{\cosh ql} \cosh q(l-a).e. \tag{44}$$

These solutions and the more general cases may be easily converted to Fourier series, if required, by a method to be explained later. In the meantime it may be noted that the general solutions (35), (36) may be got by assuming

$$V_1 = \epsilon^{qx}F + \epsilon^{-qx}G, \qquad V_2 = \epsilon^{qx}H + \epsilon^{-qx}I, \tag{45}$$

and determining the four time functions by the two intermediate conditions

$$V_1 - V_2 = e, \qquad C_1 = C_2, \tag{46}$$

at $x=a$, and by two terminal conditions. In the case of equations (37) to (44) these terminal conditions are simply either $V=0$ or $C=0$ at A and B, as may be readily tested. What the terminal conditions are in general, in relation to the reflection coefficients α, β, will now be pointed out.

The Reflection Coefficients in terms of the Terminal Resistance Operators.

§ **261.** The reflection coefficients are usually operators themselves. There is no difficulty in finding them. Let $V=Z_1C$ at B. This says that Z_1 is the resistance operator of the terminal arrangement at B. Now let v_1 be a wave incident upon B, and v_2 be the reflected wave. Their sum is the real potential. So

$$v_1+v_2=Z_1C. \tag{47}$$

Also, we multiply by q/R to get the current belonging to v_1, and by $-q/R$ to get that belonging to v_2, because the first is a positive and the second a negative wave, going to and coming from B respectively. So

$$q(v_1-v_2)=RC. \tag{48}$$

Eliminating C between (47), (48) by division, we get

$$\frac{v_1-v_2}{v_1+v_2}=\frac{R}{qZ_1}, \text{ therefore } \beta=\frac{v_2}{v_1}=\frac{Z_1-R/q}{Z_1+R/q}, \tag{49}$$

giving β in terms of Z_1.

Similarly, if Z_0 is the resistance operator of the terminal arrangement at A, we shall have

$$v_1+v_2=-Z_0C, \tag{50}$$

if v_1 is the incident and v_2 the reflected wave. Now v_1 is a negative wave, to be multiplied by $-R/q$ to get the current to correspond, whilst v_2 is a positive wave, to be multiplied by R/q; so

$$-q(v_1-v_2)=RC. \tag{51}$$

Eliminating C, we obtain

$$\frac{v_1-v_2}{v_1+v_2}=\frac{R}{qZ_0}, \text{ therefore } \alpha=\frac{v_2}{v_1}=\frac{Z_0-R/q}{Z_0+R/q}, \tag{52}$$

giving α in terms of Z_0. Using these expressions for α and β in the solutions (35), (36), they are made completely expressive. The terminal conditions they satisfy are $V = -Z_0C$ at $x=0$ and $V=Z_1C$ at $x=l$.

Cases of Vanishing or Constancy of the Reflection Coefficients.

§ 262. By inspection of (49), we see that β vanishes when $Z_1=R/q$. This is because R/q is the resistance operator of an infinitely long cable line. To say that it equals Z_1, asserts that the cable either does not stop at β, but goes on to infinity; or else that if it does stop, there is a terminal arrangement which is exactly equivalent to the infinitely long continuation, so far as the real cable itself is concerned.

In this case $\beta=0$ reduces (35) (36) to

$$V_1=(\epsilon^{qa}-\alpha\epsilon^{-qa})\epsilon^{-qx}\tfrac{1}{2}e, \tag{53}$$

$$V_2=-(\epsilon^{qx}+\alpha\epsilon^{-qx})\epsilon^{-qa}\tfrac{1}{2}e, \tag{54}$$

showing two waves only, for there is no reflection at B, and therefore can be but one reflection at A.

Again, if $Z_0=R/q$ then $\alpha=0$. There is no reflection at A, because either the cable is continued past A indefinitely, or else there is a terminal arrangement copying the continuation. We now have

$$V_1=\epsilon^{qa}(\epsilon^{-qx}+\beta\epsilon^{-q(2l-x)})\tfrac{1}{2}e, \tag{55}$$

$$V_2=-\epsilon^{qx}(\epsilon^{-qa}-\beta\epsilon^{-q(2l-a)})\tfrac{1}{2}e, \tag{56}$$

showing two waves again, the reflection being at B.

Finally, if $R/q=Z_0=Z_1$, both α and β vanish, and we reduce to

$$V_1=\epsilon^{-q(x-a)}\tfrac{1}{2}e, \tag{57}$$

$$V_2=-\epsilon^{-q(a-x)}\tfrac{1}{2}e, \tag{58}$$

which are simply the primary waves from the source, without subsequent interference.

It will be observed that the expressions for α, β contain q, involving $p^{\frac{1}{2}}$. In order, therefore, to have the coefficients freed from $p^{\frac{1}{2}}$ (besides in the previous ways), we must introduce $p^{\frac{1}{2}}$ in Z_1 and Z_0. Say, for example,

$$Z_0=\left(\frac{R_0}{S_0p}\right)^{\frac{1}{2}}, \qquad Z_1=\left(\frac{R_1}{S_1p}\right)^{\frac{1}{2}}. \tag{59}$$

These mean that the cable AB is put between two other cables of different types; of the type R_0, S_0 on the left of A, and of the type R_1, S_1 on the right of B. The differentiator p then disappears from α and β, which reduce to

$$\alpha = \frac{(R_0/S_0)^{\frac{1}{2}} - (R/S)^{\frac{1}{2}}}{(R_0/S_0)^{\frac{1}{2}} + (R/S)^{\frac{1}{2}}}, \qquad \beta = \frac{(R_1/S_1)^{\frac{1}{2}} - (R/S)^{\frac{1}{2}}}{(R_1/S_1)^{\frac{1}{2}} + (R/S)^{\frac{1}{2}}}. \qquad (60)$$

These are constants, and may evidently have any values between (and including) -1 and $+1$. We can get rid of the reflection at A by having $R_0/S_0 = R/S$, and of that at B by $R_1/S_1 = R/S$. This is somewhat more general than the previous way of having continuations of the same type as the real line.* If we do not abolish the reflections, the whole series of waves summed up in (35), (36) are in action, only with α and β constants instead of, as in the general case, operators containing p.

General Case of an Intermediate Source of Electrification subject to any Terminal Conditions.

§ 263. In conection with the preceding, the other kind of source should be mentioned. The treatment is similar when the source at $x = a$ is not impressed force, which creates a discontinuity in the potential, though not in the current, but is a source of current, creating a discontinuity in the current, though not in the potential. Thus, let h be the strength of a source of current at $x = a$ (say led to the cable by an auxiliary wire), then the current in the line increases by the amount h in passing from left to right past the source. This case has been already briefly considered (§ 250). The primary waves are

$$c_1 = \epsilon^{-q(x-a)}\tfrac{1}{2}h, \qquad \text{on the right,} \qquad (61)$$

$$c_2 = -\epsilon^{-q(a-x)}\tfrac{1}{2}h, \qquad \text{on the left,} \qquad (62)$$

---

* But it is not necessary for the terminal cables to be homogeneous. For example, if we want $R/S = R_1/S_1$, we may make it go by having any number of cables in sequence, in which the value of R_1/S_1 is constant, so far as the reaction on the real cable is concerned. Compare with the corresponding property in the diffusion of heat, as described in § 233.

and the potentials to correspond are

$$v_1 = \epsilon^{-q(x-a)}\frac{Rh}{2q}, \quad \text{on the right,} \tag{63}$$

$$v_2 = \epsilon^{-q(a-x)}\frac{Rh}{2q}, \quad \text{on the left.} \tag{64}$$

When h is steady or impulsive we have already indicated the results. The point at present in question is the extension to include terminal influences. Have the same reflection coefficients as above, and apply them to the present case. The initial wave to the right is made the same as in § 259, equation (23), by substituting Rh/q for e, whilst the initial wave to the left is made the same by writing Rh/q for $-e$. There is no other difference. So we may employ the previous results fully. In particular, the solutions (35), (36) become

$$V_1 = \frac{\epsilon^{qa} + \alpha\epsilon^{-qa}}{\epsilon^{ql} - \alpha\beta\epsilon^{-ql}}\left(\epsilon^{q(l-x)} + \beta\epsilon^{-q(l-x)}\right)\frac{hR}{2q}, \tag{65}$$

$$V_2 = \frac{\epsilon^{qx} + \alpha\epsilon^{-qx}}{\epsilon^{ql} - \alpha\beta\epsilon^{-ql}}\left(\epsilon^{q(l-a)} + \beta\epsilon^{-q(l-a)}\right)\frac{hR}{2q}, \tag{66}$$

which express the potentials V_1 on the right side and V_2 on the left side of the source h. Test that

$$V_1 = V_2, \qquad \text{and} \qquad -\frac{dV_1}{dx} + \frac{dV_2}{dx} = Rh, \tag{67}$$

at $x = a$.

Take $h = pQ$ when the source is impulsive. Then (65), (66) represent the potential due to a charge Q which is initially all at the point $x = a$, as modified by the terminal influences. The conversion to a Fourier series of this result leads to the expansion of an arbitrary function in all sorts of Fourier series, not merely the periodic case which rigourists have tried so hard* to demonstrate, but to the numberless other expansions which occur naturally in the physics of the matter, associated with different forms of the coefficients α, β regarded as functions of p. It is no easy matter from the restricted

* I do not mean that they have not succeeded, but that the rigorous demonstrations are, from a physical point of view, hard to follow and not very convincing.

rigorous mathematical point of view to answer the question, Why should an arbitrary function be capable of expansion in such or such a series? But from the physical point of view, the question is rather, Why shouldn't it? This matter, however, must come later.

The Two Ways of expressing Propagational Results, in terms of Waves, or of Vibrations.

§ 264. Having accumulated a stock of formulæ in the last few sections, it becomes necessary to explain their connection with Fourier series. Given an electromagnetic operational equation, say $e = \mathrm{ZC}$. Here C is some particular effect due to a cause e, as, for instance, the current in one part of an electrical system due to an impressed force in another part, though it is not necessary to restrict their meanings in this way. The operator Z is to be constructed in the way previously explained in § 245 (or in any equivalent way), that is, in the same way as if the elements of the combination were mere resistances subject to Ohm's law, to be generalised in the final result to the functions of p, the time differentiator, which are appropriate to the real nature of the elements.

Some ways of algebrising such operational equations have been already given, especially applicable to diffusion problems, though they have a wider application. One way in particular should be noted and remembered, namely, the resolution of the operator Z by algebraical division into wave operators. This is a very simple and powerful method, which applies very generally in physical problems concerning continuous media. The effect is to express the solution in the form of the sum of a series of waves. These may be either simultaneous or successive, according to circumstances. There may be but one wave, or two, or an infinite number. If the method had no other recommendation, it would have this important one—that in considering the effect due to a source it imitates nature by directly expressing the course of events in the way it happens in actual fact, whereas an alternative and equivalent formula might completely disguise it.

But there is another very different way of resolving an operator into other more elementary ones, which leads to a strikingly different functional expression of the developed solution. The results are so entirely unlike the wave results that their equivalence produces algebraical identities which are rather astonishing at first. Only by the familiarity with them, which their universal existence in physical problems involving propagation (subject to certain restrictions) makes possible, do they become commonplace. We may express our results either in terms of the waves or in terms of the vibrations of normal systems of disturbances. Or, in the latter case, if the vibrations be frictionally resisted, in terms of the subsidences or decadences of the normal systems. We may compare the two developments to the branches and the roots of a tree. They are widely separated, but have a common bond in the trunk which joins them. The trunk corresponds, of course, to the operational solution. No doubt the analogy will fail if pushed much further. A perfect analogy in every respect would require an identity—for nothing is wholly like anything but itself—and an identity would be useless for an analogy. The present one is good enough as far as it goes here.

It is not a matter of indifference which way of development is employed. It may be that in some particular problem one of the two is far more manageable than the other, or more amenable to numerical calculation. Apart from this, it depends upon circumstances which of the two ways is to be preferred by its natural recommendations. Take, for instance, the case of a long stretched cord, fixed at its ends. If we disturb it so as to make a hump or a number of humps run along the string, that is, if we produce evident and visible progressive waves, it is natural to express the mathematical results in the form of waves. Again, if we displace the whole cord to the form of a sine curve, and let it go, it will vibrate in the same form over and over again, whether the sine curve be only half a complete wave length, or have intermediate nodes. Here it is obviously natural to express mathematically the visibly evident simple vibrations as vibrations simply.

But in the former case the progressive wave may be expressed entirely in the form of normal vibrations, and in the latter the vibrations may be expressed entirely in the form of progressive waves. The methods are perfectly equivalent quantitatively. We see at once, however, that it is not natural to express the simple progressive wave in the form of normal vibrations, or the normal vibration in the form of progressive waves, because the results would disguise the reality. The examples used here are extreme. In intermediate cases it may be quite difficult to say which way is preferable, owing to both ways being complicated. There may, however, be numerical advantage in using one form ; or, more likely, one form may be useful in one part of the history of events, and the other form later, of which there are many examples in diffusion problems. The normal vibrations may involve much labour in calculation for some particular value of the time, when, under the same circumstances, the waves are easily done ; and conversely.

We should remark, however, that in pure diffusion problems we are not concerned with true vibrations, as of a string. The resistance stops the vibrations, so the disturbances simply subside or decay. Thus, if the potential in a submarine cable, imagined to be quite free from self-induction, be distributed according to a sine curve, say $V = V_0 \sin ax$, with nodes at beginning and end, if earthed terminally, and be left to itself, the curve of potential will preserve its sine form, though continuously falling to equilibrium. But self-induction will make it pass through the equilibrium position, and the potential will vibrate, though decaying at the same time. This is oscillatory subsidence. By reducing the resistance, or increasing the self-induction, or in both ways together, we make the vibrations last longer, and resemble those of a stretched cord. The limit would be reached if there were no resistance. The vibrations would continue for ever, without loss of intensity, like those of the string in acoustical theory, when friction is ignored. The stretched cord makes by far the best analogy for a telegraph circuit when it is desired to have a simple analogy, because every one knows something about how a cord vibrates, and how pulses are transmitted along it, and reflected, and so on until finally killed by friction. They are visibly evident. Now all these things have their close representatives in a telegraph circuit, which makes the simple analogy be very useful. Of course, it is an entirely different matter when the etherial theory is in question ; then suitable analogies are of a different nature, which may, perhaps, be as difficult to follow as the electromagnetic theory itself, as, for instance, the continuous medium first imagined by MacCullagh, in which rotational elasticity is involved. In one respect, however, the stretched cord fails. It does not, at least in any simple manner, represent the effect of leakage on a telegraph circuit. Prof. FitzGerald,* however, by employing the vibrations of air in a pipe in the analogy, was able to include leakage. But the pipe analogy is not so easily followed in most respects as the stretched cord, since compressions and condensations of the air are less easily pictured than the motions of a cord.

In speaking of the diverse modes of representation as being in terms of waves and of vibrations, of course progressive waves are referred to, whether undistorted or distorted as they progress. Progression is the essence of a wave. "Standing" waves are somewhat deceptively so called, if simple normal vibrations be included therein. Now the reader must be cautioned against supposing that every operational equation is convertible, as described, into waves or normal vibrations. Some operational equations refuse to go more than one way, save perhaps by artificial expedients. They make waves only, or vibrations only, but not both. Why these failures occur is evident in practice by physical considerations. Mathematically, it is due to peculiarities in the form of the Z operator. As an example of waves only, there can be at most only two waves from one source situated in an infinitely long cable, as in § 255. Then we cannot have a vibrational form of solution (if we include subsiding normal systems under vibrations), except artificially. And, of course, when there is only a limited number of degrees of freedom in an electrical arrangement, as in the theory of condensers and coils, we have vibrations or vibrational subsidences or pure subsidences of normal systems in limited number which do not admit of expression in the form of continuous waves, except perhaps very artificially. Generally speaking we may have both, when there is a continuous medium for propagation in some part of the system. The distinction is usually connected with the presence or absence of boundaries. If we create a disturbance in an elastic medium which is quite homogeneous and is

* *The Electrician*, May 25, 1894, p. 106.

unbounded, the disturbance goes out to infinity in wave fashion, and no free vibrations in definite and separate modes are possible. But if the medium is bounded, the wave cannot dissipate itself freely, but is reflected and re-reflected any number of times. We can now express the disturbance in terms of the vibrations of definite normal systems of a complicated nature, depending upon the nature of the medium and on the form, &c., of the boundary. The possibility arises from the superposition of an infinite number of progressive waves, and we may express our results either in terms of waves or of vibrations.

This brings us to the reservation made about an artificial way of representing a progressive wave by vibrations. If in the last case we imagine the boundary shifted further and further away, however far we go the normal systems remain distinct and separate, and the two modes of representation are in force. Now, in the limit, when the boundary is removed to an infinite distance, the expression for the sum of the normal systems becomes a definite integral. Thus a single progressive wave may be expressed by a definite integral as the sum of vibrations of infinitely numerous normal systems differing infinitely little from one to the next. Instead of a definite sequence of separated periods, the periods run into one another. Of course it is simpler to think of the progressive wave than of the vibrations in the definite integral which equivalently expresses it, though not in a desirable manner, and to derive it directly from the operational solution when possible, instead of through the integral.

After these general remarks about the chief peculiarities, we must proceed to show practical methods of manipulating the Z operator so as to convert operational solutions to vibrating or subsiding normal systems. It is not difficult by the general method to be explained. Perhaps it is too easy. That is, too easy in execution, for the theory thereof is more difficult. As, however, it applies to all sorts of normal systems besides those concerned in Fourier series, it will perhaps be best at first to show the connection of Fourier series with the operational forms in more special ways, which, though longer, may be more immediately intelligible to the unpractised in operational methods.

Conversion of Operational Solutions to Fourier Series by Special Ways.

(1). Effect due to e at A, when earthed at A and B.

§ 265. Take, to begin with, an interesting special case, namely, the effect due to steady impressed force e at A, where $x=0$, when earth is on at both A and B, where $x=l$. As before shown, the operational solution is

$$\mathrm{V}=\frac{\operatorname{shin} q(l-x)}{\operatorname{shin} ql}e. \tag{1}$$

The conversion to a Fourier series in this and many other simple cases can be done by using well-known trigonometrical identities, assisted by an elementary operational result. Thus, we have

$$\frac{ql}{\operatorname{shin} ql}=1-\frac{2}{1+\left(\frac{\pi}{ql}\right)^2}+\frac{2}{1+\left(\frac{2\pi}{ql}\right)^2}-\frac{2}{1+\left(\frac{3\pi}{ql}\right)^2}+\dots, \tag{2}$$

by trigonometry. Here noting that q^2 means RSp, we see that we have merely to algebrize $(1+\mathrm{P}/p)^{-1}1$, where P is a constant. Thus, in full,

$$\frac{1}{1+\frac{\mathrm{P}}{p}}=1-\frac{\mathrm{P}}{p}+\left(\frac{\mathrm{P}}{p}\right)^2-\left(\frac{\mathrm{P}}{p}\right)^3+\dots=1-\mathrm{P}t+\frac{(\mathrm{P}t)^2}{\lfloor 2}-\frac{(\mathrm{P}t)^3}{\lfloor 3}+\dots;$$

or, in the usual brief expression of the exponential function,

$$\frac{1}{1+\frac{\mathrm{P}}{p}}=\epsilon^{-\mathrm{P}t}. \tag{3}$$

This result should be noted, as it often turns up. Applying it to (2), we obtain

$$\frac{ql}{\operatorname{shin} ql}e=e\{1-2\epsilon^{p_1t}+2\epsilon^{p_2t}-2\epsilon^{p_3t}+\dots\}, \tag{4}$$

where the constants p_1, p_2, &c., are given by

$$p_n=-n^2\pi^2/\mathrm{RS}l^2. \tag{5}$$

Comparing (1) with (4), we see that V will be found by operating on (4) by

$$\frac{\operatorname{shin} q(l-x)}{ql}. \tag{6}$$

This is a function of q^2 also, so involves only complete differentiations. They may be carried out at length if desired, but that is unnecessary. For the power of p in the operator (6) on ϵ^{Pt} is simply P. It is therefore zero on the first constant term on the right of (4), then p_1 on the second term, and so on. That is, these values are to be put in (6) for p to express its effect. So (1) is converted to

$$V = e\left(1 - \frac{x}{l}\right) - 2\frac{\text{shin}\, q_1(l-x)}{q_1 l}\epsilon^{p_1 t} + 2\frac{\text{shin}\, q_2(l-x)}{q_2 l}\epsilon^{p_2 t} - \dots, \quad (7)$$

where the q's are the constant values corresponding to the p's, through $q^2 = \text{RS}p$. Equation (7) is the full solution, but for trigonometrical convenience, since the values of q^2 are negative, it is best to put $q^2 = -s^2$. This makes

$$V = e\left(1 - \frac{x}{l}\right) - 2\frac{\sin s_1(l-x)}{s_1 l}\epsilon^{p_1 t} + 2\frac{\sin s_2(l-x)}{s_2 l}\epsilon^{p_2 t} - \dots; \quad (8)$$

or, since $\sin sl = 0$ for every s,

$$V = e\left(1 - \frac{x}{l}\right) - \left(2\frac{\sin s_1 x}{s_1 l}\epsilon^{p_1 t} + 2\frac{\sin s_2 x}{s_2 l}\epsilon^{p_2 t} + \dots\right), \quad (9)$$

where the s's are the roots of $\sin sl = 0$, which is the same as saying that the p's are the roots of $\text{shin}\, ql = 0$. In the customary form (9) is the same as

$$V = e\left(1 - \frac{x}{l}\right) - \frac{2e}{\pi}\Sigma_1^\infty \frac{1}{n}\sin\frac{n\pi x}{l}\epsilon^{-n^2\pi^2 t/\text{RS}l^2}. \quad (10)$$

This is the equivalent of the wave solution before got.

When $t = \infty$, the time factors vanish, and there is left the final state expressed by the outside term. On the other hand, when $t = 0$, the time factors are unity, and

$$V = e\left(1 - \frac{x}{l}\right) - \frac{2e}{\pi}\Sigma_1^\infty \frac{1}{n}\sin\frac{n\pi x}{l}. \quad (11)$$

We know that V is initially zero everywhere except at A, where it is e. Therefore

$$1 - \frac{x}{l} = \frac{2}{\pi}\Sigma_1^\infty \frac{1}{n}\sin\frac{n\pi x}{l}, \quad (12)$$

except at $x = 0$, where the right member gives zero. That is, the steady state, which is $e(1 - x/l)$, has become expanded in a series of sines. But we have got more than we wanted. For the Fourier series is periodic, with period $2l$. So the right member represents $1-x/l$ between 0 and $2l$; then the same values are repeated from $2l$ to $4l$; and so on. Similarly on the left side of the origin. Equation (11) is, in fact, the solution of the problem of finding V due to an infinite number of equal impressed forces of strength $2e$ acting all the same way at the points 0, $2l$, $4l$, &c., $-2l$, $-4l$, &c., in an infinitely long cable. This was shown before as regards the operational and wave solutions.

(2). Modified Way of doing the Last Case.

§ 266. There is another similar way of getting (9) which is worth noticing. Use the trigonometrical identity

$$ql \coth ql = 1 + \frac{2}{1+\left(\frac{\pi}{ql}\right)^2} + \frac{2}{1+\left(\frac{2\pi}{ql}\right)^2} + \frac{2}{1+\left(\frac{3\pi}{ql}\right)^2} + \dots, \tag{13}$$

differing from (2) on the right side only in the signs of the terms. Let the operand, as before, be **1** in the usual manner, and we obtain

$$ql \coth ql \,.\, e = e(1 + 2\epsilon^{p_1t} + 2\epsilon^{p_2t} + 2\epsilon^{p_3t} + \dots), \tag{14}$$

instead of (4) above, with the same values of the p's.

Now (1) is the same as

$$\begin{aligned} \mathrm{V} &= (\cosh qx - \coth ql \,\text{shin}\, qx)e \\ &= \cosh qx \,.\, e - \frac{\text{shin}\, qx}{ql} e(1 + 2\epsilon^{p_1t} + 2\epsilon^{p_2t} + \dots), \end{aligned} \tag{15}$$

by (14). Here the power of p is zero in the cosh function, and also in the other so far as the constant term goes, and p_1 on ϵ^{p_1t}, &c. So, putting $q^2 = -s^2$, we get

$$\mathrm{V} = e\left(1 - \frac{x}{l}\right) - 2e\Sigma \frac{\sin sx}{sl} \epsilon^{pt}, \tag{16}$$

which is the same as (9).

Since circular functions are finally to be employed, we may put $q^2 = -s^2$ at the beginning if we like, and write

$$\mathrm{V} = \frac{\sin s(l-x)}{\sin sl} e, \tag{17}$$

instead of (1). Here of course s^2 is a differentiator, namely, $-\mathrm{RS}p$. It finally takes the meaning of a constant, or rather of an infinite series of constants, merely because p has the power of a constant on the exponential function.

The determinantal equation $\text{shin}\, ql = 0$, or, more strictly, $(\text{shin}\, ql)/ql = 0$, is really the equation finding the admissible values of p in the normal systems. To every p belongs an s^2, so if we choose to consider the determinantal equation in the form $(\sin sl)/sl = 0$ as finding the values of s, we need only attend to the positive values.

Notice also that the operator in (1) does not contain $p^{\frac{1}{2}}$, but only the complete p. The elementary waves which make up the solution (1) do depend on $p^{\frac{1}{2}}1$, but the union of all the wave operators to make the solution (1) causes $p^{\frac{1}{2}}$ to be eliminated. The determinantal equation for a cable of finite length—that is, bounded both ways by some restraint, is always a rational equation in p, provided the terminal arrangements themselves are finite* combinations whose resistance operators are themselves rational functions of p.

(3). Earth at Both Terminals. Initial Charge at a Point. Arbitrary Initial State.

§ 267. A more comprehensive case than the last, bringing a fresh peculiarity into view, is that of the effect due to an initial charge at a single place when the cable is earthed at both ends. There are two solutions in terms of waves, one for the region to the left, and the other for that to the right side of the source. To find them put $\alpha = -1 = \beta$ in (65), (66), § 263, and also $h = p\mathrm{Q}$, as there described. Then we obtain

$$\mathrm{V}_1 = \frac{\text{shin}\, qy}{\text{shin}\, ql}\, ql\, \text{shin}\, q(l-x)\, \frac{\mathrm{Q}}{\mathrm{S}l}, \tag{18}$$

on the right side of the source, which is a charge Q initially at the place $x = y$. To obtain V_2, the potential on the left side, interchange y and x in the last equation.

We see that we have the same denominator $\text{shin}\, ql$ as before. This is clearly because the terminal conditions, earth

* There is an exception to finiteness in the case of a distortionless circuit used for a terminal arrangement, because it behaves like a mere resistance when it is infinitely long.

on at both ends, are unchanged. The denominator sums up the reflective action of the ends. But it is not to be supposed that changing the terminal conditions necessarily changes the determinantal equation. For that depends upon both terminal conditions, and a change in one may be balanced by a change in the other. Thus, if both ends were cut, the denominator would be unaltered, since the signs of both a and β would be reversed. It is their product that determines the subsidence rates, as we see by (65), (66), § 263, which show that, in the general case,

$$\epsilon^{2ql} = a\beta \tag{19}$$

is the determinantal equation.

Returning to (18), and using the result (4), we get

$$V_1 = \text{shin}\, qy\, \text{shin}\, q(l-x)\, \frac{Q}{Sl}(1 - 2\epsilon^{p_1 t} + 2\epsilon^{p_2 t} - \ldots), \tag{20}$$

with the same values of p as before. The final steady V is zero obviously, so there is no outside turn, and we get at once by putting $q^2 = -s^2$,

$$V = \frac{2Q}{Sl}\Sigma \pm \sin sy \sin s(l-x)\, \epsilon^{pt}; \tag{21}$$

or, since $\sin sl = 0$,

$$V = \frac{2Q}{Sl}\Sigma \sin sy \sin sx\, \epsilon^{pt}. \tag{22}$$

We have in the last two equations written V for V_1, because the interchange of x and y makes no difference. That is, (22) represents V at x due to Q initially at y, whether x be to the right or left of the source. Only one Fourier series is required to represent the essentially different results to the right and left of the source.

Putting $t=0$ and dismissing constant factors, we see that if

$$u = \frac{2}{l}\Sigma_1^\infty \sin\frac{n\pi x}{l}\sin\frac{n\pi y}{l}, \tag{23}$$

the summation including all integral values of n from 1 to ∞, then u represents 0 everywhere between A and B, except at the point $y = x$, where it is infinite. But its space total is 1, because the space total of SV in (22) is Q at the first moment, before any of the charge has left the cable.

The function u therefore expresses an impulsive function, like the $p1$ with which we were concerned before, only now the variable is x instead of t, or the function is distributed in length instead of time. Such functions can be represented in various ways. Every Fourier series involves one form.

If we multiply a continuous function of y, say $f(y)$, by u, an impulsive function which exists only at the point $y=x$, the product is obviously zero except at that point, where it is infinite. But if we take the space total of the product $uf(y)$, the result is $f(x)$. For u only exists at x, and its total is 1. Thus,

$$\int uf(y)dy=f(x), \tag{24}$$

if the limits include the point x. If not, the result is zero. This is the property made use of in Fourier and other series when employed to express arbitrary functions. The function u, in the above or other special form, spots a single value of the arbitrary function in virtue of its impulsiveness.

Suppose, however, that the given function $f(y)$ had no definite value at a particular place, as happens when there is a discontinuity, or sudden jump from one value to another. Here $f(y)$ has any value between the two extremes. Under these circumstances, what should we expect the impulse to do, if it can only, through (24), lead to one value? To be quite fair, it should show the mean value. This is an excellent reason why it ought to do so, though not a proof that it does. Why it must may become clear later on.

Applying (24) to the problem in hand, with u as in (23), we see that

$$f(x)=\frac{2}{l}\Sigma_1^\infty \sin\frac{n\pi x}{l}\int_0^l f(y)\sin\frac{n\pi y}{l}dy \tag{25}$$

expands the arbitrary function $f(x)$ in a series of sines, so as to vanish at $x=0$ and l (where, of course, if $f(x)$ does not vanish, the formula fails). We also see that

$$\mathrm{V}=\frac{2}{l}\Sigma_1^\infty \sin\frac{n\pi x}{l}\epsilon^{pt}\int_0^l \mathrm{V}_0\sin\frac{n\pi y}{l}dy \tag{26}$$

represents the potential at x at time t after the moment when the potential was represented by V_0. The former Q is represented now by SV_0dy, that is, the charge on the element of length dy.

The formula (25) may, as we have seen, fail either at the terminals or intermediately at a discontinuity. But the formula (26) does not fail at all for any finite value of the time. Initially it is the same as (25), and may fail then. But the presence of the time functions makes V continuous. No matter how small t is taken, if not actually zero, there is no discontinuity. Sharp corners in V_0 are instantly rounded off by the time factors, though if graphically represented there might be no observable difference. We may take t so small that the time functions differed from 1 insensibly for millions of terms, yet go far enough, and we must come to time-functions differing insensibly from 0 with the same value of t. So V is necessarily made continuous in the smallest interval of time after the first moment, if it be initially discontinuous.

Another way of getting (22) is by using the coth function. Thus, by (18),

$$V_1 = ql \operatorname{shin} qy \left(\cosh qx - \frac{\operatorname{shin} qx}{ql} ql \coth ql \right) \frac{Q}{Sl}. \tag{27}$$

Noting that the first part is inoperative, the result (13) turns the last to

$$V = -\operatorname{shin} qx \operatorname{shin} qy \frac{Q}{Sl} (1 + 2\epsilon^{p_1 t} + 2\epsilon^{p_2 t} + \ldots); \tag{28}$$

and now, noting that the power of p is 0 in the first term, p_1 in the second, and so on, and putting $q^2 = -s^2$, we make

$$V = \frac{2Q}{Sl} \Sigma \sin sx \sin sy \, \epsilon^{pt}, \tag{29}$$

as before.

Finally, a third way. We have, by (18),

$$V_1 = ql \operatorname{shin} qy \times \frac{\operatorname{shin} q(l-x)}{\operatorname{shin} ql} \frac{Q}{Sl}. \tag{30}$$

Now the part after the × was algebrized in §265; *see* equations (1) and (9) there. Therefore

$$V_1 = ql \operatorname{shin} qy \frac{Q}{Sl} \left\{ \left(1 - \frac{x}{l}\right) - 2\Sigma \frac{\sin sx}{sl} \epsilon^{pt} \right\}; \tag{31}$$

and now, on effecting the differentiations, which is done at sight by the power of p being constant in every term, we obtain (29) again. These variations will, if practised, enable the reader to operate quickly and safely.

(4). Line Cut at Both Terminals. Effect of Impressed Force.

§ 268. Next, in further illustration of simple cases, let the line be cut at both ends. First, with an impressed force e at the point $x=y$. Then, by equations (39), (40), § 259, we have

$$V_1 = \frac{\operatorname{shin} qy}{\operatorname{shin} ql} \cosh q(l-x) \,.\, e, \tag{32}$$

$$V_2 = -\frac{\cosh qx}{\operatorname{shin} ql} \operatorname{shin} q(l-y) \,.\, e. \tag{33}$$

To convert to Fourier series, put $q^2 = -s^2$ in (32), and reject the part involving $\sin sx$. Then

$$V_1 = \frac{\sin sy \cos sx}{sl} \, sl \cot sl \,.\, e. \tag{34}$$

In this use (14), making

$$V_1 = e\, \frac{\sin sy \cos sx}{sl} (1 + 2\epsilon^{p_1 t} + 2\epsilon^{p_2 t} + \dots),$$

which finally makes, by the power of p being a constant in every term,

$$V_1 = \frac{ye}{l} + \Sigma \frac{2e}{sl} \sin sy \cos sx\, \epsilon^{pt}. \tag{35}$$

This is on the right side of e. On the left side, (33) makes

$$V_2 = -e + e\, \frac{\cos sx \sin sy}{sl} (1 + 2\epsilon^{p_1 t} + 2\epsilon^{p_2 t} + \dots) \tag{36}$$

$$= -e + \frac{ey}{l} + \Sigma \frac{2e}{sl} \sin sy \cos sx\, \epsilon^{pt}, \tag{37}$$

which only differs from (35) in the outside term expressing the steady state. The values of s are as before, but the expansion is in cosines, not sines. The summation represents the negative of the final state, which is discontinuous.

As regards the final state, we have two condensers in sequence, of permittances Sy and $S(l-y)$, with the impressed force between them. The effective permittance is the reciprocal of the sum of the reciprocals of the separate permittances—that is, $Sy(1-y/l)$. Multiplying by e we obtain the charge, and then dividing by $S(l-y)$, we obtain

ey/l, the final potential on the right side of the force. On the left side it is lower by the amount e.

(5). Line Cut at Terminals. Effect of Initial Charge.

§ 269. We may do the effect of a single charge at y in the same way. Let it be Q, then

$$V_1 = ql\frac{\cosh qy}{\text{shin } ql}\cosh q(l-x)\,.\,\frac{Q}{Sl}, \tag{38}$$

by (65), § 263, with $h = pQ$. This is on the right side of Q, and on the left side the formula is obtained by interchanging x and y. Put $q^2 = -s^2$, and reject the part involving $\sin sx$. Then

$$V_1 = \cos sy \cos sx\,.\,sl \cot sl\,\frac{Q}{Sl} \tag{39}$$

$$= \frac{Q}{Sl}\cos sy\cos sx\,(1 + 2\epsilon^{p_1 t} + 2\epsilon^{p_2 t} + \ldots) \tag{40}$$

$$= \frac{Q}{Sl}\{1 + 2\Sigma \cos sy \cos sx\,\epsilon^{pt}\}. \tag{41}$$

In passing from (39) to (40) the result (14) is used, and the further passage to (41) involves only a formal change by the power of p being a constant, as in former cases.

By the last formula we see that

$$u = \frac{1}{l} + \frac{2}{l}\Sigma_1^\infty \cos\frac{n\pi x}{l}\cos\frac{n\pi y}{l} \tag{42}$$

is a unit impulsive function existing at the point $y = x$, like (23) in fact, so that the formula (24) applies, and an arbitrary function may be expanded in cosines thus :—

$$f(x) = \frac{1}{l}\int_0^l f(y)\left\{1 + 2\Sigma_1^\infty \cos\frac{n\pi x}{l}\cos\frac{n\pi y}{l}\right\}dy. \tag{43}$$

The introduction of the time factor ϵ^{pt}, where $p = -n^2\pi^2/RSl^2$, in the summation will, by the preceding, represent what the initial potential $f(x)$ becomes at time t later by diffusion in the cable with its ends insulated. The final state is a uniform mean potential.

(6). Earth at A and Cut at B. Effect of Initial Charge.

§ 270. Next take a case involving a different determinantal equation. Say there is earth on at A and disconnection at B.

Then $\alpha=-1, \beta=1$. So, with an initial charge Q at $x=y$, we have, by (65), § 263,

$$V_1=\frac{\text{shin}\, qy}{\cosh ql}\cosh q(l-x)\,.\,ql\frac{Q}{Sl}, \tag{44}$$

in which the interchange of x and y will give V_2. Or, putting $q=si$,

$$V_1=sl\sin sy\,(\cos sx+\sin sx\tan sl)\,\frac{Q}{Sl}. \tag{45}$$

The first part is inoperative as regards a Fourier series. So

$$V_1=(\sin sy\sin sx)(sl\tan sl)\,\frac{Q}{Sl}. \tag{46}$$

Here we see that we have a different operator to produce the series of exponentials, viz., $sl\tan sl$ or ql than ql. The trigonometrical expansion of this function is

$$ql\text{ than }ql=\frac{2}{1+\left(\frac{\pi}{2ql}\right)^2}+\frac{2}{1+\left(\frac{3\pi}{2ql}\right)^2}+\frac{2}{1+\left(\frac{5\pi}{2ql}\right)^2}+\dots. \tag{47}$$

Remembering the meaning of q^2, namely, RSp, this makes

$$ql\text{ than }ql\,.\,1=2\epsilon^{p_1t}+2\epsilon^{p_2t}+2\epsilon^{p_3t}+\dots, \tag{48}$$

in the way (14) was got, only now the p's are a different set, namely, the roots of

$$\cosh ql=0,\qquad\text{or}\qquad \cos sl=0; \tag{49}$$

say $sl=\frac{1}{2}\pi,\ \frac{3}{2}\pi,\ \frac{5}{2}\pi$, &c.

So, by (46),

$$V_1=\frac{2Q}{Sl}\sin sy\sin sx\,\Sigma\epsilon^{pt}, \tag{50}$$

which is at once transformable to

$$V=\frac{2Q}{Sl}\Sigma\sin sy\sin sx\;\epsilon^{pt}, \tag{51}$$

by making s be constant instead of a differentiator. In this example the normal functions of the type $\sin sx$ vanish at A only. At B they have maxima, so that their slopes vanish. This is to make the current vanish at B, where there is insulation. The change in the first normal function is very great. Being $\sin(\pi x/2l)$, it is just one half of the first normal function for a line twice as long, earthed at both ends. The corresponding p is the same as that for the line of double length, so that the subsidence of the first normal system in the present case of insulation at B takes four times as long to reach a given stage, as the subsidence of the first normal system of the same line does when it is earthed at both ends.

The first normal system is, of course, the most important. When t is big it is practically the only one worth counting. When t is smaller, the second one acquires significance, then the third, and so on, up to the first moment, when the whole series is required. But these Fourier series are frequently very unpractical for numerical purposes when t is taken small on account of the slow convergency.

Periodic Expression of Impulsive Functions. Fourier's Theorem.

§ 271. The above examples are sufficient to show the connection between the operational solutions and the corresponding Fourier series in the simple cases of terminal direct earth connection and complete disconnection, at least so far as the particular method employed is concerned, bearing in mind that principles are often as well illustrated by simple as by more complicated problems, if not better. The above process consists in first converting the reciprocal of the determinator shin ql or cosh ql, &c., to the sum of partial fractions by using known trigonometrical formulæ. These are easily algebrized, and the result is to give V or C or other quantity at some particular place. Thirdly, we operate on this result directly by whole differentiations, which merely means giving p a constant value in any term, and the result is the solution at any place in the form of a Fourier series.

Before passing on to another way of considering Fourier series operationally, some remarks about the functions called u above may be useful. They were considered merely as expressing unit impulsive functions located at a point $x=y$ situated between 0 and l. So far as that goes, they are quite similar. But in other respects they differ. They are both periodic, however. Every time x is increased by the amount $2l$, the sines u and the cosines u repeat themselves. But the sines u is an odd function of x, whilst the cosines u is an even function. So the sines u is negatived by changing x to $-x$. The complete meaning of the sines u is therefore an infinite row of positive unit impulses at the points $y=x$, $x+2l$, $x+4l$, &c., $x-2l$, $x-4l$, &c.; together with an infinite row of negative unit impulses at the points $y=-x$, $-x+2l$, $-x+4l$, &c., $-x-2l$, $-x-4l$, &c. On the other hand, the cosines u represents positive unit impulses at all the above places. Neither of the u's is a periodic repetition of a single impulse.

But if we take half the sum of the sines u and the cosines u, the negative impulses will be cancelled. There is left simply an infinite row of positive unit impulses at the points $y=x\pm 2nl$, where n is any integer. The diagram will serve to fix this plainly:—

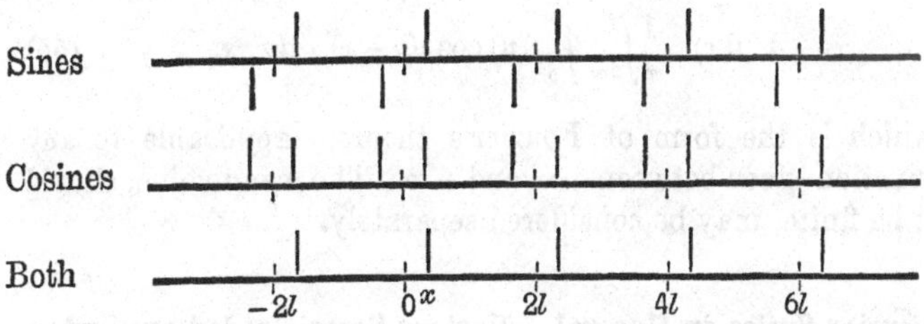

The horizontal straight line is divided into equal lengths $2l$, and the vertical lines at x and $x\pm 2nl$, $-x$ and $-x\pm 2nl$ represent the distribution of the impulses in the three cases of the sines u, the cosines u, and half their sum. By (23) and (42), the new u is represented by

$$u=\frac{1}{2l}\left\{1+2\Sigma_1^\infty \cos\frac{n\pi}{l}(y-x)\right\}, \tag{52}$$

and its meaning for our purpose is a function of y which is zero everywhere except at $x\pm 2nl$, where it is infinite, but so that its total is unity. This u is the periodic impulsive function concerned in the periodic form of Fourier's theorem. For, using (52) in (24), we have

$$f(x)=\frac{1}{2l}\int_0^{2l}\left\{1+2\Sigma_1^\infty \cos\frac{n\pi}{l}(y-x)\right\}f(y)\,dy, \tag{53}$$

which expands an arbitrary function $f(x)$ whose values are given between the limits 0 and $2l$ into the sum of a series of sines and cosines with constant coefficients, and with a constant term expressing the mean value. Besides that, it makes the given values repeat themselves over and over again, with the period $2l$. That is, the impulse operator $\int u \ldots dy$, in virtue of its periodic nature, makes $f(x)$ periodic as well.

When the period is infinitely long, if we put $s = n\pi/l$, we have $ds = \pi/l$ to represent the step from one s to the next, which is infinitely small in the limit, when the impulsive function becomes

$$u = \frac{1}{\pi}\int_0^\infty \cos s(y-x) \,.\, ds, \tag{54}$$

and its meaning is a unit impulse at the point $y = x$ only. So, applying it through (24) to a function of y, we get

$$f(x) = \frac{1}{\pi}\int_{-\infty}^{\infty}\int_0^\infty f(y) \cos s(y-x) \,.\, dy\, ds, \tag{55}$$

which is the form of Fourier's theorem applicable to any function given between $-\infty$ and $+\infty$. The mean value, should it be finite, may be considered separately.

Fourier Series in General. Various Sorts Needed even when the Function is Cyclic and Continuous. Expansions of Zero.

§ 272. Fourier's theorem as exhibited by equation (53) applies to the diffusion of charge in a cable in two cases only. First, let the cable be infinitely long, all in one piece, and let it be charged initially in a periodic manner, say arbitrarily between $x = 0$ and $2l$, and with repetitions of the same state in all the other sections of length $2l$. Then the expansion (53) is obviously the proper one. The function $f(x)$ represents the initial state of potential. The state at any time t later is got by introducing the time factor ϵ^{pt}, as in previous cases, in the summation on the right side. This allows the harmonic terms to subside or decay, leaving only the mean value of the initial state behind finally.

But, secondly, it answers the purpose equally well to have a finite cable, of length $2l$, with the ends joined to make a closed circuit. We may take the origin ($x = 0$) anywhere, and

$x=2l$, $4l$, &c., mean the same as $x=0$. Equation (53) is the appropriate expansion to employ, with the time factor introduced to indicate the resulting state at any time.

The action of the time functions is always to make the later terms in the expansion insignificant compared with the earlier, so that we may generally say that the successive terms of the summation, not counting the outside constant term, are of regularly decreasing importance. And, indeed, in making practical applications, it will be found that this is usually true for the initial state itself. We usually find that the natural order of the harmonic terms in the expansion of a function is the order of their relative importance inversely. But this property is by no means a necessary one. One or any number of terms may be wholly absent, or if not absent may be very small, and followed by bigger ones. There is then an exceptional state at the first moment, and for a certain time after. But when the decadence has progressed sufficiently we must arrive at a state of things in which the order of the terms that exist at all is also the order of their magnitude.

When a cable is closed upon itself, say as described above, the real potential, whatever it may be, is necessarily a periodic function of x. Now Fourier's theorem in the form (53) is the fully periodic form. It might therefore seem that the mere fact of the periodic character of the potential in the case of a closed circuit made the periodic form of Fourier's theorem obligatory. But this is only accidentally true, as it were. The expansion (53) is only one of an infinity of expansions applicable to a closed cable, each in its proper place. For (53) to be applicable, it is not only necessary for the cable to be closed, but also for it to be self-contained—that is, there must be no external connections imposing some restraint upon the potential or the current at a particular place. The "terminal" conditions are that V at $x=0$ and V at $x=2l$ are identical, and that C is also continuous at those places, because they are united, and in such a manner as not to interfere with the continuity of V and C. We may, indeed, choose the initial state to violate these conditions. But the violation is only momentary, for continuity is instantly established by the action of the time functions.

But suppose we put a simple leak on at the place $x=0$, everything else being unchanged. The charge in the cable will redistribute itself now in an entirely different manner. For one thing, it will all leak out of the cable, instead of settling down to a state of uniform distribution. It is obvious that the periodic form of Fourier's expansion of the initial state is unsuitable, although the real potential of the line is periodic. There is no difficulty in obtaining the proper series, though it would be too great an interruption of continuity to give it just now. The peculiarity is that although V is continuous at $x=0$ and $2l$ (except, it may be, at the first moment), C is not continuous. The discontinuity in C at the joined terminals is proportional to the potential there, by Ohm's law applied to the leak. The proper Fourier series must be found to satisfy this condition as well as that of continuity of V in order that it shall, when the time factors are introduced, represent the potential at any moment. The initial state may be chosen arbitrarily and so as to violate the conditions stated, but this is of no consequence at all in the complete solution for V at any time.

The simple leak may be generalised to any combination having a definite resistance operator, by substituting the latter for the resistance of the leak. Thus we obtain any number of Fourier series for a closed circuit, which may represent one and the same arbitrary function (the initial state). All are continuous as regards V, but discontinuous at the junction $x=0$ as regards C according to the special law imposed by the nature of the leak. That is, when t is finite; for, going right back to the moment $t=0$, there may be failure then, viz., by choosing the arbitrary initial state so as to fail to comply with the conditions.

Instead of a discontinuity in the current (or in the derivative of the potential), we may make it be a discontinuity in the potential inself. Thus, we may put a coil in circuit with the cable, doing away with the leak. And, as before, this may be variously generalised. Or we may have the potential and the current both discontinuous at the junction of the beginning and end of the cable by suitable electrical devices. In every case there is one and only one appropriate Fourier series, and there is no difficulty (save complication in detail) in finding it by seeking the solution of the physical problem of decadence of an initial state subject to the imposed conditions. On the other hand, if we eliminate the physical ideas and also the time variation, and look at the matter from a purely mathematical point of view, there are serious difficulties introduced.

The initial state may be one of equilibrium, say $V = 0$ everywhere. That is, we may expand zero in any number of ways in a Fourier series. To do this, we require an external source of electric or magnetic energy. We may, for example, insert a charged condenser and coil in sequence between earth and the cable at the junction of $x = 0$ and $x = 2l$. Let the initial state be a charge in the condenser, but none in the cable. The initial arbitrary function to be expanded in a F. series of a very special kind is zero. The time factors make the series be finite, and express the potential at time t due to the condenser's charge. Finally, if the condenser is leaky, we come to zero potential again in the cable.

We may also require to determine a Fourier series not merely so as to express $f(x)$ (including zero as a special case) between certain limits, but also so that certain functions of the Fourier series may have special values in addition. And then connected systems of Fourier series are sometimes required, as when the line is divided into connected sections, with external conditions imposed at the junctions. Or we may have a network of cables, of the same type or different, with imposed conditions, and have to find a set of Fourier series (or other kind of series if a cable varies in its resistance, &c.), to satisfy all the conditions. And so on to any extent.

There is no difficulty in obtaining the operational solutions by the method of resistance operators in a clear and definite manner, and when that is done we are virtually in possession of the real algebraical solutions in the cases of simply periodic impressed forces, and steady and impulsive forces, and arbitrary initial states, for the conversion from operational to algebraical form involves only formal transformations. But we must now return to Fourier's theorem. The above remarks originated in the departures from Fourier's theorem required even when the function under consideration is everywhere continuous and periodic. In the treatment of Fourier series to be found in certain works, the reader can hardly find a glimmer of a notion of the subject in its general and comprehensive aspect.

Special Forms of Fourier Integrals. Interchangeable Property. Use in Transforming Definite Integrals.

§ 273. There are two special forms of Fourier's theorem in the form of a double integral which should be noted. Referring to the diagram of impulses, § 271, we see that when l is ∞ the sines u, say u_2, is reduced to represent merely a positive unit impulse at $y=x$, and an equal negative one at $y=-x$. The cosines u, say u_1, on the other hand, represents a positive impulse both at y and at $-y$. The formulæ are

$$u_1=\frac{2}{\pi}\int_0^\infty \cos sy \cos sx\, ds, \tag{56}$$

$$u_2=\frac{2}{\pi}\int_0^\infty \sin sy \sin sx\, ds. \tag{57}$$

So, if we use these in the fundamental formula (24), or

$$f(x)=\int f(y)u\,dy, \tag{58}$$

the result on the left side, which is $f(x)$ when u is a single impulse within the limits as before stated, will depend upon the limits employed. If the limits in (58) are complete, viz., $-\infty$ to ∞, and we use u_1, we shall obtain $f(x)+f(-x)$. But if we use u_2, we shall obtain $f(x)-f(-x)$. For the impulses are double, and positive or negative as described.

We may, of course, choose the limits so as to exclude both x and $-x$. Then the result is zero. Or we may include just one of them, and get either $f(x)$ or $\pm f(-x)$. The most useful way is to exclude the whole of the negative values of x. Then (58) applied to u_1 and u_2 makes

$$f(x)=\frac{2}{\pi}\int_0^\infty\int_0^\infty f(y)\cos sy \cos sx\, ds dy, \tag{59}$$

$$f(x)=\frac{2}{\pi}\int_0^\infty\int_0^\infty f(y)\sin sy \sin sx\, ds dy. \tag{60}$$

These are very useful in the mathematics of physics, perhaps more so than the complete formula (55). The following way of looking at them is interesting. Given that

$$F(x)=\left(\frac{2}{\pi}\right)^{\frac{1}{2}}\int_0^\infty f(y)\cos xy\,dy. \qquad (61)$$

Then the symbols F and f are interchangeable. That is, we also have

$$f(x)=\left(\frac{2}{\pi}\right)^{\frac{1}{2}}\int_0^\infty F(y)\cos xy\,dy. \qquad (62)$$

There is a corresponding interchangeability in the other case. That is, if

$$G(x)=\left(\frac{2}{\pi}\right)^{\frac{1}{2}}\int_0^\infty g(y)\sin xy\,dy, \qquad (63)$$

then

$$g(x)=\left(\frac{2}{\pi}\right)^{\frac{1}{2}}\int_0^\infty G(y)\sin xy\,dy. \qquad (64)$$

The corresponding property connected with the zeroth Bessel function is that if

$$F(x)=\int_0^\infty f(y)J_0(xy)y\,dy, \qquad (65)$$

then the symbols F and f are interchangeable. To prove them, put say (62) in (61), and we get (59). That is, we get an equivalent, the notation being changed. We have dy both in (61) and (62), but of course in the double integral it is desirable to have them different.

The above is the way Fourier's theorem often turns up in practice. We have an integral of the form (61) say, and we can at once deduce another one. It is not necessary to go through the work of making up the double integral, and so using Fourier's theorem explicitly. But it may not be convenient to have the $(2/\pi)^{\frac{1}{2}}$ coefficient. Then, partially sacrificing the perfect interchangeability, we may put it thus. If

$$F(x)=\frac{2}{\pi}\int_0^\infty f(y)\cos xy\,dy, \qquad (66)$$

then it follows that

$$f(y)=\int_0^\infty F(x)\cos xy\,dx. \qquad (67)$$

Similarly when we write sin for cos in these formulæ. Examples are numberless. One or two will suffice here to illustrate. Say that we know that

$$J_0(sx) = \frac{2}{\pi}\int_s^\infty \frac{\sin xy}{(y^2 - s^2)^{\frac{1}{2}}}\,dy. \tag{68}$$

It follows that

$$\frac{1}{(y^2 - s^2)^{\frac{1}{2}}} = \int_0^\infty J_0(sx) \sin xy\,dx. \tag{69}$$

Observe that the limits in (68) are not complete. From $y = 0$ to $y = s$, the $f(y)$ function integrated has no existence—that is, y must be greater than s. Consequently, in (69), y must be greater than s. If y is less than s in (69) the value of the integral is zero. (We have assumed that x is positive in (68). If negative, the $-$ sign must be inserted.)

If we reversed the process and passed from (69) to (68), we should require to know that the (69) integral is true as given when $y > s$, but is zero when $y < s$. This will exclude the region 0 to s from the first integral (68).

Similarly, if $G_0(sx)$ is the companion function to $J_0(sx)$, the second kind of oscillating function, and we know that

$$\frac{1}{(y^2 - s^2)^{\frac{1}{2}}} = \int_0^\infty G_0(sx) \cos xy\,dx, \tag{70}$$

when y is greater than s, but that it is zero when y is less than s, then we deduce

$$G_0(sx) = \frac{2}{\pi}\int_s^\infty \frac{\cos xy}{(y^2 - s^2)^{\frac{1}{2}}}\,dy, \tag{71}$$

the region from $y = 0$ to $y = s$ being excluded. This is like going from (67) to (66). Reversing the process, we observe that the condition $y > s$ in (71) has to be preserved in (70). Else we get zero.

As an extreme case of (61), if we take $f(y)$ to be a single unit impulse at the point $y = z$, we obtain $F(x) = (2/\pi)^{\frac{1}{2}} \cos xz$. Next, using this function in (62), we come back to the impulsive function, in accordance with (56) above. There is a similar property involved in all normal functions, since they are capable of expressing arbitrary functions, and, therefore, a single impulse. Observe that the (69) integral contains two normal functions, so that we may not only deduce (68) from it, but also another integral, by employing the interchangeable property involved in (65).

Continuous Passage from Wave Series to Fourier Series, and its Reversion.

§ **274.** Now, a few words on the reversibility of the process of getting Fourier series operationally. Given a Fourier series, say $f(x) = \Sigma \mathrm{A} \sin sx$, can we find its meaning operationally? Of course, its immediate meaning is given by the F. series itself. But what the resultant meaning of the sum of the harmonic functions may be is not evident, and it may be rather a serious matter to find it by numerical calculation, though after all the meaning may be quite simple, a straight line, for example. There is matter for regret here, for whilst the solution of problems by F. series may be easily elaborated, the ease is confined to the formulæ, not to their calculation, which gets more and more difficult.

One way of finding the meaning may be briefly mentioned here. By the introduction of a time-factor to every term of the F. series we may make it (if of a suitable kind) represent a real physical problem in diffusion. Then $f(x)$ means the initial state. Now I have already pointed out the identity of the wave solutions and F. series. It will have been observed that we can pass continuously from the waves to the F. series. For we can construct the wave solution of a diffusion problem expressing the effect due to a given source, subject to terminal reflections, by writing down the waves themselves in operational form, first the initial waves and then their consequences in order, as done in §§ 255 to 260. Adding these together, we obtain the condensed operational solution, which may be independently obtained without thinking out the waves in detail. Finally, we can convert the result to a F. series, as explained in §§ 265 to 270 in special cases, and in a general way that will come later.

Is this process reversible? It is certainly reversible on one side, for we convert the operational solution to the wave form by algebraical division. It is not only in the simple cases of terminal earth or disconnection that the meaning of the individual waves can be ascertained, though I have not given any advanced examples in detail yet. As regards the other side, we can plainly pass from the diffusion solution in a F. series to the operational solution by simply reversing all the steps. Referring to examples before given—first, the constancy of power of the time-differentiator on ϵ^{Pt} enables us to put the circular function occurring in the F. series outside the summation sign, by taking s^2 to mean $-\mathrm{RS}(d/dt)$. Then the summation that is left, or $\Sigma\epsilon^{Pt}$, may be converted to the operational form $\Sigma(1-\mathrm{P}/p)^{-1}$. This series of fractions may then be summed. The final result is an operational formula with unit operand.

But if the reader will apply this reversed process to the examples in detail hitherto given, he will find that he will not be able to do it safely with his eyes shut. This is because in going from the operational equation to the F. series, certain operations are impotent, and are omitted. So in the reversed process, they should be restored. This can only be done by careful inspection of the problem, and is a question of experience and judgment. The terminal conditions satisfied by the solution in F. series should be examined. They must be satisfied all the way through, and this will enable the complete reversal to be effected. But, though there may be troubles of the above kind to be surmounted, it is not a useless way of transforming from the F. series to waves when the sources are of a simple nature. On the other hand, should the sources be not simple, as in the case of subsidence from an "arbitrary" state, the case is altered, on account of the integration which is involved in the determination of the coefficients of the F. series, which complicates the matter considerably. We may therefore leave this question on one side.

How to find the Meaning of a Fourier Series Operationally.

§ **275**. Another way of regarding F. series operationally will be conveniently introduced by considering an elementary example. Given

$$f(x)=\frac{2}{\pi}\,\Sigma_1^\infty\frac{1}{n}\sin\frac{n\pi x}{l}. \tag{72}$$

Find what $f(x)$ means, without numerical calculation, and also without introducing time functions to make a diffusion problem. We see that $f(x)$ is an odd function of x, and that it is periodic. It is therefore sufficient if we know its meaning from $x=0$ to $x=l$.

Regard the right member of (72) operationally as a function of x in the same way as our previous functions of t. Let Δ stand for d/dx. This replaces the former p, the time-differentiator. Then first we have to convert the series to operational form as a function of Δ. This is easy. For we have

$$\sin sx = sx - \frac{(sx)^3}{\lfloor 3} + \frac{(sx)^5}{\lfloor 5} - \dots$$

$$= \frac{s}{\Delta} - \left(\frac{s}{\Delta}\right)^3 + \left(\frac{s}{\Delta}\right)^5 - \dots = \frac{s/\Delta}{1+(s/\Delta)^2}. \tag{73}$$

Similarly, in passing, we see that $\cos sx$ = same with 1 for numerator.

Next, using (73) in (72) we produce

$$f(x) = \frac{2}{\pi}\Sigma\frac{1}{n}\frac{\frac{n\pi}{l\Delta}}{1+\left(\frac{n\pi}{l\Delta}\right)^2} = \frac{1}{l\Delta}\Sigma\frac{2}{1+\left(\frac{n\pi}{l\Delta}\right)^2}. \tag{74}$$

We have already had occasion to employ this formula, equation (13), with ql instead of $l\Delta$. So

$$f(x) = \coth l\Delta - \frac{1}{l\Delta}, \tag{75}$$

with unit operand understood. That is, in getting (73), we imagine the operand to start at $x=0$, and be 1 afterwards.

Next, expand the coth function in (75) by division. Thus,

$$f(x) = (1 - x/l) + 2\epsilon^{-2l\Delta} + 2\epsilon^{-4l\Delta} + 2\epsilon^{-6l\Delta} + \dots, \tag{76}$$

where we have also algebrised the $(l\Delta)^{-1}\,1$ in (75).

Now this equation (76) expresses the full algebraical meaning of $f(x)$ on the assumption made, namely, that it begins at $x=0$. The value of $f(x)$ is explicitly represented by (76) from $x=0$ up to $x=\infty$. It is $1-x/l$ from $x=0$ to $x=2l$. All the following terms are then zero. From $x=2l$ to $x=4l$, the first auxiliary term exists. Its value is 2, and makes $f(x) = 3-x/l$. From $x=4l$ to $6l$, the second auxiliary exists as well as the first. This adds on another 2, and makes $f(x)=5-x/l$, and so on regularly. How these jumps come about through the auxiliary terms will be explained next.

Taylor's Theorem in its Essentials Operationally Considered.

§ 276. Imagine any wave form to be transmitted at a constant speed undeformed, as, for instance, a hump running along a flexible cord. Let it be as in the figure, namely, a triangular hump whose position at the moment of time $t=0$ is A, and at some later time t is B. It is to be imagined to travel at uniform speed v from left to right, so that the distance AB is vt, increasing uniformly with the time. If the base line is the axis of x, the travelling form is our "function of x," its value being the ordinate of the wave form.

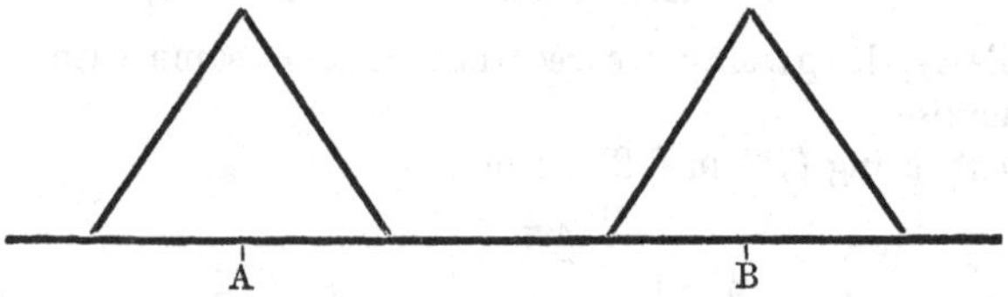

Let $f(x)$ represent the function when at A, and let it become $F(x)$ at B. They only differ in the change of origin. Their relation to one another is

$$F(x)=f(x-vt)\,; \tag{77}$$

that is, the value of F at x at the moment t is, by definition, the same as the value of f at a distance vt to the left. Equation (77) expresses the characteristic property of an undistorted and unattenuated wave when the speed is constant. It is a positive wave. When it goes the other way, it is a negative wave, and the *minus* sign must be changed to *plus*.

Differentiate equation (77) with respect to x and t separately and compare them. We see that

$$\frac{dF}{dx}=-\frac{1}{v}\,\frac{dF}{dt}. \tag{78}$$

This is the characteristic partial differential equation of a solitary undistorted wave. It is easily understood by watching the transit of a particular part of the wave past a fixed point, and considering how the speed v of transit combined with the slope determines the time rate of increase of F at the fixed spot. Equations (78) and (77) have practically the same meaning. Thus, given (78), its solution is (77), understanding that f expresses an arbitrary function of x.

Now write (78) in the form

$$\frac{dF}{dt} = -v\Delta F, \tag{79}$$

where Δ stands for d/dx, and solve it as an ordinary linear equation. This makes

$$F = \epsilon^{-vt\Delta} f, \tag{80}$$

where f is a constant with respect to t. In our present case it is any function of x. That (80) is the solution is verifiable by its satisfying (79). What f is may be seen by putting $t=0$. Then $F=f$. So f is the initial state of F, and is therefore the same as the former f. Comparing (80) with (77), we see that the operator $\epsilon^{-vt\Delta}$ is the translation operator, so that

$$\epsilon^{-vt\Delta} f(x) = f(x-vt). \tag{81}$$

This is the universal property of the operator $\epsilon^{h\Delta}$ when h is a constant. What it does, when applied to a function of x, is to translate it bodily through the distance h to the left.

Applying (81) to the result (76) which gave rise to these remarks about uniform motion of a function, we found that a certain F. series, when considered to exist only on the positive side of the origin, was identical with

$$(1-x/l) + 2(\epsilon^{-2l\Delta} + \epsilon^{-4l\Delta} + \epsilon^{-6l\Delta} + \ldots)1, \tag{82}$$

where the operand 1 is only positively existent. Denoting this operand by $f(x)$, the auxiliary terms in (82) are equivalent to

$$2f(x-2l) + 2f(x-4l) + 2f(x-6l) + \ldots. \tag{83}$$

The value of every one of these functions is 2 when it exists. But the first one only begins to exist when x reaches $2l$, the next when x reaches $4l$, and so on. The result is that the constant 2 is added on every time the variable x passes the points $x=2l$, $4l$, $6l$, &c. The function $(1-x/l)$ therefore, which falls down to -1 when $x=2l$, and which would by itself go on decreasing to $-\infty$, is lifted up by the first auxiliary to its starting value $+1$. The same series of values is then repeated between $x=2l$ and $4l$. Then the second auxiliary begins and lifts up the function to its initial value for another repetition. The F. series therefore represents $(1-x/l)$ between 0 and $2l$, followed by unending repetitions of the same series of values of the function.

If we allow the F. series to exist on the negative side also, it is sufficient to note that it is an odd function of x, to see its nature there. The negative side may be done operationally by itself by ignoring the part already done, and reckoning x positively the other way; but this is quite unnecessary, since the work has been already done equivalently on the positive side.

Equation (81), or, which is the same,

$$\epsilon^{h\Delta} f(x) = f(x+h), \tag{84}$$

is the well-known so-called "symbolical" form of Taylor's theorem. For it is merely the condensation of the form obtained by expanding the exponential, viz.:

$$\left\{1 + h\Delta + \frac{(h\Delta)^2}{\underline{|2}} + \frac{(h\Delta)^3}{\underline{|3}} + \ldots\right\} f(x) = f(x+h). \tag{85}$$

The form (84) is much more convenient. It is the form that always occurs naturally in operational mathematics. That is, we get solutions of the form $\epsilon^{h\Delta}$ acting upon something or other, and the meaning is simply a bodily translation performed upon it.

The way followed above is not the usual way of leading to Taylor's theorem, for which see works on the Calculus. It is a quasi-physical way, by which its truth becomes obvious in a general sense. It only involves the idea of translation of a form so far as the essential part goes. But now comes the question of failures. In fact, the proofs of Taylor's theorem are largely devoted to the discussion of modified forms with remainders occurring when there is some discontinuity, involving an infinite differential coefficient. When mathematicians come to an infinity they are nonplussed, and hedge round it. They would, for example, stick at the three sharp corners in the function above used, which involve discontinuity in the slope, or infinite curvature.

But there is no difficulty of the kind in the physical way of looking at the matter, so far as the act of bodily translation is concerned. One shape of the function is just as easily conceived as another, and we are not limited to the angelically perfect function which is finite and continuous itself, and has all its derivatives finite and continuous. It was long ago remarked by Sir G. Stokes that it is important for physical mathematicians to conceive of functions wholly apart from their possible symbolical expression. The remark is particularly important in the discussion of waves and their propagation, when discontinuous functions have frequently to be dealt with.

Now (77) is unexceptionable, and requires no special interpretation involving infinities. If $f(x)$ makes a jump suddenly at any place, so does $\mathrm{F}(x)$ precisely at a corresponding place. (The translation of the form should be thought of, not merely that of a value. If the curve is upright, the upright piece behaves just like the rest.) But (78), which is taken to be equivalent to (77), inasmuch as it utilises differential coefficients, may require interpretation. We should not say that the whole thing breaks down if the slope is infinite, because (78) becomes meaningless. It requires interpretation, but is not uninterpretable. The variations in $d\mathrm{F}/dx$ and those in $d\mathrm{F}/dt$ keep pace together precisely, and this tie does not break when they are momentarily infinite.

Although, of course, in a popular sense, infinities are immeasurable, they are not necessarily unmeasurable. A suitable standard is required to measure an infinity. Consider, for example, the case of electrification. Starting with a finite volume density, if we imagine it condensed on a surface, we have infinite volume density. But that causes no trouble. It is now to be measured by its surface density. Again, imagine it condensed from finite surface electrification to linear electrification. This means infinite surface density, and therefore doubly infinite volume density. But we now have a finite linear density, so there is no difficulty in measurement, or at any rate in the conception of the suitable way of measuring. Again, the notion of impulses in dynamics is exceedingly useful, though it involves the idea of an infinite force. We must not be afraid of infinity. If suitably measured, it may be no bigger than zero, or else quite small.

Now I think that similar ideas may with advantage be introduced into pure mathematics, to enlarge the scope of the mathematicians' investigations, and enable them to tackle infinities, instead of evading them. However, we are not concerned with the real or supposed failures in Taylor's theorem, perhaps rather to be considered failures in the mathematical machinery. In the operational mathematics $\epsilon^{h\Delta}$ is the translational operator, no matter how many discontinuities may be in the way. We do not have to consider the infinite values, but just jump over.

In physical applications, say to waves, the subject of operation is real and single valued, though it need not be finite and continuous. But I do not see any reason why the translational idea should not be carried further. Let it be applied to any definite series of values, or of multiple values, or perhaps to the case of a multiple pole, involving a collection of infinities.

A Fourier Series involving a Parabola interpreted by Taylor's Theorem.

§ 277. Since the investigation in § 275 shows one the working of the inner mechanism of a Fourier series in an interesting way, it will be worth while to consider another example of the same sort. We may put it thus :—

$$\coth l\Delta \, . \, f(x) = (1 + 2\epsilon^{-2l\Delta} + 2\epsilon^{-4l\Delta} + \ldots) f(x) \tag{86}$$

$$= \left\{ \frac{1}{l\Delta} + \Sigma_1^\infty \frac{2/l\Delta}{1 + (n\pi/l\Delta)^2} \right\} f(x) ; \tag{87}$$

and if we like to confine ourselves to the region 0 to $2l$, we may omit the translational auxiliaries, and write

$$f(x) = \left\{ \frac{1}{l\Delta} + \Sigma_1^\infty \frac{2/l\Delta}{1 + (n\pi/l\Delta)^2} \right\} f(x). \tag{88}$$

Take $f(x) = 1$ to make up the expansion of $1 - x/l$ considered in § 275.

Next take $f(x) = x$. Then (88) becomes

$$x = \frac{x^2}{2l} + \Sigma_1^\infty \frac{2l/(l\Delta)^2}{1 + (n\pi/l\Delta)^2}, \tag{89}$$

since $x = \Delta^{-1}1$, $\frac{1}{2}x^2 = \Delta^{-2}1$. Continuing the algebrisation we obtain

$$x = \frac{x^2}{2l} + \Sigma_1^\infty \frac{2l}{n^2\pi^2} \left(1 - \cos \frac{n\pi x}{l}\right) ; \tag{90}$$

or, which is the same,

$$x=\frac{x^2}{2l}+\frac{l}{3}-2\Sigma_1^\infty\frac{2l}{n^2\pi^2}\cos\frac{n\pi x}{l}. \tag{91}$$

This is true between $x=0$ and $2l$.

But the full meaning, according to (86), (87), including the auxiliaries, is

$$\Sigma_1^\infty\frac{2}{n^2\pi^2}\cos\frac{n\pi x}{l}=\left(\frac{x^2}{2l^2}-\frac{x}{l}+\frac{1}{3}\right)-\frac{2}{l}[x-2l]-\frac{2}{l}[x-4l]-\ldots, \tag{92}$$

where the terms with square brackets are the auxiliaries. They are to be understood in this way. In (86), $f(x)$ is to be x, so the auxiliary functions (not doubled) are $f(x-2l)$, $f(x-4l)$, &c., and these are the same as $x-2l$, $x-4l$, &c., when they exist. Now $f(x)$ is born when $x=0$, and is positively existent. So $f(x-2l)$ is born when $x=2l$, and so on. The final result is simple enough, namely, that in (92), with a particular value of x, just as many of the auxiliaries exist as have positive values. The rest are not yet born; they are zero.

In the region $x=0$ to $2l$ we have the curve

$$y=\frac{x^2}{2l^2}-\frac{x}{l}+\frac{1}{3}. \tag{93}$$

The ordinate y is $\frac{1}{3}$ at $x=0$ and $2l$, and has its minimum, namely, $-\frac{1}{6}$, midway between. After $x=2l$, it would rise up to infinity by itself, but the first auxiliary is now existent. Adding it on, we get

$$y=\frac{x^2}{2l^2}-\frac{3x}{l}+\frac{13}{3}. \tag{94}$$

This is the same as

$$y=\frac{(x-2l)^2}{2l^2}-\frac{(x-2l)}{l}+\frac{1}{3}. \tag{95}$$

Comparing the last with (93), we see that they have the same form, only x in (93) becomes $x-2l$ in (95). So the values of y between $x=0$ and $2l$ repeat themselves between $x=2l$ and $x=4l$. Then the second auxiliary comes into play, and causes another repetition of the same series of values of y. And so on to infinity.

The above refers to the positive side of the x axis. On the negative side we see the meaning of the Fourier series by observing that it is an even function of x.

The reader who is curious in this matter will find it useful to treat various simple Fourier series in the manner of § 275, to discover their meaning. Thus, given

$$f(x) = A_0 + \Sigma_1^\infty A_n \cos\frac{n\pi x}{l} + \Sigma_1^\infty B_n \sin\frac{n\pi x}{l}. \tag{96}$$

This may be converted at sight to

$$f(x) = A_0 + \Sigma_1^\infty \frac{A_n}{1+(n\pi/l\Delta)^2} + \Sigma_1^\infty \frac{B_n(n\pi/l\Delta)}{1+(n\pi/l\Delta)^2} \tag{97}$$

by the use of (73) and its cosine companion formula. Now sum up the partial fractions, if you can. The coefficients are supposed to be known functions of n. The summations may be effected by the formulæ for the expansion of coth $l\Delta$, $(\text{shin}\, l\Delta)^{-1}$, &c., in partial fractions, or by formulæ derivable from them. When this is done, comes another important step. The trigonometrical functions of $l\Delta$ may be expressed in series exponentially (*e.g.*, like $\epsilon^{-2l\Delta}$, &c.), and then Taylor's theorem, in its general translational sense, comes into play, and the full meaning of the F. series becomes evident. We determine it to represent a certain function between certain limits, and endless repetitions of the same values caused by the existence of the auxiliary terms involving the translational operators. The example in § 275 is fully descriptive in a general way, combined with the explanation in § 276, and the application may be made to a variety of formulæ. But the method is not meant for general use, because the summation of the partial fractions may be a troublesome matter.

The question arises whether a Fourier series with its constants given numerically can be done in the same way. Probably it can, by some extension of the process. But it may be necessary to discover first the law followed by the coefficients, so that they may be regarded as functions of n.

Representation of a Row of Impulses by Taylor's Theorem, leading to Fourier's.

§ 278. By equation (86), we see that the operator coth $l\Delta$, applied to a positively existent function of x, turns it into the sum of $f(x)$ and of twice the sum of its values at the points $x-2l$, $x-4l$, &c.

From this, we conclude that if $f(x)$ exists only between 0 and $2l$, meaning thereby that it is zero outside those limits, the series $\coth l\Delta \,.\, f(x)$, as in (86), represents $f(x)$ first, in its proper place, and then an infinite series of repetitions of the values of the same function doubled. For example, if $f(x)$ is x^2, we get first x^2, next $2(x-2l)^2$, then $2(x-4l)^2$, and so on, in regions of length $2l$, every one by itself.

Therefore

$$(\tfrac{1}{2}+\tfrac{1}{2}\coth l\Delta)f(x) \tag{98}$$

represents $f(x)$ and periodic repetitions of the same (not doubled), from $x=0$ to ∞, still understanding that $f(x)$ is zero except between 0 and $2l$.

And by a slight extension we see that if

$$\mathrm{F}(x)=(1+\epsilon^{-2l\Delta}+\epsilon^{-4l\Delta}+\ldots\ldots+\epsilon^{2l\Delta}+\epsilon^{4l\Delta}+\ldots)f(x), \tag{99}$$

then $\mathrm{F}(x)$ represents $f(x)$ and periodic repetitions to right and left through the whole range of x from negative to positive infinity. Only one term exists (or is finite) at a time. But if $f(x)$ be so chosen that its initial and final values do not agree, there is a jump in the manner before alluded to. The periodic property $\mathrm{F}(x+2l)=\mathrm{F}(x)$ is momentarily violated —as regards a single value of the function—unless we represent the vertical part of the curve by the mean value.

Now, the reader may amuse himself by considering the exponential expansions of other trigonometrical functions of a differentiator, as than $l\Delta$, and the reciprocals of shin $l\Delta$ and cosh $l\Delta$, and their powers as operators on a function. It will be sufficient here to derive Fourier's theorem in its general form from the above.

Go back to (99). By construction the series $\mathrm{F}(x)$ has the same meaning as Fourier's

$$\mathrm{F}(x)=\int_0^{2l} u\mathrm{F}(y)dy, \tag{100}$$

where
$$u=\frac{1}{2l}\left(1+2\Sigma_1^\infty \cos\frac{n\pi(y-x)}{l}\right). \tag{101}$$

Here u is expressive of a unit impulse at the points $x \pm 2nl$, as before explained, and (100) shows how a single value of the function is isolated.

Now $\Delta 1$ is a unit impulse at the origin of the variable. Let the variable be $y-x$. Then $\Delta 1$ signifies a unit impulse at the point $y=x$, and nothing anywhere else. So, by (99), we have

$$u=(1+\epsilon^{-2l\Delta}+\epsilon^{-4l\Delta}+\ldots+\epsilon^{2l\Delta}+\epsilon^{4l\Delta}+\ldots)\Delta 1. \qquad (102)$$

Since, by construction, this u represents the periodic unit impulse, like (101), we see that (102) must be the operational form of (101) itself.

But to algebrise (102) it is convenient to consider the positive and negative regions separately. Split u into halves, thus :—

$$u_1=\tfrac{1}{2}(1+2\epsilon^{-2l\Delta}+2\epsilon^{-4l\Delta}+\ldots)\Delta 1, \qquad (103)$$

$$u_2=\tfrac{1}{2}(1+2\epsilon^{2l\Delta}+2\epsilon^{4l\Delta}+\ldots)\Delta 1, \qquad (104)$$

Here u_1 is for the positive side, and u_2 for the negative, u_1 meaning half a unit impulse at the origin $y=x$, and whole ones at $y=x+2l$, $x+4l$, &c.; whilst u_2 means half a unit impulse at $y=x$, and whole ones at $x-2l$, $x-4l$, &c.

Now (103) is equivalent to

$$u_1=\frac{1}{2l}\coth l\Delta \,.\, l\Delta. \qquad (105)$$

This is equivalent in partial fractions to

$$u_1=\frac{1}{2l}\left\{1+\frac{2}{1+(\pi/l\Delta)^2}+\frac{2}{1+(2\pi/l\Delta)^2}+\ldots\right\}. \qquad (106)$$

Finally, the algebrisation is immediate, making

$$u_1=\frac{1}{2l}\left\{1+2\cos\frac{\pi}{l}(y-x)+2\cos\frac{2\pi}{l}(y-x)+\ldots\right\}. \qquad (107)$$

This result shows the same expression identically as Fourier's u in (101). But it has not the same meaning. Remember the reservation that $y-x$ ranges from 0 to ∞ only in connection with u_1. In (105), (106) the positively existent unit operand is understood. So (107) is limited to the positive side.

Similarly, from the symmetry of the complete u in (102) with respect to the origin $y=x$, or by reckoning $y-x$ positively the other way, so that u_2 is represented formally by u_1, we find that u_2 is given by the same expression (107) as u_1. Only in this case y has to be not greater than x.

So, removing the restriction, we see that the sum of u_1 and u_2 has the same expression as either. No change is needed in the trigonometrical formula because it is an even function of $y-x$. In its use to represent u_1 or u_2 alone, we merely restrict the range of its application. When unrestricted, we find that the u of (101) means the same as that in (102), the impulse periodic both ways.

It seems somewhat unnatural to consider the right and left halves separately. And yet it is necessary. This becomes evident at a discontinuity. Suppose $f(y)$ is zero on the left side of $y=x$, and finite instantly on the right side. Evidently u_2 is inoperative, and u_1 alone works. It gives only $\frac{1}{2}f(x)$, that is, $\frac{1}{2}$ the value on the right side.

Again, if at $y=x$, the function is A on the left and B on the right side, u_1 will give $\frac{1}{2}$B, and u_2 will give $\frac{1}{2}$A. The complete u will therefore spot the mean value.

Poisson's demonstration is interesting in this connection. Instead of u he used the function

$$v=\frac{1}{2l}\left\{1+2\Sigma_1^\infty h^n\cos\frac{n\pi(y-x)}{l}\right\}. \qquad (108)$$

It becomes the same as u when $h=1$. Now when h is <1, this v function is continuous. Its total between 0 and $2l$ is 1, and it has a hump at the point $y=x$. The closer h is made to approach to unity, the more the v function is heaped up at and close round the point $y=x$. In the limit it is all there, like u. Now the curve of v is symmetrical with respect to $y=x$; half its area is on one side, and half on the other. So, if h is very close to 1, and we use v in the formula (91) instead of u_1, we shall obtain $\frac{1}{2}$(A+B) approximately, and the more closely we make h approach unity the closer the approximation. I know no reason why a failure should occur just as perfection is reached.

Laurent's Theorem and Fourier's.

§ 279. I have said rather more about Fourier's theorem than I meant at first. That is a good enough reason for adding one more § about it, in order to have done with it.

The way Fourier's theorem presents itself in the theory of functions—that is, in *the* theory so-called thereof, which is manifestly confined to limited notions of a function—is worthy of notice. The complex variable has a double variability, and may be represented by the position of a point in a plane. But a circuital journey round the origin in the plane brings us back to the starting point. In this process the variable θ increases by the amount 2π. So we have necessary periodicity of functions of the position of the point, in the manner of Fourier's theorem, which is therefore essentially involved in the expansion of complex functions of position in integral powers of the variable denoting the position of a point in the plane.

Thus, there is Laurent's theorem for the expansion of a holomorphic function. Let $f(z)$ be the function, where z is the complex $r\epsilon^{i\theta}$, r and θ being the polar co-ordinates. Assume that f is capable of representation in integral powers of z, positive and negative, say

$$f(z) = A_0 + A_1 z + A_2 z^2 + \ldots + A_{-1} z^{-1} + A_{-2} z^{-2} + \ldots$$
$$= \sum_{-\infty}^{\infty} A_n z^n. \qquad (109)$$

Then we may easily determine the coefficients. The mean value of z^m, regarded as a function of θ, round a circle centred at the origin, is zero when m is any integer, except zero. In that exceptional case the result is 1. So, multiplying equation (109) by z^{-n}, and integrating with respect to θ, round a circle centred at the origin, we get

$$\frac{1}{2\pi}\int_0^{2\pi} f(z) z^{-n} d\theta = A_n. \qquad (110)$$

That is, the value of A_n is the mean value of $f(z)z^{-n}$ with respect to θ. So, by (109) and (110), we have

$$f(z) = \frac{1}{2\pi}\int_0^{2\pi} f(z') d\theta' \sum_{-\infty}^{\infty}\left(\frac{z}{z'}\right)^n. \qquad (111)$$

This is Fourier's theorem generalised by the introduction of the powers of r. To obtain it explicitly, make $r = r'$, so as to be concerned only with a function of θ. Then we get Fourier's theorem:—

$$f(\theta)=\frac{1}{2\pi}\int_0^{2\pi}f(\theta')d\theta'\sum_{-\infty}^{\infty}\epsilon^{in'\theta-\theta'}$$

$$=\frac{1}{2\pi}\int_0^{2\pi}f(\theta')d\theta'\sum_{-\infty}^{\infty}\cos n(\theta-\theta'). \qquad (112)$$

The substitution of the cosine for the exponential function occurs because the imaginary part of the latter contributes nothing to the result. The nth term cancels the $-n^{\text{th}}$ in this respect. But the imaginary part must not be omitted in the more general result (111).

For a rigorous demonstration of Laurent's theorem see works on the theory of functions. The line of integration may be any circle within the holomorphic region, so Laurent's theorem is the natural sequence to Fourier's when the distance from the origin is not invariable, and when the function is controlled by the restricted functional notions. The variations with r and with θ are not independent. The values of the function on one circle control those on another.

It would be of some interest to know how the function theorists would treat the subject of Fourier series in general, referred to the complex. In the above, the expansion is the rudimentary one in integral powers of the variable. In general, fractional powers are required, the n's following some special law. Say, for example, that $f(z)$ is given on the circle of z'. Then we require to have

$$f(z)=\sum A_n z^n, \qquad (113)$$

so as to be right for the given circle, and, in addition, to satisfy certain boundary conditions. For example, make a radial slit in the plane at $\theta=0$, and let the function have to satisfy a given linear differential equation between it and its derivatives with respect to θ on one side of the slit, where $\theta=0$, and another similar equation on the other side, where $\theta=2\pi$. This case corresponds to that of a cable subject to terminal but not to extra intermediate conditions. The infinite series of n's, which may be real or imaginary, will follow a law depending upon the conditions at the slit; and the determination of the coefficients A may be effected by a definite conjugate property, not that of the vanishing of the mean value of z^n round a circle as before mentioned, but of a more general character determined by the conditions at the slit. I have looked into Forsyth's work to see if there is anything about these expansions in fractional powers of the variable, but without success. If the matter has not been investigated, there is room for much expansion of the subject of the determination of functions, especially when one thinks of the various extensions that may be given by the introduction of intermediate differential conditions. Matters of this sort do not present much difficulty in strictly physical problems, and the corresponding cases in the function theory would, perhaps, involve little more than formal changes. At the same time, I must say that a matter which may be sufficiently clear in physical mathematics may become much obscured by a deliberate removal of the physical ideas, which is, perhaps, the reason of the existence of such elaborate and disagreeable demonstrations for the sake of rigour, of supposed or real necessity. It is a serious question whether the study of the theory of functions ought to be taken up by any ordinary physical student. Of course, some useful knowledge of the complex is necessary, and no one could fail to pick up some as he goes along by ordinary algebra; but that is quite different from the theory of functions as elaborated, the general tone of which is quite unlike that of the usual physical mathematics. The frame of mind required is not one that is conducive to progress in mathematical analysis in its physical aspect. A man would never get anything done if he had to worry over all the niceties of logical mathematics under severe restrictions; say, for instance, that you are bound to go through a gate, but must on no account jump over it or get through the hedge, although that action would at once bring you to the goal.

On Operational Solutions and their Interpretation.

§ 280. After this little excursion to the borders of the realms of duplicity and fearful rigour, we may return to the proper subject. We have to see how the general operational solutions must be treated in order to convert them to the appropriate Fourier series, by which (in one way) they become amenable to numerical calculation. It may be remembered that I have insisted upon the definiteness and fulness of meaning of an operational solution, and that it contains within itself not only the full statement of the problem, but also its solution. No external aid is therefore required to algebrise it fully; no assumption, for instance, of a special type of solution, and that the solution is the sum of a number of that type, with subsequent determination of the constants required to complete the matter. The work of satisfying the imposed conditions has been done already.

The conversion to algebraical or quantitative form may be easy or hard, self-evident or very obscure. But in any case it is possible, by the prior construction of the operational solution. Thus, the conversion furnishes a distinct subject of study which is of great practical value from the physical standpoint. As regards finding out how to effect the conversion, that is a matter principally of observation and experiment, and is in a great measure independent of logical demonstrations. It is the How, rather than the Why, with which we are mainly concerned in the first place; though, of course, parts of the Why cannot fail to be perceived in the course of examination of the How. A complete logical understanding of the subject implies the existence of a full theory to account for why certain ways of working are successful, and others not. It is important to note that it is just the same in mathematical research into unknown regions as in experimental physical research. Observation of facts and experiment come first, with merely tentative suggestions of theory. As the subject opens out, so does the theory improve. But it can only become logical when the subject is very well known indeed, and even then it is bound to be only imperfectly logical, for the reasons mentioned at the beginning of this volume. I feel inclined to be rather emphatic on the matter of the use of experiment in mathematics, even without proper understanding. For there is an idea widely prevalent (though it may not receive open expression) that in mathematics, unless you follow regular paths, you do not prove anything; and that you are bound to fully understand and rigorously prove everything as you go along. This is a most pernicious doctrine, when applied to imperfectly explored regions. Does anybody fully understand anything?

Three of the pernicious results of overmuch rigour may be mentioned here. First, its enfeebling action on the mind, sufficiently indicated by the analogy at the end of § 279. Secondly it leads to the omission from mathematical works of the most interesting parts of the subject, because the authors cannot furnish rigorous proofs. Thirdly, it leads to an inability to recognise the good that may be in other men's work, should it be unconventional, and be devoid of rigorous pretence.

It may not be necessary to algebrise an operational solution. It may not even be desirable to do so, if the meaning is sufficiently plain in operational form, and the algebrisation produces complication. This is very often true when the operations are merely direct ones, as, for example, $(a+bp+cp^2+\ldots)e$, where e is the operand and p the time differentiator. But it is also true, up a certain point, in more advanced forms. Consider, for instance $(\mathrm{XY})^{\frac{1}{2}}e$, or $(\mathrm{X}/\mathrm{Y})^{\frac{1}{2}}e$, or $\mathrm{R}^m e$, where X and Y, or R, are operators of the form $a+bp$. If the operand e is a simply periodic function of the time, we do not require to algebrise these forms in order to see their meaning. We can do that by geometrical construction in the now well-known manner founded upon the property $p^2=-n^2$, or $p=ni$, representing X and Y by rotations of vectors in a plane. But if we desire close calculation of the results, it is desirable to have the full algebraic formulæ. Again, I see that Mr. Kennelly has lately dispensed with the use of the full development in the more complicated cases of the shin and cosh of a complex (the complex corresponding to $a+bp$), by means of a specially constructed diagram. I do not doubt that certain cases in which e is impulsive or steady can be done by diagram. It is best, however, to do the work analytically first, in general, at least as far as the operational solution itself, and then see if, in a reasonable time, a geometrical method of representing it is suitable for calculation. For geometry by itself is rather a heavy and clumsy machine. Remember its history, and how it went forward with great bounds when algebra came to its assistance. Later on, the assistant became the master.

Sketch of Way of extending Fourier's Method to Fourier Series in General.

§ 281. The special ways which I have so far given of turning operational solutions of the diffusive kind to Fourier series have all been applied to elementary examples, the terminal conditions being such that the roots of the determinantal equation have been equally spaced, as π, 2π, &c., or $\frac{1}{2}\pi$, $1\frac{1}{2}\pi$, &c. Algebraic simplicity is thereby attained, and we are allowed to use well-known trigonometrical expansions. But, in general, the roots of the determinantal equation $\epsilon^{2ql}=\alpha\beta$, where α and β are functions of p depending on the terminal conditions, are not equally spaced, and they may not be all of the same kind—that is, they may be (as regards s in the circular functions) real or imaginary. The elementary expansion formulæ are useless. There are now three ways that present themselves of attacking the question.

First, to follow up Fourier's way by natural extensions of the kind to be found by experience. Assume the existence of a normal solution of the form

$$v=\mathrm{A}\sin(ax+b)\,\epsilon^{\mathrm{P}t}.$$

Apply it to the actually stated problem. Find the connection between P and a so as to satisfy the characteristic equation of diffusion. Next find, by applying the terminal conditions to the above assumed solution, what relation exists between a and b. Then find the law followed by the a's, and find their values. We thus get an elementary type solution fully determined except as regards size, and we know that every admissible value of a gives a solution.

Next, assume that

$$\mathrm{V}=\Sigma\, v_n=\Sigma\, \mathrm{A}_n\sin(a_nx+b_n)\,\epsilon^{\mathrm{P}_nt},$$

where the summation includes all the values of a_n, is the form of the solution of the special problem, namely, the sum of any number of the type solutions. Only the A's are arbitrary. To find them, observe that $t=0$ makes

$$\mathrm{V}_0=\Sigma\, \mathrm{A}_n\sin(a_nx+b_n).$$

The summation on the right therefore represents the initial state. Given the initial state V_0 as a function of x, we require to determine the A's in accordance therewith.

This may be done by finding the conjugate property possessed by the normal functions. When known, it allows us to operate on V_0 by quadratures in such a way as to isolate any single coefficient and determine its value. Finally, introducing the time factor, we arrive at the complete solution as regards the effects due to a given initial state, provided the effects are entirely due thereto. And the effect due to an impressed force may be done similarly by considering the final state it produces, and allowing that to discharge itself in the above manner.

The process of finding A_n is, when the terminal conditions are of the simple kind previously considered, sufficiently obvious. The integral of the product of two normal systems vanishes save when they are alike. But with terminal conditions in general, with energy involved in the terminal apparatus, it is quite another thing. The conjugate property is not the same, and although fundamentally simple in meaning, works out more difficultly. The determination of the A's may become a complicated process, requiring careful investigation of the terminal arrangements themselves. It depends upon the vanishing of the excess of the electric over the magnetic mutual energy of a pair of different normal systems (with different values of p, that is to say). In doing this, the two terminal arrangements have themselves to be included, if any part of the energy is located therein.

The Expansion Theorem. Operational Way of getting Expansions in Normal Functions.

§ 282. As a second way, we may follow up the special ways of operationally working already given in connection with simple terminal conditions. First, find the operational solution. Next investigate the proper expansions in partial fractions of the reciprocal of the determinator ($\epsilon^{2ql} - \alpha\beta$) which occurs in the operator, wherein p has to have all the values which make it (the determinator) vanish. The integration of the partial fractions follows. Complete the solution by carrying out what may be left in the way of direct operations. The expansions required are analogous to those of $\coth ql$ and other functions already used. But this process may be rather difficult, since the theory of partial fractions may lead one into obscure regions of the complex.

Finally, there is a third and very general way of converting operational solutions to the form of the sum of normal solutions. It does not require special investigation of the properties of the normal functions. It is very direct and uniform of application. It avoids, in general, a large amount of unnecessary work. The investigation of the conjugate property, and of the terminal apparatus in detail in order to apply it to the determination of the coefficients, is wholly avoided. It applies to all kinds of series of normal functions, as well as to Fourier series. And it applies generally in electromagnetic problems, with a finite or infinite number of variables; or, more generally, to the system of dynamical equations used by Lord Rayleigh in the first volume of his treatise on Sound, which covers the rest of the work, and upon which he bases his discussion of general properties.

The method may be briefly (though imperfectly) stated as follows: Let $e = ZC$ be the operational solution of an electromagnetic problem; say, for definiteness, that C is the current at a certain place due to an impressed force, e, at the same or some other place. Let the form of Z be such as to indicate the existence of normal solutions for C. Then, when e is steady, beginning at the moment $t=0$, the C due to e is expressed by

$$C = \frac{e}{Z_0} + e\Sigma \frac{\epsilon^{pt}}{p\dfrac{dZ}{dp}}, \tag{1}$$

to be understood thus:—

In the first place, the Z in the operational solution is an operator, a function of p the time differentiator. But in equation (1), Z is entirely algebraical. Thus, Z_0 is the algebraical function obtained by putting $p=0$ in Z. It is the effective steady resistance to e when, as supposed, e is a voltage, and it is at the place of C. Otherwise it is more general. Then, in the summation, dZ/dp is the ordinary differential coefficient of Z with respect to p as a quantity. Lastly, the summation ranges over all the roots of the algebraical equation $Z=0$, which is, in this respect, the determinantal equation, though Z itself is much more. These special values of p are to be used in dZ/dp as well as explicitly.

Instead of a C due to an e, it may be V due to e, or V or C due to an initial charge at a point, and many other variations might be mentioned. But it must not be inferred that equation (1) expresses a general mathematical property independent of the form of Z. Dynamical conditions restrict the form in certain ways, for one thing. But it is not even true that the expansion theorem above holds good in strictly dynamical problems. I have already pointed out that we require barriers or boundaries to produce definite and finitely separable vibrating normal systems. Also that in the cable problem, subject to terminal conditions, the determinator $\epsilon^{2ql} - \alpha\beta$ which occurs as a denominator in the general operational solutions arises from the coexistence of successively (though apparently simultaneously) generated waves. So all sorts of problems are immediately excluded from the expansion theorem, at least in the form of equation (1). Inspection of the operational equations will show this, or else physical considerations. Again, the real Z may be the sum of two operators, to one of which the expansion theorem applies fully, whilst the other requires a different treatment. It may involve direct operations of an obvious nature.

Now it would be useless to attempt to state a formal enunciation to meet all circumstances. Even supposing that an absolutely perfect knowledge of the subject made it possible to do so, it would be very unpractical. It would be worse—far worse—than that very lengthy enunciation of a theorem in the 5th Book of Euclid, which may be read and re-read fifty times without properly grasping its meaning, which is not much, after all; only something in compound proportion that the modern schoolboy does in a minute or two. It is better to learn the nature and application of the expansion theorem by actual experience and practice. A theorem which has so wide an application is a subject for a treatise rather than a proposition.

So, to begin with, I will give a few elementary cases concerning one or two degrees of freedom, just to show how the expansion theorem goes, and how to discriminate different forms of Z. After that, we can apply it to diffusion problems in the same way. Other than diffusion problems coming under the same theorem may occur later. And when the reader knows how to work the theorem, it may be possible to see its inner meaning. At present all that need be said is that in its theory it contains the first method of evaluating the coefficients by the conjugate property, and also the second method involving partial fractions. Only it jumps over most of the difficulties connected with both.

Examples of the Use of the Expansion Theorem:—
(1). Inductance Coil and Condenser separately.

§ 283. Start with the characteristic equation of a coil, being the form taken by the second circuital law when applied to a linear circuit of constant resistance. Say

$$e = (\mathrm{R} + \mathrm{L}p)\mathrm{C}, \tag{2}$$

where R is the resistance and L the inductance of the circuit, C the current and e the impressed voltage.

First, let it be given that C is steady from the initial moment, and zero previously. What is the corresponding e? The answer is immediate. We want two impressed forces, namely, $e_1 = \mathrm{RC}$ to keep up the current, and $e_2 = \mathrm{L}p\mathrm{C}$ to start it. Since $p1$ is a unit impulse, e_2 is an impulsive force of total LC. First we require the impulsive force establishing the momentum LC instantly, and then the steady force RC to keep it up against the resistance.

Now this is an example, not of the working of the expansion theorem, but of its failure. But the treatment is sufficiently clear without it. If we tried to force it into the expansion theorem we should want the root of $(\mathrm{R} + \mathrm{L}p)^{-1} = 0$, which has no finite root.

But now find the C due to steady e, which is quite a different problem. By (2) the answer is

$$\mathrm{C} = \frac{e}{\mathrm{R} + \mathrm{L}p}, \tag{3}$$

which is of a proper form for the expansion theorem. Here Z_0 is R, p is $-\mathrm{R}/\mathrm{L}$; $d\mathrm{Z}/dp$ is L; so, by (1),

$$\mathrm{C} = \frac{e}{\mathrm{R}} + \frac{\epsilon^{-\mathrm{R}t/\mathrm{L}}}{(-\mathrm{R}/\mathrm{L})\mathrm{L}} = \frac{e}{\mathrm{R}}(1 - \epsilon^{-\mathrm{R}t/\mathrm{L}}), \tag{4}$$

showing the gradual rise of C to its steady value.

Since this case is fundamental, the operational way of getting the result may also be indicated. We have

$$\frac{e}{\mathrm{R} + \mathrm{L}p} = \frac{e}{\mathrm{L}p(1 + \mathrm{R}/\mathrm{L}p)} = \frac{1}{\mathrm{R}}\left\{\frac{\mathrm{R}}{\mathrm{L}p} - \left(\frac{\mathrm{R}}{\mathrm{L}p}\right)^2 + \left(\frac{\mathrm{R}}{\mathrm{L}p}\right)^3 - \dots\right\}e. \tag{5}$$

This is got by expanding the fraction by division. The rest is done by

$$p^{-n}1 = \frac{t^n}{\underline{|n}}, \tag{6}$$

which makes, applied to (5),

$$C = \frac{e}{R}\left\{\frac{Rt}{L} - \frac{1}{\underline{|2}}\left(\frac{Rt}{L}\right)^2 + \frac{1}{\underline{|3}}\left(\frac{Rt}{L}\right)^3 - \dots\right\}, \tag{7}$$

which is the expansion of (4), by the meaning of the exponential function. The steps (5) and (7) to obtain (4) may look cumbrous. But there is deception here, for it is, perhaps, a shorter and more direct way than any other. The deception comes in in the customary use of the short expression of the exponential function to denote a certain infinite series, and in taking its properties for granted. However, once done, of course there is no further need for the intermediate work. It will be observed that the case is that of equation (3), § 265, slightly modified. To obtain it as in the equation referred to, let e be impulsive, say $=p\mathrm{E}$, where E is the impulse total. Then equation (3) above makes

$$C = \frac{p\mathrm{E}}{R + Lp} = \frac{\mathrm{E}}{L(1 + R/Lp)} = \frac{\mathrm{E}}{L}\epsilon^{-Rt/L}, \tag{8}$$

as in (3), § 265. The impulse E produces the equivalent momentum E = LC. There is no impressed voltage to keep it up, so it decays according to (8).

In doing (8) by the expansion theorem, notice that $p=0$ makes $C=0$, so there is no steady term. The rest is as in equation (4), only multiplying by p or $-R/L$. Thus, by (1),

$$C = \frac{p\mathrm{E}}{R+Lp} = \mathrm{E}\frac{p\epsilon^{pt}}{pL} = \frac{\mathrm{E}}{L}\epsilon^{pt}, \tag{9}$$

where p has the special value which makes $R+Lp=0$.

Next consider another fundamental case, that of a leaky condenser; that is, a conducting leyden or condenser, or one made conducting (equivalently) by a shunt. We have

$$C = (K + Sp)e, \tag{10}$$

where K is the conductance and S the permittance. The form is the same as for a coil, but current and voltage are interchanged, as well as conductance and resistance, and inductance and permittance.

In order that the voltage e should be instantly raised from zero to constancy, and maintained constant, we require the steady current Ke, and an initial impulsive current of total Se; that is, first the charge Se must be impulsively established, and then it must be maintained constant by the steady current Ke, which goes through the shunt K, of course. Without the steady current the charge of the condenser would disappear.

Now invert the problem. Say

$$e = \frac{C}{K + Sp}, \tag{11}$$

and inquire what e is required to make C jump from zero to constancy. Here the expansion theorem applies. It is like equation (3) above. So

$$e = \frac{C}{K}(1 - \epsilon^{-Kt/S}) \tag{12}$$

answers the question, either by translation of symbols or independently.

In another form, if the voltage be made to rise from zero to constancy in the way indicated by (12), the current will be constant, viz., C, during the whole time. This current is *the* current, the total current in the condenser and shunt. The part in the shunt is Ke; the rest is in the condenser. For convenience we assume that the conductance is supplied externally.

The Treatment of Simply Periodic Cases.

§ 284. As regards the simply periodic solutions in these cases, the working is perfectly simple, by means of the property $p^2 = -n^2$ which obtains in simply periodic states, or, which is the same, $p = ni$, applied to reduce the resistance or conductance operator to the standard form $a + bp$. The case

$$e = (R + Lp)C, \tag{13}$$

when it is C that is given as a simply periodic function, say, $C = c \sin(nt + \theta)$, is obvious, for (13) is in the standard form already.

In the other case, when e is given instead of C, we have

$$C = \frac{e}{R + Lp} = \frac{(R - Lp)e}{R^2 + L^2n^2}. \tag{14}$$

That is, we bring C to the standard form by multiplying the numerator and denominator by the operator conjugate to $R + Lp$, that is $(R - Lp)$ This realises the denominator. Whether this is done by $p^2 = -n^2$, or by $p = ni$, is indifferent. Only, it should be noted that i is not the imaginary of algebra, but, as I have pointed out before, the differentiator $d/d(nt)$; that is, p/n. It is a specialised differentiator, which may be treated like the imaginary because $p^2 = -n^2$ in simply periodic cases. But it must be finally interpreted correctly, as a differentiator, of course. There is sometimes, perhaps often, some confusion or indistinctness of ideas on this point.

Another point which may not be always understood is the assumption involved that both operand and resultant are simply periodic. This is essential. But it by no means follows that a simply periodic e produces a simply periodic C. It does practically, because there is resistance. But if it be given that there is no resistance, so that $e = LpC$; then if e is given, we obtain, by $p = ni$,

$$C = \frac{e}{Lp} = -\frac{pe}{Ln^2}. \tag{15}$$

This is the solution *if* C is simply periodic, as well as e. But it would never reach this state without resistance. That is, given a little resistance, e would, in a long time, lead to the state (15) approximately; but if $e = LpC$ is rigidly true, and e starts when $t = 0$, the resulting C is, if $e = e_0 \sin nt$,

$$C = \frac{e_0 \sin nt}{Lp} = \frac{e_0}{Ln}(1 - \cos nt), \tag{16}$$

which is quite different. Here, of course, the reciprocal of p signifies integration from 0 to t.

We may easily show the complete establishment of the simply periodic state from (5), by taking $e = e_0 \sin nt$, but *not* assuming C to be simply periodic. Then the powers of p^{-1} are successive integrations from 0 to t; and the result is

$$C = \frac{e_0}{R}\left\{\frac{R}{Ln}(1 - \cos nt) - \left(\frac{R}{Ln}\right)^2(nt - \sin nt)\right. \tag{17}$$

$$\left. + \left(\frac{R}{Ln}\right)^3\left(\frac{(nt)^2}{\lfloor 2} - 1 + \cos nt\right) - \left(\frac{R}{Ln}\right)^4\left(\frac{(nt)^3}{\lfloor 3} - nt + \sin nt\right) + \dots\right\},$$

where the functions of t in the brackets are the successive integrals of sin nt, which are obtainable at sight, or verifiable by differentiation.

By collection of terms, (17) may be put in the form

$$C=e_0\frac{R\sin nt-Ln\cos nt}{R^2+L^2n^2}+e_0\frac{Ln\epsilon^{-Rt/L}}{R^2+L^2n^2}. \qquad (18)$$

Here the second part cancels the first initially, though it ultimately vanishes and leaves the first in full play. But when R is zero we never arrive at the simply periodic state, but have the state (16) instead, which is the beginning part of (17), independent of R. As regards the establishment of (18), however, it is quite sufficient to first write down the final periodic state, and then add on a term to cancel it initially and subside at the proper rate. The above way of successive integration shows the internal working.

That p specialises itself to ni when there is simple periodicity of e and C rests upon $p^2=-n^2$. That the transformation used in (14) is fully valid without escape may be seen thus:—Operate on (13), which is general, by $R-Lp$, making

$$(R-Lp)e=(R^2-L^2p^2)C, \qquad (19)$$

which is also general. Now specialise, by letting C be simply periodic, and therefore, by (13), e also, though the converse is not necessarily true. We now have $p^2C=-n^2C$, and (19) becomes

$$(R-Lp)e=(R^2+L^2n^2)C, \qquad (20)$$

which is (14) slightly changed in form.

In the case of a complicated operational solution, if we choose to clear it of fractions, we can generally bring it to the usual form of a differential equation, say

$$Ae=BC, \qquad (21)$$

where A and B are operators involving only direct operations—that is, integral positive powers of p. Then, by specialising e and C to be simply periodic, we make the power of p^n be constant when n is even, and be $p\times$ constant when odd. So we reduce (21) to the form

$$(A_1+A_2p)e=(B_1+B_2p)C, \qquad (22)$$

where the A's and B's are constants. A similar treatment applies in the case of irrational operators, as regards their reduction to the standard form.

Finally, if want e in terms of C, multiply by $A_1 - A_2 p$. If C in terms of e, then by $B_1 - B_2 p$ instead. This work is just as well, and it may be much better done in the operational solution itself, without preliminary conversion to the usual stretched-out form of a differential equation, by either $p^2 = -n^2$ or $p = ni$. Of course, the structure of the formulæ is plainer and simpler in the operational form. This is a very important matter, especially to the electrician, as he can use electrical ideas throughout rather than purely mathematical.

(2). Coil and Condenser in Sequence. Also in Parallel.

§ 285. In the above cases of a coil or of a condenser, there is but one degree of freedom. To make two, we may put the coil and condenser in sequence. The resistance operators are additive, so

$$e = \left(R + Lp + \frac{1}{K + Sp}\right)C \tag{23}$$

is the equation of voltage, e being impressed on the complete arrangement. We need say nothing here about the simply periodic case. By inspection of (23) we see that the operator is peculiar, as the expansion theorem applies to one part of it, though not to the other. If C is zero before, and steady after the initial moment, we see that the corresponding e consists of three parts, namely, an impulsive voltage of total LC to generate the current instantly; a steady voltage RC to keep it up against R; and the variable voltage

$$\frac{C}{K}(1 - \epsilon^{-Kt/S}). \tag{24}$$

The impulsive voltage, and the steady $C(R + K^{-1})$ are clear enough. The exponential term is required to keep the total current constant in the condenser and shunt. The initial impulse does nothing to the condenser. It is merely operative on the coil.

The case $K = 0$ is interesting. Then, by (24), or much more simply by (23), the voltage on the condenser is Ct/S, increasing uniformly with the time.

Only one root is concerned in the above. But if we inquire into the C produced by a steady e, then two roots come into play. Thus, by (23),

$$C = \frac{e}{R + Lp + (K + Sp)^{-1}}. \tag{26}$$

Taking the denominator to be Z, and applying the expansion theorem, equation (1), we see first that Z_0 is $R + K^{-1}$, and next that

$$p\frac{dZ}{dp} = Lp - \frac{Sp}{(K + Sp)^2}. \tag{27}$$

So, if p_1 and p_2 denote the two roots of $Z = 0$, the expansion of C is

$$C = \frac{e}{R + K^{-1}} + \frac{e\epsilon^{p_1 t}}{Lp_1 - Sp_1(R + Lp_1)^2} + \frac{e\epsilon^{p_2 t}}{Lp_2 - Sp_2(R + Lp_2)^2}, \tag{28}$$

where, instead of $(K + Sp)^{-2}$, we have written $(R + Lp)^2$, because $Z = 0$ makes them equal. Into a discussion of (28) in detail it is unnecessary to enter, the object here being merely the application of the expansion theorem. The theory of circuits in detail would need a chapter to itself.

Another way is to convert (26) to

$$C = \frac{(K + Sp)e}{(R + Lp)(K + Sp) + 1}. \tag{29}$$

This will give the same result by the expansion theorem, first as regards the steady state, and next as regards the terms variable with the time. In the numerator, p has to receive the two values in succession which make the denominator vanish. That is, if $C = (Y/Z)e$, and the subsidence constants of the normal systems depend upon the vanishing of Z, then

$$C = \frac{Y_0}{Z_0}e + e\Sigma\frac{Y\epsilon^{pt}}{p\frac{dZ}{dp}} \tag{30}$$

expresses the expansion theorem, being equivalent to equation (1). Putting it in another form, it is on the infinities of Y/Z that the terms involving the time depend.

In (14), with the specialised meaning of p, the algebrisation was effected by introducing a common factor into the numerator and denominator. Comparing (29) with (26), we see that a common factor is also introduced, though unnecessarily. It does not alter the roots of the denominator. Suppose, however, that it did. Suppose, for instance, we multiply numerator and denominator in (26) by $r+lp$, thus introducing a fresh root $p=-r/l$ to the modified Z. It will make no difference in the working of the expansion theorem. The new term is of size zero, and is, therefore, automatically excluded.

There is an important principle involved here, which may be referred to in passing. Let $C=e/Z$ be the operational solution of an electromagnetic problem, say, involving a complicated combination of coils and condensers. If $Z=0$ has m roots, there are m degrees of freedom, and m terms in the C expansion, besides possibly an outside steady term. Now, by introducing special relations amongst the resistances, inductances, &c., concerned in Z, we may be able to reduce it to a simpler form, involving a smaller number of roots, even down to one root, it may be. This means that the combination behaves, towards the impressed force, under the circumstances, just as though it were some simpler combination. So far as the expansion of $C=e/Z$ goes, there is no loss of generality by the simplification of Z mentioned. The reduced form used in the expansion theorem will give the proper result. The same applies to simply periodic solutions. Nevertheless, the combination does possess a greater generality than the simplified equivalent in other respects, as regards a differently situated impressed force for instance, or as regards free subsidence from an arbitrarily given initial state. It is very convenient to be able to eliminate simply the unessential and inoperative parts of the work, and the Z operators allow us to do this.

We also see, by reversing the reasoning, that when we alter $C=e/Z$ to $C=Ye/YZ$, and introduce thereby a number of new roots of the denominator, viz., those (if any) of $Y=0$, we are really enlarging the problem, for YZ belongs to a larger combination. But then it is done in such a way that the larger combination is equivalent, as regards the relation between e and C, to the smaller. The extra terms in the expansion of C are of zero size. Thus, with unit operand,

$$\frac{1}{Z}=\frac{Y}{YZ}=\frac{Y_0}{Y_0Z_0}+\Sigma\frac{Y\epsilon^{pt}}{p(ZY'+YZ')}, \tag{31}$$

where the accent means differentiation to p. Now, with the roots of $Y=0$ only, the vanishing numerators make the terms be zero; but this does not happen with those of $Z=0$. We therefore get the original form of equation (1).

Returning to the condenser and coil, put them in parallel, and subject them to the same impressed force. The conductance operators are now additive, and therefore

$$C=\{(K+Sp)+(R+Lp)^{-1}\}e \tag{32}$$

is the equation of current.

To have e steady after the initial moment, we require an impulsive current of total Se, a steady current Ke, and a gradually rising current

$$\frac{e}{R}(1-\epsilon^{-Rt/L}),$$

by what has preceded. Only one time function is concerned.

On the other hand, if the current is steady after the initial moment, we have to expand

$$e=\frac{C}{(K+Sp)+(R+Lp)^{-1}}. \tag{33}$$

Comparing with (26), and its expansion, we see that the result in the present case is

$$e=\frac{C}{K+R^{-1}}+\frac{C\,\epsilon^{p_1t}}{p_1S-p_1(K+Sp_1)^2}+\frac{C\,\epsilon^{p_2t}}{p_2S-p_2(K+Sp_2)^2}. \tag{34}$$

The roots p_1, p_2, are as before, in (28).

(3). Two Coils under Mutual Influence.

§ 286. As another example of two degrees of freedom, consider two linear conductive circuits. The second circuital law applied to them makes

$$E=(R+Lp)C+Mpc, \tag{35}$$

$$e=(r+lp)c+MpC, \tag{36}$$

using big letters for one circuit, small for the other, except in the case of M, the mutual inductance, which is common to both. This is "ironless mathematics," of course. Young investigators need not be discouraged by the contempt which is too often poured upon "ironless mathematics" in engineering journals. It is fundamental, and very useful. Sometimes the contemners only indicate their own ignorance.

If both C and c are to suddenly become steady, we require the voltages indicated explicitly, viz., the steady RC in one and rc in the other circuit, following the initial impulses $\mathrm{LC}+\mathrm{M}c$ and $lc+\mathrm{MC}$ respectively.

If c is to be kept zero, put $c=0$, making

$$\mathrm{E}=(\mathrm{R}+\mathrm{L}p)\mathrm{C}, \qquad e=\mathrm{M}p\mathrm{C}. \tag{37}$$

That is, we require the voltage $\mathrm{M}p\mathrm{C}$ in the second circuit to prevent current in it. At the same time, the equation and behaviour of the first circuit are the same as if the second were taken away. The principle of the destruction of mutual induction by a suitable impressed force is a useful one in theoretical reasoning.

When, on the other hand, the forces are given, we should solve for the currents. Thus, by cross multiplication, we may eliminate C or c in turn between (35) and (36). The equation of C is

$$\mathrm{C}=\frac{(r+lp)\mathrm{E}-\mathrm{M}pe}{(\mathrm{R}+\mathrm{L}p)(r+lp)-\mathrm{M}^2p^2}, \tag{38}$$

which, by (1), or (30) more conveniently, expands to

$$\mathrm{C}=\frac{\mathrm{E}}{\mathrm{R}}+\frac{\mathrm{E}(r+lp_1)-\mathrm{M}p_1e}{\mathrm{L}p_1(r+lp_1)+lp_1(\mathrm{R}+lp_1)-2\mathrm{M}^2p_1^2}\epsilon^{p_1t} \tag{39}$$

$$+\text{ ditto with } p_2.$$

Here p_1 and p_2 are the roots of the denominator in (38).

The above examples, going only as far as two degrees of freedom, are sufficient to show how the expansion theorem is to be applied. They are not particularly good examples, but are chosen rather to show how to readily discriminate different operators, and see when the expansion theorem applies, and when it does not.

(4). Cable Earthed at A and B. Impressed Force at A.

§ 287. Passing to the consideration of the application of the expansion theorem to diffusion problems, take first what is one of the simplest cases, as well as one of the most interesting. Let a cable be earthed at both ends, and a steady impressed force e be inserted at $x=0$. The potential produced is given by (17), § 256, namely

$$V=\frac{\text{shin}\, q(l-x)}{\text{shin}\, ql}e=\frac{\sin s(l-x)}{\sin sl}e, \tag{40}$$

where $q^2=-s^2=\text{RS}p$. The practical way of getting this operational solution was described on the same page. Being an easy case, we can see that it is right by noting that it makes $V=0$ at $x=l$, and $V=e$ at $x=0$.

Take the form (30) of the expansion theorem. Take Y to be the numerator, and Z the denominator. The roots of Z are those of shin ql. But not the zero root, because to rationalise shin ql, which is an odd function of ql (and therefore of $p^{\frac{1}{2}}$), we must exclude it. This may be formally done by making Z be $(\text{shin}\, ql)/ql$; but it is unnecessary, for by omitting the zero root from shin $ql=0$ we come to the same result. Noting this, we may use either of the forms given in (40). The first gives the result in terms of the values of p. But these are negative, therefore the second form is more convenient. So we have

$$p\frac{dZ}{dp}=\tfrac{1}{2}s\frac{dZ}{ds}=\tfrac{1}{2}sl\cos sl. \tag{41}$$

Also, putting $p=0$, the final steady state is $e(1-x/l)$. So the expansion theorem (30) gives

$$V=e\left(1-\frac{x}{l}\right)+e\Sigma\frac{\sin s(l-x)\epsilon^{pt}}{\frac{1}{2}sl\cos sl}, \tag{42}$$

where the summation ranges over the finite roots of $\sin sl=0$; that is, $sl=\pi, 2\pi, 3\pi$, &c.; and p in the exponential function is given by $\text{RS}p=-s^2$. Owing to the dependence of p on the square of s, we see that only the positive roots of $\sin sl=0$ need be counted. That is, it is really the values of p that control matters. Equation (42) may be at once reduced to

$$V=e\left(1-\frac{x}{l}\right)-2e\Sigma\frac{\sin sx}{sl}\epsilon^{pt}, \tag{43}$$

which was obtained before.

(5). Cable Earthed at A and Cut at B. Impressed Force at A.

§ **288.** Next take the same case as last with the sole change made of insulating instead of earthing at $x=l$. Then, by (22), § 258, the potential is given by

$$\mathrm{V}=\frac{\cosh q(l-x)}{\cosh ql}e=\frac{\cos s(l-x)}{\cos sl}e, \tag{44}$$

which, as before, we see to be correct by the formula making $\mathrm{V}=e$ at $x=0$, and $\mathrm{C}=0$ at $x=l$. Now in this case, the denominator is rational, as a function of p, that is to say, so no reservation is needed, and the expansion theorem (30) makes

$$\mathrm{V}=e+e\Sigma\frac{\cos s(l-x)}{\frac{s}{2}\frac{d}{ds}\cos sl}\epsilon^{pt}; \tag{45}$$

ranging over the positive roots of $\cos sl=0$; or, $sl=\frac{1}{2}\pi$, $1\frac{1}{2}\pi$, &c. Effecting the differentiation in the denominator, and simplifying the numerator by the vanishing of $\cos sl$, we reduce (45) to

$$\mathrm{V}=e-2e\Sigma\frac{\sin sx}{sl}\epsilon^{pt}, \tag{46}$$

which is the expansion required. The summational part is of the same form as in (43). But of course there is a different series of values of sl, and also of the p's. There is no confusion between p the differentiator, and its specialised algebraical value denoted by the same symbol. The latter only occurs in algebraical summations; the former in operational solutions.

(6). Earth at A, Cut at B. Impressed Force at y.

§ **289.** The above two cases being sufficiently illustrative when the impressed force is situated terminally, and the terminal conditions are of the simplest type, making either the potential or the current vanish, take next the case of an intermediate impressed force, though still with the simple terminal conditions. Say that there is earth on at $x=0$, insulation at $x=l$, and an impressed force e at $x=y$. Then, by (41), (42), § 260, the resulting potentials, putting $q^2 = -s^2$ for convenience in the expansions to be found, are given by

$$V_1 = \frac{\cos sy}{\cos sl} \cos s(l-x) . e, \tag{47}$$

$$V_2 = \frac{\sin sx}{\cos sl} \sin s(l-y) . e, \tag{48}$$

V_1 being the potential when x is on the right side of y, and V_2 when on the left side. Here, of course, s^2 is the differentiator $-\text{RS}p$.

First, putting $s=0$, we see that the final steady state is $V_1 = e$, $V_2 = 0$. So applying (30), we obtain

$$V_1 = e - e\Sigma \frac{\cos sy}{\frac{1}{2}sl \sin sl} \cos s(l-x) . \epsilon^{pt}, \tag{49}$$

$$V_2 = 0 - e\Sigma \frac{\sin sx}{\frac{1}{2}sl \sin sl} \sin s(l-y) . \epsilon^{pt}, \tag{50}$$

subject to $\cos sl = 0$, as in the last case, or $sl = (n - \frac{1}{2})\pi$, where n is a positive integer. We may now simplify by means of $\cos sl = 0$, leaving

$$V_1 = e - 2e\Sigma \frac{\cos sy}{sl} \sin sx . \epsilon^{pt}, \tag{51}$$

$$V_2 = \quad -2e\Sigma \frac{\cos sy}{sl} \sin sx . \epsilon^{pt}, \tag{52}$$

the simplest form of the expansions.

The transition from the operational solutions (47), (48) to the expansions (49), (50) involves two formal processes, viz., the change in the denominator, done by the differentiation which occurs when the summational sign and the time factor are introduced, and the addition of the outside term. In going further, we reach (51), (52), by the omission of terms which vanish in the preceding equations. The same occurs in the previous examples. Now this omission may be done in the act of applying the expansion theorem, thereby deriving (51), (52) directly from (47), (48) without the more complicated intermediate equations. And, in fact, so far as arriving at (51), (52) is concerned, through the expansion theorem, we may make the omission in the operational solution. But the steady part of the solution must be extracted first; and in any case, the operational solution is made nonsense of (considered as a general operational solution) by the too early omission. So the omission is properly done in the act of conversion to the expanded series.

(7). Earth at A and B. Impressed Force at y.

§ 290. If both terminals are earthed, whilst the impressed force is intermediate, at the point y, we have, by (37), (38), § 260, the results

$$V_1 = +\frac{\cos sy}{\sin sl}\sin s(l-x)\,.\,e, \tag{53}$$

$$V_2 = -\frac{\sin sx}{\sin sl}\cos s(l-y)\,.\,e, \tag{54}$$

on the right and left sides of the impressed force respectively. So the expansion theorem makes

$$V_1 = e\left(1-\frac{x}{l}\right) - 2e\Sigma\frac{\cos sy \sin sx}{sl}\epsilon^{pt}, \tag{55}$$

$$V_2 = -\frac{ex}{l} - 2e\Sigma\frac{\cos sy \sin sx}{sl}\epsilon^{pt}, \tag{56}$$

where we first write down the outside terms got by putting $s=0$ in the previous equations, and then write the summational general term, changing the denominator by the operator $\frac{1}{2}s(d/ds)$, which is equivalent to $p(d/dp)$, and omitting terms which, in virtue of $\sin sl = 0$, vanish from the result with changed denominator.

The determinantal equation depends only upon the nature of the terminal conditions apart from impressed force. There may be any distribution of impressed force, intermediate or at the terminals as well, but unless we change the terminal arrangements, there will always be the same set of normal systems. In (55), (56), for example, e is entirely at the point y. But make it be a function of y, representing a distributed impressed force. Then the integration of (55), (56) with respect to y, so as to include all the impressed forces, will give the resulting potential. The outside terms are altered, and likewise the size of the normal systems in the summation, but the type of the normal systems, say $\sin sx\,\epsilon^{pt}$, is not changed, and the same series of s's and p's remain in force.

(8). General Terminal Conditions. Impressed Force at A.

§ 291. Owing to the equidistant spacing of the roots of the determinantal equation in the above cases, which arises from the fact that the terminal reflection coefficients are of size $+1$ or -1, as the case may be, the expanded solutions are readily calculable numerically at once. It is somewhat different when we depart from the simplicity of terminal earth or insulation, as the roots become unequally spaced, and require preliminary determination from trigonometrical tables before the formulæ can be subjected to calculation. The process of expansion is, however, the same. Let, for a first case, the impressed force be at the beginning, $x=0$. Then we only want the formula on the right side of the force. This is (35), § 259, in which put $a=0$, making

$$V = \frac{1-\alpha}{\epsilon^{ql} - \alpha\beta\epsilon^{-ql}}\{\epsilon^{q(l-x)} + \beta\epsilon^{-q(l-x)}\}\tfrac{1}{2}e, \tag{57}$$

which shows the potential at x due to e at $x=0$, when the terminal reflection coefficients are α and β, at $x=0$ and l respectively. In terms of the terminal resistance operators Z_0 and Z_1, we have

$$\alpha = \frac{Z_0 - R/q}{Z_0 + R/q}, \qquad \beta = \frac{Z_1 - R/q}{Z_1 + R/q}, \tag{58}$$

by § 261, equations (49) and (52). Here Z_1 is such that $V = Z_1C$ is the equation of voltage for the terminal arrangement at $x=l$, and $V = -Z_0C$ for that at $x=0$.

When the terminal arrangement is a mere resistance, we have $Z_0 = R_0$ say, and $Z_1 = R_1$, these being constants. Notice, in passing, the singular case $\alpha = 1$. By (57) it makes $V=0$; that is, the cable cannot be charged at all. This happens when Z_0 is infinite, which needs an infinite terminal resistance at every moment, and no terminal permittance.

Going further, let us expand (57). The best way is, by means of the given expressions for α and β, to convert it to circular form first, before applying the expansion theorem, although it will come to the same thing in the end if we expand (57) as it stands, and then change to circular functions. Adopting the former course, put $q = si$ in (57) and (58), and reduce by ordinary algebra. The result is

$$V = \frac{\{\sin + (Z_1 s/R) \cos\} s(l - x)}{(1 - Z_0 Z_1 s^2/R^2) \sin sl + (Z_0 + Z_1)(s/R) \cos sl} e. \qquad (59)$$

This is, by its construction, only another form of the operational solution itself, the meaning of s^2 being $-RSp$. Observe, also, that although Z_0 and Z_1 are unspecified, no further reductions of the kind just made are required for them in any case allowing of the application of the expansion theorem, because they are then rational functions of p, and therefore of s^2. So (59) is the operational solution in a form which is convenient for the immediate application of the expansion theorem.

Put $s = 0$ to find the steady state. Practically, put 1 for $\cos sl$ and $\cos sx$, and sl or sx for $\sin sl$ and $\sin sx$. The result is

$$V_0 = \frac{R(l - x) + Z_1}{Rl(1 - Z_0 Z_1 s^2/R^2) + Z_0 + Z_1} e. \qquad (60)$$

Now if the terminal arrangements are mere resistances, say R_0 and R_1, this becomes

$$V_0 = \frac{R(l - x) + R_1}{Rl + R_0 + R_1} e, \qquad (61)$$

which is obviously right, because the denominator is the total resistance of the circuit, and the numerator is that part of it which lies to the right of the point x. Similarly, if the terminal arrangements, though not simple resistances themselves, have finite steady resistances, then $s = 0$ in Z_0 and Z_1 produces them, and (61) is still the result, where R_0 and R_1 denote the effective steady resistances. But it may happen that the term containing s^2 in (60) does not vanish. It may be infinite, so caution is needed when the effective terminal resistance is infinite. Thus, (60) may be written

$$V_0 = \frac{R(l - x) + Z_1}{Rl(1 + Z_0 Z_1 Sp/R) + Z_0 + Z_1} e, \qquad (62)$$

in terms of p. Now suppose the terminal arrangement at A is a condenser, making $Z_0 = (S_0 p)^{-1}$. The p is cancelled. But at the same time the denominator is made infinite, R_0 being infinite. So if R_1 is finite, the result is $V_0=0$. That is, the cable settles down to a neutral state ultimately when charged through a condenser, provided there is not infinite resistance at the further end. On the other hand, when the condenser is at B, making $Z=(S_1p)^{-1}$, say, the result is $V=e$, the full value, provided the resistance at A is finite. Lastly, if both R_1 and R_0 are infinite, by having condensers at both ends, say $Z_1=(S_1p)^{-1}$, $Z_0=(S_0p)^{-1}$, the result of putting $p=0$ in (62) is

$$V_0=\frac{S_0e}{S_0+S_1+Sl}. \tag{63}$$

This result may be verified by the law of displacement in condensers. On one side of the impressed force is the elastance S_0^{-1}, and on the other side $(S_1+Sl)^{-1}$. So the total displacement is

$$\frac{e}{S_0^{-1}+(Sl+S_1)^{-1}}=\frac{eS_0(Sl+S_1)}{S_0+Sl+S_1}, \tag{64}$$

Divide by $Sl+S_1$, the permittance on the right side of e, and the result is V_0, as in (63), the final steady potential of the cable and further condenser. On the other side of the impressed force—that is, on the side of the condenser at A which is next the cable—the potential is negative, being lower than the above V_0 by the amount e. It is the side of the condenser at A next the earth that is charged positively. The reverse is the case in the dielectric of the cable and the B condenser. The simplest mental realisation is obtained by the use of Maxwell's displacement current.

If we compare these cases with the formula (61), understanding that R_0 and R_1 are the steady terminal resistances, we see that they are in agreement, except that when R_0 and R_1 are *both* infinite, the result (61) is ambiguous. In such a case we must go back to the more complete formula (60), or (62), and interpret it instead.*

Having thus settled V_0, the outside term to express the steady state when there is one, we may apply the expansion

* Of course condensers may occur terminally in other ways than the above, but in any case $s=0$ in (59) will lead to the required result.

theorem to (59), or to any form of the same that may be more convenient. Another form is

$$V = \frac{\cos sx(\tan sl + Z_1 s/R) + \sin sx(\tan sl \,.\, Z_1 s/R - 1)}{(1 - Z_0 Z_1 s^2/R^2)\left(\tan sl + \dfrac{(Z_0 + Z_1)\, s/R}{1 - Z_0 Z_1 s^2/R^2}\right)} e, \quad (65)$$

obtained from (59) by expanding the numerator, and then dividing the numerator and denominator in that equation by $\cos sl$.

The determinantal equation is

$$\tan sl + \frac{(Z_0 + Z_1)\, s/R}{1 - Z_0 Z_1 s^2/R^2} = 0, \quad (66)$$

or, which is the same,

$$\tan sl = - \frac{Rl(Z_0 + Z_1)\, sl}{(Rl)^2 - Z_0 Z_1 (sl)^2}. \quad (67)$$

This simplifies to

$$\tan sl = - \frac{Z_0}{Rl} sl, \quad (68)$$

when $Z_1 = 0$, and to

$$\tan sl = - \frac{Z_1}{Rl} sl, \quad (69)$$

when $Z_0 = 0$. In these equations the Z's are functions of $(sl)^2$ in virtue of being functions of p in general.

In the case of the terminal arrangements being mere resistances, producing a constant ratio of the potential and current at the terminals, we have the simple terminal conditions considered by Fourier. Z_0 or Z_1, or both, as the case may be, are constants. The values of sl are then very easily determined by means of a table of tangents. This kind of terminal condition, which includes the extreme cases of terminal earth and insulation, is the only one that admits of the expansion of an arbitrary function after the manner of Fourier—that is to say, by the use of the property of the vanishing of the integral of the product of any two different normal systems. The reason is because there is no energy in the terminal arrangements that can by itself affect the state of the cable. It is true that there is waste of energy in the terminal resistances, but that does not count at all. It is wasted and done for. So the specification of the initial state merely requires a statement of the potential in the cable itself; that implies electric energy, upon which the later state of the cable when left to itself depends, in the absence of an impressed force to introduce fresh energy. But whenever the terminal arrangement departs from the above simple type, involving merely a waste of energy from the cable, Fourier's method fails. The integral of the product of two normal systems along the cable is no longer zero. The initial state may or may not include any energy in the terminal arrangements. If it does, this terminal energy must be allowed for by widening the conjugate property, or in other ways. But even if it does not, the energy in the cable itself will set up energy in the terminal arrangements in a reversible manner, not by mere waste, so the complete normal systems must include the terminal arrangements as well as the cable. Now if the terminal arrangements consist only of condensers and resistances (or equivalently), however complicated in detail, we only introduce electric energy. This is like that in the cable (its self-induction being ignored here), viz., the energy of electric displacement. The conjugate property which has to take the place of Fourier's is then substantially the same, with a wider range, however. It is the property of the vanishing of the total mutual energy of a pair of complete normal systems. But should we introduce magnetic energy into either or both of the terminal arrangements, the last property breaks down. We have to go to a still wider one, viz., that the mutual potential energy (or electric, here), of a pair of normal systems equals the mutual magnetic energy of the same, the terminal arrangements being included. This may work out in a complicated manner.

But the expansion theorem goes straight to the final simplified result, irrespective of the absence or presence of energy or of the power of receiving and storing energy in the terminal arrangements, and of the kind of conjugate property required to effect the expansion after the manner of Fourier extended. Thus, applying it to (65), following the formula (30), and using $\frac{1}{2}s\,(d/ds)$ instead of $p(d/dp)$, we obtain

$$\mathrm{V} = \mathrm{V}_0 + 2e\Sigma \frac{\cos sx\,(\tan sl + \mathrm{Z}_1 s/\mathrm{R}) + \sin sx\,(\tan sl\,.\,\mathrm{Z}_1 s/\mathrm{R} - 1)}{sl(1 - \mathrm{Z}_0\mathrm{Z}_1 s^2/\mathrm{R}^2)\left(\dfrac{1}{\cos^2 sl} + \dfrac{d}{d(sl)}\dfrac{(\mathrm{Z}_0 + \mathrm{Z}_1)s/\mathrm{R}}{1 - \mathrm{Z}_0\mathrm{Z}_1 s^2/\mathrm{R}^2}\right)}\epsilon^{pt}, \qquad (70)$$

by performing one differentiation upon the denominator in (65). We cannot carry it further profitably without specifying what the natures of Z_0 and Z_1 really are. Since, however, the values of sl in the summation are fixed by the equation (66), that is, by the vanishing of the denominator in (65), we may substitute the right member of (67) for $\tan sl$ in (70), and similarly put $\sec^2 sl$ in terms of the Z's. But there is no particular advantage in doing so, unless the Z's are given, and we desire to make possible simplifications of expression.

In obtaining (70) by the expansion theorem, no notice was taken of the factor $(1 - Z_0 Z_1 s^2/R^2)$ in the denominator in (65). It has its roots truly, but they are inoperative. They lead to nothing. The factor may therefore be regarded as belonging to the numerator of (65), if we please. Or, keeping it in the denominator, we see that its vanishing does not make the denominator vanish, and therefore its roots are out of the question.

(9). General Case of an Intermediate Impressed Force.

§ 292. When the impressed force is intermediate, say at the point $x = y$, the treatment is quite similar, so that very little detail need be given. We have the operational solutions (35), (36), § 259, in which put y for a. Now circularise the functions by putting $q = si$. Remember in doing so that this is not really a complex transformation, because the operators are rational functions of p already, and therefore of s^2. This fact makes the transformation easy, by the cancelling of terms that occurs. The result is that

$$V_1 = \frac{(R\cos - Z_0 s \sin)sy \,.\, (Z_1 s \cos + R \sin)s(l - x)}{\cos sl\,(R^2 - Z_0 Z_1 s^2)\left(\tan sl + \dfrac{Rs(Z_0 + Z_1)}{R^2 - Z_0 Z_1 s^2}\right)} e \qquad (71)$$

expresses the potential due to e at a point x on the right side of y, where e is situated. To obtain the corresponding expression for V_2, when x is on the left side of y, refer again to (35), (36), § 259, and observe that V_2 is got from V_1 by interchanging a and β, x and $l - x$, y and $l - y$, and then negativing the result. So, doing the same to (71), we get

$$V_2 = -\frac{(R\cos - Z_1 s \sin)s(l - y) \,.\, (Z_0 s \cos + R \sin)sx}{\text{same denominator}} e. \qquad (72)$$

Now apply the expansion theorem. The outside terms for the steady state may be perhaps most easily got by elementary considerations—by Ohm's law and the condenser law, that is to say. When the steady terminal resistances (effective) are finite, say R_0 and R_1, we shall have

$$V_1 = e\frac{R(l-x)+R_1}{Rl+R_0+R_1}, \qquad V_2 = -e\frac{R_0+Rx}{Rl+R_0+R_1}, \tag{73}$$

by Ohm's law. But when R_0 and R_1 are both infinite, and there are terminal condensers concerned, we must go further, and apply the condenser law, say as in § 291, equation (64), which is a special case of the present. In general, however, should the terminal arrangements include several condensers, it is the effective steady permittances of the combinations that are to be regarded as S_0 and S_1 in the place referred to. Or, we may get the steady states directly out of (71) and (72) by seeking their limiting values when $s=0$. However the work be done, denoting the steady states by v_1 and v_2, and applying the expansion theorem to (71), we obtain at once

$$V_1 = v_1 + \Sigma \frac{e(R\cos - Z_0 s\sin)sy\,.\,(Z_1 s\cos + R\sin)s(l-x)}{\frac{1}{2}sl\cos sl\,(R^2 - Z_0Z_1s^2)\left(\sec^2 sl + \dfrac{d}{d(sl)}\dfrac{Rs(Z_0+Z_1)}{R^2 - Z_0Z_1s^2}\right)}\epsilon^{pt}, \tag{74}$$

by introducing the summation sign, the time factor, and changing the denominator by the operation $\frac{1}{2}sl(d/d(sl))$.

As regards V_2, we may get its formula by making the interchanges already indicated, in getting (72) from (71). But this is quite unnecessary. For, by inspection of the original operational formulæ (35), (36), § 259, we may see that when the denominator vanishes V_1 and V_2 become identical in those equations. That is, identical for all the values of p given by the determinantal equation, though not in the case of $p=0$, which finds the steady states on the two sides of e. So the V_2 formula is to be got from (74) by changing V_1 to V_2 and v_1 to v_2. The expansional part is the same, in virtue of the special values of sl. Of course, we may use the alternative formula if we like, making the interchanges before described. And when the Z's are explicitly given, simplifications of form may be readily carried out. This is not a matter of indifference, if numerical calculation is wanted, because, of two equivalent forms, one may be much more manageable than the other.

The Determinantal Equation.

§ 293. The determinantal equation is (66) or (67). Say,

$$\tan sl = \phi(sl) = -\frac{Rl(Z_0+Z_1)sl}{(Rl)^2 - Z_0Z_1(sl)^2}. \tag{75}$$

Here $\phi(sl)$ is a rational function of sl. Draw the curves

$$y_1 = \tan sl, \qquad y_2 = \phi(sl); \tag{76}$$

the abscissæ being sl, and the ordinates y_1 or y_2. The intersection of these curves indicates the admissible values of sl. It is sufficient to draw them very roughly, just to find the general situation and rough values of the roots. Only a few at the beginning are wanted, except for very small values of the time. I mentioned before that when the Z's are mere resistances, tangent tables readily find the roots. Use the graphically-obtained values as a first approximation. Say it is the first root, which is the most important in general. The tables may show that the rough value is several degrees wrong. Estimate the correction by rule of three. Apply the revised value to the tables. It will be nearly right, and a second revision may bring it right to a minute, and a third to a second.

Another easy case is that of terminal condensers. Also the case of condensers and resistances does not trouble much. And it may be remarked here that when the terminal arrangements involve only electric energy, without magnetic, the roots are always real, so that the above described process can be followed. But it is not so easy when magnetic energy is involved. There may then be complex roots as well. This occurs even when a single coil is concerned at either terminal, if the inductance be of a suitable value. Of course, in slow cable working, the inclusion of the effect of the self-induction of the coils is not an important matter, and the coils may be treated as resistances. The trouble with complex roots sets a limit to the desirability of elaborating formulæ beyond the point of ready systematic calculation. The formulæ, nevertheless, have an interest of a scientific kind, as showing what may be done if the necessary trouble be taken.

The determinantal equation may be written

$$\tan sl + \tan\left(\tan^{-1}\frac{sZ_0}{R} + \tan^{-1}\frac{sZ_1}{R}\right) = 0, \qquad (77)$$

which makes

$$n\pi = sl + \tan^{-1}\frac{Z_0 sl}{Rl} + \tan^{-1}\frac{Z_1 sl}{Rl}. \qquad (78)$$

This form shows how the difference between sl and the standard value $n\pi$, n being a positive integer, depends upon the terminal operators. Equation (78) may be readily converted to a power series, perhaps with occasional advantage. Special formulæ may also be constructed exhibiting the roots explicitly to any desired degree of accuracy, but it is questionable whether they repay the labour of obtaining them, when tables of circular functions are so useful. General principles may become smothered by overmuch detail.

The original form of the determinantal equation, as it shows itself in the operational solutions (35), (36), § 259, and (65), (66), § 253, may be written

$$\epsilon^{2ql} = \alpha\beta, \qquad \text{or} \qquad \alpha\epsilon^{-ql}\beta\epsilon^{-ql} = 1. \qquad (79)$$

If the left-hand operator be applied to a wave originated at a point, say at $x=0$, the resultant would represent what it became after a journey to $x=l$, reflection there, according to β, followed by a journey to $x=0$, and reflection there according to α. The right-hand member would show that the effect of the cycle of operations was to restore the disturbance to its original state. This is impossible, of course. But it is apparently done on a normal solution. Thus

$$\alpha\beta\epsilon^{-2ql}\epsilon^{Pt} = \epsilon^{Pt}, \qquad (80)$$

if the constant P is chosen to be any one of the values given by the determinantal equation. This equation does not imply any use of $p^{\frac{1}{2}}$, but only abbreviates a rational equation. Clear it of fractions, and we get

$$\left\{\left(1+\frac{qZ_0}{R}\right)\left(1+\frac{qZ_1}{R}\right)\epsilon^{ql} - \left(1-\frac{qZ_0}{R}\right)\left(1-\frac{qZ_1}{R}\right)\epsilon^{-ql}\right\}\epsilon^{Pt} = 0, \qquad (80\text{A})$$

where the operator is an even function of q, and is therefore rational.

Subsidence of Special Initial States.

§293a. In the above equation (74) we have not only solved the question of the establishment of a certain state by an impressed force, but also the connected question of the subsidence of that state to equilibrium when the force is removed. For, if we write (74) and its companion thus,

$$V = v - \Sigma u \epsilon^{pt}, \tag{81}$$

where v has to be v_1 or v_2 as the case may be, we have, initially,

$$v = \Sigma u. \tag{82}$$

That is, the state of potential v is expanded in the special Fourier series concerned. If, then, v is left to itself, unsupported by impressed force, the state at time t later will be given by

$$v_1 = \Sigma u \epsilon^{pt}, \tag{83}$$

where v_1 is what v then becomes.

There is this remark to be made about (82). It cannot be called without limitation *the* expansion of the function v in a Fourier series, subject to the terminal differential conditions, for it may be only one of many expansions expressing the same functions v, and subject to the same differential conditions. This is quite clear in the physics of the matter. For consider what the impressed force does in the act of setting up the state v. It energises the terminal arrangements as well, electrically and magnetically, except when only mere resistances are concerned. Now the same state of potential v in the cable may be accompanied by various states of energy terminally, and all due to the same impressed force. So when we take off the impressed force we not only let the charge redistribute itself, but also let the terminal energy act on the cable, and in various manners according to circumstances. Equation (83) exhibits the potential at time t due to the initial state u in the cable, *and* to the accompanying initial states of the terminal arrangements. This is true even when the Z's are only given by formulæ, and we do not analyse them to see what particular arrangements are really represented thereby, and what the terminal energies are.

It will be seen that the expansion theorem is a labour-saving agent of a remarkable character. For it is not from formulæ representing the expansion of an arbitrary initial state that we can most readily learn the general course of events in the physical problems concerned. We should rather prefer to examine the result of some special initial state, or the result of a disturbance initiated at a single spot, as an impulsive or a continued source. Now it is just in these cases that the expansion theorem shows to best advantage. We obtain our formulæ in a very ready manner, without the circumbendibus connected with arbitrary initial states and the conjugate property of the normal functions. We shall see presently how to apply the expansion theorem to arbitrary initial states. In connection therewith, it may be noted here that an integration applied to (74) and its companion enables us to express the effect due to an arbitrary distribution of impressed force along the line, all starting at the same moment in the simplest case. Put edy for e, and let e be a function of y. Then integrate with respect to y from 0 to l, using one or other formula, according as x is to right or left of the elementary impressed force concerned. But this is a mathematical development which is useless for our present purpose.

(10). General Case of an Arbitrary Initial State in the Cable.

§ 294. The expansion of an arbitrary function of x between the limits 0 and l naturally rests upon the expression of the function for a single point. That is, we require to find how a charge Q, initially all at the point y, diffuses itself when controlled by given terminal conditions. As was before pointed out, the construction of the operational solution for a point charge is like that for an impulsive impressed voltage, with a difference. The latter produces a jump in the potential, the former in the current, at the place of application. A continued impressed force e at a point has its analogue in a continued source of current, led in by an auxiliary wire. The operational solutions to right and left of the source are given by (65), (66), § 263, in which h is the current introduced, to be given as a function of the time. Taking $h=p\mathrm{Q}$ makes the source impulsive, and the case is then that of an initial charge Q. So, by

(65), § 263, the potential V_1 at x, on the right side of the point y, where Q is impulsively introduced, is given by

$$V_1 = \frac{\epsilon^{qy} + \alpha\epsilon^{-qy}}{\epsilon^{ql} - \alpha\beta\epsilon^{-ql}}\left\{\epsilon^{q(l-x)} + \beta\epsilon^{-q(l-x)}\right\}\frac{q\mathrm{Q}}{2\mathrm{S}}. \tag{84}$$

On the left side of y the potential V_2 is to be got by simply interchanging x and y in the last formula.

Converting (84) to circular functions, by means of $q = si$, makes the equivalent formula

$$V_1 = -\frac{(Z_0 s\cos + R\sin)sy \,.\, (Z_1 s\cos + R\sin)s(l-x)}{\cos sl\,(R^2 - Z_0Z_1s^2)\left(\tan sl + \dfrac{Rs(Z_0+Z_1)}{R^2 - Z_0Z_1s^2}\right)}\frac{s\mathrm{Q}}{\mathrm{S}}. \tag{85}$$

The final state must be either zero or constant. It must be zero if there is conductive connection with earth at either terminal—that is, when at least one of R_0 and R_1 is not infinite; and it must be finite when both are infinite, so that the charge Q cannot be got rid of. In the former case, the last equation shows the result at once, on account of the s which occurs as a factor. In the latter case, it is the permittance that controls matters. The charge Q is finally either uniformly distributed along the cable, if there is no terminal permittance, or is divided between the cable and the terminal condensers. Thus, the potential of the cable is

$$v = \frac{\mathrm{Q}}{\mathrm{S}l + \mathrm{S}_0 + \mathrm{S}_1}, \tag{86}$$

if S_0 and S_1 are the effective permittances of the terminal combinations. This result may be dug out of (85), but is sufficiently clear without doing that, which is, however, a useful test.

Applying the expansion theorem to (85), we obtain

$$V = v - \Sigma\frac{(Z_0 s\cos + R\sin)sy \,.\, (Z_1 s\cos + R\sin)s(l-x)}{\cos sl\,(R^2 - Z_0Z_1s^2)(\sec^2 sl - \phi'(sl))}\frac{2\mathrm{Q}}{\mathrm{S}l}\epsilon^{pt}, \tag{87}$$

where $\phi(sl)$ is as in (75) and $\phi'(sl)$ is its derivative. We have the same form of determinantal equation. This formula applies on both sides of the point y, both as regards the steady v and the summation. For the steady state is constant, and the original operational solutions (65), (66), § 263, become the same when the denominator vanishes. We may, however, exchange x and y in (87), and make other changes controlled by the determinantal equation.

Put Q/S = 1 and $t = 0$. Let u be the result. Then

$$u = u_0 + \frac{2}{l}\Sigma\frac{(Z_0 s\cos + R\sin)sy \,.\, (Z_1 s\cos + R\sin)s(l-x)}{\cos sl\,(Z_0Z_1s^2 - R^2)(\sec^2 sl - \phi'(sl))}, \qquad (88)$$

where u_0 is the value of v with Q/S = 1. Here u represents the initial state. Therefore u is the expansion of the unit impulsive function of x concentrated at the point y. It represents a great deal more than that, outside the limits 0 and l, for there are periodic repetitions, though not periodic in the usual sense, except in the elementary cases of total reflection at the terminals, when we have the usual Fourier series with regularly spaced roots.

Given, then, a function $f(x)$ representing an initial state, its expansion is

$$f(x) = \int_0^l f(y)u\,dy, \qquad (89)$$

using the previous expression for u; and, by introducing the time factor in the summation, we find what the potential $f(x)$ becomes at time t. The ultimate result is a state of uniform potential—viz., the total initial charge divided by the total permittance, including the effective terminal permittances.

As regards the meaning of the expansion for V outside the limits for x, viz., 0 and l, which are imposed in the physical problem, we get some knowledge from the determinantal equation. If V is the potential at x at time t in the cable, free from impressed force, its general characteristic is $(\Delta^2 - q^2)V = 0$, where Δ is the space differentiator and q^2 is RS × the time differentiator. Also, when the terminal reflection coefficients α and β are introduced, we make V satisfy

$$\left\{\left(1+\frac{\Delta Z_0}{R}\right)\left(1+\frac{\Delta Z_1}{R}\right)\epsilon^{l\Delta} - \left(1-\frac{\Delta Z_0}{R}\right)\left(1-\frac{\Delta Z_1}{R}\right)\epsilon^{-l\Delta}\right\}V = 0, \qquad (90)$$

where in the operator on V we write Δ for q. Compare with (80A). It is to be understood that in the two Z's the same change is made, or Δ^2/RS substituted for p. So, if V is $f(x)$, we have, by Taylor's theorem, § 276,

$$\left(1+\frac{Z_0\Delta}{R}\right)\left(1+\frac{Z_1\Delta}{R}\right)f(x+l) = \left(1-\frac{Z_0\Delta}{R}\right)\left(1-\frac{Z_1\Delta}{R}\right)f(x-l). \qquad (91)$$

This is the equation showing the functional connection between the values at points distant $2l$ from one another. When the Z's are zero, it reduces to $f(x+l)=f(x-l)$, showing the primary kind of periodicity. Further examination of (91) to see how this is modified by the terminal operators, would lead us too far away from the proper subject.

(11). Auxiliary Expansions due to the Terminal Energy. Case of a Condenser

§ 295. The expansion (89), subject to (88), has some right to be termed *the* expansion of $f(x)$ in the Fourier series in question, because it is the one that depends only upon the initial state of the cable. It gives the potential at any later time on the understanding that the terminal arrangements are initially unenergised. Unless, therefore, the Z's are mere resistances, the Fourier series (89), (88), may not be the proper one to use in the physical problem. It will still represent the initial state, but will fail later, should the terminal arrangements be initially energised. So there are, independently of the above expansion, others which represent zero initially, and which only come into operation when t is finite. The number of such auxiliaries is determined by the number of independent ways in which the terminal arrangements may be energised. It may, therefore, be finite or infinite. If, for example, there is a coil at A and a condenser at B, there are just two auxiliary expansions. For the full specification of the initial state must include not merely the state of charge in the cable, but also the charge of the condenser and the current in the coil. After the first moment, they are all connected. Initially, they are independent.

A case in which there is an infinite number of auxiliaries is got by letting the terminal arrangement at either end be a second cable, of finite length. Another one is got by putting a piece of metal inside a coil. Without the metal the coil would introduce one auxiliary. The metal, however, which will produce a quite determinate rational terminal operator, will introduce an infinite number of auxiliaries, because the initial state will require a specification of the magnetic force in all parts of the metal, and this initial state may be arbitrary. This case may seem fearfully complicated, but can be worked out (in the case of a round core), without great difficulty, by the operational method.

To show the action of a single auxiliary, let there be a terminal condenser. Say that $Z_0 = (S_0 p)^{-1}$ and $Z_1 = 0$. Then the cable is earthed direct at B, and through a condenser of permittance S_0 at A. So, by (85) above,

$$V_1 = \frac{\left(\frac{S}{sS_0}\cos - \sin\right)sy \,.\, \sin s(l-x)}{\cos sl\left(\tan sl - \frac{S}{S_0 s}\right)} \frac{sQ}{S} \tag{92}$$

represents the potential at x on the right side of y, due to the impulsive introduction of Q at y. And this expands to

$$V = \Sigma \frac{\left(\frac{S}{S_0}\cos - s\sin\right)sy \,.\, \sin s(l-x)}{\frac{1}{2}sl\cos sl\left(\sec^2 sl + \frac{Sl}{S_0 s^2 l^2}\right)} \frac{Q}{S}\epsilon^{pt} \tag{93}$$

which is valid on both sides of y. There is no outside term. The determinantal equation is

$$\tan sl = S/S_0 s, \tag{94}$$

and an integration according to (89) constructs the potential due to an arbitrarily given state in the cable, with no initial charge in the condenser.

Now as regards the condenser, put $y=0$ in (92). Then

$$V_1 = \frac{\sin s(l-x)}{\cos sl\,(\tan sl - S/S_0 s)} \frac{Q}{S_0} \tag{95}$$

represents V_1 in the cable due to Q at its beginning. That is, to an initial charge Q of the condenser. Or, if V_0 is its initial potential, we may substitute it for Q/S_0. So

$$V = V_0 \Sigma \frac{\sin s(l-x)\epsilon^{pt}}{\frac{1}{2} sl\cos sl\,(\sec^2 sl + Sl/S_0 s^2 l^2)} \tag{96}$$

represents the potential V at time t in the cable due to V_0 alone. Initially we have

$$0 = V_0 \Sigma \frac{\sin s(l-x)}{\text{same denominator}}, \tag{97}$$

for any finite x in the cable. This exhibits the auxiliary expansion, V_0 being any constant. The complete expansion, showing V due to the initial V_0 of the condenser and $V = f(x)$ in the cable, is therefore

$$V = \Sigma \sin s(l-x)\, \epsilon^{pt} \frac{V_0 + \int_0^l f(y)\left(\frac{S}{S_0}\cos - s\sin\right) sy\, dy}{\frac{1}{2} sl \cos sl\, (\sec^2 sl + Sl/S_0 s^2 l^2)} \tag{98}$$

This case is peculiar, inasmuch as the same operational solution serves for the elementary charge in the cable and for the terminal charge. The distinction is that Q is finite in one case and infinitesimal in the other. It is true that in (92) Q may be finite, but when we pass to an initial finite distribution of potential, it is replaced by Q dy, the charge on the element dy, which is therefore infinitesimal. On the other hand, the Q for the condenser is finite.

Points of Infinite Condensation. Exceptions to Fourier's Theorem.

§ 296. Another consideration presents itself here. It is a common error that only finite functions can be expanded in Fourier series. That this is wrong may be seen by the above investigation. Go back to (88). Let there be initially Q_1/S at y_1, Q_2/S at y_2, and so on up to Q_n/S at y_n, these points being separate. Then

$$v = \Sigma \frac{Q_n}{S} u_n \tag{99}$$

will, by (88), represent the initial state of the potential at time t due to the n point charges. Initially, v is zero except at the n points, but there it is infinite, with finite space totals, however. Combine with (98). If $f(x)$ is finite and continuous, $V + v$ is initially finite and continuous, except at the n points, where finite charges exist.

Nor need these condensations be themselves finite. The Q's may be infinite. It is, indeed, true that an infinite charge suddenly introduced at a point y would in an infinitely short time raise the potential of the whole cable infinitely. In this respect the solution would be of a useless nature. But we may combine the infinities so as to have finite results. The simplest case is first to have two finite point charges Q and $-$Q at distance z apart. Then bring them closer, increasing Q in the same ratio, so that Qz is finite. In the limit we have a point source involving a double infinity in Q, and a corresponding expansion of the potential in a F. series. Similarly as regards poles of higher multiplicity.

Nor does the matter end here. If the self-induction of the line be included, the initial state must include a specification of the initial current, as well as of V. Fourier series are involved, and such a series has to express the initial potential, whilst a connected series expresses the initial current. There may be points of initial infinite condensation in both series.

Another common error is that the expansion of a function in a series of sines and cosines of $n\pi x/l$, n being integral, is uniquely determined by Fourier's theorem. There are other expansions. To show this (in one way), let $Z_1 = -Z_0$ in (85). Then $\tan sl = 0$, or $sl = n\pi$, as in Fourier's theorem. But now $R^2 - Z_0 Z_1 s^2 = 0$ also, or $R^2 = Z_0^2 q^2$, determines a distinct set of extra normal systems, of the type ϵ^{hx}, where h is constant. So we can expand a given function in terms of the usual $\sin sx$ and $\cos sx$, and the extra functions. The result is, therefore, to represent another function of x expanded in terms of the usual sines and cosines, with equally-spaced roots. But it is not the expansion got by Fourier's theorem, though it is equally true. I will give one or two examples in illustration of this.

Abnormal Fourier Series.

§ 297. When a cable is earthed at its ends the type of the normal function for the potential is $\sin sx$, vanishing terminally, s being fixed by $\sin sl = 0$. Left to itself, a distribution of potential of this type becomes reduced to $\sin sx \, . \, \epsilon^{pt}$ at time t later, p being negative, fixed by $RSp = -s^2$. If the ends are both insulated, the normal function is $\cos sx$ for the potential, but with the same series of values of s, with the addition of a zero value, making 1 a special normal function. The roots of the determinantal equation are equally spaced.

Now, if one end be earthed and the other insulated, the normal functions are $\sin sx$ or $\cos sx$, as the case may be, with $sl = (n - \frac{1}{2})\pi$, n being integral. The roots of the determinantal equation are equally spaced in this case also.

In all other cases whatsoever, so far as I know, of real practical problems, the series of s's in the general normal function $\cos sx + a \sin sx$, where a is a constant, which are permissible and necessary in order that real and practical terminal conditions may be satisfied, differ from those given by $\sin sl = 0$ or $\cos sl = 0$, and are unequally spaced. It does not seem to be possible for any two arrangements of real condensers, resistances, and coils, acting at the terminals, to act in concert in such a way as to bring the determinantal equation back to the original simple type. It is perhaps for this reason, or rather for similar reasons in other physical problems involving Fourier series, that it has apparently escaped notice that Fourier's theorem, in which the s in the normal functions is $n\pi/l$, n being integral, does not uniquely determine expansions of functions. It is, indeed, said to do so, but that is another story. It seems natural, for the above-stated reason, that mathematical physicists should not come across exceptions. On the other hand, pure mathematicians would, perhaps, not arrive at them, owing to the peculiar way they have of regarding the subject of the expansion of functions. They would be concerned with professedly rigorous proofs of Fourier's theorem, and of the convergency, rather than with the discovery of exceptions.

Guided, however, by physical ideas, though applied to entirely unpractical arrangements, involving instability, to be produced by latent impressed forces, set going by the real electrical mechanism, it is quite easy to adjust matters so that the terminal arrangement at one end of the cable shall combine with that at the other in such a way as to bring the roots of the determinantal equation back to the original simple kind. We come back to $sl = n\pi$, or $(n - \frac{1}{2})\pi$, with, however, the important and essential fact that there are extra normal functions of an abnormal kind. And so we come to violations of Fourier's theorem regarded as a unique expansion.

Origin of Two Principal Abnormal Cases.

§ **298.** Thus, equation (85) shows the potential at x on the right side of y, due to Q initially at y, subject to the terminal conditions $V = -Z_0C$ at A, and $V = Z_1C$ at B. We may write it

$$V_1 = \frac{(Z_0 s \cos + R \sin) sy \,.\, (Z_1 s \cos + R \sin) s(l - x)}{(Z_0 Z_1 s^2 - R^2) \sin sl + Rs(Z_0 + Z_1) \cos sl} \frac{sQ}{S}. \qquad (100)$$

In general, with practical resistance operators, the denominator only vanishes by the sum of the terms in it vanishing; then we have the determinantal equation (75), with unequally spaced roots, save in the limiting cases of terminal earth or disconnection.

But let $$Z_0 + Z_1 = 0. \tag{101}$$

Then $$(Z_0 Z_1 s^2 - R^2) \sin sl = 0 \tag{102}$$

is the determinantal equation. So $\sin sl = 0$, giving $sl = n\pi$, n being integral; and also,

$$Z_0^2 q^2 = R^2, \tag{103}$$

where $q^2 = -s^2 = RSp$. Here, when Z_0 is a real resistance operator, the left member is an odd function of p. So there is an odd number of extra special normal functions.

Thus, in case of a resistance R_0 at A, we get

$$R^2 = R_0^2 RSp, \tag{104}$$

showing one root, positive. In case of a condenser at A, or $Z_0 = (S_0 p)^{-1}$, we get

$$R^2 S_0^2 p = RS; \tag{105}$$

again one root, positive. In case of a coil at A, with $Z_0 = R_0 + L_0 p$, we get

$$R^2 = (R_0 + L_0 p)^2 RSp. \tag{106}$$

There are now three roots, and three extra normal functions. But two of them combine to make an oscillatory function of a mixed kind. In general, the number of extras is unlimited. The physical interpretation will be considered a little later.

If, instead of (101) as the fundamental relation connecting the Z's, we substitute

$$Z_0 Z_1 s^2 = R^2, \quad \text{or} \quad Z_1 = -\frac{R^2}{q^2 Z_0}, \tag{107}$$

the determinantal equation becomes

$$\left(Z_0 - \frac{R^2}{q^2 Z_0}\right) \cos sl = 0. \tag{108}$$

So $\cos sl = 0$, or $sl = (n - \frac{1}{2})\pi$ for the regulars, just as if one end of the cable were earthed and the other insulated; and, in addition,

$$Z_0^2 q^2 = R^2 \tag{109}$$

for the extras. So, with the same Z_0, there are the same extras as in the former case. But now, if Z_0 is a resistance, Z_1 is a negative permittance, and if Z_0 is a permittance, Z_1 is a negative resistance.

First General Case: $Z_0 = -Z_1$.

§ 299. We cannot consider both cases at once very well. Therefore first examine the case $Z_0 = -Z_1$. Equation (100) reduces to

$$V_1 = \frac{(\mathrm{R}\sin + Z_0 s\cos)sy \,.\, (\mathrm{R}\sin - Z_0 s\cos)s(l-x)}{(Z_0^2 q^2 - \mathrm{R}^2)\sin sl}\frac{sQ}{S}. \tag{110}$$

Apply the expansion theorem. This had better be done separately for the regulars and the extras. Say $V = v_1 + v_2$, where v_1 is for the regulars, and v_2 for the extras. Then, working as in many previous cases, we find

$$v_1 = \frac{Q}{Sl} + \frac{2Q}{Sl}\Sigma\frac{(\mathrm{R}\sin + Z_0 s\cos)sy \,.\, (\mathrm{R}\sin + Z_0 s\cos)sx}{\mathrm{R}^2 + Z_0^2 s^2}\epsilon^{pt}, \tag{111}$$

where $sl = n\pi$. The outside term represents the mean potential—that is, $Q/(Sl + S_0 + S_1)$, where S_0 and S_1 are the effective terminal permittances, which, however, cancel one another. But the outside term is only required when the terminal resistance is infinite at A, or $x = 0$.

For the extra terms, we may transform (110) to

$$V_1 = \frac{(\mathrm{R}\,\mathrm{shin} + Z_0 q\cosh)qy \,.\, (\mathrm{R}\,\mathrm{shin} - Z_0 q\cosh)q(l-x)}{(\mathrm{R}^2 - Z_0^2 q^2)\,\mathrm{shin}\,ql}\frac{qQ}{S}, \tag{112}$$

in which it is to be noted that the p's determine the normal systems. We have

$$p\frac{d}{dp}(Z_0^2 q^2 - \mathrm{R}^2) = \mathrm{R}^2\left(1 + \frac{2p}{Z_0}\frac{dZ_0}{dp}\right), \tag{113}$$

when $\mathrm{R}^2 = Z_0^2 q^2$. Hence, using (113), and putting $\mathrm{R} = Z_0 q$ in the numerator of (112), it, so far as the extras are involved, expands to

$$v_2 = \frac{Q}{S}\Sigma\frac{q\epsilon^{q(x+y-l)}\epsilon^{pt}}{\mathrm{shin}\,ql \,.\, \left(1 + \frac{2p}{Z_0}\frac{dZ_0}{dp}\right)}, \tag{114}$$

ranging over the roots of $Z_0^2 q^2 = \mathrm{R}^2$.

Finally, as before said, $V = v_1 + v_2$ is the complete expansion. The unit impulsive function is got by dividing by Q/S, and omitting the time factor, making

$$u = \frac{1}{l} + \frac{2}{l}\Sigma\frac{(\mathrm{R}\sin + \mathrm{Z}_0 s\cos)sy\,.\,(\mathrm{R}\sin + \mathrm{Z}_0 s\cos)sx}{\mathrm{R}^2 + \mathrm{Z}_0^2 s^2}$$

$$+\Sigma\frac{q\epsilon^{q(x+y-l)}}{\mathrm{shin}\,ql\,.\left(1+\dfrac{2p}{\mathrm{Z}_0}\dfrac{d\mathrm{Z}_0}{dp}\right)}, \qquad (115)$$

where the extras are in the second line alone. Herein Z_0 may be any rational resistance operator. It follows that

$$f(x) = \int_0^l uf(y)dy \qquad (116)$$

is the expansion of an arbitrary function $f(x)$.

Equation (115) being general, has the failings as well as the merits of general results. But we can easily interpret specially simplified cases.

Equal Positive and Negative Terminal Resistances.

§ 300. The simplest case of all is that of a positive resistance at A, and an equal negative resistance (to be interpreted later) at B. There is but one extra, as per (104). Denote the extra q by h for distinctness. Then $h = \mathrm{R}/\mathrm{R}_0$, the reciprocal of the effective length of the terminal resistance R_0 in terms of the cable. Therefore, since $d\mathrm{Z}_0/dp = 0$,

$$u = \frac{h\epsilon^{h(x+y-l)}}{\mathrm{shin}\,hl} + \frac{2}{l}\Sigma\frac{(h\sin + s\cos)sy\,.(h\sin + s\cos)sx}{h^2 + s^2} \qquad (117)$$

expresses the unit impulsive function.

Let $f(x) = 1$ for example; that is, start with a uniform potential in the cable. Thus an easy integration gives us the following expansion:—

$$1 = \frac{2\epsilon^{hx}}{1+\epsilon^{hl}} + \frac{2}{l}\Sigma\frac{h}{s}\frac{(h\sin + s\cos)sx}{h^2 + s^2}(1 - \cos sl), \qquad (118)$$

where $sl = \pi$, 2π, 3π, &c., the terms with even multiples having zero coefficients. The introduction of the time factor to every normal function shows the result later as usual. The summational part ultimately vanishes. The outside extra term increases with the time.

We may write (118) thus, if sl/π is confined to the odd integers :—

$$\frac{2\epsilon^{hx}}{1+\epsilon^{hl}} = 1 - \frac{4}{l}\Sigma\frac{h^2 \sin sx}{s(h^2+s^2)} - \frac{4}{l}\Sigma\frac{hs\cos sx}{s(h^2+s^2)}. \tag{119}$$

The function on the left side is apparently expanded in a Fourier series. But it is *not* a Fourier series in the ordinary sense. It absolutely violates the conditions settling the size of the coefficients according to Fourier's theorem. It is only one example in a million, however. It illustrates a property of the wonderful function ϵ^x.

We may verify (118) by the conjugate property of normal systems. This is an absolutely sound method, provided we have *all* the normal functions at command. One normal function is ϵ^{hx}, the type of the rest is $(h\sin + s\cos)sx$. That the extra one is conjugate to the rest we see thus :—

$$\int_0^l \epsilon^{hx}(h\sin + s\cos)sx \,.\, dx = \int_0^l \frac{d}{dx}(\epsilon^{hx}\sin sx)\, dx = 0. \tag{120}$$

Also, if w_1 and w_2 denote any two of the regulars, we may verify that

$$\int_0^l w_1 w_2\, dx = 0. \tag{121}$$

If, therefore, we assume that

$$f(x) = 1 = A_0\epsilon^{hx} + \Sigma\, A_n w_n, \tag{122}$$

we can isolate any coefficient by quadratures. Thus

$$A_0 = \frac{\int_0^l f(x)\epsilon^{hx}dx}{\int_0^l \epsilon^{2hx}dx} = \frac{2}{1+\epsilon^{hl}} \tag{123}$$

finds A_0, and any one of the rest is found by

$$A = \frac{\int_0^l uf(x)dx}{\int_0^l u^2 dx} = \frac{2h}{sl(h^2+s^2)}(1-\cos sl). \tag{124}$$

Comparing with (118), we verify that result. The quantity h may be positive or negative. Changing its sign merely makes the positive and negative resistances change places. Between the two, with $h=0$, we have the case of terminal disconnection. The conjugate property may be readily applied to the unit impulsive function (117) itself, so as to verify that formula.

As regards the expansion of the left member of (119), if done according to Fourier's theorem, the physical assumption would be such as to make the normal systems be $\cos sx$ and $\sin sx$, and this would correspond to an entirely different state of things. Out of physics came the subject of expansion in normal functions. Unto physics should we return for fresh inspiration.

Equal Positive and Negative Terminal Permittances.

§ 301. The next simplest case is that of a terminal condenser. Say that $Z_0 = (S_0 p)^{-1}$ at A. There is but one extra q. Denoting it by h, its value is S/S_0. Also, by using the new Z_0 in (113), we see that the result is $-R^2$ instead of $+R^2$, as in the last case. So, by (111) and (114), we find that the new unit impulsive function is

$$u = \frac{1}{l} + \frac{2}{l}\Sigma\frac{(R\sin + Z_0 s\cos)sy \;.\; (R\sin + Z_0 s\cos)sx}{R^2 + Z_0^2 s^2} - h\frac{\epsilon^{h(x+y-l)}}{\operatorname{shin} hl}, \tag{125}$$

where the extra function is the last term. Here $sZ_0/R = -h/s$, and we may write (125) more simply,

$$u = \frac{1}{l} + \frac{2}{l}\Sigma\frac{(s\sin - h\cos)sy \;.\; (s\sin - h\cos)sx}{h^2 + s^2} - \frac{h\epsilon^{h(x+y-l)}}{\operatorname{shin} hl}. \tag{126}$$

Applying this u to the special initial state $f(x) = 1$, we obtain

$$f(x) = 1 = 1 - \frac{2\epsilon^{hx}}{1 + \epsilon^{hl}} + \frac{2}{l}\Sigma\frac{(s\sin - h\cos)sx}{h^2 + s^2}(1 - \cos sl), \tag{127}$$

which gives another expansion of ϵ^{hx} not conforming with Fourier's theorem. In (127) sl has the values π, 2π, 3π, &c.; and the coefficients of the terms belonging to 2π, 4π, &c., vanish as before. The formula fails at the terminals, on account of the condensers.

The normal functions are now ϵ^{hx}, the sole extra function, and an infinite series of the type $(s\sin - h\cos)sx$, including the case $s = 0$, or a constant term. But to verify (127) by the conjugate property, we must take count of the terminal condensers. Thus, if

$$w = (s \sin - h \cos) sx \tag{128}$$

be taken as the normal function along the cable it must be supplemented by

$$w_0 = -h, \qquad w_1 = -h \cos sl, \tag{129}$$

at the terminals. Similarly, the normal function ϵ^{hx} must be supplemented by 1 at A and ϵ^{hl} at B. The mutual energy of ϵ^{hx}, &c., and w, &c., is therefore

$$S_0 w_0 + \int_0^l S\epsilon^{hx} w\,dx + S_1 \epsilon^{hl} w_1, \tag{130}$$

which, by the previous two equations, may be seen to vanish, thus proving the independence of the system ϵ^{hx}, &c., and any one of the w systems.

Similarly, we may show that the mutual energy of two w systems is zero. That is,

$$0 = S_0 \underset{(x=0)}{w_m w_n} + \int_0^l S w_m w_n\,dx + S_1 \underset{(x=l)}{w_m w_n}, \tag{131}$$

if w_m and w_n are any two different regular normal functions. I have not done it, but have no reason to doubt it. Then, all the normal systems being proved to be independent, quadratures find the coefficients when we assume

$$f(x) = A_0 + \Sigma A_n w_n + B\epsilon^{hx}. \tag{132}$$

But there are different ways of doing it, according to whether $f(x)$ is the complete initial state, or whether the condensers are initially charged as well. In the latter case, if V_0 and V_1 are their initial potentials, we find the coefficient of any normal system by forming the expression for the mutual energy of the initial state and the particular normal state, using (130) or (131) to isolate the coefficient. Thus

$$A_n = \frac{S_0 V_0 w_{0n} + \int_0^l S w_n f(x)\,dx + S_1 V_1 w_{1n}}{S_0 w_{0n}^2 + \int_0^l S w_n^2\,dx + S_1 w_{1n}^2} \tag{133}$$

finds A_n, if w_{0n} and w_{1n} are the terminal values of w_n. And

$$B = \frac{S_0 V_0 + \int_0^l S f(x) \epsilon^{hx}\,dx + S_1 V_1 \epsilon^{hl}}{S_0 + \int_0^l S\epsilon^{2hx}\,dx + S_1 \epsilon^{2hl}} \tag{134}$$

finds B. The formula for A_n also answers for A_0, if it be taken to mean the coefficient of $-h$, which is what w reduces to when $s = 0$.

Now, in the case of (127), $f(x)$ alone is existent, and we must make V_0 and V_1 zero in (133), (134). Doing this, and developing the results, we shall arrive at (127), and verify it. We may also use (133) to obtain the effects due to V_0 and V_1. Initially we must have zero expansions—that is to say, expansions which represent zero, except at the terminals. To find the proper expansion to represent V in the cable due to the initial charge S_0V_0 of the real condenser at A, and the charge S_1V_1 at B, put $f(x)=0$ in (133), (134), and evaluate. The results are

$$A=\frac{2}{l}\frac{V_1\cos sl-V_0}{s^2+h^2},\quad A_0=\frac{V_0-V_1}{hl},\quad B=2\frac{V_0-V_1\epsilon^{hl}}{1-\epsilon^{2hl}},\qquad(135)$$

to be used in (132). The result is therefore

$$0=2\frac{V_0-V_1\epsilon^{hl}}{1-\epsilon^{2hl}}\epsilon^{hx}+\frac{V_0-V_1}{hl}+\frac{2}{l}\Sigma(V_1\cos sl-V_0)\frac{(s\sin-h\cos)sx}{s^2+h^2}.\qquad(136)$$

The 0 on the left side expresses $f(x)$, the potential of the cable everywhere, except at the terminals, where, on account of the condensers, special interpretation is needed.

If we assume $V_0=V_1$, the mean term disappears, and we reduce the last result to

$$0=\frac{2V_0\epsilon^{hx}}{1+\epsilon^{hl}}-\frac{2V_0}{l}\Sigma\frac{(s\sin-h\cos)sx}{s^2+h^2}(1-\cos sl),\qquad(137)$$

where only the odd values of n in $s=n\pi/l$ are effective. This corroborates (127).

If we multiply (127) by V_0, it expresses V_0 all along the cable. Adding the result to (137), we obtain $1=1$. This is obvious by arithmetic. But it means more than that here—namely, that if the initial state is V_0 constant in the cable and both condensers as well, it will remain constant, because there is nothing to disturb it.

Physical Interpretation of the Abnormal Case of § 300.

§ **302.** If we say that the equation to be obeyed at one end of the cable, say at B, is $V=R_1C$, and further say that R_1 is a positive constant, the most obvious interpretation is that R_1 is a mere resistance. The equation then asserts Ohm's law simply. But this is not necessarily the interpretation.

What the equation asserts is that the ratio of V to C is positive and constant. There may therefore be any electrical arrangement that will produce this result. A distortionless circuit will do it, and in many ways. Electric and magnetic energy are then involved, either with or without waste of energy by conductive resistance. The distortionless circuit will behave to an impressed force precisely like a resistance.

Similarly, if we say that the terminal condition is $V = -R_1C$, where R_1 is still a positive constant, we do not assert that $-R_1$ stands for a negative resistance. That would make nonsense, according to the commonly understood electrical law. Any arrangement that will, when acted upon by an impressed voltage, compel the current to obey the law $V = -R_1C$, is what is implied by that equation of condition. We are not obliged to enter any further into detail as to how it is to be done. A source of energy is involved, of course, which, however, does not require any specialisation, since the resultant effect is embodied in the law.

To save circumlocution, it is obviously convenient to speak of a negative resistance, to be understood as above. Now put a positive resistance at one end of a cable, say at A, and an equal negative resistance at B, and let the cable be initially charged at one place, say with a charge Q at y. This charge instantly begins to spread, and raises the potentials at A and B slightly. There is, therefore, a current *from* the cable at A, but *into* the cable at B. The former is normal, the latter abnormal, due to the terminal resistance being negative at B. The B end of the cable therefore becomes charged faster than the A end, and the higher its potential is raised the stronger becomes the current into the cable. At the A end, on the other hand, where things are normal, the current due to the initial charge would first rise to a maximum and then fall slowly to zero. But the continuous entry of fresh charge at B alters matters. It spreads all along the cable, from B to A, raising the potential of the whole cable. The ultimate result, therefore, when the normal effects due to the initial charge have subsided, is a state of positive charge increasing continuously with the distance from the A end, and increasing continuously with the time. This is the meaning of the extra normal function ϵ^{hx} for the potential. The current to match is directed from B to A. Both are associated with the time factor ϵ^{pt}, where $p = h^2/RS$.

The equilibrium state of zero potential is accordingly unstable. An infinitesimal charge arriving at B is sufficient to disturb the equilibrium. Which way the current will set in at B will depend on the sign of the initial point charge. If V becomes positive current will enter the cable, and continue to do so. If negative, current will leave the cable at B, and the cable will become charged negatively.

An arbitrary initial state of electrification in the cable, left to itself, will in time either wholly disappear, or will be replaced by the distribution $\epsilon^{hx}\epsilon^{pt} \times$ constant. Equation (117) determines the behaviour fully, when applied to the initial state. An initial state of the type

$$w = (h \sin + s \cos)sx$$

subsides to equilibrium, as may be readily seen by its satisfying the abnormal terminal condition at B, the current and the potential having opposite signs there. Similarly, the sum of any number of such distributions, of any sizes, will subside to equilibrium. But they cannot make up an arbitrary distribution, because the abnormal distribution is left out. An initial distribution of the type ϵ^{hx}, instead of subsiding, increases with the time. In order that no term of this sort should enter, the initial state must be specialised so as to exclude it. Thus, the first equation in (123) settles the coefficient of ϵ^{hx}. It is zero if the integral of the product of the initial state and ϵ^{hx} is zero. This is the case when the initial state is the sum of a number of systems of the regular type w. The simplest case, however, is to have two point charges, one positive, the other negative, with their sizes suited to their distances from B. Thus, if

$$Q_1\epsilon^{hy_1} + Q_2\epsilon^{hy_2} = 0,$$

Q_1 and Q_2 being the charges, at y_1 and y_2 respectively, the expression for u in (117) shows that the abnormal term ϵ^{hx} does not exist.

After so much detail concerning this simple abnormal case, others of a similar character may be very briefly treated. As for why they are considered at all, an anecdote about Dr. Elliotson comes in useful. One of his students said he did not see the use of studying morbid physiology; it was so unnatural. The doctor told him he was a blockhead, adding, "It is only by studying the morbid that the true conditions of health can be ascertained."

Positive Terminal Resistance and Negative Terminal Permittance.

§ 303. In the other way of simplifying the general determinantal equation described at the close of § 298, we reduce it to the form (108), which indicates that we have the set of normal systems corresponding to $sl = (n - \frac{1}{2})\pi$, and in addition, the set corresponding to $(Z_0 q)^2 = R^2$. So (100), the solution for a point charge, is reduced to

$$V_1 = \frac{(R\sin + Z_0 s\cos)sy \,.\, (R\cos + Z_0 s\sin)s(l-x)}{(q^2 Z_0^2 - R^2)\cos sl}\frac{sQ}{S}, \tag{138}$$

showing the potential at x on the right of y due to Q initially at y itself. Let $V = U + W$, where U is the regular part, and W the extra part. Then by the expansion theorem (30), we obtain

$$U = \frac{2Q}{Sl}\Sigma\,\epsilon^{pt}\frac{(R\sin + Z_0 s\cos)sy.(R\sin + Z_0 s\cos)sx}{R^2 + (Z_0 s)^2}, \tag{139}$$

subject to $\cos sl = 0$. And, by the same process, we find

$$W = -\frac{Q}{S}\Sigma\,\epsilon^{pt}\frac{q\epsilon^{q(x+y-l)}}{\left(1 + \frac{2p}{Z_0}\frac{dZ_0}{dp}\right)\cosh ql}, \tag{140}$$

ranging over the roots of $R^2 = (Z_0 q)^2$. The sum of U and W is the expression for the potential at time t on either side of the source.

As an example, take $Z_0 = R_0$, a mere resistance. Then there is just one extra q, of value $R/R_0 = h$, say. To balance the resistance R_0 at $x = 0$, there is a permittance of amount $-S/h$ at $x = l$.

Now, if we put $Q/S = 1$, and $t = 0$, in $U + W$, it represents u, the appropriate unit impulsive function. It need not be written out, as its form is obvious. Integrating the product of u and $f(y)$ with respect to y therefore expands $f(x)$ in the normal functions in question. In particular, if $f(x)=1$, we obtain the expansion

$$1 = -\frac{\epsilon^{hx}(1-\epsilon^{-hl})}{\cosh hl} + \frac{2}{l}\Sigma\left(\frac{h}{s}\pm 1\right)\frac{(h\sin + s\cos)sx}{h^2+s^2}, \tag{141}$$

in which the + sign is to be used in the odd terms, and the − in the evens. Introducing the time factors, as usual, shows what the initial uniform state changes to later on.

The expansion (141) may be verified by the conjugate property of the normal systems. Remember that there is one terminal condenser. So, assuming

$$1 = A_0\epsilon^{hx} + \Sigma\, A(h\sin + s\cos)sx, \tag{142}$$

we must have the identity

$$S\int_0^l \epsilon^{hx}dx + S_1\epsilon^{hl}\times 0 = A_0\left\{S\int_0^l \epsilon^{2hx}dx + S_1\epsilon^{2hl}\right\}, \tag{143}$$

which gives A_0, and also

$$S\int w dx + S_1 h \sin sl \times 0 = A\left\{S\int w^2 dx + S_1 h^2 \sin^2 sl\right\}, \tag{144}$$

which finds A. Here w stands for the general normal function in (142). The 0 which occurs stands for the initial potential of the terminal condenser. S_1 is its permittance, $= -S/h$. Carrying out (143), (144), we shall arrive at the coefficients in (141), and verify that expansion.

An interesting modification may be mentioned. Put a real condenser at A, and a negative resistance at B. Say $Z_0=(S_0 p)^{-1}$, which makes the one extra q be $h=S/S_0$. Then, instead of the above-used unit impulsive function, we shall find that

$$u = \frac{h\epsilon^{h(x+y-l)}}{\cosh hl} + \frac{2}{l}\Sigma\frac{(s\sin - h\cos)sx\,.\,(s\sin - h\cos)sy}{h^2+s^2} \tag{145}$$

is the proper one to use.

Equal Positive and Negative Terminal Inductances.

§ 304. Returning to the more interesting case in which $Z_0=-Z_1$, treated in § 299, where (115) shows the proper unit impulsive function, let Z_0 be $L_0 p$, indicating an inductance coil without resistance. The determinantal equation of the extras is a cubic, and the assumption of no resistance makes it manageable. Thus, (106) shows that

$$p^3 = \frac{\mathrm{R}}{\mathrm{SL}_0^2} = g^3, \text{ say,} \tag{146}$$

finds the three extra p's—that is, $p = g$, g_1, and g_2, where

$$g_1 = g(-\tfrac{1}{2} + i\sqrt{\tfrac{3}{4}}), \qquad g_2 = g(-\tfrac{1}{2} - i\sqrt{\tfrac{3}{4}}).$$

Also, in virtue of (146), we may write R/L_0p for q. So, by (115), the part of the unit impulsive function depending upon the extras is given by

$$\frac{\mathrm{R}}{3\mathrm{L}_0 g} \frac{\epsilon^{\mathrm{R}z/\mathrm{L}_0 g}}{\operatorname{shin} \dfrac{\mathrm{R}l}{\mathrm{L}_0 g}} + (\text{same with } g_1) + (\text{same with } g_2). \tag{147}$$

The time factors to be introduced later are ϵ^{gt}, $\epsilon^{g_1 t}$, $\epsilon^{g_2 t}$. The part of u depending on the regulars is given by the first line in (115), omitting the $1/l$ term, however, because there is terminal earth connection.

Of the three terms in (147) only the first increases with the time. The other two unite to make a real function indicating oscillatory subsidence, but it is of a complicated mixed kind, because it depends on two inductances separated by the permittance of the cable. The full development need not be given.

Impressed Force in the Case $\mathrm{Z}_0 = -\mathrm{Z}_1$.

§ 305. The same extra functions are, of course, concerned when the source is an impressed force. For example, let it be e at $x = 0$. Equation (59), § 291, shows the V that results. Put $\mathrm{Z}_0 = -\mathrm{Z}_1$ in it. Then

$$\mathrm{V} = \frac{\{\sin - (\mathrm{Z}_0 s/\mathrm{R})\cos\} s(l - x)}{\{1 + (\mathrm{Z}_0 s/\mathrm{R})^2\} \sin sl} e \tag{148}$$

gives V in terms of e.

Specialise, by making $\mathrm{Z}_0 = \mathrm{R}_0$, a resistance. Then

$$\mathrm{V} = h \frac{(h \sin - s \cos) s(l - x)}{(h^2 + s^2) \sin sl} e, \tag{149}$$

where h denotes R/R_0.

Put $s=0$ to get the part independent of the time, and then develop by the expansion theorem (30), as in previous cases. The result is

$$V = e\left(1-\frac{x}{l}-\frac{1}{hl}\right)+\frac{e\,\epsilon^{h(x-l)}}{\text{shin}\,hl}-\Sigma\frac{2he}{sl}\frac{(h\sin+s\cos)sx}{h^2+s^2}, \qquad (150)$$

without the time factors. That is, it is the initial state, representing $V=0$ everywhere between $x=0$ and l, and at the terminals as well. This is because there is terminal resistance. In the summation, we have $sl=n\pi$.

We may verify the result by the conjugate property. Assume that

$$\frac{e}{hl}-e\left(1-\frac{x}{l}\right)=A_0\epsilon^{hx}+\Sigma A(h\sin+s\cos)sx \qquad (151)$$

Multiply by ϵ^{hx} and integrate along the cable to find A_0. Multiply by $(h\sin+s\cos)sx$ and integrate to find A. The results are as in (150).

I give this case for the sake of a result which will be useful a little later. It is got by differentiating (150) to x. We obtain

$$\frac{\epsilon^{h(l-x)}}{\text{shin}\,hl}=\frac{1}{hl}+2\Sigma\frac{(hl\cos+sl\sin)sx}{(hl)^2+(sl)^2}, \qquad (152)$$

which will be used in another problem.

Singular Extreme Case of Z_0 or $-Z_1$ being a Cable equivalent to the Main One.

§ 306. It will be seen from the preceding that the subject of abnormal Fourier series is quite a large one. We find that we require to use series that look like Fourier series, inasmuch as the wave lengths are the same as in real Fourier series, but that the complete set of normal functions includes extra functions of the type ϵ^{hx}, where h is a constant, real or complex, as the case may be, the number of such functions depending upon the terminal conditions. These abnormal series are merely special cases of the general series got previously by leaving Z_0 and Z_1 independent of one another. If any one of the electrical constants involved in these operators, which constants are real and positive in real problems, be made negative, we introduce something abnormal, because there is a reversal of Ohm's law, or of the law of displacement, or of induction, in some part of the electrical system concerned, and, as a result, time functions, as ϵ^{pt}, occur in which p is positive. Some cases of this kind were considered by me in 1882, and earlier ("Electrical Papers," Vol. I.). The peculiarity of the cases above treated lies in the simplicity produced by reducing the wave lengths of the regular normal functions to be the same as in the original Fourier series. I think I have given sufficient information to enable any competent person to follow up the matter in more detail if it is thought to be desirable. It is obvious that the methods of the professedly rigorous mathematicians are sadly lacking in demonstrativeness as well as in comprehensiveness.

But it would be out of place to elaborate further in this work. I will therefore conclude the present remarks on abnormal series with the consideration of a very singular case indeed. In none of the previous cases did we cause the terminals to be nodes of the regular normal functions, or make $V = 0$ terminally in them. In the following we shall do this.

Go back to (59), § 191, and make $Z_0 = -Z_1$. This produces (148) above. So far it expresses V due to e at $x = 0$ with any terminal Z_0, provided there is its negative at $x = l$.

Next, let

$$Z_0 = \frac{R}{s} \tan sl. \tag{153}$$

This says that Z_0 is a cable identical with the one in question, or equivalent thereto, earthed at its end. We have therefore

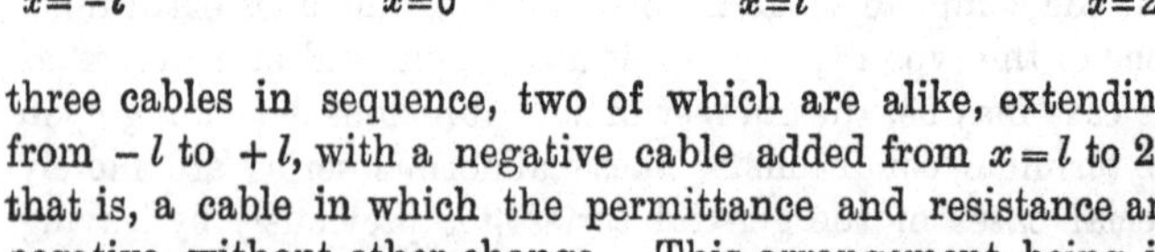

three cables in sequence, two of which are alike, extending from $-l$ to $+l$, with a negative cable added from $x = l$ to $2l$, that is, a cable in which the permittance and resistance are negative, without other change. This arrangement being in equilibrium, e is started and maintained constant. It is of little consequence to follow up the later effects. It is just at the beginning that we require the behaviour, so as to obtain the proper expansion of the initial state. The condition (153) reduces (148) to the very simple result

$$V = -\frac{\sin sx}{\tan sl}e, \tag{154}$$

which makes V permanently zero at $x=0$. This is highly anomalous. Expand (154) by the expansion theorem. The result is apparently

$$V = -\frac{ex}{l} - \Sigma \frac{2e}{sl}\sin sx\, \epsilon^{pt}, \tag{155}$$

where $\sin sl = 0$. Now this is incorrect, because the initial state is $-e$, according to (155). There is something missing.

But there is some sense in it. For it makes the final current be $C = e/Rl$. By continuity, this, if existent, should be the current through all three cables. The corresponding state of potential would be then as in the figure.

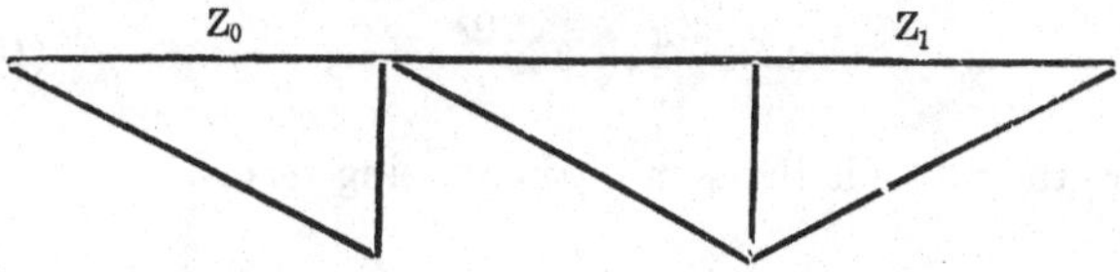

The zero potential at $x=0$ is accounted for. We may corroborate this by calculating the rise of potential in the cable Z_0 due to e, regarding the rest as a mere terminal arrangement. It will be found that the resistance operator of the two cables* from $x=0$ to $2l$, reckoned at $x=0$, is zero, which is equivalent to a short circuit, so far as the cable Z_0 is concerned. So e establishes V in the Z_0 cable just as if there were an earth on the other side (the right side) of e.

After this partial explanation return to (154) and (155), where the state of the middle cable is in question. The resistance operator to the right is zero, but then the potential at $x=0$ is zero, so there is compensation. According to (155) a complete normal system of potential would consist of $\sin sx$

* To find this operator we may use (59), § 291. It is the ratio V/C at $x=0$, when Z_0 is zero, and Z_1 is $-(R/s)\tan sl$.

from $x=-l$ up to $x=l$, combined with $-\sin sx$ from $x=l$ up to $2l$. But it is easily to be seen that we cannot expand an arbitrary function in terms of these systems alone. For instance, let $V=f(x)$ (any function) from 0 to l, zero from 0 to $-l$, and $V=f(2l-x)$ from l to $2l$. The energy is zero, so the result of expansion by quadratures is zero.

It is certain that (155) is incomplete. And, in fact, we may ask, where are the abnormal functions necessitated by the negativeness of resistance and permittance in Z_1? The answer is to be found by seeing how $\tan sl$ in (154) arose. It represents $\sec sl \times \sin sl$, and $\sec sl=0$ is the equation belonging to the extra functions. But the roots of $\sec sl$, infinite in number, are at infinity. The normal functions to correspond are therefore unmanageable from being ungetatable. Their influence, however, is in full operation. For the failure of (155) proves that the sum of the whole of the missing functions required to make it represent the initial state truly is e. That is, we should have, initially,

$$0=e-\frac{x}{l}e-e\Sigma\frac{2c}{sl}\sin sx, \tag{156}$$

where the first e is the sum of the missing terms.

More General Case to Elucidate the Last. Terminal Cable Z_0 not equivalent to Main One.

§ 307. But to make this conclusion plainer, we should bring the roots at infinity to a measurable distance (like the abolition of that fraud 4π which has been purblindly inserted in the electrical equations), so that we can manipulate them. Take, then,

$$Z_0=\frac{R_0}{s_0}\tan s_0l_0, \tag{157}$$

instead of (153). The cable Z_0 is no longer an exact or equivalent copy of the middle one. Then (148) makes

$$V=\frac{(\sin - c^{-1}\tan s_0l_0\cos)s(l-x)}{(1+c^{-2}\tan^2 s_0l_0)\sin sl}\,e, \tag{158}$$

instead of (154). Here $c=Rs_0/R_0s$ or $(RS_0/R_0S)^{\frac{1}{2}}$. The abnormal roots are now finite.

Let $V = U + W$, U being the part of V depending upon $\sin sl = 0$, and W the extra part. Develop U in the usual way by the expansion theorem. We get (with $t=0$)

$$U = e\left(1 - \frac{x}{l} - \frac{1}{cf}\right) - e\Sigma \frac{\sin sx + c^{-1}\tan f^{-1} sl \cos sx}{\frac{1}{2} sl(1 + c^{-2}\tan^2 f^{-1} sl)}, \tag{159}$$

where f is the constant $(RSl^2/R_0S_0l_0^2)^{\frac{1}{2}}$. This U evidently reduces to the right member of (154) when $c=1, f=1$.

The development of W requires more care. The extra roots are given by

$$\tan^2 s_0 l_0 = -c^2; \qquad \text{or} \qquad \text{than}^2 q_0 l_0 = c^2. \tag{160}$$

Whether we take c or $-c$ to be than $q_0 l_0$ is indifferent, provided we work consistently. Say, then, that

$$\text{than } q_0 l_0 = c, \tag{161}$$

or, which is the same,

$$\epsilon^{2q_0 l_0} = \frac{1+c}{1-c} = g^2, \text{ say,} \tag{162}$$

which makes

$$q_0 l_0 = \lambda + in\pi, \quad \text{if} \quad \lambda = \log g, \tag{163}$$

provided c is less than 1. But if c is >1, then

$$q_0 l_0 = \lambda + i(n - \tfrac{1}{2})\pi \tag{164}$$

instead, where λ is the same real constant. The form assumed by W when expanded is, therefore, quite different in the two cases. At the point $c=1$ there is a jump, and special care is required.

Let us use (161), with $c<1$. Then

$$p\frac{d}{dp}\tan^2 s_0 l_0 = s_0 l_0 \tan s_0 l_0 \,.\, (1 + \tan^2 s_0 l_0)$$
$$= -q_0 l_0 c(1 - c^2). \tag{165}$$

The application of the expansion theorem to (158), so far as the extras are concerned, therefore makes

$$W = \Sigma \frac{2ce\,\epsilon^{qx}}{q_0 l_0(1-c^2)(\epsilon^{2ql} - 1)}. \tag{166}$$

In this, the relation between q and q_0 is $ql = fq_0 l_0$, and in (163) n has to receive all integral values, positive and negative, including zero. The nth term of (166) pairs with the $-n^{\text{th}}$ to make a real function. But, owing to the exponential function in the denominator of (166), the resultant takes a rather complicated form, which need not be given. It is sufficient for our purpose to see that if f is an integer, the denominator simplifies. We then get

$$W=\frac{2ce}{1-c^2}\frac{1}{g^{2f}-1}\Sigma\frac{\epsilon^{qx}}{q_0 l_0}; \tag{167}$$

that is, by (163),

$$W=\frac{2ce}{1-c^2}\frac{\epsilon^{\lambda fx/l}}{g^{2f}-1}\left\{\frac{1}{\lambda}+2\Sigma\frac{(\lambda\cos+n\pi\sin)n\pi fx/l}{\lambda^2+(n\pi)^2}\right\}, \tag{168}$$

where the summation is with respect to n, which has all positive integral values from 1 to ∞.

Comparing the last equation with (152), we see that hl becomes λ and x becomes fx. The quantity in the big brackets in (168) is therefore

$$\frac{\epsilon^{\lambda(1-xf/l)}}{\text{shin}\,\lambda},\quad\text{or}\quad\frac{1+c}{c}\epsilon^{-\lambda xf/l}, \tag{169}$$

by (163). This reduces W to

$$W=\frac{2e}{1-c}\frac{1}{g^{2f}-1}, \tag{170}$$

provided f is an integer.

Let it be 1, *i.e.*, $RSl^2=R_0S_0l_0^2$, then

$$W=\frac{e}{c}. \tag{171}$$

Finally, $c=1$ makes $W=e$, as was to be proved.

We have proved that in the singular case of identity or equivalence between the terminal cable Z_0 and the middle one, when the extra normal functions apparently go out of existence, by the roots of the determinantal equation becoming infinite, they are nevertheless virtually existent, and must be allowed for in the expansion of the initial state.

In the more general case, in which there is not the equivalence mentioned, no such straining of the expansion theorem is needed. For instance, W in (168) may be tested in other cases in which f is integral, and be found to represent the negative of the U in (159). There is, therefore, no reason to doubt the equivalence of $-U$ with the more general form W in (166).

Arbitrary Initial State when Z_0 or $-Z_1$ is a Cable. Singular Case.

§ 308. If there is no impressed force, but the middle cable is initially charged arbitrarily instead, there is a similar curious limiting case when the terminal cable Z_0 and the middle cable are equivalent. The general solution (100), through (110), reduces to

$$V_1 = \sin s(y+l)\frac{\sin sx}{\sin sl}\frac{sQ}{S}, \tag{172}$$

indicating the potential at x, on the right of y, due to Q initially at y. When on the left side, interchange x and y to get the formula. The Z_0 operator is $(R/s)\tan sl$, as in § 306.

Developing by the expansion theorem, according to the roots of $\sin sl = 0$, and denoting the result by U, we obtain

$$U = \frac{2Q}{Sl}\Sigma \sin sy \sin sx\,\epsilon^{pt}. \tag{173}$$

Now this is exactly the same as if the cable were earthed at $x=0$ and l. See (22), § 267, for example.

But the latent auxiliary normal functions, which we know to exist in general, have not been considered. The conclusion from the above result is, therefore, that in the complete expansion of the initial state, the sum of all the extra terms is zero. Only when t is finite can they be quantitatively existent in the total. They may be infinite then, should the p in ϵ^{pt} be infinite for the extra terms; but that is of no consequence as regards the initial state.

To verify the above, we may proceed as in § 307. Let Z_0 be not equivalent to the middle cable in the first place. Then Z_0 is as in (157), and Z_1 is its negative. The general equation (110) is now

$$V_1 = -\frac{(c\sin + \tan s_0l_0\cos)sy\,(c\sin - \tan s_0l_0\cos)s(l-x)}{(c^2+\tan^2 s_0l_0)\sin sl}\frac{sQ}{S} \tag{174}$$

instead of (172) above.

Developing this by the expansion theorem, let U be the regular part, and W the extra part. Then we get

$$U = \frac{2Q}{Sl}\Sigma\frac{(c\sin+\tan s_0l_0\cos)sy\,.\,(c\sin+\tan s_0l_0\cos)sx}{c^2+\tan^2 s_0l_0}\epsilon^{pt}, \tag{175}$$

in which $\sin sl = 0$, or $sl = \pi$, 2π, &c. And, in addition,

$$W = \frac{Q}{S} \Sigma \frac{qc\,\epsilon^{q(x+y-l)}\epsilon^{pt}}{(1-c^2)\,q_0 l_0 \,\text{shin}\, ql}, \tag{176}$$

the notation being as in § 307, equations (161) or (163) finding the values of q_0. The unit impulsive function is the sum of U and W, with $t=0$, and $Q/S=1$. The U part subsides to equilibrium, whilst the W part increases with the time.

Simplify by taking $f=1$, making $RSl^2 = R_0S_0l_0^2$, and $ql = q_0l_0$. The unit impulsive function becomes

$$u = \frac{2}{l}\Sigma \sin sy \sin sx + \Sigma \frac{c}{l(1-c^2)} \frac{\epsilon^{q(x+y)}}{\epsilon^{q_0l_0}\,\text{shin}\, q_0l_0}. \tag{177}$$

Here the first part expresses the unit function concentrated at the point $x=y$, with repetitions outside the limits concerned, which do not count. We conclude that the rest represents zero between the same limits. The verification is easy, because the simplification made by assuming $f=1$ allows of the reduction of the extra part to a recognisable form. The extra part of u, say u_2, is

$$u_2 = \frac{1}{l(1+c)} \Sigma \exp\left\{(\lambda + in\pi)\frac{x+y}{l}\right\}, \tag{178}$$

where n has all integral values, positive, negative, and zero. Or, which is the same,

$$u_2 = \frac{\epsilon^{\lambda(x+y)/l}}{l(1+c)}\left\{1 + 2\Sigma_1^\infty \cos\frac{n\pi}{l}(x+y)\right\}. \tag{179}$$

This, by § 271, represents a row of impulses (not of unit size) outside the limits in question. Within the limits, therefore, u_2 is zero, as required.

Let u become v when t is finite. Then

$$v_1 = \frac{2}{l}\Sigma_1^\infty \sin\frac{n\pi y}{l} \sin\frac{n\pi x}{l}\epsilon^{-Ht}, \tag{180}$$

where H is $n^2\pi^2/RSl^2$, and the summation is with respect to n. And for the extras, we have

$$\epsilon^{q(x+y)}\epsilon^{pt} = \exp\left\{(\lambda + in\pi)\frac{x+y}{l} + \frac{t}{RSl^2}(\lambda^2 - n^2\pi^2 + 2in\pi\lambda)\right\}, \tag{181}$$

which makes

$$v_2 = \frac{\epsilon^{\mathrm{F}}}{l(1+c)}\left\{1 + 2\sum_1^\infty \cos n\pi\mathrm{G}\,.\,\epsilon^{-\mathrm{H}t}\right\}, \qquad (182)$$

where H is as before, and, for brevity, additional letters F, G are introduced, given by

$$\mathrm{F} = \frac{\lambda(x+y)}{l} + \frac{\lambda^2 t}{\mathrm{RS}l^2}, \qquad \mathrm{G} = \frac{x+y}{l} + \frac{2\lambda t}{\mathrm{RS}l^2}. \qquad (183)$$

Whilst v_1 subsides to zero, v_2 augments. In the singular case of failure, real or apparent, which occurs when $c=1$, the extra part v_2 jumps from zero to infiniteness instantly.

Real Terminal Conditions. Terminal Arbitraries. Case of a Coil. Two Ways of Treatment.

§ 309. Returning now to real electrical conditions, one of the minor matters that remains to be considered is the influence of the terminal arrangements on the state of the cable, when they are initially energised. The existence of auxiliary functions, one for every independent kind of energisation, has been pointed out. Also how, by means of the conjugate property of complete normal systems, their size may be determined. But the example given in § 295, relating to a terminal condenser, and the later examples, were not sufficiently general to illustrate the matter fully. So now take some other arrangements.

Say the $x=l$ end of the cable is earthed. Then $\mathrm{Z}_1=0$, and, by (85), § 294,

$$v_1 = -\sin s(l-x)\frac{(\mathrm{R}\sin + \mathrm{Z}_0 s\cos)sy}{(\mathrm{R}\sin + \mathrm{Z}_0 s\cos)sl}\frac{s\mathrm{Q}}{\mathrm{S}} \qquad (184)$$

is the potential at x ($>y$) due to Q at y. Interchanging x and y makes the potential at x ($<y$) due to Q at y. Differentiating the new expression with respect to x and dividing by $-\mathrm{R}$ produces the expression for the current on the left side of y. Put $x=0$ in it. The result is

$$\mathrm{C}_0 = \frac{\sin s(l-y)}{(\mathrm{R}\sin + \mathrm{Z}_0 s\cos)sl}\frac{s^2\mathrm{Q}}{\mathrm{S}}. \qquad (185)$$

This is, therefore, the total current in Z_0 due to Q, reckoned from left to right.

We conclude, by reciprocity, that a current momentarily established in Z_0 will produce potential at y, determined by the same operator as is concerned in (185), when there is magnetic inertia in Z_0. This is easily corroborated. Put an impressed voltage e at $x=0$. The C due to it, on the spot, that is, in Z_0, is

$$C=\frac{e}{Z_0+(R/s)\tan sl}=\frac{e}{Z} \text{ say,} \tag{186}$$

because the denominator is the sum of the resistance operators to right and left of e. This expands, when e is steady, to

$$C=\Sigma\frac{e\epsilon^{pt}}{p\dfrac{dZ}{dp}}, \tag{187}$$

(and a steady term), ranging over the roots of $Z=0$.

But let e be impulsive, say $=pL_0C_0$, so that L_0C_0 is the momentum generated. Then

$$C=\frac{p}{Z}L_0C_0=L_0C_0\Sigma\frac{\epsilon^{pt}}{dZ/dp}. \tag{188}$$

Putting $t=0$ makes the initial state. So

$$C_0=\Sigma\frac{L_0C_0}{dZ/dp} \tag{189}$$

must be the expansion of C_0, regarded as initially given to be the current in Z_0.

Also, the potential at x due to e is got by putting L_0pC_0 for e in (59), and $Z_1=0$. This makes

$$V=\frac{p\sin s(l-x)}{(Zs/R)\cos sl}L_0C_0, \tag{190}$$

operationally, or

$$V=\Sigma\frac{L_0C_0R\sin s(l-x)}{s\cos sl\,(dZ/dp)}\epsilon^{pt}, \tag{191}$$

by the expansion theorem. From this, again, (189) may be derived. It will be convenient to consider Z_0 to be R_0+L_0p in the above, meaning a coil of resistance R_0 and inductance L_0. At the initial moment, therefore, (191) is the zero expansion required to suit the case of an energised coil, an expansion in a Fourier series representing zero within the limits, such that V shows what it becomes at time t later, as the coil discharges itself into the cable.

Next consider the same matter from another point of view, viz., that of independent normal systems. Put

$$v=\sin s(l-x) \qquad c=\frac{s}{R}\cos s(l-x). \tag{192}$$

Since $v=0$ at $x=l$, these are the proper normal functions for V and C in the cable, provided s is suitably determined to satisfy $v/c=-Z_0$ at $x=0$. That is,

$$Z_0+\frac{R}{s}\tan sl=0=Z \tag{193}$$

is the determinantal equation. Provided, then, we include every possible normal system subject to the last equations, we see that the initial state is fully specifiable by

$$V_0=\Sigma Av=\Sigma A\sin s(l-x), \tag{194}$$

$$C_0=\Sigma Ac=\Sigma A\frac{s}{R}\cos sl, \tag{195}$$

where A is a constant. Given V_0 and C_0, the conjugate property finds A. That property asserts that

$$S\int_0^l v_m v_n dx - L_0 c_m c_n=0, \tag{196}$$

where v_m, v_n are any two normal systems of potential, and c_m, c_n the currents to match at $x=0$, because there is magnetic energy in the coil, and electric in the cable. It follows that

$$A=\frac{S\int V_0 v dx - L_0 C_0 c}{S\int v^2 dx - L_0 c^2} \tag{197}$$

finds the coefficient A belonging to any system v, c when V_0 is given as a function of x, and C_0 has any value we like.

Attending only to the part dependent upon C_0, the result is

$$A=\frac{-L_0 C_0\frac{s}{R}\cos sl}{S\left(\frac{l}{2}-\frac{\sin 2sl}{4s}\right)-L_0\left(\frac{s}{R}\cos sl\right)^2}, \tag{198}$$

which, used in (194), completes the solution so far as C_0 is concerned, since the state at time t is deducible by the introduction of the time factor. It will be found that the solution thus got agrees with the former one, viz. (191), on expanding dZ/dp. Also, that (195) agrees with (189). As regards the conjugate property (196) which has been used, its proof is to be found in testing that it is true, when the proper expressions for v and c are used. This remark applies in all expansions in normal functions. I shall, however, give later a general proof of the property.

In the numerical execution of the above case, it may be as well to remark that there is sometimes a pair of imaginaries concerned in the determinantal equation, which might be overlooked, if we merely drew the curves

$$y_1 = \tan sl, \qquad y_2 = (-s/R)\,Z_0, \tag{199}$$

and determined the roots by their intersections.

Terminal Coil and Condenser in Sequence.

§ 310. Now take a case in which both kinds of terminal energy are concerned. Say

$$Z_0 = R_0 + L_0 p + \frac{1}{S_0 p}, \qquad Z_1 = 0, \tag{200}$$

so that there are a coil and a condenser in sequence at $x=0$. The determinantal equation is

$$Z = 0 = \frac{R}{s}\tan sl + R_0 + L_0 p + \frac{1}{S_0 p}, \tag{201}$$

where $p = -s^2/RS$, as usual. Taking the normal functions as before, equation (192), the current at $x=0$ is the current in the coil. Therefore (195) is the expression for one of the terminal arbitraries—viz., the initial current in the coil. The other arbitrary is the charge (or the potential) of the condenser. Say that the condenser is next the earth, and that its potential is v_1 in a normal system and V_1 altogether. We have

$$v_1 - v = (R_0 + L_0 p)\,c, \tag{202}$$

where v and c belong to the terminal of the cable, and also

$$v_1 = -\frac{c}{S_0 p}, \tag{203}$$

these being the equations of voltage for the coil and condenser respectively. So

$$v_1 = \sin sl + (R_0 + L_0 p)\frac{s}{R}\cos sl \tag{204}$$

is the normal function for v_1. This may, by (201), or directly by (203), be simplified to

$$v_1 = \frac{S}{S_0 s} \cos sl. \tag{205}$$

The initial state is therefore fully expressible by

$$V_0 = \Sigma\, Av, \qquad C_0 = \Sigma\, Ac, \qquad V_1 = \Sigma\, Av_1, \tag{206}$$

where A is a constant, the same constant for every three connected functions, but differing in the different complete systems. Given, then, V_0 as a function of x, and C_0 and V_1 arbitrarily, we evaluate any coefficient A by

$$A = \frac{S\int V_0 v\,dx + S_0 V_1 v_1 - L_0 C_0 c}{S\int v^2 dx + S_0 v_1^2 - L_0 c^2}, \tag{207}$$

because the conjugate property is

$$0 = S\int v_m v_n dx + S_0 v_{1m} v_{1n} - L_0 c_m c_n, \tag{208}$$

the terminal energies being of different kinds. When $L_0 = 0$, the value of C_0 ceases to have any influence on the value of A, as we see by (207), although the expression

$$C_0 = \Sigma\, Ac\, \epsilon^{pt} \tag{209}$$

gives the coil current for any finite value of the time.

Coil and Condenser in Parallel.

§ 311. If the condenser and the coil are in parallel, the case is somewhat different. We now have

$$Z = 0 = \frac{R}{s} \tan sl + \left(S_0 p + \frac{1}{R_0 + L_0 p}\right)^{-1}, \tag{210}$$

because the resistance operator Z_0 is the reciprocal of the sum of the reciprocals of the resistance operators of the coil and condenser; or the conductance operator is the sum of the separate conductance operators.

The potential v at the terminal of the cable in a normal system is the potential v_1 of the condenser, so

$$V_1 = \Sigma\, Av = \Sigma\, A \sin sl \tag{211}$$

expresses one of the terminal arbitraries.

But only a part of the cable current enters the coil, so it is not the full current c that is concerned in the other arbitrary. We have

$$c_0 = -\frac{v}{R_0 + L_0 p}, \tag{212}$$

if c_0 is the proper function to associate with v. Therefore

$$C_0 = -\Sigma \frac{A \sin sl}{R_0 + L_0 p} \tag{213}$$

is the expression for the current in the coil initially.

But to determine A to suit the initial circumstances, we may still use the formula (207), taking care, however, to use the proper expressions of the present case for v, v_1 and c. The last has now become the c_0 in (212), whilst v_1 is to be the v in (211).

Two Coils in Sequence or in Parallel.

§ 312. If the terminal arrangement consists of two coils in sequence, there is but one auxiliary function, because the current is constrained to be the same in both coils. Suppose e is an impressed voltage, and there is mutual inductance m between the coils r_1, l_1 and r_2, l_2. The equation of voltage is

$$e = (r_1 + l_1 p)c_1 + mpc_2 + (r_2 + l_2 p)c_2 + mpc_1, \tag{214}$$

where c_1 and c_2 are the currents. Both being equal to c, say, this reduces to

$$e = (r_1 + r_2)c + (l_1 + l_2 + 2m)pc, \tag{215}$$

the same as for a single coil. So there is nothing new here.

But if the coils are in parallel, there are two arbitraries. For we shall now have

$$e = (r_1 + l_1 p)c_1 + mpc_2, \tag{216}$$

$$e = (r_2 + l_2 p)c_2 + mpc_1; \tag{217}$$

from which the resistance operator of the combination and the separate ones have to be deduced. Solve for the currents thus:—

$$\frac{c_1}{e} = \frac{(r_2 + l_2 p) - mp}{(r_1 + l_1 p)(r_2 + l_2 p) - m^2 p^2}, \quad \frac{c_2}{e} = \frac{(r_1 + l_1 p) - mp}{(r_1 + l_1 p)(r_2 + l_2 p) - m^2 p^2}. \tag{218}$$

These are the conductance operators separately. Their sum is the conductance operator of the combination. Its reciprocal is the resistance operator. Therefore

$$Z_0 = \frac{(r_1 + l_1 p)(r_2 + l_2 p) - m^2 p^2}{r_1 + r_2 + (l_1 + l_2 - 2m)p} \tag{219}$$

is the required Z_0, to be used in the determinantal equation. Also, the previous equations for c_1 and c_2 give the proper normal functions for them in terms of e, which becomes $-v$, the normal function for the potential at the cable end. Thus determined, the two arbitraries are ΣAc_1 and ΣAc_2, to be associated with ΣAv, expressing the potential in the cable.

In determining A by the conjugate property, m must not be forgotten. Twice the energy of c_1 and c_2 is

$$l_1c_1^2 + l_2c_2^2 + 2mc_1c_2 = \mathrm{F}, \text{ say,} \tag{220}$$

or

$$(lc_1 + mc_2)c_1 + (lc_2 + mc_1)c_2 ; \tag{221}$$

and the mutual energy of C_1, C_2 and c_1, c_2 is

$$(lc_1 + mc_2)\,\mathrm{C}_1 + (lc_2 + mc_1)\,\mathrm{C}_2 = \mathrm{G}, \text{ say.} \tag{222}$$

Consequently, given V_0 along the cable, and also the values of C_0 and C_1 initially, the value of A is

$$\mathrm{A} = \frac{\mathrm{S}\int \mathrm{V}_0 v\,dx - \mathrm{G}}{\mathrm{S}\int v^2 dx - \mathrm{F}}, \tag{223}$$

which renders the solution of the problem complete, when the normal potential function v for the cable is properly determined to suit Z_0, as in (219). The principle underlying these determinations is quite a simple one, and may be applied to the most complicated cases.

A Closed Cable with a Leak. Split into Two Simpler Cases.

§ 313. The last examples sufficiently indicating the connection between the main solution for the cable itself and the terminal arbitraries, we may pass on to another minor matter. If the cable ends A and B are joined, either directly or through apparatus, with or without an earth connection at the same place, the two conditions expressing V/C at the terminals independently of one another are replaced by two conditions of a different kind. Some difference in the treatment is, therefore, needed. It is not great, and may be readily inferred from the preceding.

Take an explicit example. Let the length of the cable be $2l$. At B, where $x = l$, there is simple continuity, without imposed condition. At A, where $x=0$ or $2l$, there is a shunt or leak to earth. It is any sort of leak, defined by a resistance operator Z_0. We have to see how this arrangement will

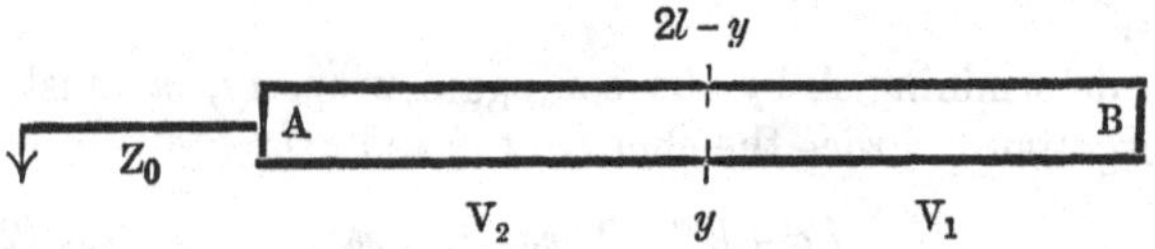

behave. We may consider three sorts of problems: the diffusion of a point charge initially at y, which will give, by integration, the effect of any initial state of the cable; the effect of an impressed force anywhere; and the effect of terminal energisation, when Z_0 is of a suitable kind. The last, however, may be inferred from the solution for an initial state in the cable, so we need not consider it specially.

As regards Q at y. By inspection of the above diagram, and a little thought, we may split the problem into two, of which the solutions are already known. Split Q into halves. Pair one of them with an equal charge at the corresponding point $2l-y$. Pair the other with an equal charge of the opposite sign at the corresponding point. The sum of the four charges is simply Q at y, so the solution is the same for the four as for the original Q.

Consider the positive pair alone, that is, $\frac{1}{2}$Q at y and $\frac{1}{2}$Q at $2l-y$. By symmetry they cause no current at B, and behave similarly in the upper and lower cables. At A the currents going to Z_0 are equal. They therefore unite and leak out through Z_0. Thus, each member of the positive pair discharges in its own cable in the same way as if it were insulated at B, and were earthed at A through the arrangement whose resistance operator is $2Z_0$. This case has been already considered. $2Z_0$ may be constructed by putting Z_0 in sequence with another Z_0, or in other ways.

The positive and negative pair, on the other hand, produce no potential at either A or B. There is no leakage current in Z_0, and consequently continuity of current between the upper and lower cables. So the $\frac{1}{2}$Q in the lower cable discharges as if it were earthed at A and B, and so does the $-\frac{1}{2}$Q in the upper cable. This part of the complete solution is also known.

Putting these together, we obtain the complete result, and can use previous formulæ without more work.

In the case of the positive pair we have this arrangement

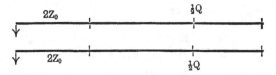

diagrammatically. Use (85), § 294. Put $Z_1 = \infty$, and $2Z_0$ for Z_0, and $\frac{1}{2}Q$ for Q. Then

$$v_1 = \frac{(2Z_0 s\cos + R\sin)sy}{(2Z_0 s\sin - R\cos)sl}\cos s(l-x)\,.\,\frac{sQ}{2S} \tag{1}$$

expresses the potential at x, on the right of y, in the lower cable. Interchange x and y when $x<y$.

In the case of the positive and negative pair we have this arrangement,

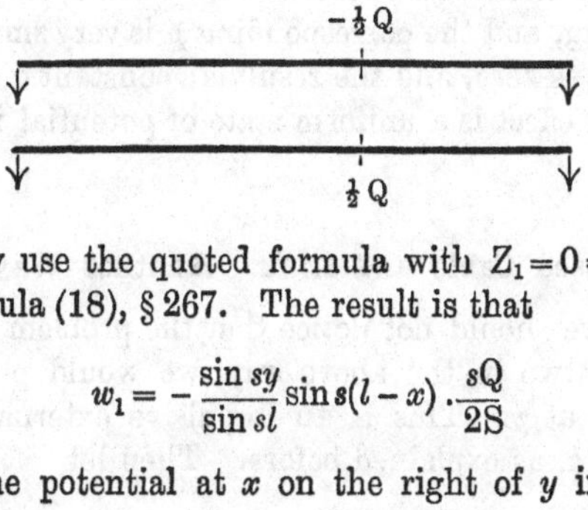

and we may use the quoted formula with $Z_1 = 0 = Z_0$, or the special formula (18), § 267. The result is that

$$w_1 = -\frac{\sin sy}{\sin sl}\sin s(l-x)\,.\,\frac{sQ}{2S} \tag{2}$$

expresses the potential at x on the right of y in the lower cable.

In the upper cables the potentials are the same at corresponding points in the first case, and their negatives in the second. Uniting the two formulæ, we see that

$$V_1 = v_1 + w_1 \tag{3}$$

is the potential at x in the lower cable, when greater than y, and the same with x and y interchanged when x is less than y. The same formula is valid all the way from y to $x = 2l$, as we may see by altering x to $2l - x$, and observing that v_1 is unchanged, whilst v_2 is negatived.

The conversion to Fourier series is to be done as usual by the expansion theorem. Doing it in the most simple case of $Z_0 = R_0$, a constant, we obtain

$$V = \frac{Q}{Sl} \Sigma \sin sy \sin sx \epsilon^{pt}$$

$$+ \frac{Q}{Sl} \Sigma \frac{(2 R_0 s \cos + R \sin) sy . \cos s(l - x)}{2 R_0 s \cos sl + (R + 2R_0/l) \sin sl} \epsilon^{pt}, \tag{4}$$

which is now valid on both sides of y. In the first summation $\sin sl = 0$. In the second, we have

$$\tan sl = R/2R_0 s \tag{5}$$

for determinantal equation. The result for the upper line need not be written separately, being the same formula.

When we make $R_0 = \infty$, we have the two cables connected together without external constraint. We then reduce the above to Fourier's periodic formula. Notice the way the constant term arises. The last equation has a very small root when R_0 is big, and the corresponding p is very small. In the limit the root is zero, and the result is a constant term showing that the final effect is a uniform state of potential in the complete cable.

Closed Cable and Leak. Another Way.

§ 314. If we should not notice that the problem admitted of splitting into two in the above way, we would proceed thus. Start with Q at y. This is an impulsive external source of current $pQ = h$, as explained before. Then let

$$V_1 = \epsilon^{qx} F + \epsilon^{-qx} G, \qquad V_2 = \epsilon^{qx} H + \epsilon^{-qx} I, \tag{6}$$

where V_1 is the potential at x when on the right side of Q, and V_2 when on the left side. There are four time functions to be determined, and four conditions which find them.

(1). Continuity of potential at A makes

$$H + I = \epsilon^{2ql} F + \epsilon^{-2ql} G. \tag{7}$$

(2). Continuity of potential at y makes

$$\epsilon^{qy} F + \epsilon^{-qy} G = \epsilon^{qy} H + \epsilon^{-qy} I. \tag{8}$$

(3). Discontinuity in the current at A makes

$$V = Z_0(C_1 - C_2), \tag{9}$$

at $x=0$. Or

$$\frac{H+I}{Z_0} = -\frac{q}{R}\left\{\epsilon^{2ql}F - \epsilon^{-2ql}G - H + I\right\}. \tag{10}$$

(4). Discontinuity in the current at y (regarding h as a function of the time) makes

$$h = C_1 - C_2 = pQ, \tag{11}$$

at $x=y$. Or

$$pQ = -\frac{q}{R}\left\{\epsilon^{qy}F - \epsilon^{-qy}G - \epsilon^{qy}H + \epsilon^{-qy}I\right\}. \tag{12}$$

These four conditions determine the time functions and make, finally,

$$V_2 = \frac{\frac{R}{2Z_0}\sin s(2l-y)\sin sx + s\sin sl\cos s(l+x-y)}{\sin sl\left(\sin sl - \frac{R}{2Z_0 s}\cos sl\right)}\frac{Q}{2S}, \tag{13}$$

$$V_1 = V_2 + \sin s(x-y)\,.\,sQ/S. \tag{14}$$

That the determinantal equation splits into two distinct equations is shown by the denominator in (13), and it may be readily shown that the operator in (13) may be represented as the sum of two operators, one having the denominator $\sin sl$, the other the bracketed part of the denominator in (13). One of these partial operators produces v_1 and v_2, the other w_1 and w_2. The expanded solution in Fourier series is the same as before got, of course.

The difference between V_1 and V_2 is operationally exhibited in equation (14). It is equivalent to exchanging y and x in the preceding formula, and taking the difference. But this difference has no existence in the Fourier series. We can see that by putting the extra part of (14) in the numerator, when it becomes multiplied by the denominator, and therefore gives zero terms when the expansion theorem is applied. We also see the same by observing that only even powers of p are involved in the extra operator in (14). But this extra operator is not of no moment in general. If, instead of the impulsive source of current pQ, we have h, any function of the time, the solutions to left and right of the source are got by writing h/p for Q in (13), (14), and the difference between the forms of V_1 and V_2 cannot be neglected in general; for example, when h is simply periodic, and the algebrisation is effected by $p^2 = -n^2$.

When we have an impressed force e at y instead of h or pQ, we can treat the matter similarly by either of the above ways, noting now that C is continuous and V discontinuous at y. Since $\frac{1}{2}e$ at y, together with $\pm\frac{1}{2}e$ at the corresponding point in the upper cable, produce zero current at B, or else zero potential, with similar effects at A, we see that the problem splits into two connected simpler cases.

Closed Cable with Intermediate Insertion. Split into Two Simpler Cases.

§ **315.** In the above, Z_0 produced a discontinuity in the current, though none in the potential. But if we introduce Z_0 in circuit with the double cable, and have no earth connection, it is the current that is continuous, and the potential discontinuous at A. The question arises whether the problem

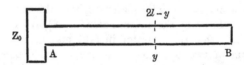

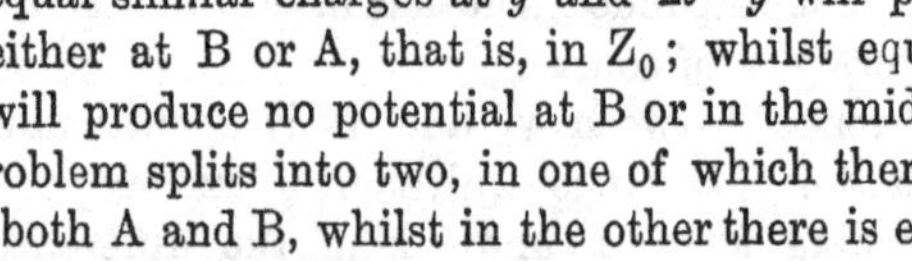

admits of splitting. It clearly does when Z_0 is a resistance, because equal similar charges at y and $2l-y$ will produce no current either at B or A, that is, in Z_0; whilst equal unlike charges will produce no potential at B or in the middle of Z_0. So the problem splits into two, in one of which there is insulation at both A and B, whilst in the other there is earth at B, and earth through resistance $\frac{1}{2}R_0$ at A.

That a similar split takes place when Z_0 is a condenser is also evident on consideration. The same may be said of any simple electrical arrangement which is divisible into similar halves. But it is not immediately evident that a split is possible when Z_0 is any electrical arrangement, which may have no sort of symmetry with respect to the cable terminals. It is sufficient to think of a coil and a condenser in sequence. Given a charge at y, and an equal charge at the corresponding point. If it be said that if the charges are unlike there is zero potential in the middle of Z_0, it may be asked, Where is the middle? The condenser and coil are not comparable in the way suggested. At the same time, the property must be true in a certain sense, because the combination of condenser and coil behaves to external voltage as a single whole. The same is true of any regular combination. It may be symbolised by an operator, without entering into its detailed structure in the form of coils and condensers.

The proper interpretation is to substitute for the combination symbolised by Z_0 two combinations in sequence, each symbolised by $\frac{1}{2}Z_0$. This is easily done. The condenser and coil may be replaced by two condensers and two coils, all in sequence. Then one of the new condensers and one of the new coils in sequence makes $\frac{1}{2}Z_0$, provided the resistance, inductance, and elastance of the original coil and condenser are all halved. Similarly as regards the remaining new condenser and coil. Z_0 has now a middle in reality, as well as in imagination. A similar process is applicable to any combination. So, without actually doing it in detail, we may say that equal unlike charges at y and the corresponding point redistribute themselves in the same way as if earth were on at B, and earth on in the middle of Z_0; or, without speaking of the middle, as if we substitute for Z_0 two arrangements each equivalent to $\frac{1}{2}Z_0$, and then put on earth between them.

So let there be a charge $\frac{1}{2}Q$ at y and $\frac{1}{2}Q$ at $2l-y$. They move as if the ends A, B were insulated. If v_1 is the resulting potential at x on the right of y, we have

$$v_1 = \frac{\cos sy}{\sin sl} \cos s(l-x) \cdot \frac{sQ}{2S}, \tag{15}$$

either by (85), § 294, with $Z_0 = \infty = Z_1$ and $\frac{1}{2}Q$ instead of Q; or by the elementary theory in § 269.

Also, if w_1 is the potential at x on the right of y due to $\frac{1}{2}Q$ at y and $-\frac{1}{2}Q$ at $2l-y$, we have

$$w_1 = -\frac{(2R\sin + Z_0 s\cos)sy}{(2R\sin + Z_0 s\cos)sl} \sin s(l-x) \cdot \frac{sQ}{2S} \tag{16}$$

by the same formula (85), with $Z_1 = 0$, $\frac{1}{2}Z_0$ instead of Z_0, and $\frac{1}{2}Q$ instead of Q.

It follows that if there is initially only Q at y, the resulting potential V_1 at x on the right of y is

$$V_1 = v_1 + w_1. \tag{17}$$

The expansions of v_1 and w_1 are to be effected as usual. I have given so many examples that it is unnecessary to exhibit the special expansions required here. The expansion theorem (30) is very easy to remember and to apply when required. On the other hand, the operational formulæ are much easier to manipulate. Note in the present case that there is a constant term in V_1. It arises from v_1, and its value is $Q/2Sl$, the mean potential which is finally existent because there is no leak anywhere.

Same as last without Initial Splitting.

§ **316.** But as the above reasoning about the splitting of Z_0 may not be wholly convincing, we may justify its accuracy by a direct investigation. Say there is initially Q at y. Let

$$V_1 = \cos sx \,.\, F + \sin sx \,.\, G \tag{18}$$

at x on the right of y, that is, from y up to $x = 2l$. Also, let

$$V_2 = V_1 + \sin s(x-y) \,.\, K \tag{19}$$

on the left of y, V_1 meaning the expression in the previous equation.

The condition of continuity of the potential at y is already satisfied. We also require

$$pQ = C_1 - C_2 \tag{20}$$

at y, C_1 and C_2 being the currents corresponding to V_1 and V_2, obtained by the operation $-(d/dx)/R$. Doing this, we get

$$pQ = sK/R, \tag{21}$$

making
$$C_2 = C_1 - \cos s(x-y) \,.\, pQ, \tag{22}$$

and
$$V_2 = V_1 - \sin s(x-y) \,.\, \frac{sQ}{S}. \tag{23}$$

There are now only the two time functions F, G to be found. The condition of continuity of current at the ends of Z_0 gives

$$\frac{s}{R}(\sin 2sl \,.\, F - \cos 2sl \,.\, G) = -\frac{Gs}{R} - \cos sy \,.\, pQ; \qquad (24)$$

and the condition

$$V_1 - V_2 = Z_0 C, \qquad (25)$$

which is the equation of voltage for Z_0, gives

$$\cos 2sl \,.\, F + \sin 2sl \,.\, G - F - \sin sy \,.\, \frac{sQ}{S} = Z_0\left(-\frac{Gs}{R} - \cos sy \,.\, pQ\right). \qquad (26)$$

These equations find F and G by the usual algebraical work. Using the results in (18) and (23), we obtain

$$V_1 = \frac{2\sin sl \cos s(l-x+y) + (Z_0 s/R)\cos sy \cos s(2l-x)}{\sin sl\,(2\sin sl + \cos sl \,.\, Z_0 s/R)}\,\frac{sQ}{2S} \qquad (27)$$

$$V_2 = V_1 - \sin s(x-y) \,.\, sQ/S. \qquad (28)$$

These are the complete operational solutions without initial splitting. We see that the determinantal equation does split into two, giving

$$\sin sl = 0, \quad \text{and} \quad \tan sl = -Z_0 s/2R, \qquad (29)$$

proving definitely the validity of the previous reasoning.

Now, if we expand the above V_1 by the expansion theorem, attending only to $\sin sl = 0$, we shall obtain the expansion of the v_1 in (15). And if we expand according to the other condition in (29) we shall obtain the expansion of w_1 in (16). Thus $V_1 = v_1 + w_1$, when in the fully expanded form.

But the equivalence is also true in the operational form. To prove this, add together the v_1 and w_1 of (15) and (16), uniting the two fractions to make one with a denominator which is the product of the former ones. The result is (27), as required.

The expanded formula for V_2 is the same as for V_1. The complete solution for the whole cable from $x=0$ to $2l$ is therefore given by a single Fourier series, although it is separable into two of different types. The same is therefore true for any initial state of the cable. The reason is that there is not (except at the initial moment of the instantaneous introduction of the charge) any sort of constraint all the way from $x=0$ to $2l$. If there were, say a constraint at $x=z$, we should want two distinct Fourier series, one for each side of z. The determinantal equation would be the same for both sides, but there would be a break in the normal systems.

Closed Cable with Discontinuous Potential and Current.

§ **317**. We have done one case involving a discontinuity in the current, and another involving a discontinuity in the potential. But in general both will be discontinuous. Let, for example, the connection between $x=0$ and $x=2l$ be made through Z_1+Z_2 in sequence, with a shunt to earth put on at

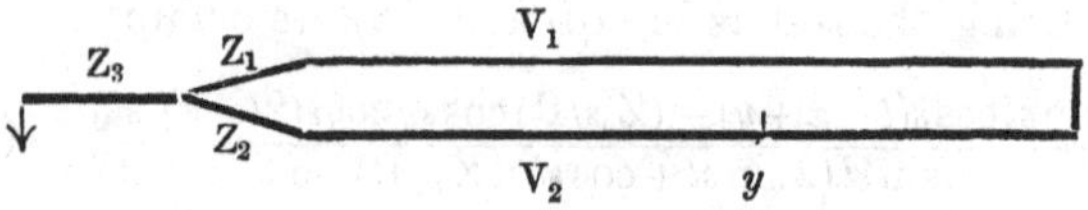

their junction. Let V_0 be the potential at the junction. Then we have these voltage equations to satisfy:—

$$V_1-V_0=Z_1C_1, \quad V_0-V_2=Z_2C_2, \quad V_0=Z_3(C_1-C_2); \qquad (30)$$

and therefore have the two terminal conditions

$$V_1=(Z_1+Z_3)C_1-Z_3C_2, \qquad (31)$$

$$V_2=Z_3C_1-(Z_2+Z_3)C_2. \qquad (32)$$

In these equations V_1 and C_1 mean the potential and current at $x=2l$; V_2 and C_2 those at $x=0$.

If, then, there is an initial charge Q at y, we can obtain the operational solution by means of equations (18) and (23) above, applied to (31), (32), to determine F and G. The result is, as before, a single Fourier series to represent both V_1 and V_2. It sometimes divides into two sets. For example, if Z_1 and Z_2 are equivalent, we get symmetry.

A Cable in Closed Circuit without Constraint.

§ **318.** The case of a cable forming a closed circuit in itself without external constraint deserves some notice. Fourier's periodic theorem is involved, of course, but it has a somewhat different application to the one before made. If we have an infinitely long cable containing an infinite series of uniformly spaced point sources of equal size, we produce periodicity by symmetry. Every source produces a double wave, diffusing to the right and left respectively; and the effect produced at any point is the sum of the effects due to the individual sources.

But in the closed cable, in the corresponding case, there is one source only, and only two waves. The apparent multiplicity arises from overlapping. The matter is best understood by introducing self-induction to such a small extent as not to alter sensibly the shape of the waves, whilst making the speed of propagation be finite, say v. Then, if there is initially a charge Q at the point y, it splits into halves, and two diffusive waves result, one to the right and the other to the left, each containing $\frac{1}{2}$Q of electrification. The manner of this diffusion has been described before. The point here is that the two waves have fronts, which are at distance vt from the source, and which therefore rush round and round the cable in opposite directions. There is no reflection anywhere, so that the effect at any point x is the resultant of the overlapping of the right wave on itself, and of the left wave on itself. The potential at x makes a very little jump when either of the wave-fronts passes the place. There are two jumps in every interval of time equal to $2l$ v, the time of transit of a wave-front round the circuit. Between the jumps, the potential changes slowly (relatively) by the natural diffusive process. As time goes on, the jumps decrease infinitesimally, and the steady state is approximated to. This is $\mathrm{SV} = \mathrm{Q}/2l$, the mean value.

If we remove the self-induction, v becomes infinite. The initial waves therefore overlap instantly, and the little jumps disappear. We now have continuous variation of the potential, which is that due to the initial waves in reality, but with the proper allowance made for the overlapping.

The contrast with the behaviour in a distortionless circuit is striking. After the initial splitting, the two half-charges would go on travelling round and round at constant speed without any diffusion, though attenuating by leakage.

But keeping to the diffusion problem, the charge Q represents the result of an impulsive current pQ at y, which divides equally to the right and left. Therefore

$$c_1 = \epsilon^{-q(x-y)}\tfrac{1}{2}pQ, \qquad v_1 = \epsilon^{-q(x-y)}\frac{RpQ}{2q}, \tag{33}$$

represent the current and potential at x, on the right of y, in the initial wave. To allow for overlapping, the factor for a circuital journey is ϵ^{-2ql}. That is, the second term in V_1 is $\epsilon^{-2ql}v_1$, the third is $\epsilon^{-4ql}v_1$, and so on. The total is

$$V_1 = \frac{\epsilon^{-q(x-y)}}{1-\epsilon^{-2ql}}\frac{RpQ}{2q}, \tag{34}$$

representing V at x so far as the wave to the right is concerned.

Now, the same point x is distant $2l-x+y$ from y, when reckoned the other way round. Otherwise it is the same. So

$$V_2 = \frac{\epsilon^{-q(2l-x+y)}}{1-\epsilon^{-2ql}}\frac{RpQ}{2q} \tag{35}$$

represents V at x so far as the wave to the left is concerned. The complete potential V is $V_1 + V_2$. That is,

$$V = \frac{\cosh q(l-x+y)}{\text{shin}\, ql}\frac{RpQ}{2q} = \frac{\cos s(l-x+y)}{\sin sl}\frac{sQ}{2S}. \tag{36}$$

Algebrising this by the expansion theorem, we obtain

$$V = \frac{Q}{S}\left\{\frac{1}{2l} + \frac{1}{l}\Sigma_1^\infty \cos\frac{n\pi}{l}(x-y)\,.\,\epsilon^{pt}\right\}, \tag{37}$$

where we recognise the unit impulsive function of Fourier's theorem, when we put $t=0$. There is only one such function now, instead of an infinite row; because x and $x \pm 2nl$ mean the *same* point.

Theory of a Leak. Normal Systems.

§ 319. Passing now to a third minor matter, something should be said about the treatment of cases in which there are two or more connected Fourier series concerned. These cases can be constructed to any extent by means of intermediate conditions, and by combinations of cables. But a whole book would be required for a full treatment, and there are more interesting matters to be considered. So all that will be done here will be to take two or three cases of the kind, to be suggestive of what is necessary in other cases, and to exhibit matters of principle as well as practical working.

Say there is a leak at z. It may be any sort of leak, to be denoted by its resistance operator r. Let the

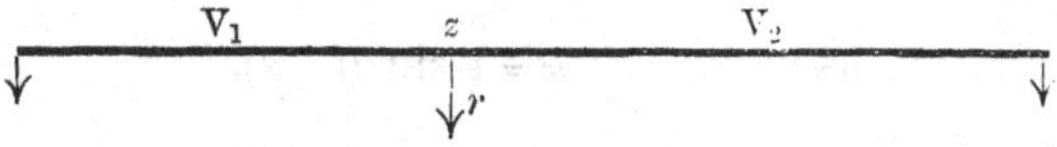

cable be earthed at $x=0$ and l.

There is one thing that can be done at once—viz., to write down the determinantal equation. For the sum of the currents leaving the point z is zero, and each current is the potential at z multiplied by the conductance operator of the path concerned. Therefore

$$0=(Y_1+Y_2+Y_3)V_z \tag{38}$$

is the differential equation of the potential at z, where the three conductance operators are

$$Y_1=\frac{s}{R}\cot sz, \quad Y_2=\frac{s}{R}\cot s(l-z), \quad Y_3=\frac{1}{r}. \tag{39}$$

If there are terminal arrangements at $x=0$ and l, Y_1 and Y_2 must be suitably modified. It follows that

$$0=Y_1+Y_2+Y_3 \tag{40}$$

is the general differential equation of the complete combination satisfied by all subsidence solutions, and the values of p satisfying it, when regarded algebraically, are the values of p in the time function ϵ^{pt} of the normal systems. That is, the solutions are of the form $\Sigma\, a\epsilon^{pt}$, where the p's are known. But the a's are functions of x, and require further investigation.

The above process of finding the determinantal equation applies when any number of circuits of any kind meet at a point. But it is important to note that any point in the combination may be taken for origin. The difference may be very great in the form of the resulting equation, which may be simple at one place and complicated at another; but all the forms are intrinsically the same in containing a common factor whose vanishing determines the p's in the time functions.

Going further, consider the normal systems of potential. Let

$$V = \sum Av, \qquad V_2 = \sum Aw, \tag{41}$$

on the left and right of z. Here v and w are the normal functions. Since V vanishes at $x=0$ and l, we require

$$v = \sin sx, \qquad w = B \sin s(l-x). \tag{42}$$

To find B, make continuity, or $v=w$ at z. Then

$$w = \sin sz \frac{\sin s(l-x)}{\sin s(l-z)}. \tag{43}$$

Two things remain—namely, s and A. To find s, we have the final condition

$$C_1 - C_2 = V/r \tag{44}$$

at z. Applying this to a normal system makes

$$\frac{\sin sz}{r} = -\frac{s}{R}\left(\cos sz + \sin sz \frac{\cos s(l-z)}{\sin s(l-z)}\right), \tag{45}$$

or

$$\frac{R}{rs}\sin sz + \frac{\sin sl}{\sin s(l-z)} = 0. \tag{46}$$

This is the determinantal equation, and is equivalent to (40) above, subject to (39), in a changed form.

Lastly, to find the A's to suit any initial state, this requires the construction of A in the case of a point charge. Say there is Q at y, less than z. Then, by the conjugate property,

$$A = \frac{Qv_y}{S\int_0^z v^2 dx + S\int_z^l w^2 dx + X}. \tag{47}$$

The numerator is the mutual energy of Q and the normal system, that is, Q multiplied by the potential of the normal system at Q. The denominator is twice the excess of the electric over the magnetic energy of the complete normal system itself. The part for the cable is explicitly shown. The part for the leak is denoted by X, to be found as before explained in special examples.

But X vanishes when r is a mere resistance. Then A becomes

$$A = \frac{(2Q/S)\sin sy}{z - \dfrac{\sin 2sz}{2s} + \left(\dfrac{\sin^2 sz}{\sin^2 s(l-z)}\right)\left(l - z - \dfrac{\sin 2s(l-z)}{2s}\right)}. \tag{48}$$

This completes the problem so far as any initial state of V from 0 to z is concerned, by an integration. But when Q is on the other side of z, we must use w_y instead of v_y in the numerator. Else it is the same.

Theory of a Leak. Operational Solution.

§ 320. It will be seen that there is considerable facility in obtaining the normal systems, and also in evaluating A, provided that there is no allowance to be made for the leak itself, which may be a troublesome matter by the above process. But there is another way, whereby the evaluation of the energy difference is done by a differentiation without detailed examination of r. This will be explained presently. The same remarks about facility and the reservation apply in more complicated cases. But there is, nevertheless, a rather bad failing. We cannot, from the above normal systems, safely deduce the general operational solutions. If, on the other hand, we get the operational solutions first, we can at once deduce the Fourier series, and other sorts of solutions as well.

We may, therefore, now exhibit the more general way. Put a charge Q at y, and see the effect.

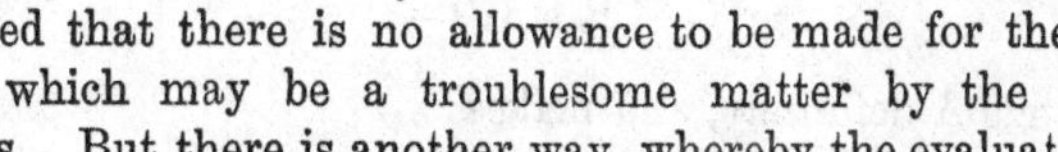

Start at $x=0$. First, we have

$$V_1 = \sin sx \,.\, A, \tag{49}$$

satisfying the terminal condition. Next, as there is to be continuity in V at y, we require

$$V_3 = V_1 + \sin s(x-y) \,.\, B. \tag{50}$$

Here B is found by

$$C_3 - C_1 = pQ \tag{51}$$

at y, and makes

$$V_3 = V_1 + \sin s(x-y) \,.\, sQ/S. \tag{52}$$

Then there is to be continuity in V at z, so

$$V_2 = V_3 + \sin s(x-z) \,.\, D, \tag{53}$$

and $V_2 = 0$ at $x = l$ finds D, and makes

$$V_2 = V_3 - V_{3l} \frac{\sin s(x-z)}{\sin s(l-z)}, \tag{54}$$

where V_{3l} means V_3 at $x = l$. Finally, we have the discontinuity in the current at z, or

$$\frac{V_{3z}}{r} = -\frac{s}{R} \frac{V_{3l}}{\sin s(l-z)}. \tag{55}$$

This finds A definitely, and makes

$$V_1 = -\sin sx \frac{\sin s(z-y) + \dfrac{sr}{R}\dfrac{\sin s(l-y)}{\sin s(l-z)}}{\sin sz + \dfrac{sr}{R}\dfrac{\sin sl}{\sin s(l-z)}} \frac{sQ}{S}. \tag{56}$$

This is the complete operational solution. Q is h/p, where h is an externally introduced source of current, which may in general be any function of the time. It is made impulsive specially to suit the treatment of an initial state.

As regards V_3, it is found by (52). It will usually differ from V_1, though not in the Fourier series solution for an initial state. It will be found that V_3 differs from V_1, comparing (52) with (56), merely in the interchange of x and y. Finally, we may use the modified V_3 in (54), and obtain

$$V_2 = -\sin sy \frac{\dfrac{sr}{R}\dfrac{\sin s(l-x)}{\sin s(l-z)}}{\sin sz + \dfrac{sr}{R}\dfrac{\sin sl}{\sin s(l-z)}} \frac{sQ}{S}. \tag{57}$$

There are two distinct expansions, one for V_1 and V_3, the other for V_2. They follow from (56), (57), in the usual way, by the expansion theorem. But, denoting the common denominator $\sin sz + \ldots$ by D, so that $D = 0$ is the determinantal equation, we can simplify the numerators by use of this relation. Then the expansion theorem makes

$$V_1 \text{ and } V_3 = \Sigma \frac{-\frac{2Qs}{SR}\sin sx \sin sy}{\sin sz \frac{d}{ds}\left(\frac{1}{r}+\frac{s}{R}\cot sz+\frac{s}{R}\cot s(l-z)\right)} \qquad (58)$$

Similarly for V_2. The form of the denominator was first changed in order to bring the relation (40) above into view. It may be verified that (58) agrees with (48) when the leak is a mere resistance. But (58) is far more general. The leak may be of any kind, r being a rational function of p, and therefore of s^2. When r is not a rational function of p, the expansion (58) fails. A definite integral is required. But the operational solution (56) remains true.

Evaluation of Energy in Normal Systems.

§ **321.** The object of altering the form of the denominator of (57), so as to make the function in the big brackets in (58) be the active determinantal factor, as in (38), (40), was to introduce a point not noticed in the previous. According to (58), the value of the coefficient A is given by

$$A = \frac{Q \sin sy}{\sin^2 sz \times -\frac{RS}{2s}\frac{d}{ds}\left(\frac{1}{r}+\dots\right)}. \qquad (59)$$

This is the same as

$$A = \frac{Qv_y}{v_z^2 \frac{dY_z}{dp}}, \qquad (60)$$

if Y_z is the conductance operator at the point z, *i.e.*, the sum of the conductance operators of the three paths meeting there.

Now the general formula to find A from the conjugate property goes thus. Let there be selected any two normal systems, belonging to p_1 and p_2, say; let U_{12} be their mutual electric, and T_{12} their mutual magnetic energy. Then

$$U_{12} - T_{12} = 0 \qquad (61)$$

expresses the conjugate property. From which it follows that

$$A_1 = \frac{U_{01} - T_{01}}{U_{11} - T_{11}} \qquad (62)$$

is the coefficient for the term involving p_1; if U_{01} is the mutual electric, and T_{01} the mutual magnetic energy of the given initial state and the normal state in question, whilst U_{11} and T_{11} are double the electric and magnetic energy of the normal state itself.

The numerator in (60) agrees with (62), of course, because of the confinement of Q to a point, and because there is no mutual magnetic energy. There would be, if the initial state involved magnetic energy in the leak r. The denominator of (60), therefore, represents the denominator in (62). That is,

$$U_{11} - T_{11} = v_z^2 \frac{dY_z}{dp_1} \tag{63}$$

for any normal system, which is here the first one. We need not use the suffixes, which may confuse, but, keeping to one normal system, write

$$U - T = v^2 \frac{dY}{dp}, \tag{64}$$

understanding that Y is reckoned at the place of v. Now, this property is general; z may be any point in the system, v its potential there, and Y the corresponding conductance operator. The variations in v will be exactly compensated by those in Y.

To illustrate, divide Y into its three members; then

$$U - T = v^2 \frac{d}{dp}(Y_1 + Y_2 + Y_3). \tag{65}$$

Here Y_1 is the ratio (conductance operator) c/v for the first path, Y_2 for the second, Y_3 for the third, all with the same v, but with different c's. And

$$v^2 \frac{d}{dp}\frac{c}{v} = v\frac{dc}{dp} - c\frac{dv}{dp} = -c^2 \frac{d}{dp}\frac{v}{c}. \tag{66}$$

We may therefore write

$$-(U - T) = c_1^2 \frac{dZ_1}{dp} + c_2^2 \frac{dZ_2}{dp} + c_3^2 \frac{dZ_3}{dp}, \tag{67}$$

if Z_1, &c., are the resistance operators.

At the point z—that is, at the leak, there are three c's and Z's to be allowed for. But the last two c's may be expressed in terms of the first, bringing us to a result of the form $c_1^2 dZ/dp$.

If, however, any other point in the cable be taken, there are only two c's, two Y's, or two Z's, and the c's are equal. Then

$$-(\mathrm{U}-\mathrm{T})=c^2\frac{d\mathrm{Z}}{dp}, \tag{68}$$

where Z is the sum of the resistance operators to right and left of the point in question, which may be any point. $d\mathrm{Z}/dp$ is an inductance operator, and $d\mathrm{Y}/dp$ a permittance operator.

It will be as well to illustrate these remarkable, though somewhat abstruse, properties explicitly. Take then the simple case of terminal earth connection and no intermediate

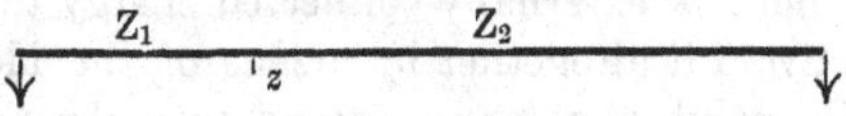

condition, so that the normal system is simply

$$v=\sin sx, \qquad c=-\frac{s}{\mathrm{R}}\cos sx, \tag{69}$$

s having any value making $\sin sl=0$. Here

$$\mathrm{Z}_1=\frac{\mathrm{R}}{s}\tan sz, \qquad \mathrm{Z}_2=\frac{\mathrm{R}}{s}\tan s(l-z), \tag{70}$$

are the resistance operators to left and right of the point z, and their sum is

$$\mathrm{Z}=\frac{\mathrm{R}}{s}\frac{\sin sl}{\cos sz\cos s(l-z)}, \tag{71}$$

so that e/Z would be the current there due to e on the spot. Also $d/dp=-(\mathrm{RS}/2s)d/ds$, since $\mathrm{RS}p=-s^2$; so

$$\frac{d\mathrm{Z}}{dp}=-\frac{\mathrm{R}^2\mathrm{S}l}{2s^2}\frac{\cos sl}{\cos sz\cos s(l-z)}=-\frac{\mathrm{R}^2\mathrm{S}l}{2s^2}\frac{1}{\cos^2 sz}, \tag{72}$$

because $\sin sl=0$. Therefore

$$-c^2\frac{d\mathrm{Z}}{dp}=\frac{s^2}{\mathrm{R}^2}\frac{\mathrm{R}^2\mathrm{S}l}{2s^2}=\frac{\mathrm{S}l}{2}. \tag{73}$$

Consequently, if there is initially Q at y, it expands to

$$\mathrm{V}=\Sigma\frac{\mathrm{Q}\sin sy\sin sx}{\frac{1}{2}\mathrm{S}l}, \tag{74}$$

which we know to be right. Of course the best place for z is at a terminal. Then only one resistance operator is concerned, and it goes simply.

Again, do it in terms of the conductance operators. Here

$$v^2 \frac{dY}{dp} = \sin^2 sz \left\{ -\frac{RS}{2s} \frac{d}{ds}\left(\frac{s}{R}\cot sz + \frac{s}{R}\cot s(l-z)\right)\right\}, \qquad (75)$$

which also works out to the result $\frac{1}{2}Sl$, independent of the position of z. Here Y is the sum of the two conductance operators to left and right of z.

The practically useful point is this. If the evaluation of U − T by integration is troublesome, which is particularly the case when there is externally connected energy to be allowed for, we can avoid it altogether by means of the identity (64), choosing the point of reference so as to make the work as simple as possible. We do not, therefore, need to form the operational solutions and apply the expansion theorem, if the only object is to develop the proper Fourier series.

It may be objected in the above that in taking $v = \sin sx$, and assuming v to be potential, or the transverse voltage, we are violating dimensional properties. The objection is valid, though rather superficial. To be very particular, we may let $A \sin sx$ be the normal system of potential, A being the same as above, and so have a symbol to put the dimensions in (though they are not known, by the way). But this will make no difference in the end, and will complicate the formulæ somewhat. Or we may let $a \sin sx$ be the normal system, and let a be represented by a unit factor. For unity, though constant in a certain sense, may have any size and any dimensions.

A more important objection is that I have given no proof of the general properties used, except what may be contained in the actual verification in the examples used, which is sufficient as far as they are concerned. But the fact is, that the expansion theorem, and the conjugate property, and the equivalence of the integration and the differentiation ways of working it, being general matters, are best proved in a general manner, out of the general equations of dynamics, which, in their application to electromagnetics, will be considered separately.

Initial States in Combinations.

§ 322. The following way of regarding the matter is useful in combinations. It has the advantage of compactness. Imagine any combination to be disturbed by the impression of a current upon it. Let the current be h, and its point of entry be denoted by y. Then

$$h = Y_y V_y, \qquad \text{or} \qquad V_y = \frac{h}{Y_y}, \tag{76}$$

is the connection between h and V_y, the resulting potential at y, if Y_y is the conductance operator there, or the sum of the conductance operators of the paths open to the impressed current. The path of entry is not counted.

If Y_y is known, (76) may be algebrised as it stands. But to represent an initial state of charge or electrification, let $h = pQ$, where Q is constant. Then

$$V_y = \frac{pQ}{Y_y} \tag{77}$$

expresses the potential at y due to the impulsive entry of the charge Q; and its expansion is

$$V_y = \Sigma \frac{Q\epsilon^{pt}}{Y_y'} + \text{steady term}, \tag{78}$$

ranging over the roots of $Y = 0$, the accent meaning differentiation to p. There is a steady term if p/Y_y is finite when $p = 0$. This is exceptional.

Now to find V_x, the potential at some other point x in the combination, we require to multiply V_y by the operator V_x/V_y; as in

$$V_x = \frac{pQ}{(V_y/V_x)Y_y}. \tag{79}$$

But in a normal state, such as the general term of the expansion refers to, we must multiply by w_x/w_y, if w is the normal function. This makes

$$V_x = \Sigma \frac{Qw_y w_x}{u_y^2 Y_y'} \epsilon^{pt} = \Sigma\, Aw_x \epsilon^{pt}. \tag{80}$$

Now, the coefficients A are subject to the conjugate property, and determined by (62). The numerator Qu_y agrees. So the denominator is U − T. Or

$$V_x = \Sigma \frac{Qu_y u_x}{U - T} \epsilon^{pt}. \tag{81}$$

But U − T represents an integration extended over the whole system. It is fixed when u is fixed at any point, and has nothing to do with the value of y, the position of Q. Therefore the alternative form of U − T shown in (80) must be independent of the position of y, so that

$$V_x = \Sigma \frac{Qu_y u_x}{u_z^2 Y_z} \epsilon^{pt}, \tag{82}$$

where z is any point, and Y_z the conductance operator there. (But do not choose a place where u vanishes, as that will require a compensating infinity.) Lastly, if there are only two paths meeting at z, we have the equivalent form

$$V_x = \Sigma \frac{Qu_y u_x}{- c_z^2 Z_z'} \epsilon^{pt}, \tag{83}$$

by (66), (68), where c_z is the normal current at z, corresponding to u_z, and Z_z the resistance operator there. This form is perhaps more generally useful.

An integration over the whole combination will determine the value of A so far as it depends upon initial charge. If further, there is initial magnetic energy, it must be allowed for as in (62). It is important to think of the normal function u as a single function. It may, indeed, have various forms of algebraical expression in different parts of the combination, but that is of no consequence in the present connection.

Two Cables with different Constants in sequence, with an insertion.

§ 323. Example should supplement precept, if it does not precede it, which may be better still. Therefore, let us apply the above to the case of two cables of different types joined

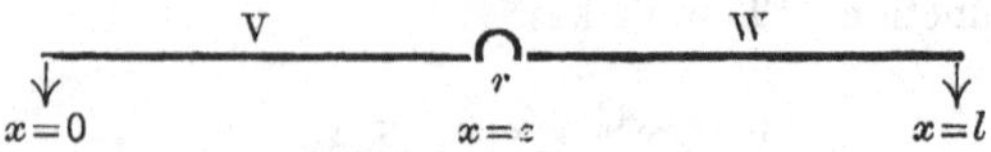

in sequence, with an intermediate insertion. Let r be the resistance operator of the insertion (containing no leak), which is made at the distance z from $x=0$. The resistance operator of the combination, reckoned at z, is

$$Z = r + \frac{R_1}{s_1}\tan s_1 z + \frac{R_2}{s_2}\tan s_2(l-z), \tag{84}$$

the cables having different R and S. Here Z is such that e/Z is the current at z due to e there.

Now put Q at y, less than z. If the normal potential function u of the last section is v in the first cable and w in the second, and the resulting potentials are V and W, we have

$$V = \Sigma \frac{Qv_y}{U-T} v\epsilon^{pt}, \qquad W = \Sigma \frac{Qv_y}{U-T} w\epsilon^{pt} \tag{85}$$

But if Q is in the second cable, then

$$V = \Sigma \frac{Qw_y}{U-T} v\epsilon^{pt}, \qquad W = \Sigma \frac{Qw_y}{U-T} w\epsilon^{pt}. \tag{86}$$

In these, v, w, V and W refer to the variable point x, whilst Q refers to y.

We have next to specify the nature of v, w. Since the first cable is earthed at $x=0$, and the second at $x=l$, we may put

$$v = \sin s_1 x, \qquad w = a \sin s_2(l-x), \tag{87}$$

provided a is settled so as to harmonise them. This is to be done by the continuity of the current at z; that is,

$$\frac{s_1}{R_1}\frac{dv}{dx} = \frac{s_2}{R_2}\frac{dw}{dx}, \quad \text{at} \quad x = z. \tag{88}$$

This makes

$$w = -\frac{R_2 s_1}{R_1 s_2}\frac{\cos s_1 z}{\cos s_2(l-z)} \sin s_2(l-x). \tag{89}$$

The normal function being complete, it remains to evaluate $U-T$ for it. Since we know Z at the point z, the shortest way is by $-c^2 Z'$. Thus

$$V = \Sigma \frac{Q \sin s_1 x \sin s_1 y}{-\left(\dfrac{s_1}{R_1}\cos s_1 z\right)^2 \dfrac{dZ}{dp}} \epsilon^{pt} \tag{90}$$

is the potential at x due to Q at y, when both x and y are in the first cable, Z being given by (84). The summation ranges over the roots of $Z=0$. There is no steady term, because of the earth connection. For d/dp may be substituted differentiation to s_1 or s_2 for trigonometrical convenience, and so we come to the fully-developed formula, suitable for calculation if required. In general r is a function of p. But we do not need to allow for that unless there is initial energy in r itself.

If x is in the second cable, y being still in the first, we must put w for v in (90). If x and y are in the second cable, use w_x and w_y in the numerator. If y is in the second cable and x in the first, use w_y and v_x. In short, follow (85), (86).

If the effect of a localised impressed voltage is wanted, we may use the resistance operator at the place. Thus, when e is at Z, then

$$C = \frac{e}{Z} = \frac{e}{Z_0} + \Sigma \frac{e}{pZ'} \epsilon^{pt}, \tag{91}$$

where Z is as in (84), and Z_0 is the steady resistance. To find C at any other place x, introduce the factor c_x/c_z in the general term. It is the ratio of the normal current functions at x and z, and is known.

If e is at $x=0$, a different Z is needed. We may regard it as the resistance operator of the first cable with a terminal arrangement, say Z_1, given by

$$Z_1 = r + \frac{R_2}{s_2} \tan s_2(l-z). \tag{92}$$

It is then a special case of (59), § 291. Again, if e is at $x=l$, we may bring the case under the same formula by treating it as a cable (the second one) with a terminal arrangement

$$Z_2 = \frac{R_1}{s_1} \tan s_1 z. \tag{93}$$

It may also be noted that the same formula allows us to construct the resistance operator of any number of cables of the same or different types put in sequence, with intermediate insertions. For instance, we know the resistance operator at

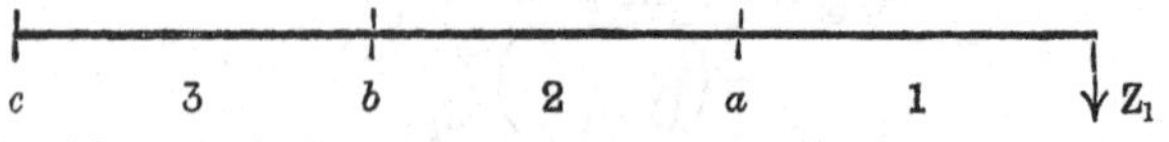

a of cable 1 with Z_1 at its end. Call it Z_2. It is

$$Z_2 = \frac{R}{s} \frac{(R\sin + Z_1 s\cos)sl}{(R\cos - Z_1 s\sin)sl}, \tag{93A}$$

using the proper R and s for the first cable. Then, by the same formula, we know the resistance operator at b of cable 2 with Z_2 at its end. Call it Z_3. Then we know the resistance operator at c of cable 3 with Z_3 at its end. That is, we know the resistance operator at c of the complete combination. This process may be continued to any extent, and results in a continued fraction, which soon assumes gigantic size. There may be intermediate insertions (without leaks), without essential change. But leaks require a little change of treatment. Say there is a leak at a, then Z_2 is to be the resultant resistance operator of the leak and that of the cable 1 with Z_1 at its end; that is, the reciprocal of the sum of their reciprocals. Using this modified Z_2, we may proceed as before till we come to the next leak, say at b, where we make a similar modification.

If an intermediate resistance operator is wanted, that is merely the sum of the resistance operators to right and left, to be got by the same process, which applies when the cables are not each of uniform resistance and permittance, and when the inductance and leakance are not ignored. Of course, however, when R and S vary in any one cable, its resistance operator requires special investigation.

Cable involving the zeroth Bessel function.

§ 324. Still keeping to two cables in sequence, for simplicity of illustration of broad principles, let one be an ordinary cable, the other with R and S varying in such a way as to produce another kind of normal function. In passing, it may be observed that by causing R and S to change gradually from uniformity of distribution to various other arrangements of the same total resistance and permittance, we can continuously deform the simple normal function $(\sin sx + a \cos sx)$ so as to make it assume the shape of any other kind of normal function. In case of two kinds of cables, the results will be exhibited in two series, say a Fourier series for one cable, and a Bessel series for the other, properly harmonised.

In order to have the zeroth Bessel normal function, we need only let the conductance and permittance per unit length of cable both vary directly as the distance from $x=0$. There are other ways, but this is the simplest. Thus, the circuital equations

$$-\frac{dV}{dx}=RC, \qquad -\frac{dC}{dx}=SpV, \tag{94}$$

unite to make the characteristic

$$\frac{d}{dx}\left(\frac{1}{R}\frac{dV}{dx}\right)=SpV, \tag{95}$$

without assumption of constancy of R and S. Therefore, if

$$\frac{1}{R}=\frac{x}{R_0}, \qquad S=S_0x, \tag{96}$$

R_0, S_0 being constant, we produce the characteristic

$$\frac{1}{x}\frac{d}{dx}x\frac{dV}{dx}=RSpV=q_0^2V, \tag{97}$$

where RS is the constant R_0S_0.

The proper solution for our purpose is

$$V=I_0(q_0x)\,.\,A=\left(1+\frac{(q_0x)^2}{2^2}+\frac{(q_0x)^4}{2^24^2}+\ldots\right)V_0, \tag{98}$$

where V_0 is a time function, and I_0 is defined by the expansion given. The corresponding C is

$$C=-\frac{1}{R}\frac{dV}{dx}=-\frac{q_0}{R}I_1(q_0x)\,.\,V_0, \tag{99}$$

where I_1 is the derivative of I_0.

The meaning of V_0 is clearly the potential at $x=0$ itself, given as a function of the time. But we do not want that

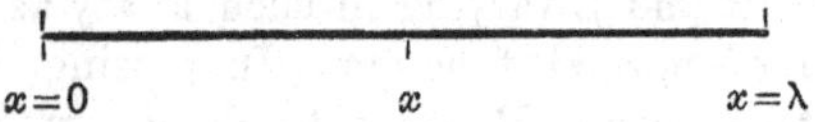

particularly. In fact, since the conductance is zero at $x=0$, there is no current there due to any finite impressed voltage, and we should avoid the point $x=0$ for our immediate purpose. We have

$$\frac{V}{C}=-\frac{R}{q_0}\frac{I_0(q_0x)}{I_1(q_0x)} \tag{100}$$

at x. Therefore, if e is impressed at λ, the resistance operator is

$$Z=\frac{R_0}{q_0\lambda}\frac{I_0(q_0\lambda)}{I_1(q_0\lambda)}, \tag{101}$$

if we now reckon the current positive from λ to 0. That is, e/Z is the current entering at λ, if there is earth on the other side of e. Or, more generally, V/Z is the current at λ when the potential there is V variable anyhow, due to sources on the side of λ beyond the cable in question.

The formula for the potential at x is simply got by (98), observing that it must reduce to e at $x=\lambda$. Thus,

$$V = \frac{I_0(q_0 x)}{I_0(q_0 \lambda)} e, \tag{102}$$

from which C follows by a differentiation according to Ohm's law.

By (102), we see that

$$I_0(q_0\lambda) = 0, \qquad \text{or} \qquad J_0(s_0\lambda) = 0, \tag{103}$$

is the determinantal equation, if $q_0^2 = -s_0^2$. Here J_0 is an oscillating function analogous to the cosine. Why we should have it so will be seen on remembering that there is no current at $x=0$, so that, if R and S were constant, the normal function would be $\cos sx$.

Applying the expansion theorem to (102), when e is steady, V expands to

$$V = e + e\Sigma \frac{I_0(q_0 x) \,.\, \epsilon^{pt}}{p\frac{d}{dp}I_0(q_0\lambda)} = e + e\Sigma \frac{J_0(s_0 x) \,.\, \epsilon^{pt}}{\frac{1}{2}s_0\frac{d}{ds_0}J_0(s_0\lambda)},$$

or

$$V = e - e\Sigma \frac{J_0(sx) \,.\, \epsilon^{pt}}{\frac{1}{2}s_0\lambda\, J_1(s_0\lambda)}, \tag{104}$$

if J_1 is the negative of the derivative of J_0. This shows the manner of establishment of the final state of V, which is $V=e$ all over, owing to the vanishing conductance at the far end.

We can, of course, put on any terminal Z_0 in place of earth on the other side of e, and solve in a similar way, but that would be premature at present.

A Fourier and a Bessel Cable in sequence.

§ 325. Pass to the matter immediately in question—namely, the harmonisation of the solutions for two cables of different natures. Let the cable on the left be of the regular kind, and that on the right be of the kind considered in the last section. Earth is on at $x=0$ in the first cable, but the $z=0$ end of the second cable is virtually insulated, if that is the point of vanishing conductance. We measure x from left to right in

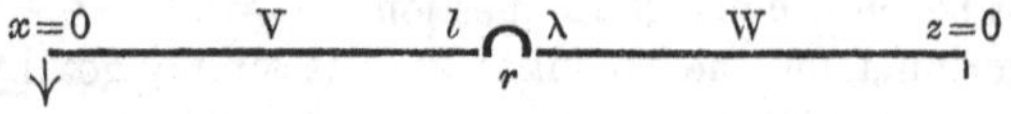

the left cable, in which the potential is V, and z from right to left in the other, where the potential is W. Between them, at $x=l$ in one and $z=\lambda$ in the other cable, is inserted any combination r without leakage. The complete resistance operator at r is

$$Z = r + \frac{R}{s}\tan sl + \frac{R_0\lambda}{q_0}\frac{I_0(q_0\lambda)}{I_1(q_0\lambda)}, \tag{105}$$

by (101), using the notation of the last section for the Bessel cable, except that x is now z, to distinguish it from x in the first cable. So the impressed voltage e at r produces the current

$$C_\lambda = \frac{e}{Z} = \frac{e}{Z_0} + \Sigma\frac{e\,\epsilon^{pt}}{p\,Z'}, \tag{106}$$

subject to $Z=0$, with the form of Z shown by (105).

This being merely a special case, let us examine the result of charging both cables initially in any given way, and leaving them to themselves. The formulæ are (85) and (86), provided we harmonise v and w, and evaluate U − T. The form of v is obvious, so

$$v = \sin sx, \qquad w = bI_0(q_0z), \tag{107}$$

where b has to be found so as to make w fit v properly. This may be done by the voltage equation at the junction, or

$$v - w = rc, \tag{108}$$

where $x=l$ in v and $z=\lambda$ in w, and c is the current corresponding to v. It is also the current corresponding to w, and their equalisation produces the determinantal equation (105) already found. Now (108) is the same as

$$\sin sl - bI_0(q_0\lambda) = -\frac{rs}{R}\cos sl, \tag{109}$$

which gives b, and makes

$$w = \frac{I_0(q_0z)}{I_0(q_0l)}\left(\sin sl + \frac{rs}{R}\cos sl\right). \tag{110}$$

This w is the other part of the normal system u, for v and w together make it complete, except as regards the internal variation in r itself, which we do not want.

There is only left the evaluation of $U - T$. As before, the resistance operator at r being known, we may employ the $-c^2Z'$ form, and therefore get, subject to $Z = 0$ in (105),

$$V = \Sigma \frac{Q \sin sx \sin sy}{-\left(\frac{s}{R} \cos sl\right)^2 \frac{dZ}{dp}} \epsilon^{pt}, \tag{111}$$

when x and y are in the first cable; it being $v_x v_y$ that is used in the numerator. Substitute $w_z w_y$ when they are in the Bessel cable, using (110); and $v_x w_y$ or $v_y w_z$ when x or z is in one and y in the other cable. Also, instead of $I_0(q_0 z)$, use $J_0(s_0 z)$.

When the initial state is given to be that the potential is U, the coefficient of the normal function is got by an integration, thus:

$$A = \frac{\int SUu\,dx}{U - T} = \frac{\int SVv\,dx + \int S_0 z W w\,dz}{U - T}, \tag{112}$$

where in the first form u is the complete normal function, whilst in the second it is separately exhibited for the two cables, the initial U becoming the initial V and W respectively.

Construction of a Normal System in General.

§ 326. The above case of two cables of the ordinary kind having different uniform resistance and permittance, and more particularly the case in which one cable is uniform whilst the other is of a variable type, are meant to lead easily to a broad understanding of a normal distribution in general. It should be regarded as a single function of position in the combination, no matter how many forms it may assume in different parts. To illustrate this, imagine any number of cables put in sequence, with any terminal arrangements, any way of variation of resistance and permittance, and any number of intermediate insertions or leaks. A normal system of potential is one which, when left to itself, preserves its form when subsiding. To find it, we may first divide the whole cable into sections containing no intermediate insertions or leaks. Then, if R, S vary in any section each according to a single formula, it may be treated by itself. But if there are two or more formulæ concerned, we must subdivide into smaller sections. So we come down to a section having a single characteristic. It is equation (95) above. Regarding p as a constant, there are two independent solutions, say $u = av + bw$, a and b being any constants, v and w the two solutions, functions of x. Every section has its u of this kind, with its own special v and w. They have next to be fitted together. This is to be done by the intermediate and the terminal conditions, introducing all the necessary continuities and discontinuities. The result is a single normal function u, represented by u_1 in the first section, u_2 in the second, and so on, but of arbitrary size. At the same time the conditions furnish a general determinantal equation. It is the characteristic of the whole combination, say $Z = 0$, and the values of p satisfying it are *the* values of p permissible. Thus $Au\epsilon^{pt}$, where A is a constant, represents a complete normal system at time t, so far as the cable is concerned, and therefore $\Sigma Au\epsilon^{pt}$, including all the normal systems corresponding to the different p's, represents the solution arising from any initial state. The normal u may, if we please, be continued over all the intermediate and terminal insertions and leaks, to make it fully complete. The size of the normal systems is then settled by their conjugate property, which leads to the formula (62), and other forms. If any part of the combination contain no energy initially, its detailed consideration may be omitted, and its influence on the rest allowed for by a resistance operator. This applies to parts of the cable itself as well as to insertions and leaks.

The characteristic of the whole combination, whose vanishing settles the admissible rates of subsidence, may be taken to be the conductance operator at any point, specialised by regarding p as a constant, all other forms being derivable from it. The normal functions are always entirely real when electric energy alone is concerned, and also when magnetic energy alone is concerned, though the latter does not occur in the above diffusive applications. In intermediate cases, the p's may be either real or complex. There is more to be said about these things in other electrical applications, but the present remarks are sufficient for the immediate purpose.

There is no material alteration when, by means of cross connections, we convert the above arrangement of cables entirely in sequence into a network of cables, of any degree of complication. To make the normal system complete, it must be continued over all the branches. We may use different symbols to indicate position, or we may use the same symbol with a different range. These things are immaterial. We still have to regard the normal system, though made up of many parts, as a single system, self-contained. At the same time, as before, any parts thereof may be left out, provided the proper operators are employed to sum up their reaction upon the retained parts.

Construction of Operational Solutions in a Connected System.

§ 327. We should now point out how a normal solution, as above specified, differs from what I have usually called the operational solution. There are likenesses and differences. Let the electrical arrangement be the same, either a single circuit of cables or a network; but now consider how to find the operational solution expressing the effect all over the combination due to impressed force at any point therein. It may be impressed voltage e, or impressed gaussage h. The former will create a discontinuity in the potential, the latter in the current, and these constitute the effective sources of disturbance, whereby energy is brought into the combination. We have here a difference from the self-sufficiency of a normal system.

We have to find V (and C) all over due to e or h; say $V = Xe$, for example. The form of X is wanted. Since e is at a definite point, whilst V is anywhere, X is a function of x, specifying position in the combination. It is not the same function (in its expression) in different branches, but since all the different forms have to be harmonised, we may still regard X as a whole, changing its form of expression as we pass over the combination. In this respect it is like the normal function. But at the same time, it is a function of p, the time differentiator, and varies in form as concerns p as well as x, because p enters along with the electrical constants.

To find X fully, we proceed as in the case of a normal function, to divide the combination into sections involving only one characteristic. But we do not assume that p is constant in the characteristic. It remains what it is fundamentally, a differentiator. So, in the general solution of any section, say $V = av + bw$, the a and b are not now constants, but time functions, and v, w are not merely algebraical functions, but differential operators. We may write $V = va + wb$, if we wish to explicitly indicate that a and b are operated upon by u and w. But this is a matter of convention. A cart may be pulled or pushed. Another form is $V = (v + jw)\,a$. Here there is but one time function explicitly. The other is ja, where j is an operator.

Now supposing we have found the general solutions of V for every one of the sections, we have next to harmonise them by the conditions at the junctions. This process is, up to a certain point, formally like that of harmonising the different parts of a normal function. The difference comes in when, say finally (though it may be done initially) we treat the section containing the impressed force. There is nothing special there in a normal solution. But in the operational case we have an additional split. The section is made two sections, and there are two auxiliary conditions, as, continuity in C, and a jump in V produced by e. The result is that all the time functions in all the branches become known in terms of e. Or, in another form, in $V = Xe$, we know the operator X completely throughout the combination, and there is nothing left over to be fixed by extra data. In the case of the normal system, on the other hand, its size is left quite indefinite. Apart from this, the normal system constitutes a succession of values making it be a numerical function of x. It is one of a set, for the constant p in it may have any one of a particular series of values. It is different in an operational solution. The operand e is any function of the time, and X by itself is not numerical. But the determinantal equation is implicitly involved. For if we put $e = 0$, producing $0 = X^{-1}V$, the latter is the general characteristic of the solutions when free from impressed force. If it is, as in most of our applications, a rational equation, then $X^{-1} = 0$ finds special solutions, of the type $\epsilon^{pt} \times f(x)$. This brings us to the normal functions again. I have shown how to expand $V = e/X^{-1}$ in normal functions when they are appropriate. From the examples given, it will be seen that the operator X itself may be regarded as expressing the type of the normal function (apart from size), viz., by making p in it become constant and assume one of the critical values. It then also simplifies in form, in virtue of the determinantal equation. The greater generality of the form of X is essential. The normal system cannot be more than it is. The other has to cover all possible cases, and is constructed to do it.

Similar remarks apply when the source is h, impressed gaussage, equivalent to an external source of current. This brings us to the normal functions again from another point of view, viz., when there are no impressed forces given whose effect is required, but the data consist in the specification of an initial state of V and C, from which the subsequent states arise. We convert this general problem to a special case of the former by the use of impulses. There are two sorts. Let $e = pP$ and $h = pQ$, where P and Q are constants, or, more strictly, constant functions of the time when t is positive, and zero when t is negative. Then e is merely the impressed voltage in an impulse at the moment $t = 0$, the total of the impulse being P. Similarly, h is the impressed gaussage in an impulse of total Q. That is, as before explained, the charge Q is instantly introduced, and then left to itself. Similarly as regards P. It represents magnetic momentum. The analogue of $Q = SV$, where S is the concrete permittance connecting the charge Q and the voltage V, is $P = LC$, where L is the concrete inductance connecting the momentum P and the current C. But in the case of a cable, if S and L belong to unit length, Q and P are the length integrals of SV and LC. Now the initial state is fully specified by V and C. That is, P and Q are given, or which is the same thing, e and h. In our diffusion problems, we do not require to know C at all, that is, e equivalently, because the assumption $L = 0$ destroys the momentum. That is why only $h = pQ$ has been employed in discussing initial states, and the expansions required to represent them. But, in general, when L is included, we must use both e and h to obtain the proper expansions. The addition of the impulsive e is important because it enables us to treat the electromagnetic problems with generality as regards initial states, and from the operational standpoint.

Remarks on Operational and Normal Solutions. Connection with the Simply Periodic.

§ **328.** The object of the mathematical investigation of physical matters being to ascertain what happens under given circumstances, and map out the field of knowledge, so to speak, the method of normal functions, or modes of vibration, or of subsidence, or of time variation in more mixed manners, is one of the most important ways of working to the desired end. But whether it should be used, and if so, how it should be used, depends upon circumstances. There are other methods. It may be that only very partial knowledge is wanted—*e.g.*, the general nature of normal modes—and then it may be quite unnecessary to resort to operational methods, though their use will often settle matters which are obscure without them. On the other hand, operational ways are much the best in dealing with the effects of localised impressed forces, and the study of such cases is often most instructive. The most serious drawback to the method of analysis into normal systems is that, beyond a certain point, the numerical examination of general results becomes impossible in any reasonable time. This is of no consequence in theory, but is a great practical objection. We should go to the operational solutions for information. We may turn them to normal systems if we want to, and if we do not, we can apply other methods to them. The most important is usually the simply periodic, from the wide application possible, involving the reduction of $V = Xe$ to $V = (X_0 + X_1 i)e$, where e is simply periodic and i is the differentiator $d/d(nt)$. Some other special ways of working have been given, and I will give more later on. The discovery of practical methods of manipulating operators is a matter of importance to the future of physical analysis. Objections founded upon want of rigour seem to be narrow-minded, and are not important, unless passive indifference should be replaced by active obstructiveness. In making them rigorists make confession of ignorance. It would be more useful for them to try to extend the matter, and remove the want of rigour, if they want to. But this is merely parenthetical.

Returning to the simply periodic solution just mentioned, it may be worth while to point out two ways of connecting it with solutions in series of normal functions. Suppose we have a telegraph line of any sort, and desire to express the current C at the distant end (or anywhere else) in terms of e at its beginning. We may do this operationally, of course. Say we come to $C = Ye$. Now, given that e is impulsive, we can convert the result to a series of normal functions. Next, observing that if e is given variable in any way with the time, it may be regarded as made up of a succession of infinitesimal impulses, we see that the last solution in normal functions, by being used as the element of a time integral, enables us to express C when e is variable. Lastly, by choosing e to be simply periodic, the result must be equivalent to that which may be obtained directly from $C = Ye$, by the transformation $p = ni$. The method described is a very complicated way of getting the simply periodic solution, but its execution and verification is quite feasible in some cases.

The process is reversible, if we are in possession of all the steps. But if not, if we only know the simply periodic solution, obtained directly from the operational, there is another and much shorter way of passing to the solution when e varies anyhow—namely, by analysing an impulse into simply periodic variations. Thus, the impulsive voltage $e = pE$ is equivalent to

$$pE = \frac{E}{\pi}\int_0^\infty \cos nt\, dn, \tag{113}$$

by (54), § 271. The simply periodic solution for the element $(E/\pi) \cos nt\, dn$ is obtained by operating on it by Y. So

$$C = \frac{E}{\pi}\int_0^\infty Y \cos nt\, dn = \frac{E}{\pi}\int_0^\infty \left(Y_0 + Y_1 \frac{d}{d\,nt}\right) \cos nt\, dn \tag{114}$$

expresses C due to an impulsive voltage, and therefore, by a time integration, as before, we express C due to any varying e by the second method. This process can also be practically carried out, in some cases, by the evaluation of the integral. But I am afraid that, in general, the evaluation of the definite integral will present much difficulty. Considering the extreme complication introduced into normal systems by the inclusion of gradual propagation into the wires as well as along them, the process (114) seems at first sight rather enticing. It looks so simple. So it is, in idea; but if we cannot effect the integration, the integral does not tell us any more than the operational solution itself. So I found it some years ago, when I desired to trace the progress of an elastic wave along a resisting wire. The integral was unmanageable. But the operational solution itself was amenable to treatment by a specially-invented process, which had its foundation in the physical ideas involved, and which proved to be relatively simple in execution. In connection with this matter, the converse process should be noticed, viz., the transformation of definite integrals to operational form, with a view to their evaluation by any means that may then present itself. This is often a very useful practice.

But the transformation from $C=YpE$ to (114) requires to be done with caution. It may be the case that the physics is such as to show that the process is legitimate. But the legitimacy cannot be asserted when Y is an arbitrary operator, of unknown meaning. The resulting integral may not be equivalent, or may be equivalent only within a certain range. That there should be failures in this application of (113) will be understood, at least partly, by considering the application or misapplication of Fourier's theorem in the definite integral form (55), § 271, to a function which does not vanish at infinity, particularly if it is infinite itself there.

A Bessel Cable with one Terminal condition. Two Bessel Cables in Sequence.

§ 329. There is little more to be said about the submarine cable without self-induction, which has been principally useful as a convenient basis upon which to rest the theory of

diffusion. The application, however, of the zeroth Bessel function may be continued in a generalised form, since the

Bessel functions are of great importance in mathematical physics, and have many applications in electromagnetism. Take the problem of § 324, but now put a terminal arrangement Z_1 at $x=\lambda$, instead of having a short-circuit there. The resistance operator at the point λ is now

$$Z = Z_1 + \frac{R_0}{q_0\lambda}\frac{I_0(q_0\lambda)}{I_1(q_0\lambda)}, \tag{115}$$

by (101), using the same notation, and it follows that if the initial state is Q at the point y, the resulting potential is

$$V = \Sigma \frac{Q\, I_0(q_0 x)\, I_0(q_0 y)\, \epsilon^{pt}}{-\left(\frac{q_0}{R} I_1(q_0\lambda)\right)^2 \frac{dZ}{dp}} \tag{116}$$

by the formula (83), § 322, applied to the present case, Z being given by (115), and $Z=0$ being the determinantal equation.

Now choose Z_1 to be another Bessel cable exactly like the first, but turned the other way, so that the zero conductance is at the far end. Then we make a single cable A B C, whose

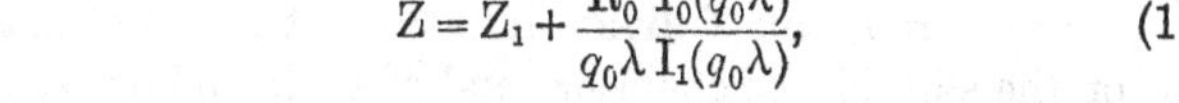

conductance and permittance are greatest at B, and fall off linearly to zero at A and C. The solution (16) relates to the left one, the right one being uncharged. The resistance operator at B is twice that of either cable, or

$$Z = \frac{2R_0}{q_0\lambda}\frac{I_0(q_0\lambda)}{I_1(q_0\lambda)}. \tag{117}$$

This is such that e/Z is the current produced by e at the point B.

But it is clear on consideration that if this Z be used in (116), that result is erroneous, if limited to the roots of $I_0(q_0\lambda) = 0$. For every normal system then has a node at B. This is perfectly right as regards the normal systems concerned when e is at B, because it charges the two cables symmetrically but oppositely, so that $V=0$ at B when e is removed. But it cannot be right in general, for the cables may be charged symmetrically so as to produce no current at B. Besides $I_0(q_0\lambda)=0$, we require another condition giving nodes to the current at B in a set of normal systems.

In fact, if in (115) we let Z_1 be nearly, but not quite, the same as the other term, say by the length of the right cable being slightly different from that of the left one, it will be found that the roots of Z divide into two sets, approximately those of the sum of the operators and of their difference. In the limit, the first lot lead to $I_0(q_0\lambda)=0$, with potential nodes at B, and the second lot to $I_1(q_0\lambda)=0$, with current nodes at B, the I_1 function being the derivative of I_0. The first set of normal systems is solely concerned when the initial state is such as to produce no potential at B, and the second set when such as to produce no current there.

I have introduced this example to illustrate the need of caution in the treatment of normal systems in peculiar cases. It is a necessity for the validity of the expansion of an arbitrary initial state in normal systems that every possible normal system should be included. But it usually happens in these peculiar cases that the peculiar results can be foreseen. It may be noticed that in the denominator in (116), representing $-c^2Z'$, the quantity c vanishes at the point B for all the missing normal functions. The reader has already been cautioned not to choose a point where v vanishes in the alternative form v^2Y'. If we chose Z at some other point, between B and A or C, there would be no failure.

In the treatment of this problem by the operational solution, there is no room for hesitation. Put Q at y in the left cable, and find the resulting V in the usual way. We require, of course, to use the second solution, $K_0(qx)$, along with $I_0(qx)$. The result is, writing q instead of q_0, for convenience, and R instead of R_0,

$$V_2=I_0(qy)\left\{\frac{\pi}{2}K_0(qx)-\frac{I_0(qx)}{2q\lambda\, I_0(q\lambda)\, I_1(q\lambda)}\right\}RpQ, \qquad (118)$$

at x, on the right side of y; and the same expression with x, y interchanged serves when x is on the left side. This result is explicit in making

$$I_0(q\lambda)I_1(q\lambda)=0 \qquad (119)$$

be the determinantal equation when (118) is expanded in a

series of the normal functions $I_0(qx)$, and shows the two sets. The $K_0(qx)$ function, though required in general, then disappears altogether. Developing (118) by the expansion theorem, we get

$$V_2 = V_0 - \Sigma\frac{RpQ\,I_0(qx)\epsilon^{pt}}{\{q\lambda\,I_1(q\lambda)\}^2} - \Sigma\frac{RpQ\,I_0(qx)\epsilon^{pt}}{(q\lambda)^2\,I_0(q\lambda)I_1'(q\lambda)}, \qquad (120)$$

where the first summation is subject to $I_0(q\lambda)=0$, and the second to $I_1(q\lambda)=0$. The accent means differentiation to $q\lambda$. There is a steady term V_0 because there is no earth connection. It is obtained as usual by making $p=0$ in the operational solution (118). The result is

$$V_0 = \frac{Q}{S\lambda^2} = \frac{Q}{\text{total permittance}}, \qquad (121)$$

if the permittance per unit length is Sx in the first cable, and symmetrically in the second. The change to the oscillating functions $J_0(sx)$ and $J_1(sx)$ by means of $q^2 = -s^2$ is obvious. But this is a special case of the next investigation, so it is only needful to record the results as above.

General Solutions for Sources in a Bessel Cable with two Terminal conditions.

§ 330. Passing now to the general case involving the zeroth Bessel functions of both kinds we can obtain our results very shortly and symmetrically by the operational method. Let terminal arrangements Z_0 and Z_1 be put on at the points $x=\lambda$ and l, in a cable subject to the circuital laws

$$-\frac{dV}{dx} = \frac{Z}{x}C, \qquad -\frac{dC}{dx} = YxV, \qquad (1)$$

V and C being the voltage (transverse) and current at x, where the resistance and leakance operators per unit length are Z/x and Yx. Here Z and Y are independent of x, and are preferentially of the forms

$$Y = K + Sp, \qquad Z = R + Lp, \qquad (2)$$

R, L, K, and S being constants, being the values at the point $x=1$ of the resistance, inductance, leakance and permittance per unit length. It is merely the introduction of variation with x in the manner specified that introduces Bessel functions.

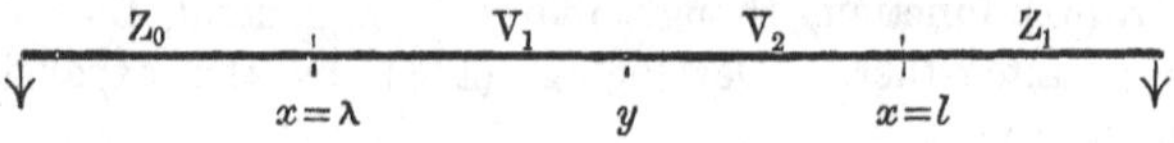

The characteristic of V, obtained by eliminating C from (1), is

$$\frac{1}{x}\frac{d}{dx}x\frac{dV}{dx} = YZV = q^2V. \tag{3}$$

Later on, we can give different meanings to q, according to those assumed for Y and Z.

The two solutions of the characteristic are

$$I_0(qx) = 1 + \frac{(qx)^2}{2^2} + \frac{(qx)^4}{2^2 4^2} + \frac{(qx)^6}{2^2 4^2 6^2} + \ldots,$$

$$K_0(qx) = \frac{2}{\pi}\left\{ -I_0(qx)(\log\frac{qx}{2} + \gamma) + \frac{q^2x^2}{2^2} + (1+\tfrac{1}{2})\frac{q^4x^4}{2^2 4^2} + (1+\tfrac{1}{2}+\tfrac{1}{3})\frac{q^6x^6}{2^2 4^2 6^2} + \ldots \right\}, \tag{4}$$

where $\gamma = 0{\cdot}5772$ is a certain constant introduced to make $K_0(qx)$ vanish at infinity. That these are solutions of the characteristic is easily verified; and since they are not in constant ratio, it follows that they are sufficient to express all solutions directly or indirectly. That is,

$$V = \{I_0(qx) - jK_0(qx)\}A, \tag{5}$$

where A and jA are any time functions, is the general solution of the characteristic, j being any operator.

When qx is numerical and positively real the I_0 function is like ϵ^{qx}, finite at the origin and increasing continuously to infinity at infinity. The K_0 function, on the other hand, is like ϵ^{-qx}, in falling continuously to zero at infinity. It is, however, not finite, but infinite at the origin. Both I_0 and K_0 are always positive when qx is positive.

Now construct the operational solutions for e and h at the point y. The characteristic fails at that point, so we have two forms, say

$$V_1 = \{I_0(qx) - \alpha K_0(qx)\}A, \tag{6}$$

$$V_2 = \{I_0(qx) - \beta K_0(qx)\}B, \tag{7}$$

where V_1 holds when x is on the left, and V_2 when on the right side of y.

In the case of e, we have

$$e = V_2 - V_1, \qquad C_1 = C_2, \tag{8}$$

at the point y. These conditions, applied to (6), (7) determine A and B, and produce

$$V_1 = \frac{\pi q y}{2} e \frac{(I_{0x} - \alpha K_{0x})(I_{1y} - \beta K_{1y})}{\alpha - \beta}, \tag{9}$$

$$V_2 = \frac{\pi q y e}{2} \frac{(I_{0x} - \beta K_{0x})(I_{1y} - \alpha K_{1y})}{\alpha - \beta}, \tag{10}$$

which are quite general, as due to the source e. The operators α and β enable us to introduce terminal conditions. Say

$$V = Z_1 C \text{ at } x = l, \qquad V = -Z_0 C \text{ at } x = 0, \tag{11}$$

These determine α and β, thus,

$$\alpha = \frac{I_{0\lambda} - (q\lambda Z_0/Z) I_{1\lambda}}{K_{0\lambda} - (q\lambda Z_0/Z) K_{1\lambda}}, \qquad \beta = \frac{I_{0l} + (qlZ_1/Z) I_{1l}}{K_{0l} + (qlZ_1/Z) K_{1l}}. \tag{12}$$

Equations (9), (10) are now made fully explicit in terms of the terminal operators, and may be developed in Bessel series.

As regards the notation, I_1 and K_1 are the derivatives of I_0 and K_0; and, to ease matters, $I_0(qx)$ is denoted by I_{0x}, and similarly for the rest.

When h is the source, we still use (6), (7), but instead of (8) we have the conditions

$$V_1 = V_2, \qquad C_2 - C_1 = h, \tag{13}$$

at y. These find A and B, and make

$$V_1 = \tfrac{1}{2}\pi Z h \frac{(I_{0x} - \alpha K_{0x})(I_{0y} - \beta K_{0y})}{\alpha - \beta}, \tag{14}$$

$$V_2 = \tfrac{1}{2}\pi Z h \frac{(I_{0x} - \beta K_{0x})(I_{0y} - \alpha K_{0y})}{\alpha - \beta}. \tag{15}$$

The expressions for α and β are the same as before. The determinantal equation of normal systems is

$$\alpha = \beta. \tag{16}$$

The h solutions are required to represent initial states of potential, and the e solutions those of current. In obtaining them use was made of the identity

$$I_0(z)K_1(z) - I_1(z)K_0(z) = -\frac{2}{\pi z}, \tag{17}$$

which often turns up in these investigations.

Conjugate Property of Voltage and Gaussage Solutions.

§ 331. It is often to be observed that a property, first noticed in a special case, then in others, if it be really a general property, admits of simple proof. Perhaps the more general the property the simpler the proof. The last equation is an example. It is a particular case of the known conjugate relation between the two solutions of the ordinary equation of the second order. The proof is simplest in terms of V and C. Thus, let

$$-\frac{dV}{dx} = \rho C, \qquad -\frac{dC}{dx} = \kappa V, \tag{18}$$

be the connections. Here ρ and κ are, in the simplest case, the resistance of the wires and the conductance of the insulator, per unit length of circuit. They may be any functions of x. More generally they are operators, functions of p the time-differentiator, when variable states are concerned, as well as functions of x. The characteristic of V is

$$\frac{1}{\kappa}\frac{d}{dx}\frac{1}{\rho}\frac{dV}{dx} = V. \tag{19}$$

Now let v_1, c_1 and v_2, c_2 be any two solutions subject to (18). Then, since

$$-\frac{d}{dx}(v_1 c_2) = -v_1\frac{dc_2}{dx} - c_2\frac{dv_1}{dx},$$

we have, by (18),

$$-\frac{d}{dx}(v_1 c_2) = v_1 \kappa v_2 + c_2 \rho c_1. \tag{20}$$

Similarly,

$$-\frac{d}{dx}(v_2 c_1) = v_2 \kappa v_1 + c_1 \rho c_2 \tag{21}$$

Now if we make p be a constant, and the *same* constant in the two systems, κ and ρ become identical functions of x in the last two equations, so that the right members are the same. Therefore

$$\frac{d}{dx}(v_1 c_2 - v_2 c_1) = 0, \tag{22}$$

and consequently

$$v_1 c_2 - v_2 c_1 = m, \tag{23}$$

where m is constant. This is the conjugate property.

If the two systems coexist, making $v=v_1+v_2$, and $c=c_1+c_2$ be the voltage and current, the product vc is the total energy flux, and its rate of decrease with x is its leakage, or the activity per unit length of circuit. It consists of the sum of the leakages from v_1c_1, v_2c_2, v_1c_2 and v_2c_1; and the reciprocal property in the form (22) asserts that the parts depending on cross products are equal. When p is real, the systems are real, and their energy, &c. When p is imaginary, the best way is to imagine it to be real, for the sake of the electrical argument. It is easy to generalise the property by extension to space distributions of electric and magnetic force, but we do not want that at present.

That p is to be the same constant in the two states, limits us to any two solutions of the ordinary equation which the characteristic (19) becomes, when the differentiator p is replaced by a constant. For example, ϵ^{qx} and ϵ^{-qx}, or combinations; $\mathrm{I}_0(qx)$ and $\mathrm{K}_0(qx)$, or combinations; and so on to other kinds. In terms of the v's only, the conjugate property is

$$v_1v_2' - v_2v_1' = -m\rho, \tag{24}$$

where the accent indicates differentiation to x; and, in terms of the c's only, it is

$$c_1c_2' - c_2c_1' = m\kappa. \tag{25}$$

Thus (24) relates to the two solutions of (19), and (25) to those of the corresponding characteristic for c, which is not the same. If the two v solutions taken specially are in constant ratio, m vanishes, of course. When not in constant ratio, they are said to be independent. The independence is, however, of a limited nature, since, when one is given, another can be deduced by the conjugate property.

Operational Solutions for Sources in the General Case.

§ 332. By means of the conjugate property we are enabled to obtain the operational solutions for sources e and h. It will be convenient to denote two independent solutions by u and w, instead of v_1 and v_2, so that the property is

$$uw' - wu' = -m\rho. \tag{26}$$

Now there is a discontinuity at a source, so two forms of solution for the potential are required, say

$$V_1 = (u_x - \alpha w_x)A, \qquad V_2 = (u_x - \beta w_x)B, \tag{27}$$

where V_1 is on the left side of the source and V_2 on the right. These are operational, p meaning the time-differentiator, α and β unknown operators, A and B unknown time-functions.

Specialise the source to be h at y. This makes $V_1 = V_2$, and $h = C_2 - C_1$ at y. Applying these conditions to (27), using the conjugate property (26), we produce

$$V_1 = \frac{(u_x - \alpha w_x)(u_y - \beta w_y)}{m(\alpha - \beta)} h, \tag{28}$$

$$V_2 = \frac{(u_x - \beta w_x)(u_y - \alpha w_y)}{m(\alpha - \beta)} h, \tag{29}$$

which are explicit when α and β are known.

When the source is e at y, which makes $C_1 = C_2$ and $e = V_2 - V_1$ at y, we similarly produce the results

$$V_1 = \frac{(u_x - \alpha w_x)(u'_y - \beta w'_y)}{mp(\alpha - \beta)} e, \tag{30}$$

$$V_2 = \frac{(u_x - \beta w_x)(u'_y - \alpha w'_y)}{mp(\alpha - \beta)} e. \tag{31}$$

The value of the constant m depends upon the standardisation of the functions u, w. It can be found as soon as they are given, by inserting them in the conjugacy equation. Only the commencement of the series need be used.

The operator α is determined by V/C at any point on the left of the source, and β similarly by V/C at any point on the right. Say $V/C = -Z_0$ at λ, and $= Z_1$ at l, then

$$\alpha = \frac{u_\lambda - (Z_0/\rho)u'_\lambda}{w_\lambda - (Z_0/\rho)w'_\lambda}, \qquad \beta = \frac{u_l + (Z_1/\rho)u'_l}{w_l + (Z_1/\rho)w'_l}. \tag{32}$$

By means of Z_0 and Z_1, we allow for the reaction of all the parts of the electrical system beyond the limits upon the part between them. That is to say, the operational solutions are complete for the region between λ and l, so far as the source e or h is concerned. The conversion to series of normal solutions is to be done in the usual way, and initial states of potential and current by $e=pP$, $h=pQ$, as before described.

Comparing the present investigation with that in § 330, we see that u is represented by $I_0(qx)$ and w by $K_0(qx)$, and that the constant m is $2/\pi Z$. It is constant as regards x. It contains p (or may do so).

Conversion of Wire Waves to Cylindrical Waves. Two Ways.

§ 333. Returning to the Bessel investigation of § 330, which contains most of the fundamental formulæ involving the zeroth Bessel functions of both kinds, by means of certain equivalent transformations we can develop the whole body of results, ordinary and extraordinary. But before doing that, it is desirable to point out that the formulæ are not merely those concerned in the theory of a telegraph line of a variable nature, of no practical use, but belong to cylindrical electromagnetic waves in general. Moreover, they do not do so by mere analogy, but, by proper interpretation of symbols, the two theories may be seen to be one and the same essentially.

Thus, I have shown in § 206, Vol. I., how, in the transmission of plane electromagnetic waves through a conducting dielectric, any tube of energy flux may be isolated from the rest and made independent, by enclosing it in a conducting tube made of two materials. The electric displacement in the tube must terminate perpendicularly upon perfect electric conductors forming two sides of the tube, and the magnetic induction upon perfect magnetic conductors forming the other two sides of the tube. Also, the electric force is tangential to the two magnetic conductors, and the magnetic force to the two electric conductors, whilst the two forces cross one another perpendicularly within the tube. Under these circumstances the internal plane waves are made self-supporting, guided by the tube. The internal state is like a beam of light, with lateral spreading prevented.

We may conveniently take the tube to be of square section for plane waves. Then the lines of force are straight. But in the present application, the tube must be imagined to be of the shape of a wedge. It increases in section regularly as we pass along it, but two of its flat sides remain equidistant, whilst the other two separate. Looking at the tube in section along its length, A B and A C represent the traces of the separating sides, to be called the top and bottom, whilst the others are parallel to the plane of the paper, and may be called simply the sides.

Now let straight lines of **H** join the sides, or be perpendicular to the plane of the paper. The **E** lines must then join the top and bottom, and be circular, centred at A, as at E in the diagram. Also, let E and H, the intensities of **E** and **H**, be each the same at the same distance from A. Under these circumstances we shall have propagation of electromagnetic waves in the tube according to the formulæ given.

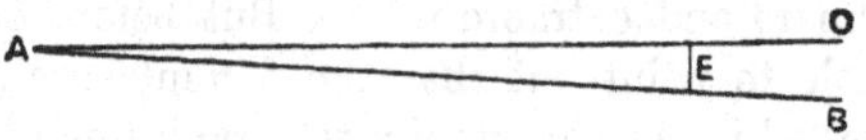

For, consider how the inductance and permittance of the tube vary. The current C is the same as H × depth of tube, being the transverse gaussage. The "potential" V, with an enlarged meaning of potential, is E × by length of arc concerned, or the transverse voltage. The inductance (per unit length of tube) varies directly as x, the distance from A, because the section of the tube of induction belonging to unit length of the tube of energy flux varies as x, whilst its length is constant. On the other hand, the permittance varies inversely as x, because the section of the corresponding tube of displacement is constant, whilst its length varies as x. The circuital equations are therefore

$$-\frac{dV}{dx} = \mathrm{L}xp\mathrm{C}, \qquad -\frac{dC}{dx} = \frac{\mathrm{S}}{x}p\mathrm{V}, \tag{33}$$

if Lx and S/x are the inductance and permittance per unit length of tube, L and S being constants, the values of the same at $x=1$. The above makes no allowance for any conductivity within the tube, which is itself merely a guide.

Next, let the medium in the tube be uniformly conducting, both electrically and magnetically. Then the magnetic conductance will vary like L, that is, as x; and the electric conductance will vary like S, or as x^{-1}. We shall therefore make the circuital equations take the form

$$-\frac{dV}{dx}=(R+Lp)xC, \qquad -\frac{dC}{dx}=(K+Sp)\frac{1}{x}V, \qquad (34)$$

if Rx and K/x are the magnetic and electric conductances per unit length of tube. So far is an exact theory. The reader should study Chapter IV., or the part relating to the transformation of the circuital equations in general to the special forms for plane waves along wires, and the interpretation of the quantity called magnetic conductivity. It is the same theory now, only applied to a tube of energy flux of variable section.

If we remove R from the inside of the tube, and substitute equivalent electric resistance in its top and bottom, the equations remain approximately true when the angle of the wedge is small, and the distance x not too great. It is now the theory of waves along a pair of finitely resisting electric conductors through an electrically conducting dielectric medium. If we remove K too, we must substitute equivalent magnetic resistance in the other pair of conductors, the side pair. But since K expresses a really existent property, it is no use doing that. Therefore retain K within the tube.

The depth of the wedge is immaterial. Make it infinite, and do away with the side magnetic conductors altogether. The wedge now consists of two planes (top and bottom) inclined at an angle. The magnetic flux is endless, which is equivalent to being circuital. If, now, in addition R is zero, the angle of the wedge becomes immaterial. This is also the case when R is not zero, provided it be in the tube of energy flux, and not at top and bottom. Increase the angle until it becomes 360deg. We may then remove the top and bottom plates altogether. The result is complete cylindrical waves in a conducting dielectric, the conductivity being of either kind or of both kinds, as we please. The voltage V is now circuital, and finite. The measure of C, on the other hand, is infinite. We need only, however, deal with a finite depth along the axis, when C will be finite too.

But in § 330 the resistance of the wires varied as x^{-1}, and the permittance as x. To obtain this case, we have merely to interchange the electric and magnetic forces in the above.

The magnetic force must be in the plane of the paper in the figure, and parallel thereto, and be circular. The electric force must be straight and perpendicular to the paper. So, to begin with, the sides of the tube must be perfect electric conductors, and the top and bottom perfect magnetic conductors. Then, if the medium in the tube is uniformly conducting in both ways, the quantities K and S will vary as x, and the quantities R and L as x^{-1}.

The same substitutions of resistance in the bounding conductors for the internal conductance may be made as before, allowing for the change in position of the electric and magnetic forces. Also, the bounding conductors may be wholly removed, leaving complete cylindrical waves in a homogeneous dielectric, conducting in either or both wavs. The gaussage C is now reckoned once round the circle of magnetic force. To make V finite we may consider only a finite depth along the axis.

We see that the problem of a telegraph circuit has real application to electromagnetic waves of much greater generality than plane waves, by letting the "constants" of the circuit be variable. The process employed is not confined in its application to cylindrical waves. Spherical waves may be similarly treated, by employing a tube whose section varies as x^2, both pairs of sides separating. Both **E** and **H** will be along arcs of circles, centred at the axis. In fact, the theory of this case is much simpler than the other, but we must not run away from the matter immediately in hand.

Special Cases of Zeroth Bessel Solutions.

§ 334. After the enlargement of ideas of the last section, we may partially discuss some of the operational results. Notice first that the results for e in § 330 are produced from those for h, by substituting $eqy(d/dqy)$ for Zh. We may keep at present to the h set. Thus, take equation (14), or

$$V_1 = \tfrac{1}{2}\pi Zh \frac{(I_{0x} - \alpha K_{0x})(I_{0y} - \beta K_{0y})}{\alpha - \beta}, \qquad (35)$$

giving V_1 at $x<y$ due to h at y. Interchange x and y when V_2 is wanted.

If l is infinity, K_{0l} is zero, and β is infinite, by (12). This reduces (35) to

$$V_1 = \tfrac{1}{2}\pi Zh(I_{0x} - aK_{0x})K_{0y}. \tag{36}$$

We see that the terminal condition at l becomes impotent, and disappears.

Again, if $\lambda = 0$, $K_{0\lambda}$ is infinite, and $a = 0$, by (12). So

$$V_1 = \tfrac{1}{2}\pi ZhI_{0x}\left(K_{0y} - \frac{I_{0y}}{\beta}\right). \tag{37}$$

Thus the terminal condition at λ becomes impotent and disappears by shifting it to the origin. Also note that in the region reaching to the origin, K_{0x} does not appear. It is usual to consider that we cannot have the K_0 function in solutions extending to the origin (the axis of the cylinder in cylindrical waves), unless there is a source at the origin, because to do so would make V infinite there. That this reason is inadequate and unsatisfactory will be evident if it be allowed that V can be infinite at the origin without having a source there, of which I will give examples. That is, we obtain infinite results at the origin without using K_{0x}. Nevertheless, there is no doubt that the K_0 function is excluded unless there is a source at the origin, as we saw above, unless there can be something very peculiar about the Z_0 operator, making a, which is given by

$$a = \frac{I_{0\lambda} - (q\lambda Z_0/Z)I_{1\lambda}}{K_{0\lambda} - (q\lambda Z_0/Z)K_{1\lambda}}, \tag{38}$$

be finite when $\lambda = 0$.

If both $\lambda = 0$ and $l = \infty$, then $a = 0$, $\beta = \infty$. Both terminal conditions are impotent, and

$$V_1 = \tfrac{1}{2}\pi ZhI_{0x}K_{0y}, \qquad V_2 = \tfrac{1}{2}\pi ZhI_{0y}K_{0x}, \tag{39}$$

which represent an important fundamental solution.

If $y = \lambda$, we have only the V_2 solution under control, and

$$V_2 = \tfrac{1}{2}\pi Zh\frac{I_{0x} - \beta K_{0x}}{a - \beta}(I_{0\lambda} - aK_{0\lambda}). \tag{40}$$

If $y = \lambda = 0$, we reduce the last to

$$V = \tfrac{1}{2}\pi Zh(K_{0x} - \frac{I_{0x}}{\beta}) \tag{41}$$

The source is at the origin, and the Z_0 condition is impotent.

Finally, if $l = \infty$, $\beta = \infty$, and (41) reduces to

$$V = \tfrac{1}{2}\pi Z h K_{0x}. \qquad (42)$$

Thus, with the source at the origin, we have the K_0 function alone if there is no reflection, but the I_0 function as well when there is reflection at a finite distance, as in (41).

Similarly as regards e. By (9),

$$l = \infty \quad \text{makes} \quad V_1 = \tfrac{1}{2}\pi q y e (I_{0x} - \alpha K_{0x}) K_{1y}, \qquad (43)$$

$$\lambda = 0 \quad \text{makes} \quad V_1 = \tfrac{1}{2}\pi q y e I_{0x}(K_{1y} - \beta^{-1} I_{1y}), \qquad (44)$$

$$l = \infty, \ \lambda = 0 \quad \text{make} \quad V_1 = \tfrac{1}{2}\pi q y e I_{0x} K_{1y}, \qquad (45)$$

$$y = \lambda = 0 \quad \text{make} \quad V_2 = 0. \qquad (46)$$

In the last case the finite e produces no disturbance beyond the origin, because of the vanishing conductance there. In (42), on the other hand, h is given to be finite. That it needs infinite impressed voltage is immaterial.

Numerical Interpretation of Formulæ. The Divergent Series.

§ 335.   The above results are usually operational. But they are strictly numerical results in certain cases, without change of form, or perhaps with only nominal change. Thus when the source gives rise to a steady state we can use the above at once. Putting $p = 0$ makes $q^2 = RK$, which is constant. Then I_{0x}, &c., are all algebraical functions and are numerical when x is given. So, knowing the general nature of the curves I_0 and K_0, we can see by inspection in some cases the nature of the steady states on both sides of the source. We have, of course, when α and β are concerned, to make Z_0 and Z_1 represent the steady resistances terminally at λ and l.

In these direct numerical applications, use the formulæ (4) when qx is small. But they are quite unsuitable when qx is big.   Then use the approximate formulæ

$$I_0(qx) = \frac{\epsilon^{qx}}{(2\pi qx)^{\frac{1}{2}}}, \qquad K_{0x} = \left(\frac{2}{\pi qx}\right)^{\frac{1}{2}} \epsilon^{-qx}, \qquad (47)$$

which show plainly how the functions behave ultimately. The quantity $8qx$ should be several times unity to make (47) be practically true. More generally, use the formulæ

$$I_0(qx) = \frac{\epsilon^{qx}}{(2\pi qx)^{\frac{1}{2}}}\left(1 + \frac{1^2}{8qx} + \frac{1^2 3^2}{\lfloor 2(8qx)^2} + \frac{1^2 3^2 5^2}{\lfloor 3(8qx)^3} + \ldots\right), \quad (48)$$

$$K_0(qx) = \frac{\epsilon^{-qx}}{(\frac{1}{2}\pi qx)^{\frac{1}{2}}}\left(1 - \frac{1^2}{8qx} + \frac{1^2 3^2}{\lfloor 2(8qx)^2} - \frac{1^2 3^2 5^2}{\lfloor 3(8qx)^3} + \ldots\right). \quad (49)$$

The first terms represent (47). These are divergent formulæ, and, to obtain the proper values, stop when you reach the bottom, and ignore the rest. That is, seek the point of convergence of the series you are calculating, or the place where the smallest term occurs. The error is limited by the size of the smallest term, and may be far less, especially if half the smallest term is counted. The above are the practical formulæ except when qx is so small that the smallest term may be big enough to introduce a large error. Even when qx is as small as 1, the error is not very great. It is when the point of convergence, which is at a distance along the series when qx is big, comes to the beginning of the series, that we are obliged to go to the convergent formulæ (4). The difference between the divergent and convergent series is, for numerical calculation, only one of degree, and the degree varies. In the convergent series, the point of convergence is always at the end of the series, which cannot be reached. But the terms at the end tend to zero in size, so that by taking enough trouble we may reduce the error to be smaller than any quantity that can be named. But then, when qx is big, a very large number of terms must be counted before we come to the convergent region, so the convergent formulæ are unpractical. On the other hand, if $8qx$ is greater than 1, the divergent formulæ are convergent from the very beginning, and allow of rapid calculation with an accuracy which is practically unexceptionable when qx is a large number, and of fair approximativeness so long as the point of convergence is not too near the first term. It is a question of practice which formulæ to use. The connections between the convergent and divergent formulæ will be given in Chapter VII., as well as their connections with other formulæ. There is a table of the I_0 and I_1 functions (and others) in Gray and Mathew's work on Bessel functions, p. 282, going up to $qx = 5{\cdot}1$. After that (48) may be used, or even (47), noting that I_1 is the derivative of I_0. But I cannot find any table of the important K_0 function, and therefore use (49) myself, or else (4), to get rough values.

Since, as has been said, the above formulæ become numerical when steady states are concerned, it is not necessary to rewrite them, with $\sqrt{RK}$ substituted for q, to indicate the special results. Observe, however, that should either R or K vanish, or both vanish, the formulæ simplify considerably, when steady states are concerned, and it will not be a bad thing for the reader to look into a few cases in detail, both with q finite first, and then with $q=0$, by the vanishing of R or K.

Look at (39) for example. The constant impressed current h at y produces V as there given, with $q^2 = RK$ in the I_0 and K_0 functions, and $Z = R$. Now, if R is zero, V is zero on both sides. The corresponding current is $C_1 = 0$, on the left, and $C_2 = h$, on the right side of the source. That is, when $Z = Lp$ and $Y = Sp$, representing the case of no waste of energy by resistance, h all goes to the right ultimately, and there is no electrification. The absence of electrification is the striking point. It is really a case of statics, or of motional statics. The setting up of the steady state involves electromagnetic waves, of course, but they disappear ultimately. [Not by actual cessation, but by transference to a great and continuously increasing distance of the variable state. Details later.]

The divergent Formulæ are fundamental. Generation of Waves in a Medium whose Constants vary as the nth power of the distance.

§ 336. The next thing to be done is to show how the Bessel solutions in general are generated by waves. This was done before in the case of a uniform cable, leading to Fourier series. We take a source at a point, and construct the double wave it generates, going to right and left. Then we add on, one after another, terms representing the reflected waves generated at the terminals by the incident first waves and their successors and overlappers. The complete set of waves, when in operational form, makes a geometric series, which, when summed, expresses the operational solution in condensed form. Conversion to series of normal functions then follows by the expansion theorem. So we prove that the series of waves and the series in terms of normal functions are equivalent.

To do this for the Bessel series is not a useless mathematical complication, but is of fundamental importance, both from the physical and the mathematical side. From the physical side because in Maxwell's theory all disturbances of electric or magnetic force are propagated at a speed v settled by the inductivity and permittivity, which property is not interfered with by the simultaneous existence of conductivity (electric or magnetic). All solutions involving change are, therefore, ultimately of wave nature. This is also true when, by artificial restrictions, we make v be infinite. In such limiting cases we do not see any succession of waves in time, for they are made simultaneous, but there is no essential difference in the principle. The waves are there, all the same. Also, from the mathematical side, the question is important in casting light upon the connection between the two kinds of Bessel formulæ, the convergent and the divergent.

The wave analysis and synthesis is nearly as easily done for the Bessel waves of mth order as for the zeroth, and is made plainer by the extension. Therefore, let

$$-\frac{dV}{dx}=\frac{Z}{x^n}C, \qquad -\frac{dC}{dx}=Yx^nV, \tag{1}$$

be the circuital connections of voltage and current, Y and Z being preferentially $K+Sp$ and $R+Lp$, so that the resistance and inductance per unit length of circuit vary as x^{-n}, and the leakage and permittance as x^n. The characteristic of V is

$$\frac{1}{x^n}\frac{d}{dx}x^n\frac{dV}{dx}=q^2V, \quad \text{or} \quad \frac{d^2V}{dx^2}+\frac{n}{x}\frac{dV}{dx}=q^2V, \tag{2}$$

where $q^2=YZ$.

This is the theory of the propagation of plane waves in a medium whose specific properties vary as the nth or $-n^{\text{th}}$ power of the distance from a fixed plane; or of cylindrical waves when the variation is as the $(n-1)^{\text{th}}$ power, and so on to other dimensions. But n is not restricted to be an integer; it may be fractional if we like.

The first step is to find the wave operators. Let

$$\mathrm{H}_m(qx) = \left(\frac{2}{\pi qx}\right)^{\frac{1}{2}} \epsilon^{qx} \left\{1 + \frac{1^2 - 4m^2}{\lfloor 1 \; 8qx} + \frac{(1^2 - 4m^2)(3^2 - 4m^2)}{\lfloor 2 \; (8qx)^2} + \ldots\right\}, \tag{3}$$

$$\mathrm{K}_m(qx) = \left(\frac{2}{\pi qx}\right)^{\frac{1}{2}} \epsilon^{-qx} \left\{1 - \frac{1^2 - 4m^2}{\lfloor 1 \; 8qx} + \frac{(1^2 - 4m^2)(3^2 - 4m^2)}{\lfloor 2 \; (8qx)^2} - \ldots\right\}, \tag{4}$$

These are to define the functions H and K. It may be readily tested that $x^{-m}\mathrm{H}_m(qx)$ and $x^{-m}\mathrm{K}_m(qx)$ satisfy the characteristic (2), provided $n = 1 + 2m$, or $m = \frac{1}{2}(n-1)$.

Now H and K are divergent series, that is, when qx is numerical. They are, nevertheless, actually the fundamental Bessel formulæ, because they are the wave generators. It is not difficult to see this. Suppose there is no waste of energy by resistance. Then $q = p/v$, and ϵ^{qx} is a mere translation operator, turning t to $t + x/v$ in any time function on which it works. Similarly ϵ^{-qx} turns t to $t - x/v$. Therefore, by § 276, the H operator belongs to an inward wave, and the K operator to an outward wave, if there is nothing else to interfere with this property. Now, in the case of a homogeneous medium $n = 0$, and $m = -\frac{1}{2}$. Then H is simply proportional to ϵ^{qx} and K to ϵ^{-qx}, indicating propagation at constant speed without distortion. In other cases, H is $\epsilon^{qx}\mathrm{F}$, and K is $\epsilon^{-qx}\mathrm{G}$, where F, G are as in (3). The translation is still done by the exponential operators, but the operand is altered first by F or G. So these operators do the distortion, or deformation of the waves.

When there is resistance, q is not p/v. But the exponential operators may still be reduced to $\epsilon^{px/v} f$ and $\epsilon^{-px/v} g$, when f and g are other operators also concerned in producing distortion. Thus, from the fundamental fact that disturbances travel at speed v, the forms of H and K prove that they are the wave operators, H for an inward, and K for an outward wave. If this general reasoning is not convincing, I may add that the algebrisation of (3), (4) with an operand added is quite easily effected in various simple cases (and the extension to the general case is only a matter of complication) with the results as above described. Some cases will be given in Chap. VII.

Having settled this point, use H and K to show the primary waves generated by a source. Let there be a source h at y. Then

$$V_1=\frac{\pi}{4}Zh\frac{H_{mx}}{x^m}\frac{K_{my}}{y^m},\qquad V_2=\frac{\pi}{4}Zh\frac{H_{my}}{y^m}\frac{K_{mx}}{x^m},\tag{5}$$

show the waves, V_1 to the left of y, and V_2 to the right. To prove this, note first that the proper operators containing x are used for the waves. Next, that $V_1=V_2$ at y. Thirdly, that $C_2-C_1=h$ at y. The last is done by the conjugate property of the H, K functions, which is

$$(HK'-H'K)_x=-\frac{4}{\pi x},\tag{6}$$

if the accent means differentiation to x.

In (5) the operand is h, which may be any function of the time, but if we take it to be constant, beginning at the moment $t=0$, then V_1 and V_2 represent waves of V starting from y at that moment, and travelling to right and left at speed v.

Construction of General Solution by Waves.

§ 337. If there are no terminal impositions—that is, if the characteristic is valid (except at the source) all the way from $x=0$ to ∞, the V_2 wave goes on to infinity. But the V_1 wave, when it reaches the origin, generates a reflected wave, which also travels out to infinity. There are no more waves than these. But should there be a boundary on the right of y, say at l, both the outward waves, primary and secondary,

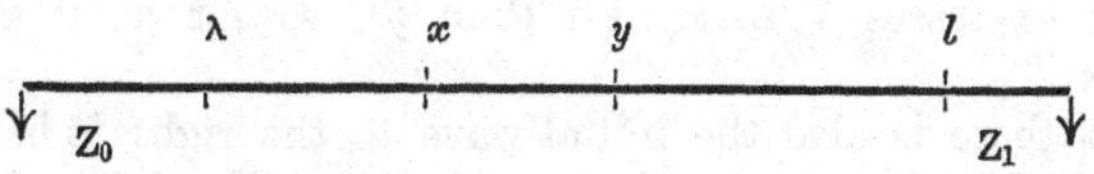

generate reflected waves there, and both these inward waves generate outward waves when they reach the origin, and so on for ever. It is the same when the origin is not in question, but the place of inner reflection is at $x=\lambda$, on the left of x and y. It is now easy to construct the wave series.

Let $x=\lambda$ and l be the boundaries, and the point x where V is wanted be on the left side of the source h at y. The first wave is

$$v_1=\frac{\pi}{4}\,\mathrm{Z}h\cdot\frac{\mathrm{K}_{my}}{y^m}\frac{\mathrm{H}_{mx}}{x^m}, \tag{7}$$

by the first of (5). Put $x=\lambda$ to produce its value at λ. Multiply by a reflection coefficient, say a', to produce the value of the reflected wave at the same place. Then multiply by $(x^{-m}\,\mathrm{K}_{mx})/(\lambda^{-m}\,\mathrm{K}_{m\lambda})$ to express the reflected wave at any point, because this operator is 1 at λ, and has the right x function in it. The result is

$$v_2=\frac{\pi}{4}\,\mathrm{Z}h\,\frac{\mathrm{K}_{my}}{y^m}\left(a'\frac{\mathrm{H}_{m\lambda}}{\mathrm{K}_{m\lambda}}\right)\frac{\mathrm{K}_{mx}}{x^m}=\frac{\pi}{4}\,\mathrm{Z}h\,\frac{\mathrm{K}_{my}}{y^m}\,a_1\frac{\mathrm{K}_{mx}}{x^m}, \tag{8}$$

if a_1 is the operator in the brackets.

Put $x=l$ to give the value of v_2 at l. Then multiply by a reflection coefficient, say b', to give the value at the same place of the third (inward wave) generated by v_2. Finally, multiply by $(x^{-m}\,\mathrm{H}_{mx})/(l^{-m}\,\mathrm{H}_{ml})$ to produce the complete third wave anywhere. The result is

$$v_3=v_1\left(a'\frac{\mathrm{H}_{m\lambda}}{\mathrm{K}_{m\lambda}}\right)\left(b'\frac{\mathrm{K}_{ml}}{\mathrm{H}_{ml}}\right)=v_1a_1b_1, \tag{9}$$

if b_1 is the operator in the second brackets. After this, it is the same over and over again. Thus,

$$v_4=v_2a_1b_1,\qquad v_5=v_3a_1b_1,\qquad v_6=v_4a_1b_1,\qquad \&c. \tag{10}$$

The sum of all these waves is

$$\Sigma v=\frac{v_1+v_2}{1-a_1b_1}, \tag{11}$$

which expresses V at x (less than y), so far as it arises from v_1.

But there is also the initial wave to the right to be considered. It does nothing at x itself directly, but only by reflection at l. Initially the primary wave is the V_2 in (5). Put $x=l$, and multiply by $b'(x^{-m}\mathrm{H}_{mx})/(l^{-m}\mathrm{H}_{ml})$, to produce the first inward reflected wave, say w_1, which can act at x. It is

$$w_1=\frac{\pi}{4}\,\mathrm{Z}h\frac{\mathrm{H}_{my}}{y^m}\,b_1\frac{\mathrm{H}_{mx}}{x^m}. \tag{12}$$

After this it is to be treated in the same way as v_1 was, for both v_1 and w_1 are inward waves. Therefore

$$w_2 = \frac{\pi}{4} Zh \frac{H_{my}}{y^m} b_1 a_1 \frac{K_{mx}}{x^m} \tag{12A}$$

is the next outward wave, and the rest are

$$w_3 = w_1 a_1 b_1, \qquad w_4 = w_2 a_1 b_1, \qquad w_5 = w_3 a_1 b_1, \qquad \&c., \tag{13}$$

Consequently the sum of the w waves is

$$\Sigma w = \frac{w_1 + w_2}{1 - a_1 b_1}, \tag{14}$$

and the complete V_1 at x is $\Sigma v + \Sigma w$; that is

$$V_1 = \frac{\frac{1}{4}\pi Zh}{1 - a_1 b_1} \left(\frac{H_{mx}}{x^m} + a_1 \frac{K_{mx}}{x^m} \right) \left(\frac{K_{my}}{y^m} + b_1 \frac{H_{my}}{y^m} \right). \tag{15}$$

It only remains to find a_1 and b_1, or a' and b', the reflection coefficients. Let $V = Z_1 C$ at l, and consider any outward wave and the reflected wave superposed. They are represented by

$$\frac{K_{mx}}{x^m} + b_1 \frac{H_{mx}}{x^m}, \tag{16}$$

multiplied by a factor not containing x. The corresponding currents are

$$-\frac{x^{m+1}q}{Z} \left(K_{m+1,x} - b_1 H_{m+1,x} \right), \tag{17}$$

by the first of (1), and using these properties of the functions,

$$\frac{1}{q} \frac{d}{dx} \frac{H_m}{x^m} = \frac{H_{m+1}}{x^m}, \qquad \frac{1}{q} \frac{d}{dx} \frac{K_m}{x^m} = -\frac{K_{m+1}}{x^m}, \tag{18}$$

which are easily verified by the definition of H and K above. The ratio of (16) to (17) is therefore Z_1 when $x = l$. So

$$Z_1 = \frac{K_{ml} + b_1 H_{ml}}{(l^n q/Z)(K_{m+1} - b_1 H_{m+1})}, \tag{19}$$

and therefore

$$b_1 = \frac{-K_{ml} + BK_{m+1,l}}{H_{ml} + BH_{m+1,l}}, \qquad \text{if } B = \frac{l^n Z_1 q}{Z}. \tag{20}$$

In a similar manner, let $V = -Z_0C$ be the condition at the other boundary λ, then we shall find by the proper change of symbols that

$$a_1 = \frac{-H_{m\lambda} + AH_{m+1,\lambda}}{K_{m\lambda} + AK_{m+1,\lambda}}, \qquad \text{if } A = \frac{\lambda^n Z_0 q}{Z}. \tag{21}$$

We may therefore write (15) symmetrically thus:—

$$V_1 = \frac{\pi Z h}{4x^m y^m} \frac{(H_{mx} + aK_{mx})(H_{my} + bK_{my})}{b - a}, \tag{22}$$

where a stands for a_1 as in (21), and b for the reciprocal of b_1 in (20), so that a is derived from b by turning l to λ and Z_1 to $-Z_0$. We have only to interchange x and y to obtain V_2 on the right side of the source, and the results are explicit in terms of the terminal resistance operators. They are derived entirely from the initial waves from the source, by the addition of the reflected waves. The divergent series are therefore essential to a proper understanding of the matter.

Construction of General Solution by the Convergent Formulæ.

§ 338. Now consider a radically different way of viewing the subject. Besides the divergent series there are convergent series satisfying the characteristic. Thus $x^{-m} I_{mx}$ and $x^{-m} I_{-mx}$ are convergent solutions of the characteristic (2), if

$$I_m(qx) = \frac{(\frac{1}{2}qx)^m}{\underline{|0}\,\underline{|m}} + \frac{(\frac{1}{2}qx)^{m+2}}{\underline{|1}\,\underline{|m+1}} + \frac{(\frac{1}{2}qx)^{m+4}}{\underline{|2}\,\underline{|m+2}} + \dots, \tag{23}$$

and I_{-mx} or $I_{-m}(qx)$ is got by turning m to $-m$. These I functions are always finite at the origin when m is positive or zero, and infinite when m is negative. It follows that

$$V_1 = \frac{I_{mx} + rI_{-mx}}{x^m} A_1, \qquad V_2 = \frac{I_{mx} + sI_{-mx}}{x^m} B_1, \tag{24}$$

are general solutions, and represent V on left and right of a source h provided r, s, A_1 and B_1 are properly found.

First, $V_1 = V_2$ at y, the place of the source h, makes

$$B_1 = \frac{I_{my} + rI_{-my}}{I_{my} + sI_{-my}} A_1, \tag{25}$$

so B_1 is eliminated. Next, A_1 is to be found by $C_2 - C_1 = h$ at y. Now the C's corresponding to (24) are

$$C_1 = -\frac{q}{Z}x^{m+1}(I_{m+1,x} + rI_{-m-1,x})A_1, \qquad C_2 = -\frac{q}{Z}x^{m+1}(I_{m+1,x} + sI_{-m-1,x})B_1, \tag{26}$$

because

$$\frac{1}{q}\frac{d}{dx}\frac{I_m}{x^m} = \frac{I_{m+1}}{x^m}, \qquad \frac{1}{q}\frac{d}{dx}\frac{I_{-m}}{x^m} = \frac{I_{-m-1}}{x^m}, \tag{27}$$

as may be quickly tested by the series (23). So, applying the h condition just mentioned, we find

$$A_1 = \frac{hZ(I_m + sI_{-m})}{qy^{m+1}(r-s)(I_m I_{-m-1} - I_{-m}I_{m+1})} \tag{28}$$

at the point y. In the denominator a conjugate property is involved of voltage and current, in the manner of § 331, and the value of the quantity in the second brackets in the denominator is easily seen to be $x^{-1} \times$ constant. To find the constant, use the series (23), writing down the first terms of I_m, I_{-m-1}, &c.; and pick out the terms involving x^{-1}. There is only one. The result is

$$I\;I_{-m-1} - I_{-m}I_{m+1} = \frac{2}{qx}\frac{1}{\underline{|m}\,\underline{|-m-1}} = -\frac{2}{\pi qx}\sin m\pi, \tag{29}$$

because

$$\underline{|-m} = -m\,\underline{|-m-1}, \quad \text{and} \quad \underline{|m}\,\underline{|-m} = \frac{m\pi}{\sin m\pi} \tag{30}$$

are elementary properties of the m function, which we shall come across again in Chapter VII., to be there proved. Thus

$$A_1 = \frac{\frac{1}{2}\pi Zh}{y^m} \cdot \frac{I_{my} + sI_{-my}}{(s-r)\sin m\pi}, \tag{31}$$

and therefore, finally,

$$V_1 = \tfrac{1}{2}\pi Zh\frac{(I_{mx} + rI_{-mx})(I_{my} + sI_{-my})}{x^m y^m (s-r)\sin m\pi}, \tag{32}$$

in a symmetrical form, is another way of expressing V_1. If we determine s and r by the same terminal conditions as before, namely $V = Z_1C$ at l, and $V = -Z_0C$ at λ, we find

$$s = -\frac{I_{ml} + BI_{m+1,l}}{I_{-ml} + BI_{-m-1,l}}, \qquad r = \frac{-I_{m\lambda} + AI_{m+1,\lambda}}{I_{-m\lambda} - AI_{-m-1,\lambda}}, \tag{33}$$

where A and B are as before, (20) and (21) above. So now the solution is, like the old one, explicit in terms of the terminal operators.

Comparison of Wave and Vibrational Solutions to deduce Relation of Divergent to Convergent Bessel functions.

§ 339. We have thus, in (22) and (32), obtained two entirely different forms of solution of the same problem. In one way we built it up with the primary waves and their reflections—and it is certainly right (barring possible working errors). The other way ignores waves altogether, or indeed any sort of elementary component solutions, but is entirely operational, using formulæ which are convergent, when numerical. But the form (32) may be expanded by the expansion theorem into a series of normal functions of the subsiding or vibrating kind. We may therefore for the present regard (32) as representing this normal expansion, in the same way as we may regard (22) as representing the series of waves, for every wave given in operational form may be algebrised if we like. We therefore prove the strict equivalence of the series of normal solutions and the series of waves, arising out of (32) and (22) respectively.

But, besides that, we may regard the investigations algebraically and numerically. For q may be a positive constant, namely, when $Z=R$ and $Y=K$. It is then the steady state due to h that is in question throughout, which is instantly assumed when h varies, because the speed v is infinite, and there are no time differentiations concerned in the various operators, which become constants or functions of x. So we prove the numerical equivalence of (22) and (32), when qx is positive number, apart from their equivalence as operational formulæ. What are then the relations between the divergent functions H, K and the convergent functions I_m and I_{-m}? They are involved in (22) and (32) of course, but it is not clear at first how they are to be exhibited. We must either rearrange (32) to show identically the same form as (22), or else the other way. But a trial with the sum and difference of I_m and I_{-m} used in (22) shows the way. Thus, keeping entirely to (22) at present, let

$$H_{mx}=I_{mx}+I_{-mx}, \qquad K_{mx}=\frac{I_{-mx}-I_{mx}}{\sin m\pi}. \tag{34}$$

These are merely to define H and K, without present reference to the former meanings. Then (32) becomes

$$V_1 = \frac{\frac{1}{2}\pi Z h}{x^m y^m (s-r)\sin m\pi}\{\tfrac{1}{2}(1+r)H_{mx} + \tfrac{1}{2}(r-1)\sin m\pi\, K_{mx}\}$$
$$\times\{\tfrac{1}{2}(1+s)H_{my} + \tfrac{1}{2}(s-1)\sin m\pi\, K_{my}\}, \qquad (35)$$

which is expressible as

$$V_1 = \frac{\frac{1}{2}\pi Z h}{x^m y^m}\frac{(H_{mx}+aK_{mx})(H_{my}+bK_{my})}{(s-r)\dfrac{4\sin m\pi}{(1+r)(1+s)}}, \qquad (36)$$

if $$a = \frac{r-1}{r+1}\sin m\pi, \qquad b = \frac{s-1}{s+1}\sin m\pi, \qquad (37)$$

in the numerator of (36). Here r and s are given in (33), according to which (37) become

$$a = \frac{-H_{m\lambda}+AH_{m+1,\lambda}}{K_{m\lambda}+AK_{m+1,\lambda}}, \qquad b = -\frac{H_{ml}+BH_{m+1,l}}{K_{ml}-BK_{m+1,l}}. \qquad (38)$$

Next, there is the denominator in (36). Putting it in terms of a and b in (37), we find its value is $2(b-a)$. So

$$V_1 = \frac{\frac{1}{4}\pi Z h}{x^m y^m}\frac{(H_{mx}+aK_{mx})(H_{my}+bK_{my})}{b-a}. \qquad (39)$$

This being merely a modification of (32), compare it with (22). They are identical; for the present a, b given by (38) are identical with the a, b of the divergent investigation, viz., a_1 in (21) and the reciprocal of b_1 in (20). But in (22), H and K are defined by the divergent series, whilst in (39) they are defined in terms of the convergent series, through (34). It follows apparently that (34) express the equivalence between the divergent and convergent formulæ.

But it is not a rigorous proof. For there is just this possibility in a proof by comparison. However improbable it may be, it is possible (unless proved to be impossible) for some other combination of the functions I_m and I_{-m} to behave in the same way as regards reducing (32) to identically the form (22). If it did, we should soon find out something anomalous by the impossibilities which would arise on further pursuit. However, I may mention here that in Part 3 of my Paper on "Operators in Physical Mathematics" (May, 1894),

I have given an investigation which transforms the I_m and I_{-m} functions to H_m and K_m functions according to (34). It is an entirely different process to the above, effecting the transformation by algebra alone, without any differentiations or integrations. Of this I will give some account in Chapter VII.

The above suggested ambiguity does occur. For if we put $H_m = 2I_m$ instead of $I_m + I_{-m}$, and still use the second of (34), we shall arrive at the same result, equation (39).

Going further, if we use

$$I_m = \rho H_m + \sigma \sin m\pi K_m, \tag{40}$$

leaving ρ and σ arbitrary, in place of the first of (34), still using the second, the result is that the right member of (39) is multiplied by 2ρ, so we require $\rho = \frac{1}{2}$ to harmonise the convergent and divergent formula. As for σ, it does not appear in the result at all, so it looks at first as if it were indeterminate, and that

$$H_m = 2I_m - 2\sigma \sin m\pi K_m, \tag{41}$$

with any value of σ. But there is another consideration. The last formula must not contradict the second of (34). Now H_m and K_m are even functions of m, so the last equation makes

$$H_m = 2I_{-m} + 2\sigma \sin m\pi \,.\, K_m. \tag{42}$$

By addition, we obtain the first relation in (34); and by subtraction the second relation, provided $\sigma = -\frac{1}{2}$, and only then. So the matter is made square.

Nature of Algebraical Transformation from Divergent to Convergent Formulæ.

§ 339. But I have previously given an investigation which covers the case of integrality of m. I have shown that the operational solution of a certain physical problem, when algebrised in one way leads to the convergent form of the zeroth Bessel function, and in another way leads to the divergent form; thus, $H_0 = 2I_0$. Now, given a Bessel function of any order, all those differing from it in order by an integer may, as is known, be derived by complete differentiations, as in (18) and (27) above. Therefore $H_m = 2I_m$, when m is any integer.

To see what the rigorous mathematicians have to say on this matter, I have referred to the latest treatise (Gray and

Mathews). On p. 68, equation (142) is a result given, which, allowing for the difference in notation, harmonises with the second of (34). But the other result given, equation (143), is discordant. It is equivalent to $H_{mx} = 2I_{mx}$ in my notation, and is therefore true only when m is an integer, in which case the functions I_m and I_{-m} become identical, as (23) shows. No proof is given in either case.

As regards the algebraical transformation from the divergent to the convergent series, it goes thus. Let

$$B_m(qx, r) = \ldots + \frac{(\frac{1}{2}qx)^{m+2r-2}}{\underline{|m+r-1}\ \underline{|r-1}} + \frac{(\frac{1}{2}qx)^{m+2r}}{\underline{|m+r}\ \underline{|r}} + \frac{(\frac{1}{2}qx)^{m+2r+2}}{\underline{|m+r+1}\ \underline{|r+1}} + \ldots \quad (43)$$

This B_m function is the generalised Bessel function of the mth order. In it, r is any number positive or negative. If we increase r by 1 it reproduces itself, so it is a periodic function of r. The series is to be continued both ways, unless it stops.

Now in the H, K formulæ, put for ϵ^{qx} and ϵ^{-qx} the following generalised expressions,

$$\epsilon^{qx} = \ldots + \frac{(qx)^{r-1}}{\underline{|r-1}} + \frac{(qx)}{\underline{|r}} + \frac{(qx)^{r+1}}{\underline{|r+1}} + \ldots, \quad (44)$$

$$\epsilon^{-qx} \cos r\pi = \ldots - \frac{(qx)^{r-1}}{\underline{|r-1}} + \frac{(qx)^r}{\underline{|r}} - \frac{(qx)^{r+1}}{\underline{|r+1}} + \ldots, \quad (45)$$

where r is as before. In (44) the signs are all +; in (45) they are alternately + and −. On performing the multiplications, the H, K functions are turned into B functions according to the following:—

$$B_m(qx, r) = \tfrac{1}{2}H_m(qx) - \tfrac{1}{2}K_m(qx) \sin \pi(m + 2r). \quad (46)$$

Change the sign of m to obtain a second formula. The two formulæ then give H_m and K_m in terms of B_m and B_{-m}.

When r is zero or any integer, B_m reduces to I_m, and we obtain (34) above. In the generalised formulæ (43) to (46), qx should be a real positive quantity, except in special cases.

The identity of I_m and I_{-m} when m is integral, makes the second of (34) assume the 0/0 form. Then take the limit. Thus,

$$K_{mx} = -\frac{1}{\pi} \frac{1}{\cos m\pi} \left(\frac{dI_{mx}}{dm} - \frac{dI_{-mx}}{dm} \right), \quad (47)$$

when m is integral. Or, by (23),

$$-\mathrm{K}_{mx}\cos m\pi=\frac{1}{\pi}\left\{2\mathrm{I}_{mx}\log\frac{qx}{2}+\frac{(\frac{1}{2}qx)^m f'(m)}{\lfloor 0}+\frac{(\frac{1}{2}qx)^{m+2}f'(m+1)}{\lfloor 1}\right.$$
$$\left.+\frac{(\frac{1}{2}qx)^{-m}f'(-m)}{\lfloor 0}+\frac{(\frac{1}{2}qx)^{-m+2}f'(-m+1)}{\lfloor 1}+\dots\right\}, \quad (48)$$

if $f(m)=(\lfloor m)^{-1}$ and $f'(m)$ is its derivative. The convergent formula before given for K_{0x} is a special case of this. The general case presents no difficulty, but requires $f(m)$ to be explained, which belongs to Chapter VII., along with related matters concerning the development of wave formulæ.

The reader should be cautioned against concluding that equivalence, as of a divergent and a convergent formula, means identity. The fact that they are different shows that they are not alike in *all* respects, and cannot be interchanged under *all* circumstances. I am inclined to think that this is true even when the equivalence exists between two convergent formulæ of different types, in fact, what rigorous mathematicians call an identity. Or there may be equivalence when the argument is real, but not when it is imaginary or even negative. The extent to which equivalence persists is an interesting matter, but is better observed in the practical concrete examples than theorised about upon incomplete data. Experience and experiment must precede theory.

Rationality in p of Operational Solutions with two Boundaries. Solutions in terms of I_m and K_m.

§ 341. In order to convert the operational solution to a series of Bessel normal functions we naturally use in the first place the form (32), involving the convergent functions We are virtually in possession of the unit impulsive function, —that is, by putting $h=p\mathrm{Q}$, and developing by the expansion theorem, the result is a formula showing how the charge Q, initially at y, subsequently behaves. But to allow of this development, (32) should be a rational differential equation. Two necessary failures are obvious. First, if $l=\infty$, which does away with the infinite series of reflections, leaving in general only two waves, these waves themselves constitute the practically significant result. The set of normal functions with distinctly separated periods or rates of subsidence no longer exists, although the series of normal functions appropriate when l is finite has its ultimate representative in a definite integral, the limit of the series. Secondly, if either of the terminal operators be irrational, we have a similar failure, and a similar resulting definite integral.

But, assuming that l is finite, and that both the terminal operators are rational (as well as Y, Z in the circuital equations), we may confidently expect that the operational equation (32) contains p the time differentiator rationally, in spite of the presence in the functions concerned of q^m, where m may be fractional, or of $\log q$ in some cases. As a matter of fact, such irrationalities are inoperative by appearing in a suitable manner for cancellation. Thus, in (32) it will be found that r and s both have the factor q^{2m}, whilst I_m has q^m, and I_{-m} has q^{-m}. So the numerator and denominator in (32) both have the factor q^{2m}, and, therefore, (32) is a rational function of q^2, itself a rational function of p.

This being true for any value of m, it follows that in the case of m being integral, when we need to employ the 0/0 form of the K_m function, which brings in the logarithm of q, there is a similar elimination of this logarithm. It goes out in this fashion:— $\log \frac{1}{2}ql - \log \frac{1}{2}q\lambda = \log l/\lambda$. In the results we have only logarithms of numbers.

The alternative form of (32), in terms of I_m and K_m instead of I_m and I_{-m}, is got by using the second of (34), or

$$I_{-m} = I_m + K_m \sin m\pi \tag{49}$$

in (32). This brings us to

$$V_1 = \frac{\frac{1}{2}\pi Z h}{x^m y^m} \frac{(I_{mx} - fK_{mx})(I_{my} - gK_{my})}{f - g}, \tag{50}$$

where
$$f = \frac{I_{m\lambda} - AI_{m+1,\lambda}}{K_{m\lambda} + AK_{m+1,\lambda}}, \tag{51}$$

and g is got by turning λ to l and A to $-$B. Here the convergent K_m function may be understood, though the same result is true with the divergent form.

If l is made infinite, it will be found, by using the divergent series, that g becomes either $+\infty$, or $-\infty$ according to the nature of Z_1. In either case Z_1 is impotent, and (50) reduces to

$$V_1 = \frac{\frac{1}{2}\pi Z h}{x^m y^m} (I_{mx} - fK_{mx}) K_{my}, \tag{52}$$

showing just the primary wave and a reflected one. This is the form suitable for simply periodic states when fully established, as to which more later.

The conversion of (50) to normal functions is merely a formal one by the expansion theorem, but there may be a good deal of work needed sometimes to bring the result to a form suitable for calculation by such tables as exist. In general, the roots of the determinantal equation $f=g$ are complex. That is too bad. But there are important cases of a relatively simple nature. Say $L=0$, retaining R, S, K finite. Then only electric energy is concerned in the "medium," whether with wires or not. If also there is only electric energy concerned at the terminals, the roots (for p) of the determinantal equation are all real and negative. In like manner, if $S=0$, retaining R, L, K finite, only magnetic energy is concerned in the medium; and if this is also the case terminally, we have again roots of the same nature. These are two extreme cases of diffusion, with infinite speed v of propagation, though the practical result may be slow enough. Also, if R, K are zero, and L, S finite, we have finite speed of propagation without waste by resistance; therefore undamped vibrations can occur. In this case the roots are $p=\pm\theta i$, where θ is real, and corresponding terms can be paired to make real vibrations. To this may be added that if R and K are not zero, but are properly balanced, we have the last case again, with attenuation due to R and K superimposed. This nearly exhausts all the applications of a relatively simple nature; though if we do away with the terminal considerations, by taking Z_0 and Z_1 to be either zero or infinity, which makes either the current or the potential vanish terminally, we can extend the matter further.

The convergent Oscillating Bessel functions, and Operational Solutions in terms thereof.

§ 342. In all such cases, it is convenient to convert the functions from I_m and K_m to the oscillating functions J_m and G_m, or perhaps to J_m and J_{-m} instead. The cases of integrality of m are perhaps more common; then it is better to use K_m or G_m, and ignore I_{-m} and J_{-m}.

The oscillating functions are got by putting $q=si$. Thus,

$$\mathrm{I}_m(qx)=i^m\mathrm{J}_m(sx),\qquad \mathrm{I}_{-m}(qx)=i^{-m}\mathrm{J}_{-m}(sx), \tag{53}$$

where

$$\mathrm{J}_m(sx)=\frac{(\frac{1}{2}sx)^m}{\underline{|0}\ \underline{|m}}-\frac{(\frac{1}{2}sx)^{m+2}}{\underline{|1}\ \underline{|m+1}}+\frac{(\frac{1}{2}sx)^{m+4}}{\underline{|2}\ \underline{|m+2}}-\dots, \tag{54}$$

and J_{-m} is the same with $-m$ put for m. When the argument sx is real, this is an oscillating function, the original Bessel function in fact. But the functions involving q are more primitive.

Now as regards K_m. Use the second of (34), with $q=si$. Then,

$$\mathrm{K}_m(qx)=i^{-m}\frac{\mathrm{J}_{-m}(sx)-i^{2m}\mathrm{J}_m(sx)}{\sin m\pi}. \tag{55}$$

Here $i^{2m}=(\cos+i\sin)\,m\pi$. So we may write

$$\mathrm{K}_m(qx)=i^{-m}\{\mathrm{G}_m(sx)-i\mathrm{J}_m(sx)\}, \tag{56}$$

where G_m is defined by

$$\mathrm{G}_m(sx)=\frac{\mathrm{J}_{-m}(sx)-\cos m\pi\,\mathrm{J}_m(sx)}{\sin m\pi}. \tag{57}$$

Or, if we expand i^{-m} in (55) as well, we get

$$\mathrm{K}_m=\frac{\mathrm{J}_{-m}-\mathrm{J}_m}{2\sin\frac{1}{2}m\pi}-i\frac{\mathrm{J}_{-m}+\mathrm{J}_m}{2\cos\frac{1}{2}m\pi}. \tag{58}$$

But the form (56) is the one to take note of.

So now, by the use of (53), we convert the equation (32) to

$$\mathrm{V}_1=\frac{\frac{1}{2}\pi\mathrm{Z}h}{x^m y^m}\frac{(\mathrm{J}_{mx}-\rho\mathrm{J}_{-mx})(\mathrm{J}_{my}-\sigma\mathrm{J}_{-my})}{(\rho-\sigma)\sin m\pi}, \tag{59}$$

where

$$\rho=\frac{\mathrm{J}_{m\lambda}+\mathrm{A}'\mathrm{J}_{m+1,\lambda}}{\mathrm{J}_{-m\lambda}-\mathrm{A}'\mathrm{J}_{-m-1,\lambda}},\qquad \mathrm{A}'=\frac{\lambda^n s\mathrm{Z}_0}{\mathrm{Z}}, \tag{60}$$

and σ is got by turning λ to l and A' to $-\mathrm{B}'$ in ρ. We have $\mathrm{A}=i\mathrm{A}'$, and $\mathrm{B}=i\mathrm{B}'$.

Also, by the use of (56), we convert the alternative equation (50) to the form

$$\mathrm{V}_1=\frac{\frac{1}{2}\pi\mathrm{Z}h}{x^m y^m}\frac{(\mathrm{J}_{mx}-\alpha\mathrm{G}_{mx})(\mathrm{J}_{my}-\beta\mathrm{G}_{mx})}{\alpha-\beta}, \tag{61}$$

where

$$\alpha=\frac{\mathrm{J}_{m\lambda}+\mathrm{A}'\mathrm{J}_{m+1,\lambda}}{\mathrm{G}_{m\lambda}+\mathrm{A}'\mathrm{G}_{m+1,\lambda}}, \tag{62}$$

and β is got by turning λ to l and A' to $-\mathrm{B}'$ in α.

Observe, comparing the forms of (50) and (61), and of α and f, that the transformation is the same in effect as turning I_m to J_m and K_m to G_m, with the change from A to A′ in the denominator, and to − A′ in the numerator of f, to turn it to α. Similarly, as regards the transformation from (32) to (59). The symbol i does not appear in (50), but when we put $q = si$, it appears in K_m through (56), and therefore in f and g. It cancels out on further reduction, and then the strikingly similar form (61) results. That i ought to cancel out is clear enough, because since (50) is rational in q^2, (61) must be rational in s^2. Nevertheless the way the symbol i goes out is somewhat remarkable, depending, as it does, upon the existence of two boundaries. It will not take place when there is only one. Say $l = \infty$, then g^{-1} is zero, and the form (52) results. Put $q = si$, and we do not get a result rational in s^2, and we ought not.

In connection with this, there is sometimes a bit of hocus pocus. If we like we may make the functions J_m and G_m the primary objects of attention, so that

$$V_1 = (J_{mx} - \alpha G_{mx})X \qquad (63)$$

is the initial form of operational solution, s^2 having the meaning − YZ. Determining α and X by the conditions as regards h and terminally, we shall arrive at the result (61). This may, in fact, be the best way to work, when the ultimate results are to be simply periodic or normal solutions, and the development of waves is not in question. But there is a curious irreversibility sometimes concerned. We can always pass from the primitive I_m and K_m to J_m and G_m, but we cannot always go the other way. For instance, (61) leads to (50), by substituting $i^{-1}q$ for s, only when there are two boundaries. It fails when $l = \infty$.

For example, suppose that the condition at l is that $V_2 = 0$. Then $J_{mx} - \beta G_{mx}$ must vanish at l, and this shows that β must be J_{ml}/G_{ml}. Introduce some other kind of terminal condition, and we get some other form of β. But how find its value when $l = \infty$? If we have worked entirely with J_m and G_m, and know nothing of I_m and K_m, there is apparently nothing to show what β should be. For J_m/G_m, the ratio of two real oscillating functions when the argument is real, has no particular limiting value when l is infinite. Nevertheless, if we write $J_{ml}/G_{ml} = -i$, we shall come to the proper result. For it is the same as $\beta = -i$, which is correct.

The explanation has been already virtually given. It is the K_m function that is concerned alone when l is ∞; when $q = si$, then $(G_m - iJ_m)i^{-m}$ takes its place; that is, $-i(J_m + iG_m)i^{-m}$, or $\beta = -i$. But, quite independently of this determination of β, a physically-minded man, who was working in terms of J_m and G_m for convenience in the practical application, would arrive at the correct result by considering the flux of energy. In a simply periodic state produced between the source at y and infinity, with no reflection, the flux of energy must be outward. This necessitates $\beta = -i$.

The divergent Oscillating Bessel functions.

§ **343.** We know that the operational solution in terms of the divergent H_m and K_m is equivalent to that in terms of I_m and I_{-m} or of I_m and K_m. Therefore, the same transformation $q = si$ in the divergent operational formula should lead to a result equivalent to that in terms of J_m and G_m. For distinctness, put a bar over the divergent functions. Then, $q = si$ in $\overline{H}_m$, $\overline{K}_m$ produces

$$\overline{H}_m(qx) = i^m\{\overline{J}_m(sx) - i\overline{G}_m(sx)\}, \tag{64}$$

$$\overline{K}_m(qx) = i^{-m}\{\overline{G}_m(sx) - i\overline{J}_m(sx)\}, \tag{65}$$

when $\overline{J}_m$ and $\overline{G}_m$ are given by

$$J_m(sx) = \left(\frac{2}{\pi sx}\right)^{\frac{1}{2}}\left(P\cos + Q\sin\right)(sx - \tfrac{1}{4}\pi - \tfrac{1}{2}m\pi), \tag{66}$$

$$\overline{G}_m(sx) = \left(\frac{2}{\pi sx}\right)^{\frac{1}{2}}\left(-P\sin + Q\cos\right)(sx - \tfrac{1}{4}\pi - \tfrac{1}{2}m\pi), \tag{67}$$

in which P and Q are the divergent functions

$$P = 1 - \frac{(1^2 - 4m^2)(3^2 - 4m^2)}{1.2(8sx)^2}\left(1 - \frac{(5^2 - 4m^2)(7^2 - 4m^2)}{3.4(8sx)^2}\left(1 - \ldots,\right.\right. \tag{68}$$

$$Q = \frac{1^2 - 4m^2}{1.(8sx)}\left(1 - \frac{(3^2 - 4m^2)(5^2 - 4m^2)}{2.3(8sx)^2}\left(1 - \frac{(7^2 - 4m^2)(9^2 - 4m^2)}{4.5(8sx)^2}\right.\right. \tag{69}$$

The G_m formula is obtained from the J_m one by turning sin to cos and cos to − sin. When x is large, $P = 1$ and $Q = 0$, so

$$\bar{J}_m(sx) = \left(\frac{2}{\pi sx}\right)^{\frac{1}{2}} \cos\left(sx - \tfrac{1}{4}\pi - \tfrac{1}{2}m\pi\right), \qquad (70)$$

$$\bar{G}_m(sx) = -\left(\frac{2}{\pi sx}\right)^{\frac{1}{2}} \sin\left(sx - \tfrac{1}{4}\pi - \tfrac{1}{2}m\pi\right), \qquad (71)$$

show the ultimate nature of the oscillating functions at a considerable distance from the origin. They behave just like $\cos sx$ and $\sin sx$, but with amplitudes varying inversely as the square root of the distance from the origin.

J_m and $\bar{J}_m$ are equivalent when sx is real, and so are G_m and $\bar{G}_m$. The first was proved by Sir G. Stokes; the second I find in Gray and Mathew also, though somewhat difficult to recognise, owing to the use of several forms of the second function. It has been standardised in different ways, some of which are very inconvenient.

If in the H_m, K_m formula (22) we make the changes according to (64), (65), we shall arrive at (61) above precisely, only with $\bar{J}_m$ instead of J_m, and $\bar{G}_m$ instead of G_m. This alone would not prove the equivalence of the convergent and divergent oscillating functions. For example, if we put $\bar{H}_m = 2i^m\bar{J}_m$ in (22), instead of the proper form (64), we shall still arrive at the same result (61).

There is, in fact, an essential difference between the two divergent functions $\bar{H}_m$ and $\bar{K}_m$. Say $m=0$, for instance. This is an important case. We do have $\bar{H}_m = 2I_m$ when qx is real and positive. But it is not an equivalence when $q=si$, and sx is real. One makes $\bar{J}_0 - i\bar{G}_0$. The other makes $2J_0$. On the other hand, $\bar{K}_0 = \bar{G}_0 - i\bar{J}_0$, and $K_0 = G_0 - iJ_0$; and we have $\bar{K}_0 = K_0$, both when q is real positive, and when s is real positive.

Physical reason of the unlikeness of the two divergent functions H_m, K_m.

§ **344.** I have given elsewhere* an algebraical explanation of the distinction. But it will be more satisfactory here to regard the matter physically. We ought to have the one agreement, and we ought to have the other discrepancy, by consideration of the physics. Go back to (5), for example, expressing the initial waves, and suppose that the source h is simply periodic,

* O. in P.M., Part 2.

and that there is no waste of energy by resistance. This makes $q=p/v$, so that if the frequency of the source is $n/2\pi$, then $q=ni/v$, and $s=n/v$. The wave to the right is then

$$V_2=\frac{K_{mx}}{x^m}E=\frac{G_{mx}-iJ_{mx}}{x^m}e, \tag{72}$$

when E and e are simply periodic functions of the time. Or, if $e=e_0\sin nt$, then

$$V_2=\frac{G_{mx}\sin nt-J_{mx}\cos nt}{x^m}e_0. \tag{73}$$

This represents the ultimate result of the periodic source, and the corresponding current is

$$\begin{aligned} C_2 &= \frac{q}{Z}x^{m+1}(G_{m+1,x}-iJ_{m+1,x})i^{-1}e \\ &= -\frac{q}{Z}x^{m+1}(J_{m+1,x}+iG_{m+1,x})e \\ &= -\frac{qe_0}{Z}x^{m+1}(J_{m+1,x}\sin nt+G_{m+1,x}\cos nt). \end{aligned} \tag{74}$$

Here $q/Z=(Y/Z)^{\frac{1}{2}}=$ constant. These solutions indicate a train of waves travelling outwards. The flux of energy is V_2C_2. Its mean value over a period is

$$(V_2C_2)_{\text{mean}}=\tfrac{1}{2}\frac{qx}{Z}e_0^2\,(J_mG_{m+1}-G_mJ_{m+1})_x. \tag{75}$$

This is constant, on account of the conjugate property of J_m and G_m, which is (if the accent is d/dsx)

$$J_mG_{m+1}-G_mJ_{m+1}=-J_mG_m'+J_m'G_m=\frac{2}{\pi sx}. \tag{76}$$

That is to say, there is a steady average flux of energy from the source out to infinity, and this is as it should be, because there is no barrier.

In a similar manner, the potential on the left side is

$$V_1=\frac{H_{mx}}{x^m}F=\frac{(J_{mx}-iG_{mx})}{x^m}f, \tag{77}$$

where F and f are simply periodic, and this represents a wave train travelling to the left. The flux of energy will again be found to be constant on the average, and to be directed to the left. But this state of things is an impossible one, regarded as the ultimate state, like the other. It can only exist approximately initially. Say that h is at a great distance, and the frequency is so great that very many wave lengths can exist between h and the left barrier before it is reached. Then the above solution may be nearly true in a great part of the region occupied by the disturbance. But, whereas there is no barrier on the right side, there must be one on the left side, either at λ, as before, or at the origin itself. When the barrier is reached a new state of things will begin to prevail, first at the barrier, and then travelling out to infinity. Its nature will depend upon the kind of condition imposed at the barrier. It is then generally necessary to consider the second wave equation (8) as well as the first, equation (7), and superimpose them properly.

If the barrier should be such as not to take in energy continuously, the result between the barrier and the source must be a stationary vibration, involving no average flux of energy. The case $m=0$ is peculiar, when the barrier is at the origin. There can be no current there, and no flux of energy. So to the inward wave $H_{0x}.e$ or $(J_{0x}-iG_{0x})e$, add the outward wave $(G_{0x}-iJ_{0x})ie$. The result is $2J_{0x}e$, that is, $2I_{0x}e$. Here $2J_{0x}.e$ represents the ultimate stationary vibration which replaces the preceding state $(J_{0x}-iG_{0x})e$. The operator H_{0x} is valid at first, and then, later, the equivalent $2I_{0x}$, as soon as the origin is reached. We see that we have no right to expect that the property $H_m=I_m+I_{-m}$ should be true when s is real as well as when q is real.

The matter is made plainer by considering h to be steady, beginning at the moment $t=0$. The inward wave from h at y is calculable from H_{0x} until the origin is reached. The result is convergent. But after that, that is, after the moment $t=y/v$, it is only a partial solution, valid between y and the front of the return wave. In the region occupied by the return wave and the primary, we can calculate the wave by $2I_{0x}$ instead of H_{0x}. The two forms of solution become identical at the junction. This matter will be made plainer by one or two special examples. The present remarks are directed to the cause of the failure of $H_m=I_m+I_{-m}$ when $q=si$ and s is real, a cause which is not operative when the K_m function is in question.

Electrical Argument showing the Impotency of Restraints at the origin, unless $1>n>-1$.

§ **345.** So long as the barrier at λ is at a finite distance from the origin, there is no interference with the power of imposing any terminal condition $V=-Z_0C$ of the usual nature, because the functions concerned have finite values. But when $\lambda=0$, there are some noteworthy peculiarities. We know that in one special case (viz., $n=0$, or $m=-\frac{1}{2}$), that of plane waves in a homogeneous medium, we may impose any condition at the origin. We have also observed that in another case ($n=1$, or $m=0$), that of cylindrical waves in a homogeneous medium, or of plane waves in a medium in which the constants vary as the first (or inverse first) power of the distance from a fixed plane, the Z_0 condition is impotent, because there can be no current at its place of application under any finite voltage. We have, therefore, to inquire when in general the Z_0 condition is potent, and when impotent. This is to be done by examination of the limiting form assumed by a in (22), or by r in (32) or by f in (50), when λ is made zero, under different circumstances. This is rather tedious mathematically, from the absence of luminosity. But we can throw some light upon the matter physically, and see that the mathematical results are justifiable.

Under what circumstances can there be a current at the origin when under an impressed voltage? Plainly the resistance must not be infinite. Now the resistance per unit length is Rx^{-n}, which is infinite or zero at the origin according as n is + or −. But the resistance per unit length is not (under the circumstances) the same as the resistance of unit length at the origin. The resistance of the length from 0 to x is $\int_0^x Rx^{-n}dx$. This is finite when n is negative, and also when it is positive, up to $n=1$, when it becomes infinite, and remains infinite for all greater values of n. Here the distance 0 to x may be a very little bit at the origin. So we see that there can be no current there when n is 1 or >1. This is true also as regards the source h. The terminal condition, if applied, must be impotent when n is 1 or >1.

Next, consider the permittance of a little bit from 0 to x at the origin. This is $\int_0^x Sx^n dx$, which is finite when n is positive, and also when n is negative, down to $n=-1$, when it becomes infinite, and remains infinite for greater negative values of n. If, then, n is -1 or <-1, no finite charge can raise the potential at the origin above zero. The terminal condition must be again impotent.

Thirdly, when n is between -1 and $+1$, both the resistance and the permittance of the little bit at the origin are finite; V and C may then have any ratio, and the Z_0 condition of the usual kind is operative if applied.

In the first case, we require to use I_{mx} only, when m is 0 or $+$, or $n =$ or >1. This makes C be zero at the origin. In the next case, when n is -1 or <-1, we must use I_{-mx}, m being -1 or <-1. This makes V be zero at the origin. In the intermediate case both functions will or may occur, according to the nature of Z_0; that is, both I_m and I_{-m}, or I_m and K_m, or J_m and G_m, if we use the oscillating functions. The last is rather remarkable, it leads not only to expansions of the form ΣAJ_{mx}, but also the form $\Sigma C(J_{mx} - aG_{mx})$, when the origin is one of the barriers.

Reduced Formulæ when one Boundary is at the Origin.

§ 346. The preceding electrical reasoning will enable us to understand the results produced in the formulæ when the inner boundary is shifted to the origin. At the same time it does not absolve us from making the examination, because without it we cannot say what special form is assumed when the terminal condition is potent.

In the original formula (22) put $\lambda = 0$. There are three results. If n is not less than 1, we get $a = -\sin m\pi$. If n is not greater than -1, we get $a = +\sin m\pi$. In the intermediate case, when n is between -1 and $+1$, we get

$$a = \sin m\pi \frac{1-u}{1+u}, \quad \text{where} \quad u = \frac{Z_0 q}{Z}\left(\frac{2}{q}\right)^n \frac{\underline{|m}}{\underline{|-m-1}}. \qquad (78)$$

That is, in the first case the x function in (22) reduces to $2I_{mx}$; but in the second case to $2I_{-mx}$; whilst in the intermediate case Z_0 remains potent, and a has the special value shown in the last equation.

From the above may be derived the changes in the other formulæ, or they may be done separately. Thus, in the alternative formulæ (32) and (50), which are connected by

$$-f=\frac{r\sin m\pi}{1+r}, \qquad -g=\frac{s\sin m\pi}{1+s}, \tag{79}$$

we have $r=0$ when $n>1$; $r=\infty$ when $n<-1$, and $r=u^{-1}$ intermediately, u being as in (78). Therefore $s=0$, or $g=0$, is the reduced form of the determinantal equation when $n>1$, or $b+\sin m\pi=0$, or

$$(I_m+BI_{m+1})_l=0, \quad \text{or} \quad (J_m-B'J_{m+1})_l=0. \tag{80}$$

Also, when $n<-1$, the determinantal equation is $s=\infty$, or $g+\sin m\pi=0$, or $b-\sin m\pi=0$; or

$$(I_{-m}+BI_{-m-1})_l=0, \quad \text{or} \quad (J_{-m}+B'J_{-m-1})_l=0. \tag{81}$$

As regards a in (61) when $\lambda=0$, first we have $a=0$ when $n>1$. Then $a=-\tan m\pi$, when $n<-1$. Intermediately

$$a=\frac{-\sin m\pi}{\cos m\pi+v}, \quad \text{where} \quad v=\frac{Z_0 s}{Z}\left(\frac{2}{s}\right)^n\frac{|\underline{m}}{|\underline{-m-1}}. \tag{82}$$

So, in the first case, we have

$$V_1=\frac{\frac{1}{2}\pi Zh}{(xy)^m}J_{mx}\left(G_{my}-\frac{J_{my}}{\beta}\right), \qquad (n=\text{or}>1) \tag{83}$$

and $\beta=0$ is the determinantal equation of normal systems; the same as (80).

In the second case the x function is $J_{-mx}/\cos m\pi$, and

$$V_1=\frac{\frac{1}{2}\pi Zh}{(xy)^m}\frac{J_{-mx}}{\cos m\pi}\frac{\beta G_{my}-J_{my}}{\beta+\tan m\pi}, \qquad (n=\text{or}<-1) \tag{84}$$

whilst the determinantal equation is equivalent to (81).

Intermediately, using a as in (82), the determinantal equation $a=\beta$ is represented by

$$(J_{-m}+B'J_{-m-1})_l+v(J_m-B'J_{m+1})_l=0, \tag{85}$$

where v is as in (82), and B′ is $Z_1 sl^n/Z$, as before.

As a test of avoidance of error in the way of wrong factors, make $n=0$, or $m=-\frac{1}{2}$ in the last equation. It reduces to

$$\tan sl\,(1-Z_0Z_1s^2/Z^2)+Z_1s/Z+Z_0s/Z=0. \tag{86}$$

Comparing with (75), § 293, we find proper agreement, allowing for the present generalised meanings of Z and s compared with their meanings in that place.

The peculiarities of infinite resistance or permittance at the origin with suitable values of n place restrictions upon the restraints possible. The source at y cannot raise the potential at the origin above zero when n equals or is less than -1. We may, however, also have the potential zero there when n is greater, namely, up to just less than $+1$, but then it must be done by external restraint, through $Z_0=0$, being equivalent to a short circuit if a pair of conductors be in question. But when $n=$ or >1 it is no use trying to make the potential vanish.

Similarly, the current at the origin due to the source at y vanishes naturally when $n=$ or >1. We can also make it vanish there when n is less, down to just over -1, by external restraint, through $Z_0=\infty$. It is equivalent to a disconnection in the case of wires. But it is no use trying to make the current vanish when n equals or is less than -1.

Equation (85) only applies when n is between -1 and $+1$, but it harmonises with the proper forms outside those limits. Thus v vanishes when Z_0 does, which makes the potential be zero terminally, provided n is intermediate, and then (85) reduces to (81). Also, if Z_0 is infinite (85) reduces to (80), the other form.

If we impose the condition $V_2=0$ at l, then $B'=0$, and (85) becomes

$$(J_{-m}+vJ_m)_l=0, \qquad (-1<m<0). \tag{87}$$

If, in addition, $V_1=0$ at origin by external restraint, $(Z_0=0)$, then $J_{-ml}=0$ is the determinantal equation. This is replaced by $J_{ml}=0$ if the current is zero at the origin by external restraint, $(Z_0=\infty)$.

The Expansion Theorem and Bessel Series. The Potential due to initial Charge.

§ **347.** The development in Bessel series when h is steady or impulsive is to be done in just the same way as for Fourier series, which department has been somewhat elaborated. So little need be said about it. First find the final steady state, when there is one, as is nearly always the case in practical problems. Let this be V_0. Then apply the $p(d/dp)$ operation to the denominator $\alpha-\beta$ (or other form) in the operational solution, according to the expansion theorem. Thus, using

(61), and supposing h to be steady, beginning when $t=0$, we expand it to

$$V_1 = V_0 + \Sigma \frac{\frac{1}{2}\pi Z h}{(xy)^m} \frac{(J_m - \alpha G_m)_x (J_m - \beta G_m)_y}{p(d/dp)(\alpha - \beta)} \epsilon^{pt}. \qquad (88)$$

The values of p being the roots of $\alpha = \beta$, over which the summation ranges, we see that either α or β may be eliminated from the numerator. The interchange of x and y makes no difference. Therefore the formula for V_2, on the right side of the source at y is just the same as for V_1 as far as the summation goes. It can only differ in the expression of the steady part V_0.

In passing, it may be remarked that cases may arise in which there is no tending to a steady state. For instance, if the above refers to a circuit consisting of a pair of parallel wires, and the insulation is quite perfect intermediately and terminally, then the effect of h accumulates incessantly, and V_1 rises to infinity. The outside term then contains t. But such exceptional cases need not delay us here, but can be treated when they arise. At present assume that there is a steady state tended to. There must be one when the source is impulsive, and there is waste of energy in some part (no matter how limited) of the connected system, and there must be one with a steady source unless there is perfect insulation in the way mentioned.

Now in (88) the J and G functions concerned have sx for argument, and the values of s are settled by the determinantal equation. Also s^2 is a function of p, and so is Z. The further development therefore rests upon the nature of Y and Z in the original circuital equations, for we naturally want to have the result entirely in terms of s. If $Z = R$, and $Y = Sp$, we have diffusion, with one value of p for one of s^2. Then

$$s^2 = -RSp, \quad \text{and} \quad \frac{d}{dp} = -\frac{RS}{2s}\frac{d}{ds}. \qquad (89)$$

This brings us to

$$V_1 = V_0 - \Sigma \frac{\pi Q s}{S(xy)^m} \frac{(J_m - \alpha G_m)_x (J_m - \beta G_m)_y}{(d/ds)(\alpha - \beta)} \epsilon^{pt}, \qquad (90)$$

where the value of p in the time function is $-s^2/RS$.

If $\lambda = 0$, and $\alpha = 0$, which occurs naturally when n is 1 or more, the last reduces to

$$V_1 = V_0 + \Sigma \frac{\pi Q s}{S(xy)^m} \frac{J_{mx} J_{my}}{d\beta/ds} \epsilon^{pt}, \qquad (91)$$

where $\beta=0$ finds the values of s^2 and p. If the potential is constrained to be zero at l, then $B'=0$ in β, making $J_{ml}=0$ be the determinantal equation. Also $V_0=0$ now. So

$$V_1=\Sigma\frac{\pi Qs}{Sl(xy)^m}\frac{J_{mx}J_{my}G_{ml}}{J'_{ml}}\epsilon^{pt}=\Sigma\frac{2Q}{Sl^2(xy)^m}\frac{J_{mx}J_{my}}{(J'_{ml})^2}\epsilon^{pt}, \quad (92)$$

if the accent denotes differentiation to the argument sl. The second form is derived by the conjugate property

$$J_mG'_m-J'_mG_m=-\frac{2}{\pi sl}, \quad (93)$$

remembering that $J_{ml}=0$. Equation (92) expands an arbitrary function of x to suit the conditions stated. Put $Q=Sy^nvdy$, and integrate with respect to y from 0 to l. The result is the V arising from the initial distribution of potential v. This also applies when n is smaller, down to just over -1, provided $Z_0=\infty$ is imposed terminally.

But if n is -1 or less, the solution takes a different form, as before explained. We get, by (84) and (57),

$$V_1=-\frac{\frac{1}{2}\pi RpQ}{(xy)^m}\frac{J_{-mx}(J_m-\beta G_m)_y}{(J_{-m}+B'J_{-m-1})_l}(G_m-B'G_{m+1})_l, \quad (94)$$

with any Z_1. So $Z_1=0$, making $B'=0$, makes

$$V_1=-\frac{\frac{1}{2}\pi RpQ}{(xy)^m}\frac{J_{-mx}}{J_{-ml}}(J_m-\beta G_m)_yG_{ml}, \quad (95)$$

where β is J_{ml}/G_{ml}, which makes $V_2=0$ at l, V_2 being got by interchanging x, y. The last expands to

$$V_1=-\Sigma\frac{\pi Qs}{(xy)^mSl}\frac{J_{ml}J_{-my}J_{-mx}}{J'_{-ml}\sin m\pi}\epsilon^{pt}, \quad (96)$$

subject to $J_{-ml}=0$. The cases in which m is integral had better be kept in terms of J_m and G_m. The connection is

$$J_{-m}=J_m\cos m\pi+G_m\sin m\pi, \quad (97)$$

A companion formula is

$$G_{-m}=G_m\cos m\pi-J_m\sin m\pi. \quad (98)$$

See equations (66), (67), (70), (71).

Equation (96) does not look right. But we have the conjugate property,

$$J_mJ_{-m-1}+J_{-m}J_{m+1}=-\frac{2}{\pi sl}\sin m\pi=-J_{-m}J'_m+J_mJ'_{-m}, (99)$$

which may be derived from (29). So, since $J_{-ml}=0$ at present, J_{ml} may be eliminated from (96). This brings it to

$$V_1=\Sigma\,\frac{2Q}{Sl^2(xy)^m}\,\frac{J_{-mx}J_{-my}}{(J'_{-ml})^2}\,\epsilon^{pt}; \tag{100}$$

that is, the same form as (92), except in the reversal of the sign of m in the J functions. It is valid when m is -1 or less, but the range may be extended up to just under $m=0$ by a terminal restraint making $V_1=0$ at the origin, ($Z_0=0$).

Time Function when Self-induction is allowed for.

§ 348. In the more general case in which R, L, K, S, are all finite, and

$$q^2=-s^2=(R+Lp)(K+Sp), \tag{101}$$

there are two p's for every single s^2; thus,

$$p=-j\pm k,\quad j=\tfrac{1}{2}\left(\frac{R}{L}+\frac{K}{S}\right),\quad k=\left\{\tfrac{1}{4}\left(\frac{R}{L}+\frac{K}{S}\right)^2-\frac{RK+s^2}{LS}\right\}^{\frac{1}{2}}, \tag{102}$$

$$\frac{dq^2}{dp}=-\frac{ds^2}{dp}=S(R+Lp)+L(K+Sp)=\pm 2LSk. \tag{103}$$

The sum of the two time functions of the form $Z\epsilon^{pt}/(dq^2/dp)$, got by using the above two values of p, is therefore

$$\frac{1}{S}\epsilon^{-jt}\left\{\cosh kt+\frac{R/2L-K/2S}{k}\,\text{shin}\,kt\right\}=\frac{\phi(t)}{S}. \tag{104}$$

To show the application, take the simple case of the potential constrained to vanish both at λ and l. Then in (88) $V_0=0$. Also,

$$\frac{d}{ds}(a-\beta)=\frac{d}{ds}\left(\frac{J_{m\lambda}}{G_{m\lambda}}-\frac{J_{ml}}{G_{ml}}\right)=\frac{2}{\pi s}\left(\frac{1}{G^2_{m\lambda}}-\frac{1}{G^2_{ml}}\right), \tag{105}$$

by using the conjugate property (93), with argument $s\lambda$ or sl, as the case may be, and remembering that $a=\beta$.

Therefore (88) takes the special form

$$V_1=\Sigma\,\frac{\pi^2s^2Q}{2S(xy)^m}\,\frac{(J_{mx}-aG_{mx})(J_{my}-aG_{my})}{(G_{ml})^{-2}-(G_{m\lambda})^{-2}}\,\phi(t), \tag{106}$$

where a is $J_{m\lambda}/G_{m\lambda}$, and $\phi(t)$ is the time function in (104). Its value is 1 when $t=0$, so we can expand any initial state U of potential by (106), by putting $Q=Sy^nU\,dy$, and integrating.

The Potential due to initial Current.

§ 349. The inclusion of the self-induction in the last makes the magnetic energy operative. The specification of the initial state is therefore incomplete if the potential alone is given. We require to know the initial current as well. Instead of the above h, let the source at y be e, producing a jump in the potential, thus, $e = V_2 - V_1$ at y, but with continuity of current, or $C_2 = C_1$. To find the result, we need not go through the work in detail, but generalise the former result when the zeroth Bessel function was concerned, § 330. We shall now have

$$V_1 = \tfrac{1}{4}\pi q e \frac{y^{m+1}}{x^m} \frac{(H_m + aK_m)_x (H_{m+1} - bK_{m+1})_y}{b-a}, \quad (107)$$

$$V_2 = \tfrac{1}{4}\pi q e \frac{y^{m+1}}{x^m} \frac{(H_m + bK_m)_x (H_{m+1} - aK_{m+1})_y}{b-a}; \quad (108)$$

V_1 being on the left, V_2 on the right of y. We turn V_1 to V_2 not by interchanging x and y, but by interchanging a and b and negativing the result.

The operation $-x^n Z^{-1}(d/dx)$ finds the current. So

$$C_1 = -\tfrac{1}{4}\pi q^2 e \frac{(xy)^{m+1}}{Z(b-a)} (H_{m+1} - aK_{m+1})_x (H_{m+1} - bK_{m+1})_y, \quad (109)$$

and now C_2 is got by interchanging x and y.

It is easy to test that these satisfy the conditions at y and terminally, a and b being the same as before, by the conjugate property

$$H_m K_{m+1} + K_m H_{m+1} = \frac{4}{\pi q y} = K_m H'_m - H_m K'_m, \quad (110)$$

the argument being qy, and the accent denoting differentiation to it.

Similarly, to find the solution in terms of J_m and G_m, we may derive the results from the last, putting H, K in terms of J and G; but this is tedious, and the results are easily got independently. Thus,

$$V_1 = -\tfrac{1}{2}\pi s e \frac{y^{m+1}}{x^m} \frac{(J_m - \alpha G_m)_x (J_{m+1} - \beta G_{m+1})_y}{\alpha - \beta}, \quad (111)$$

$$V_2 = -\tfrac{1}{2}\pi s e \frac{y^{m+1}}{x^m} \frac{(J_m - \beta G_m)_x (J_{m+1} - \alpha G_{m+1})_y}{\alpha - \beta}, \quad (112)$$

show V due to e at y. The current on the left of y is

$$C_1 = -\frac{s^2\pi e}{2Z}(xy)^{m+1}\frac{(J_{m+1}-\alpha G_{m+1})_x(J_{m+1}-\beta G_{m+1})_y}{\alpha-\beta}, \quad (113)$$

and interchanging x and y produces C_2. In deriving this we use

$$\frac{1}{s}\frac{d}{dx}\frac{J_m}{x^m} = -\frac{J_{m+1}}{x^m}, \qquad \frac{1}{s}\frac{d}{dx}\frac{G_m}{x^m} = -\frac{G_{m+1}}{x^m}, \quad (114)$$

the argument being sx. The continuity in C at y is obvious. The discontinuity in V is easily tested, for we get

$$V_2 - V_1 = \tfrac{1}{2}\pi sye(J_m G_{m+1} - G_m J_{m+1}) = e. \quad (115)$$

Since α, β are as before, the same limitations mentioned apply when $\lambda = 0$.

The development in Bessel series is similar. For an initial state of current, let e be impulsive, say $e = p$P, where P is the momentum generated, as explained in § 327.

Then we get

$$V_1 = -\Sigma\,\pi s^2 P\frac{y^{m+1}}{x^m}\frac{(J_m-\alpha G_m)_x(J_{m+1}-\beta G_{m+1})_y}{\frac{ds^2}{dp}\frac{d}{ds}(\alpha-\beta)}\epsilon^{pt} \quad (116)$$

Here we see that if the potential is made to vanish terminally, and s^2 is as in the last section, making two p's to one s^2, the two time functions to be added are not the same as before. The time function is now $\epsilon^{pt} \div (ds^2/dp)$. The sum of the two is $-(\text{LS}k)^{-1}\epsilon^{-jt}$ shin kt. So, using this, and (103) and (105) again, we convert (116) to

$$V_1 = \Sigma\,\tfrac{1}{2}\pi^2 s^3\frac{P}{\text{LS}k}\frac{y^{m+1}}{x^m}\frac{(J_m-\alpha G_m)_x(J_{m+1}-\alpha G_{m+1})_y}{(G_{m\lambda})^{-2}-(G_{ml})^{-2}}\epsilon^{-jt}\text{ shin } kt, \quad (117)$$

showing the potential due to the initial momentum P at y. Put $P = Ly^{-n}c\,dy$, and integrate from λ to l to show the potential due to the initially given state c of current.

In finding the current due to the initial current, say by the operational solution (113), the presence of Z in the denominator should not be overlooked. Its vanishing sometimes introduces another term depending upon $p = -R/L$ (*see* the next Section). The interpretation is that if the initial state is c = constant, and we have also $V = 0$ imposed terminally, the result is a current $c\epsilon^{-Rt/L}$ at time t.

Uniform Subsidence of Induction and Displacement in combinations of Coils and Condensers.

§ 349A. When formulæ become somewhat intricate, there is a natural tendency to treat them mathematically only, so that the avoidance of error rests upon the mechanical accuracy of working, which can only be effectively confirmed by repetition and by the harmony of results obtained in varied ways. Under these circumstances it is satisfactory to be able to utilise some simple physical property of wide generality to test the formulæ. Such a property can be applied to the preceding formulæ with advantage.

When a simple coil, whose time-constant is L/R, has a current in it, say C_0 at time $t=0$, and is left to itself on short circuit without impressed force, the current subsides in such a way that $C=C_0\epsilon^{-Rt/L}$ is its value at time t. It is the elementary case of the destruction of momentum by a resisting force varying as the velocity. Any number of coils of different R and L, but with the same time-constant, will behave in the same way when on short circuit separately, and without mutual influence. The same is true when they are all connected in series to make a closed circuit. If the initial current is C_0 in the same sense in all, then the current in all subsides as if they were short circuited. There is no difference of potential generated between any of the terminals. There is, it is true, usually some difference of potential between parts of any one coil; but that is a residual effect, arising from the inductance not being quite the same for every turn of wire. This residual effect does not occur in the application to be made. Anyway, the terminals are at the same potential. They may therefore be joined together through any unenergised arrangement of coils and condensers (without introducing mutual influence across the air), and the current in the original circuit will behave in the same way as before described. Every coil wastes its energy against its own resistance independently of the rest. (It is also possible for the external combination to be energised in special ways without interference, but we do not want that at present.)

The above being a purely magnetic property, there is a similar one concerning electric displacement. A leaky condenser (or a condenser with a shunt), if initially charged to potential V_0, and then left to itself, discharges itself so that $V = V_0\epsilon^{-Kt/S}$ is the potential at time t. Here S/K is the time-constant of subsidence, the ratio of the permittance of the condenser to its conductance. The same is true of any number of condensers separately: if they have the same time-constant, the discharges will be alike. They may be put in parallel without any alteration if their potentials are the same, and no difference will be made by joining the various positive terminals together through any (usually unenergised) electrical arrangements, and also the negative terminals. Every condenser will still discharge itself through its own conductance, and waste its energy therein.

The two properties may be co-existent in one combination in many ways. The particular way we want now is this. First, have a long series of coils of any resistances, but all with the same time-constant L/R. Then put their junctions to earth through condensers of any permittances, but all with the same time-constant S/K. The result is a generalised telegraph circuit, in which the resistance, inductance, permittance, and leakance are collected in lumps, so to speak. But the actual distribution may be continuous, if we like, and is, in any case, quite arbitrary, subject to the constancy of the magnetic and electric time-constants.

Two rows of similar coils may be employed. Then the condensers are to be joined across from one row to the other. But this somewhat complicates the description. One series is enough.

By the preceding, it follows that if the initial state is $V = V_0$, constant, and $C_0 = 0$, where V_0 is the voltage of the condensers, and C_0 the current in the conductor consisting of the series of coils, then the state at time t later is simply $V = V_0\epsilon^{-Kt/S}$, and $C = 0$. The waste of energy is in the leakage conductance, and is quite local. There is no development of magnetic force. The true current (in Maxwell's sense) is the sum of the conduction and displacement current, and this is zero for every condenser, little or big.

The circuit may be infinitely long. But if it be only of finite length, we must take care that the terminal arrangements obey the same law. Either the ends must be insulated, making $C = 0$, or we may put the ends to earth through condensers having the same time-constant S/K, and have them initially charged to the same voltage V_0.

Similarly, if the initial state is $C = C_0$, a constant current, and $V_0 = 0$, the state at time t will be $C = C_0 \epsilon^{-Rt/L}$, and $V = 0$. The energy starts magnetic and remains magnetic. The waste of energy is in the conductor of the circuit, and is quite local. In order to complete the analogy as regards "true current" in a physical manner, that is, by making a possible case of electromagnetic wave propagation in a conducting medium, we require to introduce the idea of magnetic conductance to replace the real electric resistance of the circuit, as explained in Chapter IV. In default of that, the analogy is partly only a mathematical one. Leaving out the completion of the analogy physically, it is to be further noted that if the circuit is not infinitely long, the terminal arrangements must be suitably chosen. We require either a dead earth at the ends, making $V = 0$, or else terminal coils possessing the same time-constant L/R, and initially charged with the same current as the rest of the circuit.

It follows further, that if in the first case, where the initial energy is wholly electric, the initial state be not one of constant V_0, there must still usually be a term involving the time factor $\epsilon^{-Kt/S}$ in the resulting potential, unless the mean value of the initial potential should be exactly zero. Also, in the second case, concerning magnetic energy only initially, there must be a term involving $\epsilon^{-Rt/L}$ in the resulting current when C_0 is not constant to begin with, unless its mean value should also be zero. How to reckon the mean values will appear presently. Moreover, the determinantal equation must contain the isolated factor Y or $K + Sp$ in one case, and Z or $R + Lp$ in the other, when the above conditions are complied with.

Uniform Subsidence of Mean Voltage in a Bessel Circuit.

§ **349** B. The Bessel results previously given come under the last Section, because they involve constancy of the electric and magnetic time-constants in spite of the variation of the resistance, &c., per unit length of circuit. We may therefore test the results, and exhibit the solitary terms concerned. If the source is $h = pQ$ at y, that is, a charge Q initially at y, we may write the operational solution for the potential which results thus,

$$V = \phi^{-1}pQ, \tag{118}$$

and the values of p which make $\phi^{-1}p$ infinite are the constants p in the time factor ϵ^{pt}. If v is the partial solution for any p, then

$$v = \frac{Q}{\phi'}\epsilon^{pt}, \tag{119}$$

where ϕ' means $d\phi/dp$. Now, if we know that $Y=0$ gives an isolated root of $\phi=0$, we can write $\phi = Y\theta$, where θ does not vanish with Y. Then

$$\phi' = Y'\theta + \theta'Y = Y'\theta_0 = S\theta_0, \tag{120}$$

if θ_0 is the value of θ when $Y=0$. So

$$v = \frac{Q}{S\theta_0}\epsilon^{-Kt/S} \tag{121}$$

is the partial solution. The denominator $S\theta_0$ is evidently the permittance concerned.

The full operational solution in this case is (61), § 342, in terms of J and G functions, but it is easier to evaluate θ_0 in terms of J_m and J_{-m}. So use the equivalent form (59), § 342, or

$$V_1 = \frac{\frac{1}{2}\pi Zh}{(xy)^m}\frac{(J_m - \rho J_{-m})_x (J_m - \sigma J_{-m})_y}{(\rho - \sigma)\sin m\pi}, \tag{122}$$

where $$\rho = \frac{J_m + A'J_{m+1}}{J_{-m} - A'J_{-m-1}}(\lambda), \qquad A' = \frac{\lambda^n s Z_0}{Z}, \tag{123}$$

and σ is got by turning λ to l and A′ to $-$ B′ in ρ; B′ also containing l instead of λ and Z_1 instead of Z_0.

It does not look as if $Y=0$ were involved in (122). But then it has to be remembered that J_m has the factor s^m. So, put $J_m = s^m P_m$, and balance the powers of s in the numerator against those in the denominator of (122), on the understanding that

$$Y_0 = Z_0^{-1} = m_0(K + Sp), \qquad Y_1 = m_1(K + Sp), \tag{124}$$

where m_0 and m_1 are positive constants, so that there are terminal condensers with the proper time-constant. We shall then obtain, not the factor Z shown in (122), but Z/s^2; that is, $-Y^{-1}$, as required, making $Y=0$ give a root of the determinantal equation. Remember that x^nY and $x^{-n}Z$ are the leakance and resistance operators of the circuit per unit length.

The evaluation of θ_0 is now easy, because $Y=0$ makes $s=0$, and reduces the J functions to their first terms, or the P functions to constants free from s. The result is that (121) becomes

$$v=\frac{Q\epsilon^{-Kt/S}}{S\left(m_0+m_1+\frac{l^{n+1}-\lambda^{n+1}}{n+1}\right)}, \tag{125}$$

showing that the mean value of the potential—that is, the charge divided by the total permittance of the circuit, including the terminal condensers—subsides in the proper way.

In case of terminal insulation, $m_0=0$, or $m_1=0$, or both. These are included in (125), and give finite v. At the other extreme we have $m_0=\infty$, or $m_1=\infty$, or both, and $v=0$. But the term v will still exist finitely if the initial state includes charge of one of the terminal condensers to a finite potential, for then the mean potential will be finite. Practically, however, a short circuit will mean simply $K_0=\infty$ at the terminal, and no condenser. Then $v=0$. The special term does not exist.

Observe that if m_0 and m_1 are defined to be Y_0/Y and Y_1/Y, and if it be assumed (with or without warrant) that $Y=0$ gives a root, we shall obtain $v=0$ usually, because m_0 and m_1 are made infinite. The exception is when Y_0 and Y_1 vanish when Y does; the above case, in fact. Then $Y=0$ really gives a root.

Uniform Subsidence of Mean Current in a Bessel Circuit.

§ 349c. In the other case, if the source is $e=pP$, or the momentum P initially at y, we may write

$$C=\phi^{-1}pP, \tag{126}$$

using the proper ϕ, not the same as before, of course. Then if we know that $Z=0$ gives a root of $\phi=0$, we have $\phi=Z\theta$, not the same θ as the last. Also, if c is the partial solution depending upon $Z=0$, we have

$$c=\frac{P}{\phi'}\epsilon^{pt}=\frac{P}{L\theta_0}\epsilon^{-Rt/L}, \tag{127}$$

and now we have only to evaluate θ_0, the value of θ or ϕ/p when $s=0$, because $Z=0$ makes $s=0$.

The potential and current solutions are (111), (113). In terms of J_m and J_{-m} and the previous ρ and σ, they are

$$V_1 = -\tfrac{1}{2}s\pi e\frac{y^{m+1}}{x^m}\frac{(J_m-\rho J_{-m})_x(J_{m+1}+\sigma J_{-m-1})_y}{(\rho-\sigma)\sin m\pi}. \tag{128}$$

$$C_1 = \tfrac{1}{2}\pi Ye(xy)^{m+1}\frac{(J_{m+1}+\rho J_{-m-1})_x(J_{m+1}+\sigma J_{-m-1})_y}{(\rho-\sigma)\sin m\pi}. \tag{129}$$

This is on the left of y. The C_2 on the right side is got by interchanging x and y in C_1. The V_2 is got by interchanging ρ and σ in the numerator only of (128). The signs of ρ and σ have to be carefully attended to. The formula of derivation used is the first of (114) for J_m in V_1. But it is different with J_{-m}. The companion formulæ to (114) are

$$\frac{1}{s}\frac{d}{dx}\frac{J_{-m}\text{ or }G_{-m}}{x^m} = +\frac{J_{-m-1}\text{ or }G_{-m-1}}{x^m}. \tag{130}$$

Applying (127) to (129), we see first that Z^{-1} does not show itself. But as before, put $J_m = s^m P_m$. The factor Y then becomes Ys^{-2}, that is, $-Z^{-1}$, showing that $Z=0$ gives a root, provided

$$Z_0 = R_0 + L_0 p = n_0(R+Lp), \quad Z_1 = R_1 + L_1 p = n_1(R+Lp), \tag{131}$$

so that there are terminal coils having the proper time-constant.

Lastly, evaluating θ_0 by the first terms of the J functions as before, we obtain

$$c = \frac{P\epsilon^{-Rt/L}}{L\left(n_0 + n_1 + \dfrac{\lambda^{-2m} - l^{-2m}}{2m}\right)}, \tag{132}$$

where $2m = n-1$. The denominator is the total inductance of the circuit, including that of the terminal coils, and c expresses the mean value of the current throughout the whole circuit at every moment.

When $n_0 = 0$, or $n_1 = 0$, or both, we have one or more terminal earths or short circuits. If either of n_0 or n_1 is infinite, it expresses more than a disconnection, for L_0 (for instance) is also infinite. The result is $c=0$, but c can be finite if the initial state includes a finite current in the terminal coil. Practically, a disconnection produces $c=0$ in another way.

There is no terminal coil as a reservoir for magnetic energy, and the special term involving $Z=0$ does not exist.

It will be observed in the above evaluations that the quantities A′ and B′ which occur in ρ and σ, equation (123), are reduced to $s\times$ constant or $s^{-1}\times$ constant by the special terminal conditions. This makes $\rho-\sigma$, (apart from the special factor Z or Y above considered), become a function of s^2. In general it is not a function of s^2 but of p. So now in concluding this part of the subject it may appropriately be pointed out that this property can be generalised in many ways by appropriate terminal conditions involving electric and magnetic energy. We have merely to make Z_0/Z and Z_1/Z be functions of s^2. It is sufficient to illustrate by an easy example. Let

$$Z_0 = R_0 + L_0 p + (K_0 + S_0 p)^{-1}, \tag{134}$$

This says that Z_0 is a coil and a condenser in sequence between the terminal and earth. Divide by Z; then, if $R_0 + L_0 p = n_0 Z$ and $K_0 + S_0 p = m_0 Y$, we obtain

$$\frac{Z_0}{Z} = n_0 - \frac{1}{m_0 s^2}. \tag{135}$$

To be a function of s^2, n_0 and m_0 must be constants, that is, $n_0 = R_0/R = L_0/L$, and $m_0 = K_0/K = S_0/S$. Similarly we may make Z_1/Z a function of s^2.

In all such cases $\rho\ \ \sigma$ is made a function of s^2 (with the possible extra factor), and its roots are calculable by tables of Bessel functions. Then there are two p's to every s^2 in a known manner, so that the time functions are known, and a complete development can be obtained.

[NOTE.—In § 330 and § 334, the function $K_1(qx)$ is defined to be the derivative of $K_0(qx)$. But there are good reasons for the later notation, in § 336 and after, which makes $K_1(qx)$ be the negative of the derivative of $K_0(qx)$. See equations (3), (4), p. 240, and (18), p. 243. The function $K_m(qx)$ is always positive.

The $G_m(sx)$ function has the opposite sign to that employed in my "Electrical Papers," for good reasons. I have endeavoured to smooth matters, and from § 336 to the end have employed that standardisation which experience in the complicated relations of Bessel functions has shown me to be the best and the easiest to follow.]

APPENDIX C. RATIONAL UNITS.

In 1891 I endeavoured, to the best of my ability, to revive an old labour by directing attention to the irrational nature of the B. A. system of units (once so much praised), and advanced arguments to show that not merely should the presentation of theory be altered, but that the practical units should be reformed. (See Vol. I., Chapter II., and the Preface). In 1892 Prof. Lodge wrote asking me if I had any practical proposal to make. The following letters resulted :—

THE POSITION OF 4π IN ELECTROMAGNETIC UNITS.

[*Nature*, July 28, 1892, p. 292.]

There is, I believe, a growing body of opinion that the present system of electric and magnetic units is inconvenient in practice, by reason of the occurrence of 4π as a factor in the specification of quantities which have no obvious relation with circles or spheres.

It is felt that the number of lines from a pole should be m rather than the present $4\pi m$, that "ampere turns" is better than $4\pi n$C, that the electromotive intensity outside a charged body might be σ instead of $4\pi\sigma$, and similar changes of that sort ; see, for instance, Mr. Williams's recent paper to the Physical Society.

Mr. Heaviside, in his articles in *The Electrician* and elsewhere, has strongly emphasised the importance of the change and the simplification that can thereby be made.

In theoretical investigations there seems some probability that the simplified formulæ may come to be adopted—

$$\mu \text{ being written instead of } 4\pi\mu, \text{ and } k \text{ instead of } \frac{4\pi}{K};$$

but the question is whether it is or is not too late to incorporate the practical outcome of such a change into the units employed by electrical engineers.

For myself I am impressed with the extreme difficulty of now making any change in the ohm, the volt, &c., even though it be only a numerical change ; but in order to find out what practical proposal the supporters of the redistribution of 4π had in their mind, I wrote to Mr. Heaviside to inquire. His reply I enclose; and would merely say further that in all probability the general question of units will come up at Edinburgh for discussion. OLIVER J. LODGE.

MY DEAR LODGE,—I am glad to hear that the question of rational electrical units will be noticed at Edinburgh—if not thoroughly discussed. It is, in my opinion, a very important question, which must, sooner or later, come to a head and lead to a thoroughgoing reform. Electricity is becoming not only a master science, but also a very practical science. Its units should therefore be settled upon a sound and philosophical basis. I do not refer to practical details, which may be varied from time to time (Acts of Parliament notwithstanding), but to the fundamental principles concerned.

If we were to define the unit area to be the area of a circle of unit diameter, or the unit volume to be the volume of a sphere of unit diameter, we could, on such a basis, construct a consistent system of units. But the area of a rectangle or the volume of a parallelepiped would involve the quantity π, and various derived formulæ would possess the same peculiarity. No one would deny that such a system was an absurdly irrational one.

I maintain that the system of electrical units in present use is founded upon a similar irrationality, which pervades it from top to bottom. How this has happened, and how to cure the evil, I have considered in my papers—first in 1882-83, when, however, I thought it was hopeless to expect a thorough reform; and again in 1891, when, in my "Electromagnetic Theory," I adopted rational units from the beginning, pointing out their connection with the common irrational units separately, after giving a general outline of electrical theory in terms of the rational.

Now, presuming provisionally that the first and second stages to Salvation (the Awakening and Repentance) have been safely passed through, which is, however, not at all certain at the present time, the question arises, How proceed to the third stage, Reformation? Theoretically, this is quite easy, as it merely means working with rational formulæ instead of irrational; and theoretical papers and treatises may, with great advantage, be done in rational formulæ at once, and irrespective of the reform of the practical units. But taking a far-sighted view of the matter, it is, I think, very desirable that the practical units themselves should be rationalised as speedily as may be. This must involve some temporary inconvenience, the prospect of which, unfortunately, is an encouragement to shirk a duty; as is, likewise, the common feeling of respect for the labours of our predecessors. But the duty we owe to our followers, to lighten their labours permanently, should be paramount. This is the main reason why I attach so much importance to the matter; it is not merely one of abstract scientific interest, but of practical and enduring significance; for the evils of the present system will, if it continue, go on multiplying with every advance in the science and its applications.

Apart from the size of the units of length, mass and time, and of the dimensions of the electrical quantities, we have the following relations between the rational and irrational units of voltage V, electric current C, resistance R, inductance L, permittance S, electric charge Q, electric force E, magnetic force H, induction B. Let x^2 stand for 4π, and let the suffixes r and i mean rational and irrational (or ordinary). Also let the presence of square brackets signify that the "absolute" unit is referred to. Then we have—

$$x=\frac{[E_r]}{[E_i]}=\frac{[V_r]}{[V_i]}=\frac{[H_r]}{[H_i]}=\frac{[B_r]}{[B_i]}=\frac{[C_i]}{[C_r]}=\frac{[Q_i]}{[Q_r]};$$

$$x^2=\frac{[R_r]}{[R_i]}=\frac{[L_r]}{[L_i]}=\frac{[S_i]}{[S_r]}.$$

The next question is, what multiples of these units we should take to make the practical units. In accordance with your request I give my ideas on the subject, premising, however, that I think there is no finality in things of this sort.

First, if we let the rational practical units be the same multiples of the "absolute" rational units as the present practical units are of *their* absolute progenitors, then we would have (if we adopt the centimetre, gramme, and second, and the convention that $\mu=1$ in ether)

$$[R_r]\times 10^9 = \text{new ohm} = x^2 \text{ times old.}$$
$$[L_r]\times 10^9 = \text{new mac} = x^2 \quad ,,$$
$$[S_r]\times 10^{-9} = \text{new farad} = x^{-2} \quad ,,$$
$$[C_r]\times 10^{-1} = \text{new amp} = x^{-1} \quad ,,$$
$$[V_r]\times 10^8 = \text{new volt} = x \quad ,,$$
$$10^7 \text{ ergs} = \text{new joule} = \text{old joule.}$$
$$10^7 \text{ ergs per sec} = \text{new watt} = \text{old watt.}$$

I do not, however, think it at all desirable that the new units should follow on the same rules as the old, and consider that the following system is preferable :—

$$[R_r]\times 10^8 = \text{new ohm} = \frac{x^2}{10}\times \text{old ohm.}$$
$$[L_r]\times 10^8 = \text{new mac} = \frac{x^2}{10}\times \text{old mac.}$$
$$[S_r]\times 10^{-8} = \text{new farad} = \frac{10}{x^2}\times \text{old farad.}$$
$$[C_r]\times 1 = \text{new amp} = \frac{10}{x}\times \text{old amp.}$$
$$[V_r]\times 10^8 = \text{new volt} = x \times \text{old volt.}$$
$$10^8 \text{ ergs} = \text{new joule} = 10 \times \text{old joule.}$$
$$10^8 \text{ ergs per sec.} = \text{new watt} = 10 \times \text{old watt.}$$

It will be observed that this set of practical units makes the ohm, mac, amp, volt, and the unit of elastance, or reciprocal of permittance, all larger than the old ones, but not greatly larger, the multiplier varying roughly from $1\frac{1}{4}$ to $3\frac{1}{2}$.

What, however, I attach particular importance to is the use of one power of 10 only, viz., 10^8, in passing from the absolute to the practical units; instead of, as in the common system, no less than four powers, 10^1, 10^7, 10^8, and 10^9. I regard this peculiarity of the common system as a needless and (in my experience) very vexatious complication. In the 10^8 system I have described, this is done away with, and still the practical electrical units keep pace fairly with the old ones. The multiplication of the old joule and watt by 10 is, of course, a necessary accompaniment. I do not see any objection to the change. Though not important, it seems rather an improvement. (But transformations of units are so treacherous, that I should wish the whole of the above to be narrowly scrutinised.)

It is suggested to make 10^9 the multiplier throughout, and the results are:—

$$[R_r]\times 10^9 = \text{new ohm} = x^2 \times \text{old ohm.}$$

$$[L_r]\times 10^9 = \text{new mac} = x^2 \times \text{old mac.}$$

$$[S_r]\times 10^{-9} = \text{new farad} = x^{-2} \times \text{old farad.}$$

$$[C_r]\times 1 = \text{new amp} = \frac{10}{x} \times \text{old amp.}$$

$$[V_r]\times 10^9 = \text{new volt} = 10x \times \text{old volt.}$$

$$10^9 \text{ ergs} = \text{new joule} = 10^2 \times \text{old joule.}$$

$$10^9 \text{ ergs p. sec.} = \text{new watt} = 10^2 \times \text{old watt.}$$

But I think this system makes the ohm inconveniently big, and has some other objections. But I do not want to dogmatise in these matters of detail. Two things I would emphasise:—First, rationalise the units. Next, employ a single multiplier, as, for example, 10^8.

Paignton, Devon, July 18, 1892. OLIVER HEAVISIDE.

Nothing particular seemed to result. I do not know that there was any discussion of the matter at the Edinburgh meeting. The development was apparently only in its first stage, the Awakening.

The B. A. Committee, so far as I know, took no formal notice of a serious matter in which they should be so much interested. About 1894-5 too, they were so ill-advised (in my opinion) as to persist in their errors and announce that there did not appear to be any reason why their practical units should not be legally adopted (I have not the document by me to give the exact words). This was accordingly done, by proclamation, so to speak, under the Royal Arms, as may be seen in contemporary journals. The question of rationalisation was apparently nowhere.

In the meantime, however, between 1891 and 1895, a remarkable diffusion of knowledge on this subject, and consequent change of opinion and formation of opinion, had taken place, as some of the following will show. The discussion arose out of Prof. Lodge's Report on Magnetic Units, which was printed and circulated amongst members of the B. A. Committee and others, including myself, for opinions. It was reprinted in *The Electrician*, August

2, 1895, p. 449, along with letters from Mr. F. G. Bailey and Prof. J. D. Everett. It was not considered proper to circulate my own opinions of the Report amongst the members of the Committee. Hence their separate publication in *The Electrician*, August 16, 1895, p. 511-12, in the form of the following two letters to Dr. Lodge :

MAGNETIC UNITS.

Paignton, January 28, 1895.

DEAR LODGE : I had some idea of marking your paper all through, in the way of simplifying it mainly, but I gave it up when you got immersed in the 4π muddle. You (the B. A. Committee I mean) are in a beautiful state of muddle by reason of refusing to complete your work properly. You cannot say you did not know you were wrong till after the "legalisation" ; you cannot put it on to International Conferences ; you began it, and the blame is yours—all the more so from your refusal to put it right, or even to make the beginning of an attempt to put it right, by open admission of error, and recommendation of a revision, and by properly discussing it at your B. A. meetings and at Chicago. There is no way out of the muddle than by my radical cure, I believe. When practicians get to be a little more enlightened than they are, the B. A. system will be something for them to laugh at and damn, if it is not already. Even in pure theory, it has been the cause of much mischief, of which I could give examples in the theories of eminent men. Swinburne has suggested in *Nature* that I am very likely wrong in this matter. What is more suggestive is that Magnus Maclean, of Glasgow, who wrote on Units in the *Electrical Engineer* lately, had the assurance to dismiss my reform with the condemnation that my reasons were unwarranted. The geographical suggestiveness is obvious, though perhaps equally unwarranted. But then *you* are a rationalist, and so is FitzGerald, and Larmor, and perhaps many more. Perhaps a majority on the B. A. Committee are rationalists. Then why do they not do the proper thing, and complete their work properly? You cannot get out of the muddle in any other way.

Voltage and gaussage. I dare say practicians will not like them. Gaussage especially. (Sausage !) I do not admire them myself *very* much, on account of the "age," but I took voltage as I found it, and extended the meaning. (How about voltation and gaussation ?) Accepting these words voltage and gaussage, however (or others), it should be noted that they stand for E.M.F. and for M.M.F., *not* for falls of potential (electric or magnetic), because the latter are exceptional, and in fact often become meaningless and quite wrong. Practicians are quite up to circuitation ; the E.M.F. in a circuit, for instance, is the sum of the elementary effective parts of the real electric force. I think, then, that they should speak of the gaussage or the voltage in a circuit, or along a line, or from a to b, &c.; not gaussfall, which I do not like at all.

I think "intensity of" may be dropped altogether. I maintain that **E** and **H** *are* forces, dynamically (generalised, of course) ; specify them as the electric force and the magnetic force, and you are all right, and dynamically sound. Their factors **D** and **B** in the energy density are the corresponding fluxes.

I think your opening part could be simplified a good deal. After that, when you bring in 4π and 10^{-1} and $10/4\pi$, and so forth, I would not presume to criticise. I would rather not concern myself with such a bad job.

About the meaning to be given to inductance, permeance, &c., when you take in iron and *do not* keep to very small forces or small variations in big forces. Here the practical requirements of the practician have to be consulted undoubtedly, but if they are let alone they may do it in some way that they will be sorry for afterwards. The difficulty seems to me to be that there is no definite connection between H and B. In § 192 of "Electromagnetic Theory" I have tried to indicate how we may perhaps come to a good magnetic theory, in which, however, it would be necessary to discriminate; thus, $\mathbf{H}=\mathbf{F}+\mathbf{h}$; **F** *only* to be free, such that the curl of **F** is the current density.

But as regards the extended meaning of μ: suppose we do take a definite connection between **H** and **B**, ignoring hysteresis, and that we have

$$-\operatorname{curl}\mathbf{E}=\dot{\mathbf{B}}.$$

How put $\dot{\mathbf{B}}$ in terms of **H**? It seems to me that the best theoretical way would be

$$\dot{\mathbf{B}}=\frac{d\mathbf{B}}{d\mathbf{H}}\frac{d\mathbf{H}}{dt}=\mu\dot{\mathbf{H}}.$$

so that $\mu=\dfrac{d\mathbf{B}}{d\mathbf{H}}$ and $\mathbf{B}=\int\mu d\mathbf{H}$. (Similarly $\mathbf{D}=\int c\, d\mathbf{E}$.) (This is like saying that the volumic heat capacity of a body is

$$c=\frac{d\mathrm{H}}{dv},\qquad \text{so that}\qquad \mathrm{H}=\int c dv,\qquad \text{and}\qquad \dot{\mathrm{H}}=c\frac{dv}{dt},$$

H being the heat per unit volume, v the temperature. It goes well in the diffusion of heat.)

Then, similarly, we should have in a circuit, (N = total induction),

$$\mathrm{E}=\dot{\mathrm{N}}=\frac{d\mathrm{N}}{d\mathrm{C}}\frac{d\mathrm{C}}{dt}=\mathrm{L}\dot{\mathrm{C}};$$

so that $\mathrm{L}=\dfrac{d\mathrm{N}}{d\mathrm{C}}=$ and $\mathrm{N}=\int\mathrm{L}\,d\mathrm{C}$. The activity **HB** per unit volume would give

$$\dot{\mathrm{T}}=\mathrm{H}\dot{\mathrm{B}}\qquad\text{or}\qquad \mathrm{T}=\int\mathrm{H}\,d\mathrm{B}$$
$$=\mathrm{H}\mu\dot{\mathrm{H}},\qquad\qquad =\int\mathrm{H}\frac{d\mathrm{B}}{d\mathrm{H}}d\mathrm{H}=\int\mu\mathrm{H}\,d\mathrm{H};$$

and by volume integration, we should get for a coil,

$$\dot{\mathrm{T}}=\mathrm{C}\dot{\mathrm{N}}=\mathrm{C}\mathrm{L}\dot{\mathrm{C}},$$
$$\mathrm{T}=\int\mathrm{C}\frac{d\mathrm{N}}{dt}dt=\int\mathrm{C}d\mathrm{N}=\int\mathrm{C}\mathrm{L}d\mathrm{C}.$$

But whether this is likely to be convenient for the iron people, I would not presume to say. Perhaps they do not want any of these quasi-scientific ways of trying to represent facts which are not definite in themselves; *i.e.*, $d\mathrm{B}/d\mathrm{H}$ is not a mere function of H.

Paignton, February 3, 1895.

DEAR LODGE: In my last, commenting on your report, I had doubts as to whether the $dB/dH = \mu$ definition would be convenient for practicians. It is obviously theoretically recommendative. On consideration, I again doubt it. Further than that, I do not think that quasi-official or conferential decisions are desirable on moot points involving unsettled theory. (That, however, is not a new opinion.) For the case is peculiar. There is no theory of magnetism, but only the beginnings of one. It is an excellent theory, too, only the application is limited. Now the effect of iron on the magnetic field is in general theory only a side matter, a secondary phenomenon, like many others. Commercially, it assumes exaggerated importance, so much so, that one is apt to overlook other and more important considerations in general theory. Practicians swallow camels in their "predetermination" work. It is based on theory, in fact on the precise theory, but is modified empirically by characteristic curves, percentage allowances for waste by hysteresis and leakage (which is a big camel). They do their swallowing with complacency, so I suppose they do not suffer ill effects. If I were a practician, I would swallow camels too, if I found that they agreed with me, and sacrifice rigour to expediency. Not being a practician, though, I should very much like to see a good theory of magnetism with variable μ, and leave practicians to work any way they like, with fictitious make-believe permeances and reluctances and lumping together of independent variables.

Hysteresis is the theoretical trouble. Along with this, waste of energy. Now it is not enough to know how much waste there is in a cycle; I want to know how the waste comes in in different parts thereof, and on what it depends. Is it *invariably* associated with a change of the intrinsic magnetism (intrinsic *pro tem.*), or only accidentally, as a secondary matter? There is a curious case in last week's *Electrical Engineer*, in a paper by Mavor. The stuff is *called* steel, but is said to be chemically pure iron. It gives relatively small waste, but large hysteresis, and large μ in the ordinary sense. Say as in the figure. It is imaginable that *quite* pure iron would make the loop become of insensible area. As it is, it is suggested to ignore the loop, and take the median curve; but if we do that we have B vanishing with H, a regular $\mu = dB/dH$ system, with energy stored and no waste. But owing to the extreme steepness of the curves we see that there is a very large intrinsic B when H is zero or small, and this will be so even when the loop is made of smaller and smaller area. It seems quite absurd to take B/F to represent permeability, or dB/dF either. Here F is what the practicians call H; it belongs to the coil; curl $\mathbf{F} = \mathbf{C}$; not curl $\mathbf{H}$. But say $\mathbf{H} = \mathbf{h} + \mathbf{F}$, so that curl $(\mathbf{H} - \mathbf{h}) = \text{curl } \mathbf{F} = \mathbf{C}$. We *must* allow for distinction between $\mathbf{H}$ and $\mathbf{F}$ in theory. As B is not a function of F only, we must have *at least* one other variable quantity. Perhaps h would be enough for a practical theory under limitations. But we need to know how h varies with F, or how much of B at any moment is connected with F and how much is independent *pro tem.* I think this is the right way to look at it, because, first, this way satisfies Poisson's old theory (greatly simplified in expression) of induced magnetisation, and also the more modern view of the same,induced and intrinsic together (intrinsic constant though) (also simplified in expression), and if h could be considered given (as a function of the time for instance) with a given connection between B and H, we should still have a workable theory, a generalisation of the present. But practically we do not have h as a given datum (constant or variable). It comes in through the action of F, and it seems, not in a regular or constant way. For the symmetrical loop is only got after many reversals, putting the iron into a peculiar state. If there were no waste (or it were insensible) it might be the same, but I am inclined to think there must be waste in the initial settling, even if there is none finally (in a suggested pure iron).

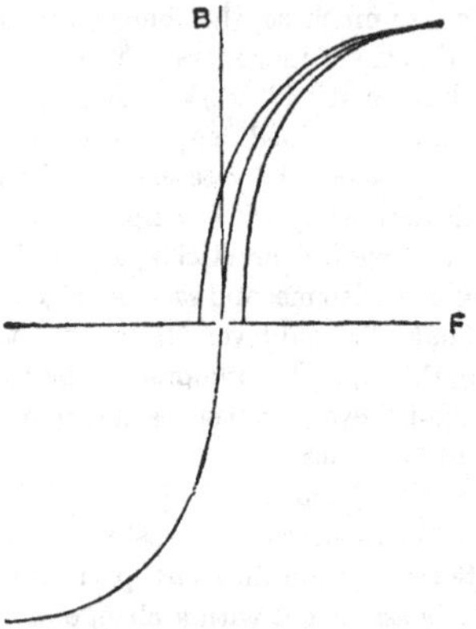

I should think Ewing ought to have the material at his disposal, and the proper realisation of the facts, to be able to discriminate between B total and intrinsic, and to get a sort of normal true inductivity curve (quite different from the commonly assumed) and so come to a sort of theory.

The Report and above letters were followed by an interesting discussion in *The Electrician* (summer and autumn of 1895), of which I give a few notes. Prof. Lodge's report mainly consisted of an attempt to systematise magnetic relations and units *without* employing rational units ; and the discussion was mainly upon it, and not about rational units. Of course, I hold that Prof. Lodge's procedure was wrong, and that the units should be rationalised first. I therefore only notice (in general) the opinions on the question of rationalisation. I condense.

Mr. J. A. Kingdon said engineers would be dismayed by the Report.

Prof. S. P. Thompson encountered the great 4π, but did not overcome it.

Mr. Sydney Evershed was apparently put in a state of fever by the Report, and seemed to be amazed at my audacity in actually proposing to abolish 4π. He also misused words.

Prof. Ewing discussed the Report. Also, he thought it impossible to formulate a theory of magnetic induction, thus practically declining my invitation.

Mr. W. B. Esson was thoroughly unsympathetic with Prof. Lodge. As for me, he called me a "ringleader," and said I ought to be both proud and happy that my work (which, he says, will endure) would never be degraded by the practician by application to things useful.

Mr. H. W. Ravenshaw thought, with regard to the 4π question, it was not generally known that $2H =$ ampere turns per inch within 2 per cent. He also thought "practician" was offensive !

Mr. F. V. Anderson remarked that the C. G. S. is not a rational system.

Prof. Lodge said the reluctance to extend to magnetic units the same sort of treatment as had proved successful for electrical units was somewhat surprising. [Not at all, remembering the two objections, that most practicians or engineers only want to be let alone ; whilst more scientific persons want to go further still, and do it rationally. See below.]

Mr. W. E. Sumpner said $4\pi/10$ can't be got rid of. No importance to practical men. Used to it. He asked whether my system was ever likely to be adopted. [There seems to be something brain-paralysing in the dynamo and transformer, producing a feeling of helplessness.]

Mr. L. B. Atkinson said he thought it was too late to discuss whether the B. A. system was best or not ; thought the new units an unmitigated nuisance ; and added that we were threatened with another complication in the adoption of a "system in which air or ether is not to be the standard subtance." [This is quite new.]

Mr. A. T. Snell thought $4\pi/10$ was really of little importance in dynamo work.

Mr. W. B. Sayers did not remark on it.

Prof. G. F. FitzGerald entirely agreed with me that it is a great misfortune that the units have been wrongly based, but did not at all agree that a change is possible, and discouraged men from wasting their time in endeavouring to bring it about when there are so many other things better worth doing. [Truly there is much to do, but there are many men to do it. And Prof. FitzGerald's argument against doing this little matter seems weak. Besides, it is not such a little matter in the long run, but a very important one.]

Mr. G. L. Addenbrooke agreed with me thoroughly on the 4π question. Said that in 10 or 15 years it will probably be found advisable to start with a complete new set of units on a rational basis. [Much easier now. if the inclination prompts to action.]

Prof. Lodge, referring to the above discussion, and before the B. A. discussion, called attention to some aspects of the matter of magnetic units which might be overlooked.

The Electrician summed up in a leader. The practical man holds up his hands in horror at my proposal, and says, No ; it shall not be, it must not be. At the same time, the writer was not sure that I am not right after all, even though it be H. v. Mundum ; and that some day we shall wonder how the B. A. system was ever blundered into, and turn to the man who tried to save us from ourselves.

The reader of the above cannot fail to notice the gradual change in tone. It is getting quite favourable. Then came the B. A. discussion, when more progress was made. "One of the most striking features was the leaning shown by many of the speakers towards rational units."

Prof. SILVANUS THOMPSON was impressed by the importance of rational units, physically and practically. Moreover, in his opinion, the change would not be very difficult. But because we might soon have to remodel the whole system was a very strong reason for taking a minimum of action now. [I think he meant as regards the proposals in the Report, without doing it rationally.]

Prof. W. E. AYRTON considered the question was not what ought to have been done 30 years ago, but what would be best under present circumstances. [Exactly so ; rationalise now, or make preparations.]

Dr. JOHNSTONE STONEY's printed remarks do not bear on the matter.

Dr. FEDERICK BEDELL said it was too soon. Not quite ready to take up the rationalisation question yet.

Prof. J. D. EVERETT agreed with me in theory, but objected that the harmony between astronomical and other [so-called] absolute systems of units would no longer hold. [So much the worse for the astronomical units ; but I am unable to see that there is much contact at present between astronomical and electromagnetic quantities. They are practically independent.]

Prof. PERRY was sorry to think that the Committee did not boldly face a difficulty which became greater by delay, and adopt at once my suggestions as to rational units. He thought it was quite possible to make the change now.

Mr. TREMLETT CARTER did not think posterity would admire the present system. "All agreed" that the rational system was better, and should be adopted. No more difficult than to introduce the metric system of weights and measures. [Much easier. Consider what heaps of old weights and measures there are, and that they enter into the daily life of the multitude!]

Dr. LODGE feared it was too late for so radical though desirable a change, but was interested in seeing how many seemed to favour it. If done at all, it should be done thoroughly, and applied to electric and magnetic and astronomical units. Perhaps the best time would be when the real nature of the ethereal constants became understood. [This is cold water indeed. It may mean the Greek Kalends. But I don't see why astronomy should be brought in. It is not necessary, if astronomers object.]

This finished the B. A. discussion. There was a little more in *The Electrician*.

Mr. W. H. PREECE, F.R.S., said some object to the presence of π, and would relegate it, "by mere artificialism," to a less intrusive place. [This reminds me of the member of the B. A. Committee who objected to $E=RC$ and said it should be $C=E/R$. He, and Prof. Maxwell, and the other members of the Committee, had so arranged the units that it should be so, without any arbitrary and unnecessary constant].

Prof. A. GRAY thought the inconvenience of 4π, though sensible, had been exaggerated.

Prof. JOHN PERRY had no doubt whatever. The change must come, and "Better soon than syne." [Bravo !]

Mr. F. G. BAILY proposed $\mu=4\pi/10^9$ for air, and thought that the simplest way out of the 4π trouble.

Mr. C. G. HAWKINS expressed his belief in the ultimate and perhaps not very far distant victory of my "radical cure." Nothing required to be added to the reasons adduced by me. He also sketched lightly his idea of the way the change should come, beginning with an international agreement. [But that is a very doubtful point. I differ. I think the original sinner should reform first. Then matters would be greatly smoothed for the others.]

Mr. W. WILLIAMS pointed out an auxiliary argument in favour of the rational system, based upon his dimensional views.

Mr. W. B. SAYERS asked whether the distinction between H and B is not a relic of the action at a distance idea, and also whether it should not be $\mu=0$ in air.

That is about all, and it is instructive as well as somewhat amusing to see how rapidly the three stages to Salvation were run through, from initial ignoration to the consideration of details of Reformation. Now, is the matter to end here? Surely not. I would say to all and sundry, do not let the matter drop after such a successful beginning, but keep pegging away till the actual demand for the reform is pressing. It is not likely that an old institution like the B. A. Committee will do anything without pressure. It meets every year.

Of course, there have been many other expressions of opinion than the above on the question of rational units. Prof. J. J. Thomson, for example, has commended the simplicity of the rational way of displaying the electromagnetic relations; but I doubt whether Cambridge men are favourable to a change. One of them advanced this argument, "But, after all, 4π must come in somewhere." As if it didn't matter *where!* It is also curious to note the action of the dynamo and transformer. Dr. Fleming, however, is a marked exception. He was an early convert, I believe.

Of the progress of rationalistic principles outside the United Kingdom, I have next to no knowledge.

It is difficult to advance any new argument. But the following may put the rationalisation question in a new light for some people. There is a natural tendency for theory and practice to diverge. To keep this divergence within bounds, the same ideas should be in action in both cases. This can only be secured by the rational system. There may then be identity of ideas, and parallel modes of expressing them. See Chap. II., Vol. I.

CHAPTER VII.

ELECTROMAGNETIC WAVES AND GENERALISED DIFFERENTIATION.

Determination of the Value of $p^{\frac{1}{2}}1$ by a Diffusion Problem.

§ **350.** At the very beginning of the treatment of the subject of diffusion there presented itself to our consideration the execution of a differentiating operation which, according to ordinary notions of differentiation, was unintelligible. This was the operation concerned in the function $p^{\frac{1}{2}}1$, where p is the time differentiator, and 1 is the special function of t which is zero before and unity after the moment $t=0$. Instead of the operand being 1, it may be any function of the time. The square root of a differentiator occurs in the fundamentals of the physical subject, namely, the generation of a wave of diffusion. It is necessary and inevitable; also, when studied, it is found to facilitate working.

In order to avoid introducing the idea of fractional differentiation from the theoretical standpoint, I took the value of $p^{\frac{1}{2}}1$ as known experimentally, § 241, equation (A). There is no question as to its value; that is settled by Fourier's investigations in the theory of the diffusion of heat in conductors. But, without this reference to a known result, we should be justified by the consistency of the results obtained by the assumption that the function $p^{\frac{1}{2}}1$ was of the form employed. For the general formulæ for diffusive waves were obtained, and then series of reflected waves, and finally these were converted to series of normally subsiding states. The same process was also carried out for Bessel waves and normal series.

There will be a good deal of use made occasionally of fractional differentiation in the following, when it turns up. The theory of the matter, also, will not be overlooked. But the primary object of the chapter now beginning is electromagnetic waves, and the generalisations will take a subsidiary place as they are suggested. Those who may prefer a more formal and logically-arranged treatment may seek it elsewhere, and find it if they can; or else go and do it themselves.

At present, merely for the sake of comple eness, I introduce one example of the experimental discovery of the meaning of $p^{\frac{1}{2}}1$, founded upon the old diffusive methods. We found in § 240, equation (7), that when an infinitely long cable, with constants R, S (resistance and permittance per unit length), is subjected at its beginning to impressed voltage e, the current produced on the spot was expressed by

$$\mathrm{C} = (\mathrm{S}p/\mathrm{R})^{\frac{1}{2}} e. \tag{1}$$

If e is constant, we have to find what $p^{\frac{1}{2}}1$ means. Now, we can work out this problem in Fourier series, first for a finite cable, and then proceed to the limit. Thus, let the cable be of length l, and be earthed there. Then, if $s^2 = -\mathrm{RS}p$,

$$\mathrm{V} = \frac{\sin s(l-x)}{\sin sl} e \tag{2}$$

is the potential V at x due to e, as in § 265, equation (1), since it makes $\mathrm{V} = e$ at the beginning and $\mathrm{V} = 0$ at the end. The algebrisation by the expansion theorem (§ 287, equation (43), or in any equivalent way) makes

$$\mathrm{V} = e\left(1 - \frac{x}{l}\right) - \frac{2e}{\pi} \Sigma \frac{\pi}{l} \frac{\sin sx}{s} \epsilon^{-s^2t/\mathrm{RS}}, \tag{3}$$

where, in the summation, s has the values π/l, $2\pi/l$, $3\pi/l$, &c.

Observing that the step from one s to the next is π/l, and that it becomes infinitely small when l is made infinitely great, we see at a glance that $l = \infty$ converts (3) to

$$\mathrm{V} = e - \frac{2e}{\pi}\int_0^\infty ds \frac{\sin sx}{s} \epsilon^{-s^2t/\mathrm{RS}}. \tag{4}$$

That is, there is a conversion of the Fourier series to a definite integral, the previous finite step π/l becoming the infinitesimal step ds. The current at the beginning, $x=0$, is got by $C = -R^{-1}(dV/dx)$, and then putting $x=0$. This makes

$$C = \frac{2e}{R\pi}\int_0^\infty \epsilon^{-s^2t/RS}ds. \tag{5}$$

Comparing with (1), and removing unnecessary constants, we see that

$$p^{\frac{1}{2}}1 = \frac{2}{\pi}\int_0^\infty \epsilon^{-s^2t}\,ds, \tag{6}$$

which is a well-known integral. Perhaps the easiest way to evaluate it (by an ingenious device, also well-known), is thus:

$$(p^{\frac{1}{2}}1)^2 = \frac{2}{\pi}\int_0^\infty \epsilon^{-x^2t}\,dx \times \frac{2}{\pi}\int_0^\infty \epsilon^{-y^2t}\,dy = \frac{4}{\pi^2}\int_0^\infty\int_0^\infty \epsilon^{-(x^2+y^2)t}\,dx\,dy$$

$$= \frac{1}{\pi^2}\int_0^\infty 2\pi r dr\,\epsilon^{-r^2t} = \frac{1}{\pi t}\Big[-\epsilon^{-r^2t}\Big]_0^\infty = \frac{1}{\pi t}. \tag{7}$$

So, taking the square root, we arrive at the required result,

$$p^{\frac{1}{2}}1 = (\pi t)^{-\frac{1}{2}}. \tag{8}$$

The above is only one way in a thousand. I do not give any formal proof that all ways properly followed must necessarily lead to the same result.

It should be noticed, in passing, that the operator C/e which is rational when l is finite, reduces to the irrational form in (1) just when the Fourier series passes into the definite integral, by making l infinite. At the same time the infinite series of waves involved in the Fourier series reduces to a single wave.

Elementary generalised differentiation. Value of p^m1 when m is integral or midway between.

§ 351. On the basis of the result just obtained, we are in possession of the value of $p^{m+\frac{1}{2}}1$, when m is any integer, positive or negative. Thus, when m is positive, we have whole differentiations to perform upon $p^{\frac{1}{2}}1$. For example,

$$\left.\begin{aligned} p^{\frac{3}{2}}1 &= pp^{\frac{1}{2}}1 = p(\pi t)^{-\frac{1}{2}} = -\frac{t^{-\frac{3}{2}}}{2\pi^{\frac{1}{2}}}, \\ p^{\frac{5}{2}}1 &= \frac{1.3}{2.2}\frac{t^{-\frac{5}{2}}}{\pi^{\frac{1}{2}}}, \qquad p^{\frac{7}{2}}1 = -\frac{1.3.5}{2^3}\frac{t^{-\frac{7}{2}}}{\pi^{\frac{1}{2}}}, \end{aligned}\right\} \tag{9}$$

and so on. When m is negative we must perform integrations from 0 to t. Thus,

$$\left.\begin{aligned} p^{-\frac{1}{2}}1 &= p^{-1}p^{\frac{1}{2}}1 = p^{-1}(\pi t)^{-\frac{1}{2}} = \frac{2t^{\frac{1}{2}}}{\pi^{\frac{1}{2}}}, \\ p^{-\frac{3}{2}}1 &= \frac{2^2 t^{\frac{3}{2}}}{1.3\,\pi^{\frac{1}{2}}}, \qquad p^{-\frac{5}{2}}1 = \frac{2^3 t^{\frac{5}{2}}}{1.3.5\,\pi^{\frac{1}{2}}}, \end{aligned}\right\} \tag{10}$$

and so on.

We also know the value of $p^{m+\frac{1}{2}}t^n$, where n is integral, or is an integer $+\frac{1}{2}$. Say it is integral and positive. Let the operand be $t^n/\underline{|n}$ where $\underline{|n}$ is the factorial function $1.2.3.\ldots.n$. Then

$$p\frac{t^n}{\underline{|n}} = \frac{t^{n-1}}{\underline{|n-1}}, \qquad p^n\frac{t^n}{\underline{|n}} = 1, \qquad p^{-n}1 = \frac{t^n}{\underline{|n}}; \tag{11}$$

and now introducing the index $\frac{1}{2}$, we get

$$p^{m+\frac{1}{2}}\frac{t^n}{\underline{|n}} = p^{m-n+\frac{1}{2}}1 = p^{m-n}\frac{1}{(\pi t)^{\frac{1}{2}}}. \tag{12}$$

Here $m-n$ is integral, so we have the former cases again, as in (9) and (10). Now the fundamental property of $\underline{|n}$ is

$$\underline{|n} = n\,\underline{|n-1}, \tag{13}$$

with the addition that its value must be fixed for any one value of n, for instance, $\underline{|1} = 1$. It follows that $\underline{|0} = 1$ also, and that $\underline{|n}$ is ∞ for all negative integral values of n. Consequently (11) and (12) are also valid when n is negative. For example, $p1 = 0$, provided t is positive. It is really an impulse at the moment $t = 0$. Also $p1 = t^{-1}/\underline{|-1}$, and this is zero, unless t is also zero.

Comparing results, it will be observed that if we use the formula (13) when n is an integer $+\frac{1}{2}$, as well as when it is an integer, and introduce the datum that

$$\underline{|-\tfrac{1}{2}} = \pi^{\frac{1}{2}}, \tag{14}$$

the above results in generalised differentiation are valid with the extended meaning of n. Thus,

$$\left.\begin{aligned} &\underline{|\tfrac{1}{2}} = \tfrac{1}{2}\underline{|-\tfrac{1}{2}} = \tfrac{1}{2}\pi^{\frac{1}{2}}, && \underline{|\tfrac{3}{2}} = \tfrac{3}{2}\tfrac{1}{2}\pi^{\frac{1}{2}}, \quad \text{\&c.} \\ &\underline{|-\tfrac{3}{2}} = -2\underline{|-\tfrac{1}{2}}, && \underline{|-\tfrac{5}{2}} = \tfrac{2\,2}{1\,3}\underline{|-\tfrac{1}{2}}, \quad \text{\&c.} \end{aligned}\right\} \tag{15}$$

We shall now have

$$p^n 1 = \frac{t^{-n}}{\underline{|-n}}, \qquad p^{-n}1 = \frac{t^n}{\underline{|n}}, \qquad p^m\frac{t^n}{\underline{|n}} = p^{m-n}1, \tag{16}$$

when n is any integer, or is midway between two consecutive integers; and the same as regards m.

The extension to the case of n being any real number, positive or negative, is not difficult, but we do not want it at present. The above can be applied to numerous electromagnetic problems without going further into the meaning of generalised differentiation. That $\underline{|-\frac{1}{2}}$ is $\pi^{\frac{1}{2}}$ is, as far as the above is concerned, merely a convention or definition, the factorial notation being convenient for showing the system involved, and for the expression of results.

Cable Problem:—$C = (K + Sp)^{\frac{1}{2}}(R + Lp)^{-\frac{1}{2}}$
(1). Elementary Cases by Inspection.

§ 352. After the above little mathematical excursion we may return to the physical problem out of which it arose, but generalised to include self-induction and leakage. Let the cable have the four constants R, K, L, S, in the notation previously employed, the additional L and K being the inductance and leakance per unit length. Then the current produced by e impressed at the beginning of an infinitely long cable is

$$C = \left(\frac{K + Sp}{R + Lp}\right)^{\frac{1}{2}} e, \tag{17}$$

that is, R is generalised to $R + Lp$ and Sp to $K + Sp$. *See* Vol. I, § 221, equation (12).

Now this is a far more developed case than the former. Ways of algebrising it have to be found. The previous mode of attack will be found to be enormously complicated. But we can find what (17) means pretty straight out from itself, without the circumbendibus involved in evaluating complicated integrals by rigorous methods.

Notice some special cases first. If only R and S are finite we have the former case. But if R and S are zero, whilst L and K are finite, we have a similar case. Thus

$$C = (K/Lp)^{\frac{1}{2}} e = 2(Kt/L\pi)^{\frac{1}{2}}, \tag{18}$$

by using the value of $p^{-\frac{1}{2}}1$. The current increases to infinity according to the square root of the time. This is a curious case of leakage conductance and inductance only, and is purely a magnetic problem.

Next, all four constants being finite, $p=0$ in (17) produces the steady ultimate state $C=(K/R)^{\frac{1}{2}}e$. Here $(R/K)^{\frac{1}{2}}$ is the effective steady resistance reckoned at the beginning of the circuit. In the extreme case $R=0$, no steady state is reached, of course.

Further, let $R/L=K/S$. Then p goes out from (17), which again reduces to $C=(K/R)^{\frac{1}{2}}e$. This is the distortionless case, § 209. It now holds good always, whether e is steady or not, the cable behaving towards the impressed voltage as though it were a mere resistance.

Again, putting $p=ni$ in (17) will produce the simply periodic current that results when e is simply periodic, by reducing it to the form $C=(K'+S'p)e$. The developed formula was given in § 221, Vol. I., and need not be repeated.

Finally, $p=\infty$ in (17) gives the *initial* value of C when e suddenly jumps from zero to a finite value, viz.,

$$C=(S/L)^{\frac{1}{2}}e=e/Lv, \tag{19}$$

if v is the speed of propagation, or $(LS)^{-\frac{1}{2}}$.

In the theory of a plane electromagnetic wave the equation corresponding to (17) is

$$H=\left(\frac{g+\mu p}{k+cp}\right)^{\frac{1}{2}}E, \tag{20}$$

where E and H are the electric and magnetic forces, c and μ the permittivity and inductivity, k and g the conductivities, electric and magnetic respectively. It can be treated in the same way, and, in fact, it represents the same problem physically, except as regards the constant g. This was explained in Chapter IV.

(2). Algebrisation when e is constant and K zero. Two ways. Convergent and divergent results.

§ 353. Now let e be constant after $t=0$, and zero before, and consider the case in which K is the only constant that vanishes. Then

$$C=\left(\frac{Sp}{R+Lp}\right)^{\frac{1}{2}}e=\frac{e}{Lv(1+2ap^{-1})^{\frac{1}{2}}}, \tag{21}$$

if $a=R/2L$.

The suggestion to employ the binomial theorem is obvious It will expand the operator in powers of p, and so substitute a series of easy operations for an unintelligible one. Thus, expanding in rising powers of a/p,

$$C = \frac{e}{Lv}\left\{1 - \frac{a}{p} + \frac{1.3}{\underline{|2}}\left(\frac{a}{p}\right)^2 - \frac{1.3.5}{\underline{|3}}\left(\frac{a}{p}\right)^3 + \dots\right\}1. \qquad (22)$$

Here we have only whole integrations. So the immediate result is, by the third of equations (11) above,

$$C = \frac{e}{Lv}\left\{1 - at + \frac{1.3}{(\underline{|2})^2}(at)^2 - \frac{1.3.5}{(\underline{|3})^2}(at)^3 + \dots\right\}, \qquad (23)$$

a convergent solution in rising powers of the time. The straightforward and rapid way of getting the result is remarkable.

But the binomial theorem furnishes another way of expanding the operator in (21), viz., in rising powers of p. Thus,

$$C = \frac{1}{Lv}\frac{(p/2a)^{\frac{1}{2}}}{(1+p/2a)^{\frac{1}{2}}}e$$

$$= \frac{1}{Lv}\left\{1 - \frac{p}{4a} + \frac{1.3}{\underline{|2}}\left(\frac{p}{4a}\right)^2 - \frac{1.3.5}{\underline{|3}}\left(\frac{p}{4a}\right)^3 + \dots\right\}\left(\frac{p}{2a}\right)^{\frac{1}{2}}e. \qquad (24)$$

Here we know already the value of $(p/2a)^{\frac{1}{2}}e$, viz., $e/(2\pi at)^{\frac{1}{2}}$. We have, therefore, merely to perform whole differentiations upon it to produce the solution with the same directness in this form :—

$$C = \frac{e}{Lv}\frac{1}{(2\pi at)^{\frac{1}{2}}}\left\{1 + \frac{1^2}{8at} + \frac{1^2 3^2}{\underline{|2}(8at)^2} + \frac{1^2 3^2 5^2}{\underline{|3}(8at)^3} + \dots\right\}. \qquad (25)$$

This is a divergent series. So much the better. It is easier to calculate except when at is so small as to bring the point of convergence too near the beginning.

Equations (23) and (25) are equivalent. Comparing (25) with (3) § 336, we see that (25) is the same as

$$C = \frac{e}{2Lv}\epsilon^{-at}H_0(at), \qquad (26)$$

where $H_0(at)$ is the divergent zeroth Bessel function which was shown to be numerically equivalent to $2I_0(at)$. Therefore (23), being convergent, should represent

$$C = \frac{e}{Lv}\epsilon^{-at}I_0(at). \qquad (27)$$

That it does may be verified by multiplying together the ordinary series for the exponential and the convergent Bessel function; (27) then becomes (23).

Besides the very direct way of getting the results (which, I may remark, are quite correct*), there are several points to be noticed. Thus, we find the use of the binomial theorem is justifiable, to substitute an infinite series of separate integrations or differentiations for the operator involving the radical. Next, that the convergency of the series in powers of p obtained does not enter into the question at all. Either (24) or (22) is divergent when a/p is numerical. But both are valid, though one gives rise to a convergent final result, the other to a divergent one. As regards the practical use of the latter, *see* § 335 for the present. The question will arise again.

Notice further that we obtained (23) from (21) through (22) without any use of fractional differentiation. But if we take the special case of the same got by making $L=0$, we reduce (21) to the form (1), and cannot now escape from $p^{\frac{1}{2}}$. This is curious. The lesser seems to contain more than the greater. The explanation is to be seen in the other way of expanding the operator. The reduced form of (24) is (1), and both involve $p^{\frac{1}{2}}$. It still remains remarkable, however, that we can escape from $p^{\frac{1}{2}}1$, and so evaluate it by generalising the problem involved in (1).

(3). Third way. Change of Operand.

§ 354. Observing that in the form (27) we have an exponential factor, a third mode of algebrising (21) is suggested—viz., by putting in the exponential factor at the beginning. This is the way. We have

$$\left(\frac{p}{p+2a}\right)^{\frac{1}{2}}1=\epsilon^{-at}\epsilon^{at}\left(\frac{p}{p+2a}\right)^{\frac{1}{2}}1, \tag{28}$$

obviously. Now, here ϵ^{at} may be shifted to the right, provided we simultaneously change p to $p-a$. This makes the last become

$$=\epsilon^{-at}\left(\frac{p-a}{p+a}\right)^{\frac{1}{2}}\epsilon^{at}, \tag{29}$$

* How *know* that (27) is right? Because, when t is turned to $(t^2-x^2/v^2)^{\frac{1}{2}}$ in the I_0 function, the result is the complete wave of current entering the cable. The partial characteristic is satisfied, as well as the terminal condition. This generalisation will occur further on.

and now it is $\epsilon^{at}1$ or ϵ^{at} that is the operand, not 1. (If an operand is always understood to be at the end, the unit operand may be omitted in general, just as in arithmetical and algebraical operations, and it is sometimes an advantage to omit it.) So far only makes a pretty change; but we can go further. For

$$\epsilon^{at} = \frac{p}{p-a}, \tag{30}$$

with unit operand understood. Compare with (3), § 265. Substituting (30) in (29), reduces it to

$$\epsilon^{-at}\left(\frac{p-a}{p+a}\right)^{\frac{1}{2}}\frac{p}{p-a} = \epsilon^{-at}\frac{p}{(p^2-a^2)^{\frac{1}{2}}} = \epsilon^{-at}\frac{1}{(1-a^2/p^2)^{\frac{1}{2}}}, \tag{31}$$

with unit operand again. Now expand by the binomial theorem and algebrise. We get

$$\epsilon^{-at}\left\{1+\tfrac{1}{2}\frac{a^2}{p^2}+\frac{1\,.\,3}{2^2\lfloor 2}\frac{a^4}{p^4}+\frac{1\,.\,3\,.\,5}{2^3\lfloor 3}\frac{a^6}{p^6}+\dots\right\}$$

$$= \epsilon^{-at}\left\{1+\frac{a^2t^2}{2^2}+\frac{a^4t^4}{2^24^2}+\frac{a^6t^6}{2^24^26^2}+\dots\right\} = \epsilon^{-at}\,\mathrm{I}_0(at), \tag{32}$$

as required. *See* (23), § 338. Introduce the omitted factor $e/\mathrm{L}v$ into equations (28) to (32) to produce the working of the electrical problem.

Here again, there are points to notice. The transformation from (28) to (30) depends upon the property $p^n\epsilon^{at} = a^n\epsilon^{at}$. That is, the potence of p is, under the circumstances, simply a, when n is integral. So

$$p^n(u\epsilon^{at}) = \epsilon^{at}(p+a)^n u, \tag{33}$$

if u is a function of t. Thus ϵ^{at} may be shifted to the left by increasing p to $p+a$. Similarly

$$f(p)(u\epsilon^{at}) = \epsilon^{at}f(p+a)u, \tag{34}$$

if $f(p)$ is a rational integral function of p. We find that this process is justifiable (by results) in the case of the irrational functions of p we have had in question.

Having changed the operand and obtained (29), we then change it to unity again, through (30). This is expressed by

$$f(p)\,1 = \epsilon^{-at}f(p-a)\epsilon^{at} = \epsilon^{-at}\frac{pf(p-a)}{p-a}1. \tag{35}$$

The result is to come to an entirely different kind of operator. Instead of the first power of p under the radical sign, we have the square of p in (31). The subsequent expansion and correct algebrisation in (32) are obvious.

But there is another thing to be noticed. Previously we had p with the + sign under the radicals. Now it has the − sign. Furthermore, it will be found that the alternative method of expanding (31) fails. This point will be noticed again.

Collecting results as regards $I_0(at)$, disconnected from unnecessary constants, we have

$$I_0(at) = \epsilon^{at}\left(\frac{p}{2a+p}\right)^{\frac{1}{2}} = \epsilon^{-at}\left(\frac{p}{p-2a}\right)^{\frac{1}{2}}$$

$$= \left(\frac{p-a}{p+a}\right)^{\frac{1}{2}}\epsilon^{at} = \left(\frac{p+a}{p-a}\right)^{\frac{1}{2}}\epsilon^{-at} = \frac{p}{(p^2-a^2)^{\frac{1}{2}}}. \quad (36)$$

The operand is ϵ^{at} or ϵ^{-at} when at the end. At the beginning they are factors only. The operand is 1, when no time function is written at the end.

So far there is only one constant essentially concerned along with p, viz., a. But in the general case, when K is not neglected, there are two constants involved. To this development I now proceed.

(4). V due to steady C when R = 0. Instantaneous Impedance and Admittance.

§ 355. A case which is similar to that of § 353 occurs when R is zero, making

$$V = \left(\frac{Lp}{K+Sp}\right)^{\frac{1}{2}} C. \quad (37)$$

The form is the same as (21). So the developed solution is the same for V due to C impressed as for C due to V impressed, provided we interchange the electric and magnetic quantities S and L, R and K. For instance, if the source is the current h, steady after $t=0$, and zero before, then (27) is transformed to

$$V = \frac{h}{Sv}\epsilon^{-bt} I_0(bt), \qquad b = K/2S, \quad (38)$$

which shows the voltage required to produce the steady current.

The Sv that occurs here is the instantaneous admittance. It is the reciprocal of Lv, the instantaneous impedance. In the distortionless circuit the impedance remains $\mathrm{L}v$ always (that is, in common units, 30 ohms × the number representing the inductance per centim., say from 10 to 30 usually, according to the size of the wires and distance apart), unless interfered with by reflection. But $\mathrm{L}v$ is the instantaneous impedance always, that is, whatever values R and K may have. In the corresponding plane electromagnetic waves, $\mathrm{E}=\mu v\mathrm{H}$, when undistorted, and the impedance is μv. It is only a different way of reckoning, using the elements of V and C instead of the totals. That is, μv is the impedance for a unit tube of flux of energy, and $\mathrm{L}v$ is the total effective impedance for the total flux. But if R and K (g and k in the proper wave problem) are not balanced, the impedance immediately begins to alter by reflection due to the unbalanced action of g and k upon the magnetic and electric fluxes.

(5). C due to steady V when S = 0. The error function again.

§ 356. If R and L are both zero, then V/C is 0, and C/V is ∞, obviously. But if K and S are both zero instead, then C/V is 0. In this case no finite voltage can produce a current, because the resistance of the conductor is infinite, and the leakage is stopped, both elastic and conductive.

If S is alone zero, then

$$\mathrm{C}=\frac{\mathrm{K}^{\frac{1}{2}}}{(\mathrm{R}+\mathrm{L}p)^{\frac{1}{2}}}\mathrm{V}=\left(\frac{\mathrm{K}}{\mathrm{R}}\right)^{\frac{1}{2}}\left(1+\frac{2a}{p}\right)^{-\frac{1}{2}}\left(\frac{2a}{p}\right)^{\frac{1}{2}}\mathrm{V}. \qquad (39)$$

This is for a leaky circuit in which the permittance is of insensible effect compared with the other influences. The initial admittance is zero, the final is $(\mathrm{K}/\mathrm{R})^{\frac{1}{2}}$, the effective steady conductance, as we see by putting $p=\infty$ and 0 in turn.

As regards the state at time t, when V is steady, observe that the method of expanding in rising powers of p appears to fail. We get a constant term, plus an infinite series of zeros. Now there are zeros *and* zeros. An absolute zero is like the point in geometry, which you cannot see even when you use a magnifying glass, as the schoolboy said. But some zeros can be magnified, and an infinite number of them might make finiteness. I do not think that is the explanation here, though. But we can pass the matter over at present, because the other way of expanding by the binomial theorem, viz., in descending powers of p, presents no difficulty. Thus,

$$\frac{\mathrm{C}}{\mathrm{V}}=\left(\frac{2\mathrm{K}}{\mathrm{R}}\right)^{\frac12}\left\{\left(\frac{a}{p}\right)^{\frac12}-\left(\frac{a}{p}\right)^{\frac32}+\frac{1\,.\,3}{\underline{|2}}\left(\frac{a}{p}\right)^{\frac52}-\frac{1\,.\,3\,.\,5}{\underline{|3}}\left(\frac{a}{p}\right)^{\frac72}+\dots\right\} \tag{40}$$

$$=\left(\frac{2\mathrm{K}}{\mathrm{R}}\right)^{\frac12}\left\{\frac{(at)^{\frac12}}{\underline{|\tfrac12}}-\frac{(at)^{\frac32}}{\underline{|\tfrac32}}+\frac{1\,.\,3}{\underline{|2}}\frac{(at)^{\frac52}}{\underline{|\tfrac52}}-\frac{1\,.\,3\,.\,5}{\underline{|3}}\frac{(at)^{\frac72}}{\underline{|\tfrac72}}+\dots\right\} \tag{41}$$

$$=\frac{2}{\pi^{\frac12}}\left(\frac{\mathrm{K}t}{\mathrm{L}}\right)^{\frac12}\left\{1-\frac{\mathrm{R}t}{3\mathrm{L}}+\frac{1}{5\underline{|2}}\left(\frac{\mathrm{R}t}{\mathrm{L}}\right)^2-\frac{1}{7\underline{|3}}\left(\frac{\mathrm{R}t}{\mathrm{L}}\right)^3+\dots\right\}. \tag{42}$$

This is complete, the integrations being done at sight by $p^{-n}1=t^n/\underline{|n}$, as explained in § 351. Or, in terms of the error function, before used,

$$\frac{\mathrm{C}}{\mathrm{V}}=\left(\frac{\mathrm{K}}{\mathrm{R}}\right)^{\frac12}\operatorname{erf}\left(\frac{\mathrm{R}t}{\mathrm{L}}\right)^{\frac12}. \tag{43}$$

It is curious in how widely different a manner this function arises now. *See* § 247.

(6). **V due to steady C when L=0. Two ways.**

§ 357. There is a similar case when L is alone zero. Then

$$\mathrm{V}=\left(\frac{\mathrm{R}}{\mathrm{K}+\mathrm{S}p}\right)^{\frac12}\mathrm{C},\ \text{or}\ \frac{\mathrm{V}}{\mathrm{C}}=\left(\frac{\mathrm{R}}{\mathrm{K}}\right)^{\frac12}\operatorname{erf}\left(\frac{\mathrm{K}t}{\mathrm{S}}\right)^{\frac12}, \tag{44}$$

when C is steady, as we see by comparing with (39) and (43), and interchanging symbols. But it will be more instructive to vary the method. We have, when C is constant,

$$\frac{\mathrm{V}}{\mathrm{C}}=\left(\frac{\mathrm{R}}{\mathrm{S}}\right)^{\frac12}\frac{1}{(p+2b)^{\frac12}}=\left(\frac{\mathrm{R}}{\mathrm{S}}\right)^{\frac12}\epsilon^{-2bt}\frac{1}{p^{\frac12}}\epsilon^{2bt}=\epsilon^{-2bt}\left(\frac{\mathrm{R}}{\mathrm{S}}\right)^{\frac12}\frac{1}{p^{\frac12}(1-2b/p)} \tag{45}$$

by changing the operand from 1 to ϵ^{2bt}, and then to 1 again, in the manner of § 354. Or

$$\frac{\mathrm{V}}{\mathrm{C}}=\left(\frac{\mathrm{R}}{\mathrm{K}}\right)^{\frac12}\epsilon^{-2bt}\left\{\left(\frac{2b}{p}\right)^{\frac12}+\left(\frac{2b}{p}\right)^{\frac32}+\left(\frac{2b}{p}\right)^{\frac52}+\dots\right\},$$

$$=\left(\frac{\mathrm{R}}{\mathrm{K}}\right)^{\frac12}\epsilon^{-2bt}\left\{\frac{(2bt)^{\frac12}}{\underline{|\tfrac12}}+\frac{(2bt)^{\frac32}}{\underline{|\tfrac32}}+\frac{(2bt)^{\frac52}}{\underline{|\tfrac52}}+\dots\right\},$$

$$=\frac{2}{\pi^{\frac12}}\epsilon^{-\mathrm{K}t/\mathrm{S}}\left(\frac{\mathrm{R}t}{\mathrm{S}}\right)^{\frac12}\left\{1+\frac13\left(\frac{2\mathrm{K}t}{\mathrm{S}}\right)+\frac{1}{3\,.\,5}\left(\frac{2\mathrm{K}t}{\mathrm{S}}\right)^2+\dots\right\}, \tag{46}$$

showing the result in terms of the product of an exponential and another series. Multiplying them together, we shall obtain the result (44) again.

(7). V due to steady C when S is zero. Two ways.

§ 358. Connected with the last examples is the inversion of the same. Say,

$$\mathrm{V}=\left(\frac{\mathrm{R}+\mathrm{L}p}{\mathrm{K}}\right)^{\frac{1}{2}}\mathrm{C}=\left(\frac{\mathrm{L}p}{\mathrm{K}}\right)^{\frac{1}{2}}\left(1+\frac{\mathrm{R}}{\mathrm{L}p}\right)^{\frac{1}{2}}\mathrm{C}. \tag{47}$$

By the process of § 356, we obtain

$$\frac{\mathrm{V}}{\mathrm{C}}=\left(\frac{\mathrm{R}}{\mathrm{K}}\right)^{\frac{1}{2}}\left\{\left(\frac{\mathrm{R}}{\mathrm{L}p}\right)^{-\frac{1}{2}}+\tfrac{1}{2}\left(\frac{\mathrm{R}}{\mathrm{L}p}\right)^{\frac{1}{2}}-\frac{1.1}{2.4}\left(\frac{\mathrm{R}}{\mathrm{L}p}\right)^{\frac{3}{2}}+\frac{1.1.3}{2.4.6}\left(\frac{\mathrm{R}}{\mathrm{L}p}\right)^{\frac{5}{2}}-\dots\right\}$$
$$=\left(\frac{\mathrm{L}}{\mathrm{K}\pi t}\right)^{\frac{1}{2}}\left\{1+\frac{\mathrm{R}t}{\mathrm{L}}-\frac{1}{3\lfloor\underline{2}}\left(\frac{\mathrm{R}t}{\mathrm{L}}\right)^{2}+\frac{1}{5\lfloor\underline{3}}\left(\frac{\mathrm{R}t}{\mathrm{L}}\right)^{3}-\frac{1}{7\lfloor\underline{4}}\left(\frac{\mathrm{R}t}{\mathrm{L}}\right)^{4}+\dots\right\} \tag{48}$$

This is straightforward enough, but there is a simpler way still. For

$$\frac{\mathrm{V}}{\mathrm{C}}=\frac{\mathrm{R}+\mathrm{L}p}{\mathrm{K}}\left(\frac{\mathrm{K}}{\mathrm{R}+\mathrm{L}p}\right)^{\frac{1}{2}}=\frac{\mathrm{R}+\mathrm{L}p}{\mathrm{K}}\left(\frac{\mathrm{K}}{\mathrm{R}}\right)^{\frac{1}{2}}\mathrm{erf}\left(\frac{\mathrm{R}t}{\mathrm{L}}\right)^{\frac{1}{2}}, \tag{49}$$

by using (43), which is the solution of (39). Only one differentiation is now concerned. When executed, the result is (48). But it is not a general truth that we may introduce $(\mathrm{R}+\mathrm{L}p)/(\mathrm{R}\cdots\mathrm{L}p)$ and consider it to represent 1. We did the integration represented by the denominator first. If we do the differentiation first, it will make a difference. Thus $pp^{-1}1=pt=1$, but $p^{-1}p1=p0=0$, unless we say $p^{-1}p1=p^{-1}\dfrac{t^{-1}}{\lfloor\underline{-1}}=\dfrac{t^{0}}{\lfloor\underline{0}}=1$. This property has to be remembered sometimes.

(8). All constants finite. C due to V varying as $\epsilon^{-\rho t}$.

§ 359. Now let all four electrical constants be finite. Then

$$\mathrm{C}=\left(\frac{\mathrm{K}+\mathrm{S}p}{\mathrm{R}+\mathrm{L}p}\right)^{\frac{1}{2}}\mathrm{V}=\frac{1}{\mathrm{L}v}\left(\frac{2b+p}{2a+p}\right)^{\frac{1}{2}}\mathrm{V}=\frac{1}{\mathrm{L}v}\left(\frac{p+\rho-\sigma}{p+\rho+\sigma}\right)^{\frac{1}{2}}\mathrm{V}, \tag{50}$$

where there are only two effective constants, a and b, or ρ and σ, connected thus,

$$a=\mathrm{R}/2\mathrm{L},\quad b=\mathrm{K}/2\mathrm{S},\quad \rho=a+b,\quad \sigma=a-b. \tag{51}$$

When $\sigma=0$ we have the distortionless case, and when $\sigma=\rho$ we have the case of no leakage, § 353.

Now we can expand each of the two radicals in (50) in powers of p^{-1}; their product is then a series in rising powers of p^{-1}. The algebrisation is then immediate, by integrations according to $p^{-n}1 = t^n/\underline{|n}$. But this is a mechanical and complicated process, without illumination, and the resulting series contains ρ and σ in a way that does not show plainly how to simplify it. So vary the method.

Introduce $\epsilon^{\rho t}$ into the operand, by simultaneously introducing the factor $\epsilon^{-\rho t}$, and turning p to $p+\rho$. Follow (35). Then (50) becomes

$$C = \frac{\epsilon^{-\rho t}}{Lv}\left(\frac{p-\sigma}{p+\sigma}\right)^{\frac{1}{2}}\epsilon^{\rho t}V = \frac{\epsilon^{-\rho t}}{Lv}\left(\frac{p-\sigma}{p+\sigma}\right)^{\frac{1}{2}}e, \tag{52}$$

if $e = V\epsilon^{\rho t}$. If e is constant, or the voltage is the special one $e\epsilon^{-\rho t}$, we can solve at once. Thus

$$\left(\frac{p-\sigma}{p+\sigma}\right)^{\frac{1}{2}} = \left(1-\frac{\sigma}{p}\right)\left(1-\frac{\sigma^2}{p^2}\right)^{-\frac{1}{2}} = \left(1-\frac{\sigma}{p}\right)I_0(\sigma t), \tag{53}$$

by (36). Here one integration upon $I_0(\sigma t)$ is wanted. The result is

$$C = \frac{e\epsilon^{-\rho t}}{Lv}\left\{I_0(\sigma t) - \left(\sigma t + \frac{(\sigma t)^2}{2\underline{|3}} + \frac{1.3(\sigma t)^5}{2.4\underline{|5}} + \dots\right)\right\}. \tag{54}$$

The initial current is e/Lv. The final current is zero, of course, because the voltage falls to zero.

(9). C due to V varying as $\epsilon^{-Kt/S}$. Discharge of a charged into an empty Cable.

§ 360. A more significant case is got by letting the voltage decay in a different way. Say $V = \epsilon^{-Kt/S}e$, where e is constant. Then the first of (52) becomes

$$C = \frac{e\epsilon^{-\rho t}}{Lv}\left(\frac{p-\sigma}{p+\sigma}\right)^{\frac{1}{2}}\epsilon^{\sigma t} = \frac{e}{Lv}\epsilon^{-\rho t}I_0(\sigma t), \tag{55}$$

by (36). Here we have a compact solution, differing from (27) in having ρ and σ instead of both σ. When $K=0$, we fall back upon the case (27).

This solution is important thus. Imagine a cable which is infinitely long both ways, to be charged initially to transverse voltage $2e$ on the left side and zero on the right side of the origin. V will instantly fall to e at the origin, and its later value will be $V = e\epsilon^{-Kt/S}$, as in the above. This is, therefore, the voltage impressed upon the initially empty cable to the right. The resulting current is (55). It will be generalised later to represent the complete wave of current.

Notice, in passing, that in the course of the transformations employed we are engaged incidentally upon other problems than those under discussion. For example, (52) shows that if

$$C = \frac{\epsilon^{-\rho t}}{Lv} e, \quad \text{then} \quad V = \epsilon^{-\rho t} \left(\frac{p+\sigma}{p-\sigma}\right)^{\frac{1}{2}} e, \tag{56}$$

which shows the voltage needed to make the current vary as $\epsilon^{-\rho t}$.

(10). C due to steady V, and V due to steady C.

§ **361.** Now let the impressed voltage V itself be constant. Then we have

$$\frac{C}{V} = \frac{K + Sp}{(R + Lp)^{\frac{1}{2}}(K + Sp)^{\frac{1}{2}}} = \frac{v(K + Sp)}{\{(p + \rho)^2 - \sigma^2\}^{\frac{1}{2}}}. \tag{57}$$

This can, by what has been done, be reduced to a single integration. First, we have

$$\frac{p}{\{(p+\rho)^2 - \sigma^2\}^{\frac{1}{2}}} = \epsilon^{-\rho t} \frac{p - \rho}{(p^2 - \sigma^2)^{\frac{1}{2}}} \frac{p}{p - \rho} = \epsilon^{-\rho t} \frac{p}{(p^2 - \sigma^2)^{\frac{1}{2}}}, \tag{58}$$

by the process (35); or, by (36)

$$\frac{p}{\{(p+\rho)^2 - \sigma^2\}^{\frac{1}{2}}} = \epsilon^{-\rho t} I_0(\sigma t). \tag{59}$$

Using this in (57) we obtain

$$\frac{C}{V} = \frac{1}{Lv}\left(1 + \frac{K}{Sp}\right) \epsilon^{-\rho t} I_0(\sigma t). \tag{60}$$

Thus the C due to V constant, when all four electrical properties are active, is expressed in terms of a known function, and its time-integral, the residual p^{-1} meaning integration from 0 to t.

Similarly, $$\frac{V}{C} = \frac{1}{Sv}\left(1 + \frac{R}{Lp}\right) \epsilon^{-\rho t} I_0(\sigma t) \tag{61}$$

is the V when C is constant (always understanding that $t = 0$ begins the operand). We get the last by interchanging symbols R and K, L and S. The symbol ρ does not change, but σ is negatived, though this makes no difference in $I_0(\sigma t)$.

(11). C due to impulsive V, and V due to impulsive C.

§ **362.** If $V = pP$, where P is a constant, the case is that of an impulsive voltage, total P, generating the momentum P, the space integral of LC.

Then, by (60),

$$C=\frac{P}{Lv}(p+\rho-\sigma)\epsilon^{-\rho t}I_0\sigma t=\frac{P\epsilon^{-\rho t}}{Lv}(p-\sigma)I_0(\sigma t). \qquad (62)$$

Or, executing the differentiations,

$$C=\frac{P}{Lv}\epsilon^{-\rho t}\sigma\{I_1(\sigma t)-I_0(\sigma t)\}. \qquad (63)$$

This shows the current due to the impulsive voltage. It is negative when σ is positive, and positive when σ is negative. That is, if R is in excess, the current following the impulse is back from the cable; if in deficit, it is forward, into the cable.

Similarly,

$$V=\frac{Q}{Sv}\epsilon^{-\rho t}(p+\sigma)I_0(\sigma t)=\frac{Q}{Sv}\epsilon^{-\rho t}\sigma\{I_1(\sigma t)+I_0(\sigma t)\}, \qquad (64)$$

obtained from (62), (63), by interchanges, represents V due to the sudden injection of the charge Q, followed by insulation. That is, $C=pQ$ is the datum, and V follows. It is positive when σ is positive and negative when σ is negative.

From these we conclude that if the cable is infinitely long both ways, and the charge Q or the momentum P be suddenly introduced at the origin, then

$$V=\frac{Q}{2Sv}\epsilon^{-\rho t}(p+\sigma)I_0(\sigma t),\quad C=\frac{P}{2Lv}\epsilon^{-\rho t}(p-\sigma)I_0(\sigma t) \qquad (65)$$

represent the resulting V at the origin in one case, and C in the other. If the line is not infinitely long, these are still true for a time, until in fact the first reflected wave arrives at the origin. The halving is done because P and Q immediately split into halves which separate. It should be noted that (65) represent V or C in the middle of the tail connecting the two heads or waves at distance $\pm vt$ from the origin. See § 208. The complete wave formulæ will follow.

Cubic under radical. Reversibility of Operations. Distribution of Operators.

§ 363. Observe that we have algebrised $(p^2+Ap+B)^{-\frac{1}{2}}p.1$. For it is the same as (59) above, if $A=2\rho$ and $B=\rho^2-\sigma^2$. That is,

$$\frac{p}{(p^2+Ap+B)^{\frac{1}{2}}}=\epsilon^{-\frac{1}{2}At}I_0\{(\tfrac{1}{4}A^2-B)^{\frac{1}{2}}t\}. \qquad (66)$$

The extension to a cubic under the radical sign is therefore suggested, say

$$u = \frac{p}{(p^3+ap^2+bp+c)^{\frac{1}{2}}}. \tag{67}$$

We can reduce it to a quadratic by making the operand be ϵ^{-xt}, where x is one of the roots of the cubic. Turn p to $p-x$, and put ϵ^{xt} at the beginning and ϵ^{-xt} at the end, or $(1+x/p)^{-1}$ at the end, or earlier. Thus

$$u = \epsilon^{xt}\frac{p^{\frac{1}{2}}}{p+x}\,\frac{p}{\{p^2+(3x+a)p+3x^2+2ax+b\}^{\frac{1}{2}}}, \tag{68}$$

that is,

$$\epsilon^{xt}\frac{p^{\frac{1}{2}}}{p+x}\,\epsilon^{-\mathrm{F}t}\mathrm{I}_0(\mathrm{G}t), \tag{69}$$

when F and G are known, by (63). This may be carried further, but it would be out of place here, having no immediate relation to the matter in hand.

The following is more to the point, concerning the reversibility of the operations used. That

$$\mathrm{I}_0(\sigma t) = \frac{p}{(p^2-\sigma^2)^{\frac{1}{2}}}1 \tag{70}$$

is clear by inspection of the form of the I_0 function, and then putting it in terms of powers of p^{-1}. We get a series which the binomial theorem allows us to write in the form (67), provided we understand that in expanding it, we are to employ integrations, not differentiations.

But it is also true that

$$\frac{(p^2-\sigma^2)^{\frac{1}{2}}}{p}\mathrm{I}_0(\sigma t) = 1. \tag{71}$$

Expand by the binomial theorem again in rising powers of p^{-1}, and execute the work on $\mathrm{I}_0(\sigma t)$. The coefficient of every power of t vanishes save the zeroth, leaving the result 1.

In connection with this, and with some of the preceding work, it is to be noted that if we have an operator θ which is the product of any number of others, say,

$$u = \theta 1 = \phi_1\phi_2\phi_3 1, \tag{72}$$

and if the type of ϕ_1 is

$$\phi_1 = a_1 + b_1 p^{-1} + c_1 p^{-2} + \ldots, \tag{73}$$

that is a series in rising powers of p^{-1}, and if all the ϕ's are of this kind, then the same is true for the resultant θ, and in whatever order the ϕ's are written. So there are all sorts of ways of algebrising (72), which are bound to lead to the same function, if properly followed, though intermediately we may be led to all sorts of functions. We may regard u as $(\phi_1\phi_2)\phi_3 1$, or $(\phi_2\phi_3)\phi_1 1$, &c. The utility of such changes is to bring operational solutions to more convenient forms for algebrisation, by changes in the operand and operator. The matter is not so limited as it was just now stated. It is not always necessary to keep to positive powers of p^{-1}, that is to integrations. Sometimes a series of differentiations may equivalently replace a series of integrations. But sufficient of theory now. Practice is more important, and the final integration involved in (60) remains to be done.

(12). Development of Equation (60). C due to steady V.

§ 364. There are several ways of obtaining a full development of the solution (60), one or two of which may be done here, with side matters. We have to find the time integral of $\epsilon^{-\rho t}\mathrm{I}_0(\sigma t)$, and a first way is to shift the exponential to the left, thus

$$\frac{1}{p}\epsilon^{-\rho t}\mathrm{I}_0(\sigma t) = \frac{\epsilon^{-\rho t}}{\rho}\frac{\rho}{p-\rho}\mathrm{I}_0(\sigma t) = \frac{\epsilon^{-\rho t}}{\rho}u, \text{ say.} \tag{74}$$

Here the operation $(p-\rho)^{-1}$ to find u must be done by integrations, like the p^{-1} from which it arose. So

$$u = \left(\frac{\rho}{p} + \frac{\rho^2}{p^2} + \frac{\rho^3}{p^3} + \dots\right)\left(1 + \frac{\sigma^2 t^2}{2^2} + \frac{\sigma^4 t^4}{2^2 4^2} + \dots\right). \tag{75}$$

This is integrable at sight, making

$$u = \rho t\left(1 + \frac{\sigma^2 t^2}{2^2 3} + \frac{\sigma^4 t^4}{2^2 4^2 5} + \dots\right) + \rho^2 t^2\left(\frac{1}{\underline{|2}} + \frac{\sigma^2 t^2}{2^2 3 . 4} + \frac{\sigma^4 t^4}{2^2 4^2 5 . 6} + \dots\right)$$
$$+ \rho^3 t^3\left(\frac{1}{\underline{|3}} + \frac{\sigma^2 t^2}{2^2 3 . 4 . 5} + \frac{\sigma^4 t^4}{2^2 4^2 5 . 6 . 7} + \dots\right) + \dots, \tag{76}$$

where the law of the coefficients is obvious, every set being the integral of the preceding one. Therefore, by (60),

$$\mathrm{C} = \frac{\mathrm{V}}{\mathrm{L}v}\epsilon^{-\rho t}\left\{\mathrm{I}_0(\sigma t) + \left(1 - \frac{\sigma}{\rho}\right)u\right\} \tag{77}$$

expresses the complete development in one form, showing the C due to steady V when all four properties of the circuit are in operation. The part involving u goes out when there is no leakage.

As a check upon (77), another way may be indicated. We have

$$C = \frac{V}{Lv} \epsilon^{-\rho t} \left(1 - \frac{\sigma}{p}\right)\left(1 - \frac{\sigma^2}{p^2}\right)^{-\frac{1}{2}} \epsilon^{\rho t}, \qquad (78)$$

by the first of (52). Here write out the operand $\epsilon^{\rho t}$ in full, then operate on it by $(1 - \sigma^2/p^2)^{-\frac{1}{2}}$ expanded in rising powers of p^{-2}, and then operate on the result by $(1 - \sigma/p)$. We shall obtain (77), after a little rearrangement of terms. There is nothing particular to notice in the process, so the details need not be given.

(13). Another Development of Equation (60).

§ 365. But with a view to finding possible better forms of the function u, do (78) without expanding $\epsilon^{\rho t}$ first, so that the result is in terms of $\epsilon^{\rho t}$ and its integrals. We then get

$$C = \frac{V}{Lv} \epsilon^{-\rho t}\left\{ I_0(\sigma t) + \left(1 - \frac{\sigma}{\rho}\right)\left(\epsilon^{\rho t} - 1 + \frac{1}{2}\frac{\sigma^2}{\rho^2}(\epsilon^{\rho t} - e_2)\right.\right.$$
$$\left.\left. + \frac{1.3}{2.4}\frac{\sigma^4}{\rho^4}(\epsilon^{\rho t} - e_4) + \frac{1.3.5}{2.4.6}\frac{\sigma^6}{\rho^6}(\epsilon^{\rho t} - e_6) + \ldots\right)\right\}, \qquad (79)$$

where e_n is the sum of the first $n + 1$ terms of the series for $\epsilon^{\rho t}$. Now, in (79) the inside exponential is cancelled by the outside one. The part of C not containing t may therefore be exhibited separately. Thus,

$$\frac{1}{Lv}\left(1 - \frac{\sigma}{\rho}\right)\left(1 + \frac{1}{2}\frac{\sigma^2}{\rho^2} + \ldots\right) = \frac{1}{Lv}\left(1 - \frac{\sigma}{\rho}\right)\left(1 - \frac{\sigma^2}{\rho^2}\right)^{-\frac{1}{2}} = \left(\frac{K}{R}\right)^{\frac{1}{2}}, \qquad (80)$$

by the definition of ρ and σ. So (79) becomes

$$C = V\left(\frac{K}{R}\right)^{\frac{1}{2}} + \frac{V}{Lv}\epsilon^{-\rho t}\left\{I_0(\sigma t) - \left(1 - \frac{\sigma}{\rho}\right)\left(1 + \frac{1}{2}\frac{\sigma^2}{\rho^2}e_2\right.\right.$$
$$\left.\left. + \frac{1.3}{2.4}\frac{\sigma^4}{\rho^4}e_4 + \ldots\right)\right\}. \qquad (81)$$

Another modification is got by arranging the function u as it appears in (79) in powers of ρt. We get

$$C = \frac{V}{Lv}\epsilon^{-\rho t}\left\{\frac{\sigma}{\rho}I_0(\sigma t) + \left(1 - \frac{\sigma}{\rho}\right)\left\{(1 + \rho t) + \left(\frac{\rho^2 t^2}{\underline{|2}} + \frac{\rho^3 t^3}{\underline{|3}}\right)f_1\right.\right.$$
$$\left.\left. + \left(\frac{\rho^4 t^4}{\underline{|4}} + \frac{\rho^5 t^5}{\underline{|5}}\right)f_2 + \left(\frac{\rho^6 t^6}{\underline{|6}} + \frac{\rho^7 t^7}{\underline{|7}}\right)f_3 + \ldots\right\}, \qquad (82)$$

where f_n is the sum of the first $n + 1$ terms in the expansion of $(1 - \sigma^2/\rho^2)^{-\frac{1}{2}}$ in rising powers of σ/ρ by the binomial theorem, and the coefficients of the f's are the successive terms, in pairs, of the expansion of $\epsilon^{\rho t}$. It is a curious form. It can be seen to be correct in the two cases $\sigma = 0$ and $\sigma = \rho$.

Comparing (82) with (77), we require this identity,

$$\mathrm{I}_0(\sigma t) + \left(1 - \frac{\sigma}{\rho}\right)u = \frac{\sigma}{\rho}\mathrm{I}_0(\sigma t) + \left(1 - \frac{\sigma}{\rho}\right)\left\{ (1 + \rho t) + \left(\frac{\rho^2 t^2}{\underline{|2}} + \frac{\rho^3 t^3}{\underline{|3}}\right)f_1 + \dots \right\}. \qquad (83)$$

or, which is the same,

$$u = -\mathrm{I}_0(\sigma t) + 1 + \rho t + \left(\frac{\rho^2 t^2}{\underline{|2}} + \frac{\rho^3 t^3}{\underline{|3}}\right)f_1 + \left(\frac{\rho^4 t^4}{\underline{|4}} + \frac{\rho^5 t^5}{\underline{|5}}\right)f_2 + \dots \qquad (84)$$

This may be verified by means of (76), rearranging terms therein.

(14). Third Development of Equation (60). Integration by Parts.

§ 366. Lastly, compare the results with those got by the process of integration by parts. We have

$$\int_0^t \epsilon^{-\rho t}\,\mathrm{I}_0(\sigma t)dt = \left[\frac{\epsilon^{-\rho t}}{-\rho}\mathrm{I}_0(\sigma t)\right]_0^t + \int_0^t \frac{\epsilon^{-\rho t}}{\rho}p\mathrm{I}_0(\sigma t)dt. \qquad (85)$$

Repeat this process again and again, and we get

$$= \left[\frac{\epsilon^{-\rho t}}{-\rho}\left(1 + \frac{p}{\rho} + \frac{p^2}{\rho^2} + \frac{p^3}{\rho^3} + \dots\right)\mathrm{I}_0(\sigma t)\right]_0^t$$

$$= \left[\frac{\epsilon^{-\rho t}}{-\rho}\frac{\mathrm{I}_0(\sigma t)}{1 - p/\rho}\right]_0^t = \frac{1}{(\rho^2 - \sigma^2)^{\frac{1}{2}}} - \frac{\epsilon^{-\rho t}}{\rho}\frac{\mathrm{I}_0(\sigma t)}{1 - p/\rho}. \qquad (86)$$

It is to be understood here that $(1 - p/\rho)^{-1}$ is done by differentiations. We therefore make (78) or (60) become

$$\mathrm{C} = \mathrm{V}\left(\frac{\mathrm{K}}{\mathrm{R}}\right)^{\frac{1}{2}} + \frac{\mathrm{V}}{\mathrm{L}v}\epsilon^{-\rho t}\,\mathrm{I}_0(\sigma t) - \frac{\mathrm{VK}v}{\rho}\epsilon^{-\rho t}\frac{\mathrm{I}_0(\sigma t)}{1 - p/\rho}. \qquad (87)$$

This agrees with (81), provided that

$$\frac{\mathrm{I}_0(\sigma t)}{1 - p/\rho} = 1 + \frac{1}{2}\frac{\sigma^2}{\rho^2}e_2 + \frac{1.3}{2.4}\frac{\sigma^4}{\rho_4}e_4 + \dots, \qquad (88)$$

and this may be verified by carrying out the differentiations on the left side. Now we also know that

$$\frac{\rho/p}{1 - \rho/p}\mathrm{I}_0(\sigma t) = u, \qquad \text{as above in (76)}, \qquad (89)$$

this being done by integrations.

So, adding the results (88) and (89), this mathematical result follows :—

$$\left(1+\frac{\rho}{p}+\ldots+\frac{p}{\rho}+\ldots\right)\mathrm{I}_0(\sigma t)=\sum_{-\infty}^{\infty}\left(\frac{p}{\rho}\right)^m\mathrm{I}_0(\sigma t)=\frac{\rho\epsilon^{\rho t}}{(\rho^2-\sigma^2)^{\frac{1}{2}}}, \qquad (90)$$

the summation including all integral values of m, including zero.

The results are all consistent. But it is unfortunate that the additional terms brought in by leakage do not seem to be reducible to a single known simple function. If it could be so, then the result could be readily generalised to express the complete wave of current. If not, then we must rest satisfied with one or other cumbrous form of series.

(15). Generalisation. The Complete Wave of C due to V at the origin varying as $\epsilon^{-\mathrm{K}t/\mathrm{S}}$.

§ 367. Having now got full results as regards V/C and C/V at the origin under different circumstances, we can go on to generalise them in certain cases. Consider (55) for example. Say,

$$\mathrm{C}_0=\frac{e}{\mathrm{L}v}\epsilon^{-\rho t}\mathrm{I}_0(\sigma t). \qquad (91)$$

This is the current produced at the origin ($x=0$) when the voltage there is $\mathrm{V}_0=e\,\epsilon^{-\mathrm{K}t/\mathrm{S}}$, e being constant, beginning when $t=0$. If the cable is short-circuited at the origin, V_0 may be regarded as a variable impressed voltage inserted. If infinitely long both ways, then the impressed voltage must be $2\mathrm{V}_0$. But we may also regard V_0 as being produced not by impressed voltage but by an initial state, namely, $\mathrm{V}=2e$ initially on the whole of the left side of the origin.

Now, at distance x, we have

$$\mathrm{C}=\epsilon^{-qx}\,\mathrm{C}_0, \qquad \text{where} \qquad q=\frac{1}{v}\{(p+\rho)^2-\sigma^2\}^{\frac{1}{2}}. \qquad (92)$$

The current must be the same at the same distance on the negative side as on the positive. We may therefore put $\cosh qx$ for ϵ^{-qx}. So

$$\mathrm{C}=\cosh qx\,\frac{e}{\mathrm{L}v}\epsilon^{-\rho t}\,\mathrm{I}_0(\sigma t). \qquad (93)$$

Shift $\epsilon^{\quad pt}$ to the left. Then

$$C = \frac{e\epsilon^{-pt}}{Lv}\cosh\frac{x}{v}(p^2-\sigma^2)^{\frac{1}{2}} \,.\, I_0(\sigma t). \tag{94}$$

This gives C by direct differentiations, since only the integral powers of $(p^2-\sigma^2)$ are involved. Now

$$(\sigma^2-p^2)\,I_0(\sigma t) = t^{-1}pI_0(\sigma t) = \sigma^2\frac{I_1(\sigma t)}{\sigma t}, \tag{95}$$

and generally,

$$(\sigma^2-p^2)\frac{I_m(\sigma t)}{(\sigma t)^m} = (2m+1)t^{-1}p\frac{I_m(\sigma t)}{(\sigma t)^m} \tag{96}$$

$$= (2m+1)\,\sigma^2\,\frac{I_{m+1}(\sigma t)}{(\sigma t)^{m+1}}. \tag{97}$$

This general property is proved by the differential equation of $x^{-m}I_m(x)$, § 336, equation (2); see also (18), § 337. Applying (96) to (94), we obtain

$$C = \frac{e\epsilon^{-pt}}{Lv}\left\{1 - \frac{1}{2}\frac{x^2}{v^2}(t^{-1}p) + \frac{1}{2.4}\,\frac{x^4}{v^4}(t^{-1}p)^2 - \frac{1}{2.4.6}\,\frac{x^6}{v^6}(t^{-1}p)^3 + \dots\right\}I_0(\sigma t), \tag{98}$$

and, if we use (97) we get

$$C = \frac{e\epsilon^{-pt}}{Lv}\left\{P_0(\sigma t) - \left(\frac{\sigma x}{v}\right)^2\frac{P_1(\sigma t)}{2^2} + \left(\frac{\sigma x}{v}\right)^4\frac{P_2(\sigma t)}{2^24^2} - \left(\frac{\sigma x}{v}\right)^6\frac{P_3(\sigma t)}{2^24^26^2} + \dots\right\}, \tag{99}$$

where $P_m(\sigma t)$ is the function got by dividing $I_m(\sigma t)$ by its first term, so that the first term of P_m itself is unity. Of course the convergent formula is used, equation (23), § 338. Thus

$$P_m(\sigma t) = 1 + \frac{(\frac{1}{2}\sigma t)^2}{1\,.\,m+1}\left(1 + \frac{(\frac{1}{2}\sigma t)^2}{2\,.\,m+2}\left(1 + \frac{(\frac{1}{2}\sigma t)^2}{3\,.\,m+3}\dots\right.\right. \tag{100}$$

Now (99) is nothing more than the expanded form of

$$C = \frac{e}{Lv}\epsilon^{-pt}I_0\left\{\frac{\sigma}{v}(v^2t^2-x^2)^{\frac{1}{2}}\right\}, \tag{101}$$

as may be seen by arranging its expansion in a series of powers of x^2. Therefore, the last equation is the generalisation required. Comparing with (91), we see that vt is turned to $(v^2t^2-x^2)^{\frac{1}{2}}$ in the I_0 function.

At the time t the region occupied by the current extends only to the distance vt from the origin. Beyond that distance the current is zero. Equation (101) therefore expresses a wave, whose front travels at speed v. At the wave front the current is $(e/\mathrm{L}v)\epsilon^{-\rho t}$. Since the I_0 function has been tabulated, the shape of the current curve along x, at successive moments of time, can be readily calculated.

(16). Summary of Work showing the Wave of C due to V at origin varying as $\epsilon^{-\mathrm{K}t}$ S.

§ 368. The above is by far the simplest way of obtaining the wave solution, without prior knowledge. Of course, if it is already known that the formula satisfies the characteristic partial differential equation of C, there is no difficulty in finding the nature of the problem of which it is the solution. But it would not be reasonable to expect people to possess the prior knowledge.

Seeing that the preceding portion of this Chapter contains the treatment of various other cases than this important one, it is desirable to exhibit collected separately the various steps in the process of deduction. First, from the connections of voltage and current,

$$-\frac{d\mathrm{V}}{dx}=(\mathrm{R}+\mathrm{L}p)\mathrm{C}, \qquad -\frac{d\mathrm{C}}{dx}=(\mathrm{K}+\mathrm{S}p)\mathrm{V}, \tag{102}$$

we deduce the characteristic

$$\frac{d^2\mathrm{V}}{dx^2}=(\mathrm{R}+\mathrm{L}p)(\mathrm{K}+\mathrm{S}p)\mathrm{V}=q^2\mathrm{V}; \tag{103}$$

from which we conclude that

$$\mathrm{V}=\epsilon^{qx}\mathrm{A}+\epsilon^{-qx}\mathrm{B}, \tag{104}$$

is the type of solution required in general when A and B are time functions, and that the ϵ^{-qx} part is the only one wanted when we send disturbances into an empty cable; so that

$$\mathrm{V}=\epsilon^{-qx}\mathrm{V}_0, \qquad \mathrm{C}=\epsilon^{-qx}\mathrm{C}_0, \qquad \mathrm{C}_0=\left(\frac{\mathrm{K}+\mathrm{S}p}{\mathrm{R}+\mathrm{L}p}\right)^{\frac{1}{2}}\mathrm{V}_0, \tag{105}$$

express V and C in terms of V_0 and C_0 at the origin, which are themselves connected by the third equation. So

$$\mathrm{C}_0=\frac{1}{\mathrm{L}v}\left(\frac{p+\rho-\sigma}{p+\rho+\sigma}\right)^{\frac{1}{2}}\mathrm{V}_0=\frac{\epsilon^{-\rho t}}{\mathrm{L}v}\left(\frac{p-\sigma}{p+\sigma}\right)^{\frac{1}{2}}\epsilon^{\rho t}\mathrm{V}_0, \tag{106}$$

by making the operand be $V_0\epsilon^{\rho t}$. Then let $\epsilon^{\rho t}V_0 = \epsilon^{\sigma t}e$, where e is constant; or $V_0 = e\epsilon^{-Kt/S}$. This makes

$$C_0 = \frac{e\epsilon^{-\rho t}}{Lv}\left(\frac{p-\sigma}{p+\sigma}\right)^{\frac{1}{2}}\frac{p}{p-\sigma} = \frac{e\epsilon^{-\rho t}}{Lv}I_0(\sigma t), \qquad (107)$$

fully realised. Finally, generalise for the complete wave,

$$C = \epsilon^{-qx}C_0 = \cosh qx\, C_0 = \frac{e\epsilon^{-\rho t}}{Lv}\cos\frac{x}{v}(\sigma^2 - p^2)^{\frac{1}{2}}\, .\, I_0(\sigma t), \quad (108)$$

which develops to

$$C = \frac{e\epsilon^{-\rho t}}{Lv}I_0\left[\frac{\sigma}{v}(v^2t^2 - x^2)^{\frac{1}{2}}\right]. \qquad (109)$$

If we do not make use of the property of equality of the current at x and $-x$, without discontinuity at the origin, we have

$$C = \cosh qx\, .\, C_0 - \text{shin}\, qx\, .\, C_0,$$

instead of as in (108), but the additional part will not bring in any new terms, because its first term will involve

$$(p^2 - \sigma^2)^{\frac{1}{2}}I_0(\sigma t) = p1, \qquad (110)$$

by (71), and the rest involve complete differentiations upon the first term.

(17). Derivation of the Wave of V from the Wave of C.

§ 369. To obtain the formula for the V wave corresponding to (109), there are many ways of working. First, we may use the C formula itself, and the second circuital law. We have

$$-\frac{dV}{dx} = (R + Lp)C = L(p + \rho + \sigma)C$$
$$= \frac{e\epsilon^{-\rho t}}{v}(p+\sigma)I_0\left[\frac{\sigma}{v}(v^2t^2 - x^2)^{\frac{1}{2}}\right], \qquad (111)$$

by (109).

Therefore V is the negative of the x integral of the right member, provided it is standardised properly, to have the value $e^{-Kt/S}$ at the origin. Now, looking at the expansion (99), we see that its x integral will vanish at the origin, So the real value at the origin must be added on. We therefore get

$$V = e\epsilon^{-\rho t}\left\{\epsilon^{\sigma t} - \left(1+\frac{p}{\sigma}\right)\left[\frac{\sigma x}{v}P_0 - \left(\frac{\sigma x}{v}\right)^3\frac{P_1}{2\cdot 3} + \left(\frac{\sigma x}{v}\right)^5\frac{P_2}{2^2 4^2 5} - \dots\right]\right\} \qquad (112)$$

for a complete development of the V wave. There is one time differentiation concerned. We have

$$\frac{p}{\sigma}\mathrm{P}_m(\sigma t) = \frac{\sigma t \mathrm{P}_{m+1}(\sigma t)}{2(m+1)}, \tag{113}$$

by (100). From this,

$$\left(1+\frac{p}{\sigma}\right)\mathrm{P}_m(\sigma t) = 2^m \lfloor m \frac{\mathrm{I}_m(\sigma t) + \mathrm{I}_{m+1}(\sigma t)}{(\sigma t)^m}. \tag{114}$$

Using this in (112), we obtain the equivalent form,

$$\mathrm{V} = e\epsilon^{-\rho t}\left\{\epsilon^{\sigma t} - \frac{\sigma x}{v}(\mathrm{I}_0 + \mathrm{I}_1) + \left(\frac{\sigma x}{v}\right)^3 \frac{\mathrm{I}_1 + \mathrm{I}_2}{\sigma t \lfloor 3} - \frac{1.3}{\lfloor 5}\left(\frac{\sigma x}{v}\right)^5 \frac{\mathrm{I}_2 + \mathrm{I}_3}{(\sigma t)^2} + \dots\right\}. \tag{115}$$

The argument of all the I functions is σt.

This is not so compact as the C formula. But there is a similar peculiarity in the corresponding formulæ in pure diffusion, the current being given by an exponential formula, the voltage by the error function. *See* §§ 246, 247.

(18). The Wave of V independently developed.

§ 370. Another way of getting the V formula is more primitive. Do not use the developed C formula, but work from the beginning. Thus,

$$\mathrm{V} = \epsilon^{-qx}\mathrm{V}_0 = e\,\epsilon^{-qx}\epsilon^{(\sigma-\rho)t}, \tag{116}$$

because the voltage at the origin is $e\epsilon^{(\sigma-\rho)t}$. Now for ϵ^{-qx} put (cosh − shin)qx, and shift $\epsilon^{-\rho t}$ to the left. Then

$$\mathrm{V} = e\epsilon^{-\rho t}(\cosh - \mathrm{shin})\frac{x}{v}(p^2-\sigma^2)^{\frac{1}{2}}\,.\,\epsilon^{\sigma t}. \tag{117}$$

Here the cosh part involves only complete differentiations upon $\epsilon^{\sigma t}$, and the potence of $p^2 - \sigma^2$ is zero. Only the first term of the cosh function does anything. So we get

$$\mathrm{V} = e\epsilon^{-\rho t}\left\{\epsilon^{\sigma t} - \frac{\mathrm{shin}\frac{x}{v}(p^2-\sigma^2)^{\frac{1}{2}}}{(p^2-\sigma^2)^{\frac{1}{2}}}(p^2-\sigma^2)^{\frac{1}{2}}\epsilon^{\sigma t}\right\}. \tag{118}$$

But here

$$(p^2-\sigma^2)^{\frac{1}{2}}\epsilon^{\sigma t} = (p+\sigma)\mathrm{I}_0(\sigma t), \tag{119}$$

by (36). So

$$\mathrm{V} = e\epsilon^{-\rho t}\left\{\epsilon^{\sigma t} - \frac{x}{v}(p+\sigma)\left[1 - \frac{x^2}{\lfloor 3 v^2}(\sigma^2 - p^2) + \dots\right]\mathrm{I}_0(\sigma t)\right\}. \tag{120}$$

Now carry out the differentiations $(\sigma^2 - p^2)$, once, twice, thrice, &c., upon $I_0(\sigma t)$, according to the formula (97) above, and the development (112) immediately results. This is the most direct way of obtaining the V formula. And, just as we derived it from the C formula by the second circuital law, so we may derive the C formula from that for V by a similar process, using the first circuital law.

(19). The Waves of V and C due to Initial Charge or Momentum at a Single Place.

§ 371. We are now in virtual possession of the complete tail formulæ arising when electrification spreads and when momentum spreads. First, let there be the charge Q initially at the origin, which may be any point in a cable continuous both ways. Then, at time t later, the voltage and current are

$$V = \frac{Q}{2Sv}\epsilon^{-pt}(p+\sigma)I_0\left[\frac{\sigma}{v}(v^2t^2 - x^2)^{\frac{1}{2}}\right], \tag{121}$$

$$C = \tfrac{1}{2}Qv\epsilon^{-pt}\left(-\frac{d}{dx}\right)I_0\left[\frac{\sigma}{v}(v^2t^2 - x^2)^{\frac{1}{2}}\right]. \tag{122}$$

Similarly, let there be initially momentum P at the origin. Then at time t later the current and voltage at x are

$$C = \frac{P}{2Lv}\epsilon^{-pt}(p-\sigma)I_0\left[\frac{\sigma}{v}(v^2t^2 - x^2)^{\frac{1}{2}}\right], \tag{123}$$

$$V = \tfrac{1}{2}Pv\epsilon^{-pt}\left(-\frac{d}{dx}\right)I_0\left[\frac{\sigma}{v}(v^2t^2 - x)^{\frac{1}{2}}\right]. \tag{124}$$

The last pair can be written down from the first pair, by interchanging V and C, P and Q, L and S, R and K.

Observe that the spreading of current due to initial charge and the spreading of charge due to initial current take place in precisely the same way, (122) and (124). But the charge due to initial charge and the current due to initial current do not behave similarly, because of the change of sign of σ in passing from (121) to (123).

If we superimpose two distributions of electrification, one given by $V = V_0$, constant on the whole of the left side of the origin and up to distance $x = \frac{1}{2}a$ on the positive side, the other given by $V = -V_0$ on the whole of the left side except from the origin to the distance $x = -\frac{1}{2}a$, the result is simply $V = V_0$

along distance a at the origin itself, or the charge SV_0a. Keeping the last constant, say Q, and decreasing a indefinitely, the result is a finite charge Q at the origin only. The waves of voltage and current which arise from it are therefore given on the positive side by the difference of the waves arising from the two uniform distributions of V before described, which neutralise one another save at the origin. In this way the solutions (121), (122) for a point-charge are derived from (109) and (111), or (115).

But the following way is interesting. We know that $C = \epsilon^{-qx}C_0$, if C_0 is the current impressed at the origin. Take $C_0 = \frac{1}{2}pQ$; then C_0 is due to the charge Q impulsively introduced and dividing into two equal parts at the origin. Therefore

$$C = \tfrac{1}{2}Q\epsilon^{-qx}p1 = \tfrac{1}{2}Qvq\epsilon^{-qx}\frac{p}{qv} = -\tfrac{1}{2}Qv\frac{d}{dx}\left(\epsilon^{-qx}\frac{p}{qv}\right). \qquad (125)$$

This is done by introducing qv/qv, and then noting that the fir-t q is equivalent to $-d/dx$.

Now the function in the brackets is already known. For

$$\frac{p}{qv} = \epsilon^{-\rho t}I_0'(\sigma t), \quad \text{and} \quad \epsilon^{-qx}\frac{p}{qv} = \epsilon^{-\rho t}I_0\{\sigma(t - x^2/v^2)^{\frac{1}{2}}\}, \qquad (126)$$

by (59) and its extension in § 367, before done, equation (91) leading to (101), or (105) to (109). So (125) produces (122).

The voltage formula to match may be derived thus,

$$V = \left(\frac{R+Lp}{K+Sp}\right)^{\frac{1}{2}}C = (R+Lp)\tfrac{1}{2}Qv\,\epsilon^{-qx}\frac{p}{qv}$$

$$= \frac{Q}{2Sv}(p+\rho+\sigma)\,\epsilon^{-qx}\frac{p}{qv} = \frac{Q}{2Sv}\epsilon^{-\rho t}(p+\sigma)I_0\{\sigma(t^2 - x^2/v^2)^{\frac{1}{2}}\}, \qquad (127)$$

which is (121).

Knowing the voltage and current due to initial charge or momentum at a single point, we can at once write down the integrals expressing the voltage and current arising from any given initial states of voltage and current. But we do not want them at present.

The Waves of V and C due to a Steady Voltage Impressed at the Origin.

§ 372. Now let us tackle the more difficult problem of the waves of voltage and current generated by a steady voltage impressed at the origin. It is the combination of resistance with leakage that makes this case complicated. But it must be done, to complete the investigation. If e is the steadily-impressed voltage, started at the initial moment, we have

$$V = e\epsilon^{-qx}1 = e\epsilon^{-\rho t}(\cosh - \text{shin})(x/v)(p^2-\sigma^2)^{\frac{1}{2}} \,.\, \epsilon^{\rho t}, \quad (128)$$

by making $\epsilon^{\rho t}$ the operand, and separating even and odd powers of x. Here the cosh operator contains only integral powers of p, and the potence of p is therefore ρ. So

$$V = e\epsilon^{-\rho t}\left\{\cosh(x/v)(\rho^2-\sigma^2)^{\frac{1}{2}} - \frac{\text{shin}(x/v)(p^2-\sigma^2)^{\frac{1}{2}}}{(p^2-\sigma^2)^{\frac{1}{2}}}(p^2-\sigma^2)^{\frac{1}{2}}\right\}\epsilon^{\rho t}. \quad (129)$$

A part has been algebrised. In the rest, the shin operator divided by the radical involves even powers of p only. Also

$$(p^2-\sigma^2)^{\frac{1}{2}}\epsilon^{\rho t} = p\left\{1 - \frac{1}{2}\frac{\sigma^2}{p^2} - \frac{1.1}{2.4}\frac{\sigma^4}{p^4} - \frac{1.1.3}{2.4.6}\frac{\sigma^6}{p^6} - \dots\right\}\epsilon^{\rho t}, \quad (130)$$

by the binomial theorem. Algebrising, we get

$$= p\left\{\epsilon^{\rho t} - \frac{1}{2}\frac{\sigma^2}{\rho^2}(\epsilon^{\rho t} - e_1) - \frac{1.1}{2.4}\frac{\sigma^4}{\rho^4}(\epsilon^{\rho t} - e_3) - \dots\right\}$$

$$= \rho\epsilon^{\rho t}(1-\sigma^2/\rho^2)^{\frac{1}{2}} + \rho\left\{\frac{1}{2}\frac{\sigma^2}{\rho^2} + \frac{1.1}{2.4}\frac{\sigma^4}{\rho^4}e_2 + \frac{1.1.3}{2.4.6}\frac{\sigma^6}{\rho^6}e_4 + \dots\right\}, \quad (131)$$

where e_n = sum of first $n+1$ terms of $\epsilon^{\rho t}$. A point to be noticed in the above that in (130) we introduce pp^{-1}. The differentiation should be done afterwards, as in (131). Let

$$U_0 = \frac{1}{2}\frac{\sigma^2}{\rho^2} + \frac{1.1}{2.4}\frac{\sigma^4}{\rho^4}e_2 + \frac{1.1.3}{2.4.6}\frac{\sigma^6}{\rho^6}e_4 + \dots; \quad (132)$$

then by (130), (131),

$$(p^2-\sigma^2)^{\frac{1}{2}}\epsilon^{\rho t} = \epsilon^{\rho t}v(\text{RK})^{\frac{1}{2}} + \rho U_0. \quad (133)$$

Now when the shin operator works on this, the potence of p is ρ again on the first term in (133) on the right side. So a second part is algebrised. Putting the two algebrised parts together we shall obtain $e\epsilon^{-rx}$, where $r = (\text{RK})^{\frac{1}{2}}$. Therefore (129) becomes

$$V = e\epsilon^{-rx} - e\epsilon^{-\rho t}\frac{\rho x}{v}\left\{1 - \frac{\sigma^2-p^2}{\lfloor 3}\frac{x^2}{v^2} + \dots\right\}U_0; \quad (134)$$

or, in a more convenient form,

$$V = e\epsilon^{-rx} - e\rho\epsilon^{-\rho t}\left\{\frac{x}{v}U_0 - \frac{x^3}{v^3}\frac{U_1}{\lfloor 3} + \frac{x^5}{v^5}\frac{U_2}{\lfloor 5} - \dots\right\}, \quad (135)$$

where the U's are time functions, of which the first is as in (132), whilst the rest are derived from it by

$$U_n = (\sigma^2 - p^2)^n U_0. \tag{136}$$

It is quite easy to derive U_n, because $p^2 e_n = \rho^2 e_{n-2}$. Thus, to illustrate, first $(\sigma^2 - p^2)^n 1 = \sigma^{2n}$, and then,

$$(\sigma^2 - p^2)e_2 = \sigma^2 e_2 - \rho^2, \qquad (\sigma^2 - p^2)e_4 = \sigma^2 e_4 - \rho^2 e_2,$$

$$(\sigma^2 - p^2)^2 e_2 = \sigma^4 e_2 - 2\sigma^2\rho^2, \qquad (\sigma^2 - p^2)^2 e_4 = \sigma^4 e_4 - 2\sigma^2\rho^2 e_2 + \rho^4,$$

$$(\sigma^2 - p^2)^3 e_2 = \sigma^6 e_2 - 3\sigma^4\rho^2, \qquad (\sigma^2 - p^2)^3 e_4 = \sigma^6 e_4 - 3\sigma^4\rho^2 e_2 + 3\sigma^2\rho^4,$$

&c., &c. So we have

$$U_1 = \frac{\sigma^2}{\rho^2}\left\{\tfrac{1}{2}\sigma^2 + \frac{1.1}{2.4}\frac{\sigma^2}{\rho^2}(\sigma^2 e_2 - \rho^2) + \frac{1.1.3}{2.4.6}\frac{\sigma^4}{\rho^4}(\sigma^2 e_4 - \rho^2 e_2) + \dots\right\},$$

$$U_2 = \frac{\sigma^2}{\rho^2}\left\{\tfrac{1}{2}\sigma^4 + \frac{1.1}{2.4}\frac{\sigma^2}{\rho^2}(\sigma^4 e_2 - 2\rho^2\sigma^2)\right.$$

$$\left. + \frac{1.1.3}{2.4.6}\frac{\sigma^4}{\rho^4}(\sigma^4 e_4 - 2\sigma^2\rho^2 e_2 + \rho^4) + \dots\right\},$$

and so on. I regret that the result should be so complicated. But the only alternatives are other equivalent infinite series, or else a definite integral which is of no use until it is evaluated, when the result must be the series (135), or an equivalent one. In it, the term $e\,\epsilon^{-rx}$ represents the steady state, all the rest ultimately vanishing.

As regards the wave of current, there are several ways of obtaining it. The one analogous to the above goes thus :—

$$C = \left(\frac{K + Sp}{R + Lp}\right)^{\frac{1}{2}} \epsilon^{-qx} e = \frac{e}{Lv}\epsilon^{-\rho t}\left(\frac{p - \sigma}{p + \sigma}\right)^{\frac{1}{2}} \epsilon^{-(p^2 - \sigma^2)^{\frac{1}{2}} x/v}\,\epsilon^{\rho t}$$

$$= \frac{e}{Lv}\epsilon^{-\rho t}(p - \sigma)(\cosh - \text{shin})\frac{x}{v}(p^2 - \sigma^2)^{\frac{1}{2}} \cdot \frac{\epsilon^{\rho t}}{(p^2 - \sigma^2)^{\frac{1}{2}}}, \tag{137}$$

by first making $\epsilon^{\rho t}$ the operand, and then introducing $(p - \sigma)/(p - \sigma)$. We now have the potence of p in the shin operator divided by the radical equivalent to ρ. This algebrises a part. For use with the cosh operator, we have

$$(p^2 - \sigma^2)^{-\frac{1}{2}}\epsilon^{\rho t} = \frac{1}{p}\left(1 + \tfrac{1}{2}\frac{\sigma^2}{p^2} + \frac{1.3}{2.4}\frac{\sigma^4}{p^4} + \dots\right)\epsilon^{\rho t}$$

$$= \frac{1}{\rho}(\epsilon^{\rho t} - 1) + \tfrac{1}{2}\frac{\sigma^2}{\rho^3}(\epsilon^{\rho t} - e_2) + \dots$$

$$= \epsilon^{\rho t}(\rho^2 - \sigma^2)^{-\frac{1}{2}} - \frac{1}{\rho}W_0, \tag{138}$$

where $$W_0 = 1 + \tfrac{1}{2}\frac{\sigma^2}{\rho^2}e_2 + \frac{1.3}{2.4}\frac{\sigma^4}{\rho^4}e_4 + \frac{1.3.5}{2.4.6}\frac{\sigma^6}{\rho^6}e_6 + \dots \qquad (139)$$

Using (138) in the previous equation (137), we see that the potence of p in the cosh operator on the first part is again ρ; and as regards the rest, we have a series of functions derived from W_0 according to

$$W_n = (\sigma^2 - p^2)^n W_0. \qquad (140)$$

So, finally, putting together separately the parts independent of the time, we obtain

$$C = e\left(\frac{K}{R}\right)^{\frac{1}{2}}\epsilon^{-rx} - \frac{e\epsilon^{-\rho t}}{Lv}\frac{p-\sigma}{\rho}\left\{W_0 - \frac{x^2}{v^2}\frac{W_1}{\lfloor 2} + \frac{x^4}{v^4}\frac{W_2}{\lfloor 4} - \dots\right\}. \qquad (141)$$

As before with the U functions, the W functions are easily developable.

Analysis of Transmission Operator to Show the Deformation, Progression and Attenuation of Waves.

§ 373. I shall now explain another way of algebrising the operational solutions, which is, perhaps, the simplest possible in the ideas concerned, and also sometimes in the execution thereof. Say we require to find the wave of voltage generated by impressed voltage at the origin; that is, $V = \epsilon^{-qx}e$, where e is a given function of the time. Practically, e is suitably selected to ease the work. As an example, in passing, take

$$e = e_0\epsilon^{-\rho t}I_0(\sigma t), \qquad (142)$$

where e_0 is constant (zero before, steady after $t=0$). We can see at once that

$$V = e_0\epsilon^{-\rho t}I_0\{\sigma(t^2 - x^2/v^2)^{\frac{1}{2}}\}, \qquad (143)$$

because we have already found the effect of ϵ^{-qx} on the function in (142). See (126) above.

Now, in the distortionless case, when $\sigma = 0$, if also $R=0$, $K=0$, we have $q = p/v$. Therefore, if $e = f(t)$, we have

$$V = \epsilon^{-px/v}f(t) = f(t - x/v), \qquad (144)$$

by Taylor's theorem operationally considered. That is, the voltage impressed at the origin travels out at speed v without any change whatever.

But if there is resistance, and also leakage to balance, so that $\sigma = 0$ still, then $qv = \rho + p$, and

$$V = \epsilon^{-qx}e = \epsilon^{-\rho x/v}\epsilon^{-px/v}e = \epsilon^{-\rho x/v}f(t - x/v), \qquad (145)$$

Thus the disturbances impressed at the origin now travel out at speed v and attenuate according to $\epsilon^{-\rho t}$ in the interval, because $t=x/v$ is the time of transit from the origin to x. At the same time the wave of current is $\mathrm{C}=\mathrm{V}/\mathrm{L}v$, the current and transverse voltage being in the same phase and in constant ratio. This is the beginning of the theory of the distortionless circuit, which I recommend every electrician to study in full detail, as an introduction to electromagnetic waves in general, since it casts light in the most obscure places. It allows us to understand electromagnetic waves mathematically not merely as a collection of formulæ, sometimes disagreeably complicated, but in terms of physical ideas of translation, attenuation, distortion, absorption, reflection, and so on. I beg to refer again to the portions of Chapter IV., Vol. I., which are devoted to the subject of waves along wires. It is obvious that an electrician who aims at telephony through very long cables by methods which violate the conditions under which it is possible, is only wasting his labour.*

There are only two ideas concerned in (145), viz., the progression of the wave, and the uniform attenuation which occurs by absorption in transit. If σ is not zero, there must be a third process involved. This is the partial reflection that occurs to every part of a wave in transit, whereby its parts are redistributed, or the wave shape is deformed or distorted. It follows from the distortionless theory that when σ is not zero, we still have propagation at speed v, and still have the attenuation factor $\epsilon^{-\rho t}$. It therefore suggested itself to me that in

$$\epsilon^{-qx}e=\epsilon^{-\rho t}\epsilon^{-(p^2-\sigma^2)^{\frac{1}{2}}x/v}\epsilon^{\rho t}e, \tag{146}$$

the operator containing p should be expressed thus,

$$\epsilon^{-(p^2-\sigma^2)^{\frac{1}{2}}x/v}=\epsilon^{-px/v}\mathrm{F}(p^{-1}), \tag{147}$$

* This refers to the proposal of Mr. W. H. Preece, in his paper at the Liverpool (1896) meeting of the B. A., "Electrical Disturbances in Submarine Cables," printed in full in *The Electrician*, September 25, 1896, p. 689. Mr. Preece seriously proposes, for the furtherance of long cable telephony, to bring the two conducting leads as near together as possible, separated by a piece of paper, in fact. This will reduce the inductance to a minimum, and increase the permittance to a maximum. Both results are entirely antagonistic to telephonic transmission in long cables. Is there a single electrician, theoretical or practical, who will support Mr. Preece? However, the paper is not more remarkable than some of his previous ones on the subject.

where $F(v^{-1})$ is a series in rising powers of p^{-1}, having unity for its first term, because when $\sigma=0$, $F=1$. Say,

$$F=1+F_1(\sigma/p)+F_2(\sigma/p)^2+F_3(\sigma/p)^3+\ldots, \qquad (148)$$

so that

$$V=\epsilon^{-qx}e=\epsilon^{-\rho t}\epsilon^{-px/v}\left(1+F_1\frac{\sigma}{p}+F_2\frac{\sigma^2}{p^2}+\ldots\right)\epsilon^{\rho t}e. \qquad (149)$$

If this is possible, the algebrisation will be immediate. First carry out the operation F. This will effect the rearrangement of parts. Then carry out $\epsilon^{-px/v}$. This will do the translation. Finally, the factor $\epsilon^{-\rho t}$ will attenuate properly.

It does not appear from the operator in (146) that the expansion in inverse powers of p is a natural one. But it goes well. The F's are simple functions of x, of which the first three are

$$F_1=\frac{\sigma x}{2v}, \qquad F_2=\frac{\sigma^2x^2}{8v^2}, \qquad F_3=\frac{\sigma x}{8v}+\frac{\sigma^3x^3}{48v^3}. \qquad (150)$$

How to obtain these will be explained separately. Supposing them known at present, we can show the working of (149).

Three Examples. The Wave of V due to impressed Voltage, varying as $\epsilon^{-\rho t}$, $\epsilon^{-Kt/S}$, or steady.

§ 374. The simplest case is $e\epsilon^{\rho t}=e_0$, or $e=e_0\epsilon^{-\rho t}$. Then $(\sigma/p)^n 1$ is the same as $(\sigma t)^n/\underline{|n}$, so the F series is algebrised. Then the next operator turns t to t_1, if t_1 means $t-x/v$. Therefore,

$$V=e_0\epsilon^{-\rho t}\left\{1+\sigma t_1F_1+\frac{\sigma^2t_1^2}{\underline{|2}}F_2+\frac{\sigma^3t_1^3}{\underline{|3}}F_3+\ldots\right\} \qquad (151)$$

expresses the complete wave of voltage. Or,

$$V=e_0\epsilon^{-\rho t}\left\{1+\sigma t\left(1-\frac{x}{vt}\right)\frac{\sigma x}{2v}+\frac{\sigma^2t^2}{\underline{|2}}\left(1-\frac{x}{vt}\right)^2\frac{\sigma^2x^2}{8v^2}+\ldots\right\}. \qquad (152)$$

Observe that V has the same value at the wave front as at the origin, at any moment. This is because we have chosen V at the origin so that it shall be so. The natural attenuation from origin to wave front is $\epsilon^{-\rho t}$, when $t=x/v$, and this is just the law assumed for the variation at the origin. In the distortionless case the curve of V is a straight line all the way up to the front, where V drops to zero suddenly. But it is not generally a straight line, even when V is the same at the extremes. It is easy to see that if e subsides faster than $\epsilon^{-\rho t}$, the voltage at the wave front will be greater than at the origin; and conversely, less.

Next, let e be constant, zero before, steady after $t=0$, then we have to integrate $\epsilon^{\rho t}$ once, twice, &c., according to (149). Thus,

$$V = e\epsilon^{-\rho t}\left\{\epsilon^{\rho t_1} + F_1\frac{\sigma}{\rho}(\epsilon^{\rho t_1} - 1) + F_2\frac{\sigma^2}{\rho^2}(\epsilon^{\rho t_1} - 1 - \rho t_1) + \dots\right\}, \quad (153)$$

where we have put t_1 for t, as before explained. Collect together the terms involving $\epsilon^{\rho t_1}$; that is, $\epsilon^{\rho t}\epsilon^{-\rho x/v}$. Then

$$V = e\epsilon^{-\rho x/v}\left\{1 + \frac{F_1\sigma}{\rho} + \frac{F_2\sigma^2}{\rho^2} + \dots\right\}$$
$$- e\epsilon^{-\rho t}\left\{F_1\frac{\sigma}{\rho} + F_2\frac{\sigma^2}{\rho^2}e'_1 + F_3\frac{\sigma^3}{\rho^3}e'_2 + \dots\right\}, \quad (154)$$

where e'_n means the sum of the first $n+1$ terms of $\epsilon^{\rho t_1}$.

This is an alternative form of the solution (135) above. We conclude that

$$\epsilon^{-rx} = \epsilon^{-\rho x/v}\{1 + F_1\sigma/\rho + F_2\sigma^2/\rho^2 + \dots\}, \quad (155)$$

so (154) may be simplified to this extent. When $\sigma=\rho$, then $r=0$; therefore,

$$\epsilon^{\sigma x/v} = 1 + F_1 + F_2 + F_3 + \dots = \Sigma F, \quad (156)$$

since $F_0 = 1$.

As a third case, let $e\epsilon^{\rho t} = e_0\epsilon^{\sigma t}$, where e_0 is constant. Then $e = e_0\epsilon^{-Kt/S}$, subsiding according to the natural leakage law. We have now to integrate $\epsilon^{\sigma t}$ in (149), instead of $\epsilon^{\rho t}$. That is, instead of (153) we have the following:—

$$V = e_0\epsilon^{-\rho t}\left\{\epsilon^{\sigma t_1} + F_1(\epsilon^{\sigma t_1} - 1) + F_2(\epsilon^{\sigma t_1} - 1 - \sigma t_1) + \dots\right\}; \quad (157)$$

or, collecting the terms involving $\epsilon^{\sigma t_1}$,

$$V = e_0\epsilon^{-Kt/S}\epsilon^{-\sigma x/v}\left\{1 + F_1 + \dots\right\} - e_0\epsilon^{-\rho t}\left\{F_1 + F_2 e_1'' + \dots\right\}, \quad (158)$$

where ϵ''_n is now the sum of the first $n+1$ terms of $\epsilon^{\sigma t_1}$.

But use (156). Then in the first portion of (158) we have

$$\epsilon^{-\rho t}\epsilon^{\sigma t_1}\Sigma F = \epsilon^{-Kt/S},$$

and therefore

$$V = e_0\epsilon^{-Kt/S} - e_0\epsilon^{-\rho t}\{F_1 + F_2 e_1'' + F_3 e_2'' + \dots\}. \quad (159)$$

This is an alternative form of the solution (115) of the same problem.

An important reduced case is $\sigma=\rho$, or no leakage. Then $e_0=e$, and is steady. We get

$$V=e-e\epsilon^{-\sigma t}\{F_1+F_2e_1''+F_3e_2''+\dots\}. \tag{160}$$

If we desire the expansions to be entirely in rising powers of $1-x/vt$, like (152) in fact, then the operand $\epsilon^{\rho t}$ or $\epsilon^{\sigma t}$ should be expanded first, and integrated term by term. Thus instead of (153) and (154) we shall get

$$V=e\epsilon^{-\rho t}\Big[1+\rho t_1+\frac{\rho^2t_1^2}{\lfloor 2}+\frac{\rho^3t_1^3}{\lfloor 3}+\dots$$
$$+\frac{\sigma x}{2v}\Big(\sigma t_1+\frac{\sigma\rho t_1^2}{\lfloor 2}+\frac{\sigma\rho^2t_1^3}{\lfloor 3}+\dots\Big)$$
$$+\frac{\sigma^2x^2}{8v^2}\Big(\frac{\sigma^2t_1^2}{\lfloor 2}+\frac{\sigma^2\rho t_1^3}{\lfloor 3}+\frac{\sigma^2\rho^2t_1^4}{\lfloor 4}+\dots\Big)+\dots\Big]. \tag{161}$$

Now rearrange in powers of t_1, and we obtain

$$V=e\epsilon^{-\rho t}\Big[1+\Big(1-\frac{x}{vt}\Big)\Big(\rho t+\sigma t\frac{\sigma x}{2v}\Big)$$
$$+\frac{1}{\lfloor 2}\Big(1-\frac{x}{v}\Big)^2\Big(\rho^2t^2+\rho t\sigma t\frac{\sigma x}{2v}+\sigma^2t^2\frac{\sigma^2x^2}{8v^2}\Big)+\dots\Big]. \tag{162}$$

This is when e is constant. And in the other case, instead of (157) or (159) we obtain

$$V=e_0\epsilon^{-\rho t}\Big[1+\Big(1-\frac{x}{vt}\Big)\sigma t\Big(1+\frac{\sigma x}{2v}\Big)$$
$$+\frac{1}{\lfloor 2}\Big(1-\frac{x}{vt}\Big)^2\sigma^2t^2\Big(1+\frac{\sigma x}{2v}+\frac{\sigma^2x^2}{8v^2}\Big)+\dots\Big], \tag{163}$$

where e_0 is constant. It follows from the previous equation by changing e to e_0, and putting $\rho=\sigma$ inside the square brackets.

Expansions of this sort, which are so easily obtained, and in a variety of cases, are useful in calculating the initial stages of the development of waves, whether due to impressed force or to an initial distribution of charge or current. It is not desirable, then, to exhibit the ϵ^{-rx} term separately.

Expansion of Distortion Operators in Powers of p^{-1}.

§ 375. In the method of § 374 we require the expansion of the operator $\epsilon^{-(p^2-\sigma^2)^{\frac{1}{2}}x/v}$ in powers of p^{-1}. This is obtainable by expanding it in powers of σ^2. Simplify by removing the $-$ signs. Put $-x/v=y$, and $-\sigma^2=s^2$. Then, by Taylor's Theorem,

$$A=\epsilon^{y(p^2+s^2)^{\frac{1}{2}}}=\epsilon^{s^2D}\epsilon^{y(p^2+s^2)^{\frac{1}{2}}}{}_{(s=0)}, \tag{164}$$

if D means d/ds^2, and we put $s=0$ in the differential coefficients after they are developed. This process will be found to be very laborious, owing to the rapid elongation of the expressions for the differential coefficients, several of which are required before the law of the expansion can be recognised. But the waste work can be avoided by observing that differentiation to p^2 produces the same result as differentiation to s^2; that is, D may signify d/dp^2. Then we see that s^2 may be put $=0$ in the operand before the differentiations. That is,

$$A = \epsilon^{s^2 D} \epsilon^{yp} = \left\{1 + s^2 \frac{d}{dp^2} + \frac{s^2}{\underline{|2}} \left(\frac{d}{dp^2}\right)^2 + \dots\right\} \epsilon^{yp}$$

$$= \left\{1 + \frac{s^2}{2p}\frac{d}{dp} + \frac{1}{\underline{|2}} \frac{s^2}{2p}\frac{d}{dp}\frac{s^2}{2p}\frac{d}{dp} + \dots\right\} \epsilon^{yp}. \qquad (164\text{A})$$

There is now no waste work, and the result is

$$A = \epsilon^{yp}\left\{1 + \frac{s^2 y}{2p} + \frac{1}{\underline{|2}}\left(\frac{s^2 y}{2p}\right)^2\left(1 - \frac{1}{yp}\right) + \frac{1}{\underline{|3}}\left(\frac{s^2 y}{2p}\right)^3\left(1 - \frac{3}{yp} + \frac{3}{y^2p^2}\right) + \dots\right\}, \qquad (164\text{B})$$

where the function in the {} contains p inversely only.

If we interchange s and p we shall obtain another expansion of A. Numerically considered, if one is convergent, the other is divergent. But we are not guided in our choice of the form (164B) by that consideration, but by its having p in the denominators, so that the integrations can be at once carried out as in § 374. Besides, we cannot use the other form of expansion in our application, because s^2 has to be negative then, which causes failure. What is important here is to recognise the law of formation of the coefficients of the powers of $s^2y/2p$. We obtain the rule by observing how a differentiation with respect to p derives any term in the expansion from the preceding one. So if we write

$$A = \epsilon^{yp}\left\{1 + \frac{s^2 y}{2p} + \frac{1}{\underline{|2}}\left(\frac{s^2 y}{2p}\right)^2 r_1 + \frac{1}{\underline{|3}}\left(\frac{s^2 y}{2p}\right)^3 r_2 + \dots\right\}, \qquad (164\text{C})$$

we shall obtain r_3, r_4, &c., by this simple process:—

r_2	1	3	3			
		3	12			
r_3	1	6	15	15		
		4	30	90		
r_4	1	10	45	105	105	
		5	60	315	840	
r_5	1	15	105	420	945	945

It almost explains itself. We start with the coefficients in r_2, and multiply them by 3, 4, 5 respectively, producing the second line, and add them as shown to produce the coefficients in r_3. These, again, multiplied by 4, 5, 6, 7 and added on, produce the coefficients in r_4. And so on. Thus

$$r_3 = 1 - \frac{6}{py} + \frac{15}{p^2y^2} - \frac{15}{p^3y^3}, \qquad r_4 = 1 - \frac{10}{py} + \frac{45}{p^2y^2} - \frac{105}{p^3y^3} + \frac{105}{p^4y^4},$$

&c., &c. Guided by this rule, we can see that the expansion (164c) is really a series of H_m functions. See equation (3) § 336. For, by that formula,

$$H_{\frac{1}{2}}(py) = \frac{\epsilon^{py}}{(\frac{1}{2}\pi py)^{\frac{1}{2}}}, \quad H_{\frac{3}{2}}(py) = \frac{\epsilon^{py}}{(\frac{1}{2}\pi py)^{\frac{1}{2}}}\left(1 - \frac{1}{py}\right), \qquad (164\text{D})$$

and so on. This makes

$$A = (\tfrac{1}{2}\pi py)^{\frac{1}{2}}\left\{H_{-\frac{1}{2}} + \frac{s^2y}{2p}H_{\frac{1}{2}} + \frac{1}{\underline{|2}}\left(\frac{s^2y}{2p}\right)^2 H_{1\frac{1}{2}} + \frac{1}{\underline{|3}}\left(\frac{s^2y}{2p}\right)^3 H_{2\frac{1}{2}} + \dots\right\} \qquad (165)$$

But the above numerical process is sufficient.

We also require the expansion of $A/(p^2+s^2)^{\frac{1}{2}}$. Call this B. Then working in the same way, we shall get B by (164A), provided we use the operand ϵ^{yp}/p instead of ϵ^{yp}. The result is

$$B = \frac{\epsilon^{yp}}{p}\left\{1 + \frac{s^2y}{2p}r_1 + \frac{1}{\underline{|2}}\left(\frac{s^2y}{2p}\right)^2 r_2 + \frac{1}{\underline{|3}}\left(\frac{s^2y}{2p}\right)^3 r_3 + \dots\right\}. \qquad (166)$$

As a verification, derive the A series (164c) from the B series by differentiation to y. Comparing the two series, we see that the series for Bp is obtained from that for A by shifting all the r functions one term to the left.

Now, in the application, for y put $-x/v$ and for s^2 put $-\sigma^2$. Then A and B are expressed by

$$\epsilon^{-(p^2-\sigma^2)^{\frac{1}{2}}x/v} = \epsilon^{-px/v}\left\{1 + \frac{\sigma^2x}{2pv} + \frac{1}{\underline{|2}}\left(\frac{\sigma^2x}{2pv}\right)^2 r_1 + \frac{1}{\underline{|3}}\left(\frac{\sigma^2x}{2pv}\right)^3 r_2 + \dots\right\}, \qquad (167)$$

$$\frac{\epsilon^{-(p^2-\sigma^2)^{\frac{1}{2}}x/v}}{(p^2-\sigma^2)^{\frac{1}{2}}} = \frac{\epsilon^{-px/v}}{p}\left\{1 + \frac{\sigma^2x}{2pv}r_1 + \frac{1}{\underline{|2}}\left(\frac{\sigma^2x}{2pv}\right)^2 r_2 + \dots\right\}, \qquad (168)$$

where

$$r_1 = 1 + \frac{v}{px}, \quad r_2 = 1 + \frac{3v}{px} + \frac{3v^2}{p^2x^2}, \quad r_3 = 1 + \frac{6v}{px} + \frac{15v^2}{p^2x^2} + \frac{15v^3}{p^3x^3}, \qquad (169)$$

&c. All signs are now positive in the r functions. Finally, if we arrange in powers of σ/p, so that

$$A = \epsilon^{-px/v}\left\{1 + F_1\frac{\sigma}{p} + F_2\frac{\sigma^2}{p^2} + \dots\right\},$$

$$B = \frac{\epsilon^{-px/v}}{p}\left(1 + G_1\frac{\sigma}{p} + G_2\frac{\sigma^2}{p^2} + \dots\right), \tag{170}$$

then the F's and G's are functions of x only, given by

$$F_1 = \frac{\sigma x}{2v}, \quad F_2 = \frac{1}{2^2\lfloor 2}\left(\frac{\sigma x}{v}\right)^2, \quad F_3 = \frac{1}{2^2\lfloor 2}\left(\frac{\sigma x}{v}\right) + \frac{1}{2^3\lfloor 3}\left(\frac{\sigma x}{v}\right)^3,$$

$$F_4 = \frac{3}{2^3\lfloor 3}\left(\frac{\sigma x}{v}\right)^2 + \frac{1}{2^4\lfloor 4}\left(\frac{\sigma x}{v}\right)^4,$$

$$F_5 = \frac{3}{2^3\lfloor 3}\left(\frac{\sigma x}{v}\right) + \frac{6}{2\lfloor 4}\left(\frac{\sigma x}{v}\right)^3 + \frac{1}{2^5\lfloor 5}\left(\frac{\sigma x}{v}\right)^5, \tag{171}$$

$$G_1 = \frac{\sigma x}{2v}, \quad G_2 = \tfrac{1}{2} + \frac{1}{2^2\lfloor 2}\left(\frac{\sigma x}{v}\right)^2, \quad G_3 = \frac{3}{2^2\lfloor 2}\left(\frac{\sigma x}{v}\right) + \frac{1}{2^3\lfloor 3}\left(\frac{\sigma x}{v}\right)^3,$$

$$G_4 = \frac{3}{2^2\lfloor 2} + \frac{6}{2^3\lfloor 3}\left(\frac{\sigma x}{v}\right)^2 + \frac{1}{2^4\lfloor 4}\left(\frac{\sigma x}{v}\right)^4, \text{ \&c.} \tag{172}$$

where the coefficients are picked out from the r functions.

Example. The Current Wave due to Impressed Voltage $e_0\epsilon^{-Kt/S}$ at the Origin.

§ **376.** It is with the expansion B that we are concerned when dealing with the wave of current generated by impressed voltage. For the current due to e at the origin is

$$C = (K + Sp)\frac{\epsilon^{-qx}}{q}e = \frac{1}{Lv}(p + \rho - \sigma)\frac{\epsilon^{-qx}}{qv}e$$

$$= \frac{\epsilon^{-\rho t}}{Lv}(p - \sigma)\frac{\epsilon^{-(p^2-\sigma^2)^{\frac{1}{2}}x/v}}{(p^2-\sigma^2)^{\frac{1}{2}}}e\epsilon^{\rho t} = \frac{\epsilon^{-\rho t}}{Lv}(p - \sigma)Be\epsilon^{\rho t}. \tag{173}$$

This is developable by the B expansion in the same way as the wave of V in § 374. One case will do to illustrate. Take $e\epsilon^{\rho t} = e_0\epsilon^{\sigma t}$, where e_0 is constant, so that the voltage impressed at $x = 0$ is $e_0\epsilon^{-Kt/S}$, such as would arise from the initial distribution of transverse voltage $V = 2e_0$ on the whole of the left side of the origin. Then we may put $p/(p - \sigma)$ for $\epsilon^{\sigma t}$ in the operand, and produce

$$C = \frac{e_0\epsilon^{-\rho t}}{Lv}Bp = \frac{e_0\epsilon^{-\rho t}}{Lv}\epsilon^{-px/v}\left\{1 + \frac{\sigma}{p}G_1 + \frac{\sigma^2}{p^2}G_2 + \dots\right\}, \tag{174}$$

which develops at once to

$$C = \frac{e_0\epsilon^{-pt}}{Lv}\left\{1 + \sigma t_1 G_1 + \frac{\sigma^2 t_1^{\,2}}{\underline{|2}} G_2 + \frac{\sigma^3 t_1^{\,3}}{\underline{|3}} G_3 + \ldots\right\}; \qquad (175)$$

or, in terms of x and t,

$$C = \frac{e_0\epsilon^{-pt}}{Lv}\left\{1 + \sigma t\left(1 - \frac{x}{vt}\right)\frac{\sigma x}{2v} + \frac{\sigma^2 t^2}{\underline{|2}}\left(1 - \frac{x}{vt}\right)^2\left(\frac{1}{2} + \frac{1}{2^2\underline{|2}}\frac{\sigma^2 x^2}{v^2}\right) + \ldots\right\}. \qquad (176)$$

This is an alternative form of the compact solution (101). We conclude that

$$1 + \sigma t_1 G_1 + \frac{\sigma^2 t_1^{\,2}}{\underline{|2}} G_2 + \ldots = I_0\{\sigma(t^2 - x^2/v^2)^{\frac{1}{2}}\}. \qquad (177)$$

Since the sign of x may be changed, the coefficients of all the odd powers of x must vanish in (176) and previous equations. We might, in fact, have started using $\cosh qx$ instead of ϵ^{-qx}.

We may also interchange $\sigma^2 t^2$ and $-\sigma^2 x^2/v^2$ in (177), or σt and $i\sigma x/v$. This makes σt_1 become $i\sigma(t + x/v)$. But for the reason just mentioned $i\sigma(t - x/v)$ will do as well. That is $i\sigma t_1$. So by (177),

$$1 + \sigma t_1 G_1 + \frac{\sigma^2 t_1^{\,2}}{\underline{|2}} G_2 + \ldots = 1 + \sigma t_1 g_1 + \frac{\sigma^2 t_1^{\,2}}{\underline{|2}} g_2 + \ldots, \qquad (178)$$

where the G's are known functions of x, as above, whilst the g's are functions of t instead. They are obtained from the functions of x by turning x/v to t/i, and then multiplying by i, i^2, &c., for g_1, g_2, &c. Thus, by (172),

$$g_1 = \frac{\sigma t}{2}, \qquad g_2 = -\frac{1}{2} + \frac{(\sigma t)^2}{2^2\underline{|2}}, \qquad g_3 = -\frac{3\sigma t}{2^2\underline{|2}} + \frac{\sigma^3 t^3}{2^3\underline{|3}}. \qquad (179)$$

So in the expansion (175) we may substitute these g functions of t for the functions G of x. This is a very curious change. The modified form of solution arises directly in another way of developing solutions, which will be referred to later.

To put connected matters in one place, it may be mentioned here that if we arrange the solution (175) in terms of powers of $\sigma x/v$ as they occur in the G functions, we shall obtain the form

$$C = \frac{e_0\epsilon^{-pt}}{Lv}\left\{I_0(\sigma t_1) + \frac{\sigma x}{v} I_1(\sigma t_1) + \frac{1}{\underline{|2}}\left(\frac{\sigma x}{v}\right)^2 I_2(\sigma t_1) + \ldots\right\}. \qquad (179\text{A})$$

Compare with the expansion (99), § 367. That is also arranged in powers of $\sigma x/v$, though only even powers. But it is the P functions of σt that occur there, whilst now we have the I functions of σt_1, or $\sigma(t-x/v)$.

Collecting identities conveniently for memory, writing t for σt and x for $\sigma x/v$, we have

$$\frac{p}{(p^2-1)^{\frac{1}{2}}}\epsilon^{-x(p^2-1)^{\frac{1}{2}}} = \mathrm{I}_0\{(t^2-x^2)^{\frac{1}{2}}\} \tag{179B}$$

$$= \mathrm{I}_0(t-x) + x\mathrm{I}_1(t-x) + \frac{x^2}{\underline{|2}}\mathrm{I}_2(t-x) + \dots \tag{179C}$$

$$= \mathrm{P}_0(t) - \frac{x^2}{2^2}\mathrm{P}_1(t) + \frac{x^4}{2^2 4^2}\mathrm{P}_2(t) - \dots ; \tag{179D}$$

and in these the signs of x and t may be changed. Also, by interchanging t and xi,

$$= \mathrm{P}_0(xi) + \frac{t^2}{2^2}\mathrm{P}_1(xi) + \frac{t^4}{2^2 4^2}\mathrm{P}_2(xi) + \dots \tag{179E}$$

$$= \mathrm{J}_0(t+x) + t\mathrm{J}_1(t+x) + \frac{t^2}{\underline{|2}}\mathrm{J}_2(t+x) + \dots, \tag{179F}$$

in terms of oscillating functions, where the signs of x and t may be changed.

Value of $(p^2-\sigma^2)^n \mathrm{I}_0(\sigma t)$ when n is a Positive or Negative Integer. Structure of the Convergent Bessel Functions.

§ 377. In the method of developing the complete wave of current from the formula for the current at the origin followed in § 367, we had occasion to use the expression $(p^2-\sigma^2)^m \mathrm{I}_0(\sigma t)$, and its result in terms of $\mathrm{P}_m(\sigma t)$, m being positive and integral. It may be asked how things work out when m is negative and integral? It happens that a comparison between two of the formulæ previously obtained furnishes the answer, so that it is worth while indicating the result in passing. We found in § 367 that

$$\epsilon^{-(p^2-\sigma^2)^{\frac{1}{2}}x/v}\,\mathrm{I}_0(\sigma t) = \mathrm{I}_0\{\sigma(t^2-x^2/v^2)^{\frac{1}{2}}\}; \tag{180}$$

and previously, equation (36), § 354, that

$$\frac{p}{(p^2-\sigma^2)^{\frac{1}{2}}} = \mathrm{I}_0(\sigma t). \tag{181}$$

Now in the last, turn σ^2 to $\sigma^2(1-x^2/v^2t^2)$. Thus,

$$\frac{p}{(p^2-\sigma^2+\sigma^2x^2/v^2t^2)^{\frac{1}{2}}} = \mathrm{I}_0\{\sigma(t^2-x^2/v^2)^{\frac{1}{2}}\}. \tag{182}$$

Comparing the last with (180), we see that the resultant functions are the same. Of course, on the left side of (182), t explicit is treated as a constant. Now we can easily make the operand on the left side of (182) be $I_0(\sigma t)$, as in (180). Thus,

$$\frac{p}{(p^2-\sigma^2+\sigma^2x^2/v^2t^2)^{\frac{1}{2}}}=\frac{p(p^2-\sigma^2)^{-\frac{1}{2}}}{\left\{1+\frac{\sigma^2x^2/v^2t^2}{p^2-\sigma^2}\right\}^{\frac{1}{2}}}=\frac{I_0(\sigma t)}{\left\{1+\frac{\sigma^2x^2/v^2t^2}{p^2-\sigma^2}\right\}^{\frac{1}{2}}}, \tag{183}$$

by using (181) again. So

$$\left\{1+\frac{\sigma^2x^2/v^2t^2}{p^2-\sigma^2}\right\}^{-\frac{1}{2}}I_0(\sigma t)=\cosh\frac{x}{v}(p^2-\sigma^2)^{\frac{1}{2}}.I_0(\sigma t), \tag{184}$$

where we put cosh for exp because the resultant function is even as regards x. Now expand. Then

$$\left\{1+\frac{1}{\lfloor 2}\frac{x^2}{v^2}(p^2-\sigma^2)+\frac{1}{\lfloor 4}\left(\frac{x}{v}\right)^4(p^2-\sigma^2)^2+\dots\right\}I_0(\sigma t)$$

$$=\left\{1-\frac{1}{2}\frac{\sigma^2x^2}{v^2t^2}(p^2-\sigma^2)^{-1}+\frac{1\,.\,3}{2^2\lfloor 2}\frac{\sigma^4x^4}{v^4t^4}(p^2-\sigma^2)^{-2}-\dots\right\}I_0(\sigma t). \tag{185}$$

So, comparing similar powers of x, we see that

$$\frac{1}{\lfloor 2n}(\sigma^2-p^2)^nI_0(\sigma t)=\frac{\lfloor 2n}{(\lfloor n)^2}\left(\frac{\sigma}{2t}\right)^{2n}(p^2-\sigma^2)^{-n}I_0(\sigma t), \tag{186}$$

which shows the relation required. For we know already that

$$\frac{1}{\lfloor 2n}(\sigma^2-p^2)^nP_0(\sigma t)=\left(\frac{\sigma}{2}\right)^{2n}\frac{P_n(\sigma t)}{(\lfloor n)^2}, \tag{187}$$

this being equivalent to (97). Therefore, equating the right members of the last two equations, we obtain

$$(p^2-\sigma^2)^{-n}P_0(\sigma t)=\frac{t^{2n}}{\lfloor 2n}P_n(\sigma t), \tag{188}$$

when n is a positive integer. Thus, whilst $(p^2-\sigma^2)^n$ leads to P_n, its reciprocal leads to $t^{2n}P_n$, n being positive in both cases.

There are very likely easier ways of getting (188). The interesting thing in the above is the way it comes out from such very dissimilar operational solutions as (180), (182). To verify our constants, take $n=1$. Then

$$\frac{\sigma^2}{p^2-\sigma^2}P_0(\sigma t)=\left(\frac{\sigma^2}{p^2}+\frac{\sigma^4}{p^4}+\dots\right)\left(1+\frac{\sigma^2t^2}{2^2}+\frac{\sigma^4t^4}{2^24^2}+\dots\right). \tag{189}$$

This is easily developed in the way done before many times, and the result is $(\sigma^2t^2/\lfloor 2)P_1(\sigma t)$, in accordance with (188). On the left side of (185) we have differentiations only, on the right side integrations only.

As regards harmonisation with the vanishing of $(p^2-\sigma^2)^{\frac{1}{2}}P_0$, which seems to assert that a repetition of the operation $(p^2-\sigma^2)^{\frac{1}{2}}$ will still produce 0, whereas $(p^2-\sigma^2)P_0$ is really $-\frac{1}{2}\sigma^2 P_1$, that is easily done. What $(p^2-\sigma^2)^{\frac{1}{2}}P_0$ really represents is the impulse $p1$; so a repetition of the operation produces

$$(p^2-\sigma^2)^{\frac{1}{2}}p1 = p^2(1-\sigma^2/p^2)^{\frac{1}{2}} = p^2 \times -\frac{\sigma^2}{2p^2}P_1 = -\tfrac{1}{2}\sigma^2 P_1, \qquad (190)$$

the third expression being got by algebrising the second through the binomial theorem.

It is worth pointing out here that these P functions furnish the readiest way of exhibiting the structure and connection of the Bessel functions. Thus, let D mean d/dy. Then

$$\epsilon^{D^{-1}} = 1 + \frac{1}{\lfloor 1 D} + \frac{1}{\lfloor 2 D^2} + \dots = 1 + \frac{y}{\lfloor 1 \lfloor 1} + \frac{y^2}{\lfloor 2 \lfloor 2} + \dots. \qquad (191)$$

Therefore, if y stands for $(\frac{1}{2}\sigma t)^2$, we have

$$I_0(\sigma t) = P_0 = \epsilon^{D^{-1}}. \qquad (192)$$

It is now easily to be seen that

$$D^n P_0 = \frac{P_n}{\lfloor n}, \qquad D^{-n}P_0 = \frac{y^n}{\lfloor n}P_n, \qquad (193)$$

when n is integral. Also

$$I_n(\sigma t) = \frac{y^{\frac{1}{2}n}}{\lfloor n}P_n = y^{\frac{1}{2}n}D^n P_0 = y^{-\frac{1}{2}n}D^{-n}P_0, \qquad (194)$$

when n is integral. But if n is not integral we do not have equivalence of differentiations and integrations in the way expressed in (194). Since, however, by (23), §338,

$$I_n(\sigma t) = y^{\frac{1}{2}n}\left(\frac{1}{\lfloor 0 \lfloor n} + \frac{y}{\lfloor 1 \lfloor n+1} + \frac{y^2}{\lfloor 2 \lfloor n+2} + \dots\right). \qquad (195)$$

generally, we do have

$$I_n(\sigma t) = \frac{y^{\frac{1}{2}n}}{\lfloor n}P_n = y^{-\frac{1}{2}n}D^{-n}P_0 = y^{-\frac{1}{2}n}D^{-n}\epsilon^{D^{-1}}, \qquad (196)$$

for any value of n from $-\infty$ to $+\infty$, provided that $D^{-n}1 = y^n/\lfloor n$ generally, as well as when n is integral. This matter will be treated separately.

Analysis Founded upon the Division of the Instantaneous State into Positive and Negative Pure Waves.

§ 378. Another way of treating the general characteristic of V and C is founded upon the division of any initial state into two waves travelling in opposite directions. If V_0 and C_0 be the initial states of V and C, and

$$U_0 = \tfrac{1}{2}(V_0 + LvC_0), \qquad W_0 = \tfrac{1}{2}(V_0 - LvC_0), \tag{1}$$

then U_0 is the positive wave, and W_0 the negative wave, at the initial moment. If $V_0 = LvC_0$ the initial state represents a positive wave only; if $V_0 = -LvC_0$, then a negative wave only.

If there is no waste of energy, that is, if $R = 0 = K$, the state at time t is got by shifting U_0 through the distance vt to the right, and W_0 through the same distance to the left. That is, if U_0 is $f(x)$ and W_0 is $g(x)$, then

$$U = f(x - vt), \qquad W = g(x + vt) \tag{2}$$

express what U and W become at time t later. Their sum and difference will show the real V and C.

Going further, if R and K are finite, but balanced, so that $\sigma = 0$, the same division into a positive and negative wave occurs, but there is, besides the translation, attenuation due to the waste of energy, so that

$$U = \epsilon^{-\rho t} f(x - vt), \qquad W = \epsilon^{-\rho t} g(x + vt) \tag{3}$$

express the waves at time t.

If σ is not zero, the theory of the distortionless circuit with resistances and leaks inserted shows that (3) always represents the phenomena in the first stage, approximately. How long it will remain fairly true depends on the amount of distortion in the interval due to the unbalanced wastes due to R and K respectively. It is clear, then, that the quantities U and W may themselves be made the objects of attention in a mathematical treatment differing from the preceding. At the same time, instead of x and t we should make $x - vt$ and $x + vt$ be the independent variables. To show how to do this, go back to the connections of V and C,

$$-\Delta V = (R + Lp)C, \qquad -\Delta C = (K + Sp)V, \tag{4}$$

where Δ and p are the x and t differentiators. These are the same as

$$-v\Delta(V\epsilon^{\rho t}) = Lv(p + \sigma)(C\epsilon^{\rho t}), \tag{5}$$

$$-v\Delta(C\epsilon^{\rho t}) = Sv(p - \sigma)(V\epsilon^{\rho t}). \tag{6}$$

Multiply the second of these by Lv, and then add and subtract these equations to form two new ones,

$$-v\Delta(V + LvC)\epsilon^{\rho t} = p(V + LvC)\epsilon^{\rho t} - \sigma(V - LvC)\epsilon^{\rho t}, \tag{7}$$

$$-v\Delta(\mathrm{V}-\mathrm{L}v\mathrm{C})\epsilon^{\rho t}=p(\mathrm{L}v\mathrm{C}-\mathrm{V})\epsilon^{\rho t}+\sigma(\mathrm{V}+\mathrm{L}v\mathrm{C})\epsilon^{\rho t} \tag{8}$$

or, which are the same,

$$(p+v\Delta)\mathrm{U}=\sigma\mathrm{W}, \qquad (p-v\Delta)\mathrm{W}=\sigma\mathrm{U}, \tag{9}$$

if $\mathrm{U}=\frac{1}{2}\epsilon^{\rho t}(\mathrm{V}+\mathrm{L}v\mathrm{C}), \qquad \mathrm{W}=\frac{1}{2}\epsilon^{\rho t}(\mathrm{V}-\mathrm{L}v\mathrm{C}),$ (10)

Going further, let

$$u=\tfrac{1}{2}\sigma(t+x/v), \qquad w=\tfrac{1}{2}\sigma(t-x/v) \tag{11}$$

be the new independent variables, then

$$\frac{d}{du}=\frac{p+v\Delta}{\sigma}, \qquad \frac{d}{dw}=\frac{p-v\Delta}{\sigma}; \tag{12}$$

consequently the equations (9) become

$$\frac{d\mathrm{U}}{du}=\mathrm{W}, \qquad \frac{d\mathrm{W}}{dw}=\mathrm{U}, \qquad \frac{d^2\mathrm{U}}{du\,dw}=\mathrm{U}. \tag{13}$$

The third of these is the characteristic of U (or of W) obtained from the previous two connecting equations.

It is easy to see that the mathematical expression is now simplified to the uttermost. All solutions have, except at sources or places of imposition of external influence, to satisfy the new characteristic. The general solution is easily found. Let a be the u differentiator and β the w differentiator, so that

$$(a\beta-1)\mathrm{U}=0. \tag{14}$$

Then

$$\mathrm{U}=\frac{0}{a\beta-1}=\frac{0/a\beta}{1-(a\beta)^{-1}}=\left(1+\frac{1}{a\beta}+\frac{1}{a^2\beta^2}+\dots\right)\frac{0}{a\beta} \tag{15}$$

Here $0/a\beta$ is *any* function of u and w which when operated upon by $a\beta$ is made to vanish. This can only be $\mathrm{F}(u)+\mathrm{G}(w)$, the sum of any function of u and any function of w. So

$$\mathrm{U}=\frac{\mathrm{F}(u)+\mathrm{G}(w)}{1-(a\beta)^{-1}}=\left(1+\frac{1}{a\beta}+\frac{1}{a^2\beta^2}+\dots\right)\{\mathrm{F}(u)+\mathrm{G}(w)\} \tag{16}$$

is one form of the general solution. We can now see how it works out and examine some special cases.

In the same way we can see that the general solution of the partial differential equation

$$d_1d_2d_3\dots d_n\mathrm{U}=\mathrm{U}, \tag{17}$$

where the d's are any number of independent differentiators, with respect to the variables $x_1, x_2, \dots x_n$, may be written thus,

$$\mathrm{U}=\frac{\Sigma\,\mathrm{F}}{1-d^{-1}}=\left(1+\frac{1}{d}+\frac{1}{d^2}+\dots\right)(\mathrm{F}_1+\mathrm{F}_2+\dots), \tag{18}$$

if d stands for the product of the d's, as in (17), and the F's are arbitrary functions of all the variables save one; thus F_1 a function of all save x_1, F_2 of all save x_2, and so on. If the operand Σ F is simply 1, we get the fundamental solution

$$U = 1 + x + \frac{x^2}{(\underline{2})^n} + \frac{x^3}{(\underline{3})^n} + \ldots, \qquad (19)$$

if x is the product of all the variables. But the treatment of (18), for the derivation of other solutions, will be sufficiently shown in discussing the case of two variables.

In (15), (16) we apparently generate U out of nothing. But nothing can really come out of nothing. And, in fact, the 0 is not absolute zero. It can be seen to represent an impulsive function. Since there are two variables, a function which is impulsive as regards both is $\beta f + \alpha g$, where f is a function of u only, and g a function of w only. Now if $f = \alpha F$ and $g = \beta G$, this impulsive function is $\alpha\beta(F + G)$. Inspection of (15), (16) will show that this is the actual form assumed there. Similarly, in solving (17), by $(d - 1)U = 0$, and therefore

$$U = \frac{0}{d-1} = \frac{d\Sigma F}{d-1} = \text{as above in (18)},$$

the 0 means the impulsive function $d\Sigma F$.

Simplest Solutions. Waves of Infinite Length, and of Length $2\pi v/\sigma$.

§ 379. Coming to the consideration of the instantaneous electromagnetic waves U and W, there are a few specially simple solutions of the characteristic which should be noted. Suppose that $U = \phi(u + w)$, then $\alpha\beta U = U$ shows that $\phi'' = \phi$. So there are just two cases, $U = V_0\epsilon^{\sigma t}$ and $U = V_0\epsilon^{-\sigma t}$, because $\sigma t = u + w$.

Considering the first, we derive the negative wave by $\alpha U = W$. This makes $W = U$. The waves are identical as regards V, and oppositely identical as regards C. That is, the actual C is zero, and the actual V is $V = V_0\epsilon^{-\rho t}\epsilon^{\sigma t} = V_0\epsilon^{-Kt/S}$. This means that the uniform distribution of transverse voltage V_0 may be regarded as the sum of two electromagnetic waves whose electric forces are similar and whose magnetic forces are opposite.

In a similar way the solution $U = LvC_0\epsilon^{-\sigma t}$ gives $W = -U$, which means $V = 0$, and $C = C_0\epsilon^{-\rho t}\epsilon^{-\sigma t} = C_0\epsilon^{-Rt/L}$. The uniform current C_0 is represented by two waves, additive as regards magnetic force, destructive as regards electric force.

From these we can pass to a case which is equally simple in expression, though far more developed in theory. Suppose that $U = \phi(u - w)$. Then $\alpha\beta U = U$ shows that $\phi'' = -\phi$. As before, there are two independent solutions. Combining them, we have

$$U = V_0 \sin(\sigma x/v + \theta), \qquad W = V_0 \cos(\sigma x/v + \theta), \qquad (20)$$

where the constant θ enables us to include the second solution. Remember that $u - w = \sigma x/v$.

Now U is a positive pure wave, and W a negative wave. Their translation in opposite directions at speed v will give the later state of things, except as regards the distortion. But the last is very important here. The waves are apparently stationary. This means that with the special wavelength in question, the distortion due to unbalanced R and K keeps the instantaneous waves unaltered in form and position. The voltage and current are

$$\left.\begin{aligned} V &= \epsilon^{-\rho t}V_0(\sin + \cos)(\sigma x/v + \theta), \\ LvC &= \epsilon^{-\rho t}V_0(\sin - \cos)(\sigma x/v + \theta). \end{aligned}\right\} \qquad (21)$$

The larger σ (and the distortion), the smaller the wavelength. When σ is quite zero we have uniform V and C subsiding.

Development of General Solution in $u^m P_m$ and $w^m P_m$ Functions.

§ **380**. Now take in hand the general solution (16). Let the operand F + G be 1, then

$$U = \left(1 + \frac{1}{\alpha\beta} + \frac{1}{\alpha^2\beta^2} + \dots\right)1 = 1 + \frac{uw}{\lfloor 1 \lfloor 1} + \frac{u^2w^2}{\lfloor 2 \lfloor 2} + \dots . \qquad (22)$$

But here $uw = \frac{1}{4}\sigma^2(t^2 - x^2/v^2)$, therefore

$$U = I_0\{\sigma(t^2 - x^2/v^2)^{\frac{1}{2}}\} = I_0(z). \qquad (23)$$

The new symbol z is introduced for clearness, because the quantity it stands for occurs so often. We see that the function $I_0(z)$ is the fundamental solution, from which to develop others, unless we use the operator $\{1 - (\alpha\beta)^{-1}\}^{-1}$ directly.

Similarly, if the operand is $u^m/\underline{|m}$, then

$$U = \frac{a^{-m}}{1-(a\beta)^{-1}} = \frac{1}{a^m} + \frac{1}{\beta a^{m+1}} + \dots$$

$$= \frac{u^m}{\underline{|0}\,\underline{|m}} + \frac{u^{m+1}w}{\underline{|1}\,\underline{|m+1}} + \frac{u^{m+2}w^2}{\underline{|2}\,\underline{|m+2}} + \dots = \frac{u^m}{\underline{|m}} P_m(z). \qquad (24)$$

And if the operand is $w^m/\underline{|m}$ or $\beta^{-m}1$, the result is got by interchanging u and w, and is, therefore,

$$U = \frac{w^m}{\underline{|m}} P_m(z). \qquad (25)$$

From the above we see that

$$U = \Sigma\left(A_m \frac{u^m}{\underline{|m}} + B_m \frac{w^m}{\underline{|m}}\right) P_m(z) \qquad (26)$$

is a comprehensive solution, not itself perfectly general, but perhaps admitting of the derivation of perfectly general results.

The manipulation of such solutions is made rather easy by these simple properties :—

$$a^m P_0(z) = \frac{w^m}{\underline{|m}} P_m(z) = \beta^{-m} P_0(z), \qquad (27)$$

$$\beta^m P_0(z) = \frac{u^m}{\underline{|m}} P_m(z) = a^{-m} P_0(z), \qquad (28)$$

the truth of which may be seen by inspecting (24).

To illustrate, let operand be ϵ^u, then

$$U = \frac{\epsilon^u}{1-(a\beta)^{-1}} = \frac{(1-a^{-1})^{-1}}{1-(a\beta)^{-1}} = \left(1 + \frac{1}{a} + \frac{1}{a^2} + \dots\right) P_0(z)$$

$$= (1 + \beta + \beta^2 + \dots) P_0(z) = P_0 + uP_1 + \frac{u^2}{\underline{|2}} P_2 + \dots, \qquad (29)$$

showing variations of method. Similarly,

$$\frac{\epsilon^w}{1-(a\beta)^{-1}} = (1 + a + \dots) P_0 = P_0 + wP_1 + \frac{w^2}{\underline{|2}} P_2 + \dots. \qquad (29\text{A})$$

Derivation of C Wave from V Wave, and Conversely, with Examples. Condition at a Moving Boundary. Expansion of $\epsilon^{\sigma t}$.

§ 381. There are two ways in which the solution (26) occurs, namely, to represent the effects due to given initial states, and to represent the waves resulting from actions impressed at a certain spot. It is with the latter that we have been principally concerned, so the treatment may be applied to these first.

We have

$$V\epsilon^{\rho t} = (1 + a)U, \quad LvC\epsilon^{\rho t} = (1 - a)U, \qquad (30)$$

by the definition of U and W. So

$$V\epsilon^{\rho t} = \frac{1 + a}{1 - a} LvC\epsilon^{\rho t}, \quad LvC\epsilon^{\rho t} = \frac{1 - a}{1 + a} V\epsilon^{\rho t}. \qquad (31)$$

The last enable us to find the C wave when the V wave is known, generated by impressed voltage at the origin, or the V wave when the C wave is known. Here it is to be understood that the disturbances are impressed upon an empty cable.

For example, let e be constant, and

$$LvC\epsilon^{\rho t} = eI_0(z), \qquad (32)$$

This is a known case. What is the expression for the V wave? Use the first of (31) to find the answer. It is

$$V\epsilon^{\rho t} = e(1 + 2a + 2a^2 + 2a^3 + \dots)P_0(z)$$
$$= e\Big(P_0 + 2wP_1 + 2\frac{w^2}{2}P_2 + 2\frac{''^3}{3}P_3 + \dots\Big)(z), \qquad (33)$$

by (27). The process is easy enough, but how make sure it is right, seeing that the fractional operator in (31) might be otherwise treated? That the result satisfies the characteristic is obvious. But it must be correct at the limits as well. To test this, put $x = vt$, then $w = 0$ and $P_0 = 1$, so we obtain $V = e\epsilon^{-\rho t}$. This indicates that the value of V at the origin was e at the moment $t = 0$, and nothing more. But also, put $x = 0$, then $w = \frac{1}{2}\sigma t$, and (33) gives

$$V_0\epsilon^{\rho t} = e(I_0 + 2I_1 + 2I_2 + \dots)(\sigma t). \qquad (34)$$

But

$$\epsilon^{\sigma t} = (I_0 + 2I_1 + 2I_2 + 2I_3 + \dots)(\sigma t) \qquad (35)$$

is an identity. Therefore $V_0 = e\epsilon^{-\rho t}\epsilon^{\sigma t} = e\epsilon^{-Kt/S}$. Now we know that this is the voltage impressed at the origin that does produce the current wave (32) above. See (101), §367. So (33) is correct, and is an alternative form of (115), §369. Perhaps (33) is the best for numerical calculation. There are tables of the I_n functions at the end of Gray and Matthews' work on Bessel functions, and the P_n functions are closely related.

We have made use of the usual principle that the characteristic must be satisfied, and also the boundary conditions, with this extension, that they have to be satisfied when a boundary is in motion. The outer boundary, $x = vt$, is at the front of the wave, and travels at speed v. But there is no distortion at the very front of a wave. The disturbance there consists merely of what has been transmitted, and is known in terms of the disturbance at the origin.

The easy way in which the V solution is derived from that for C, as above, might lead us to overlook some other considerations. For it might go thus,

$$V\epsilon^{pt} = (\beta + 1)W, \quad LvC\epsilon^{pt} = (\beta - 1)W, \tag{36}$$

therefore,

$$V\epsilon^{pt} = \frac{\beta + 1}{\beta - 1} LvC\epsilon^{pt}. \tag{37}$$

Now, if we make β act by integrations, as shown by

$$\frac{\beta + 1}{\beta - 1}P_0 = \frac{1 + \beta^{-1}}{1 - \beta^{-1}}P_0 = (1 + 2\beta^{-1} + \dots)P_0, \tag{38}$$

we shall come to the same result (33) precisely. But in getting (33) we made α act by differentiations. That is the same as β integrations, we know. But if we, instead of as in (38), use β differentiations, we get

$$\frac{\beta + 1}{\beta - 1}P_0 = -(1 + 2\beta + \dots)P_0 = -(P_0 + 2uP_1 + 2\frac{u^2}{\underline{|2}}P_2 + \dots). \tag{39}$$

This is wrong, not merely because the sign is wrong, but because we have the factors u, u^2, &c., instead of w, w^2, &c. Now u does not vanish at the wave front on the + side, but it does on the − side. So (39) belongs to a problem concerning affairs on the left side of the origin, due to a negative voltage impressed there. We are not concerned with that. The practical note to make is that we must use α differentiations or β integrations, as in (33) or (38), in order to introduce w^n and not u^n, when we are concerned with the wave sent from the origin to the right side.*

* There are two points in question. First, α and β^{-1} are equivalent on the operand $P_m(z) \times u^m$ or $\times w^m$. Secondly, which way of expanding the fractional operators should be employed? Answer, so as to get w^m factors to the $P_m(z)$ functions.

Nevertheless there is some further interest in the matter. Thus, given $(\beta - 1)\,W = P_0$. What is W? It is clearly indeterminate in its expression, without further information. But what is the connection between the two principal solutions, say W_1 and W_2, got by using β integrations or β differentiations? Thus,

$$W_1 = \frac{P_0}{\beta - 1} = \frac{\beta^{-1}P_0}{1-\beta^{-1}} = \left(\frac{1}{\beta} + \frac{1}{\beta^2} + \dots\right)P_0 = wP_1 + \frac{w^2}{\underline{|2}}P_2 + \frac{w^3}{\underline{|3}}P_3 \dots, \tag{40}$$

$$W_2 = -\frac{P_0}{1-\beta} = -(1 + \beta + \beta^2 + \dots)P_0 = -\left(P_0 + uP_1 + \frac{u^2}{\underline{|2}}P_2 + \right. \tag{41}$$

To see the difference between W_1 and W_2, expand them. The result is

$$W_1 = \epsilon^{u+w} - \epsilon^u - w\left(u + \frac{u^2}{\underline{|2}} + \dots\right) - \frac{w^2}{\underline{|2}}\left(\frac{u^2}{\underline{|2}} + \frac{u^3}{\underline{|3}} + \dots\right) - \dots; \tag{42}$$

and for W_2 we obtain the same without the first term. So $W_1 - W_2 = \epsilon^{\sigma t}$, or, which is the same,

$$\epsilon^{\sigma t} = P_0 + (u + w)P_1 + \frac{u^2 + w^2}{\underline{|2}}P_2 + \frac{u^3 + w^3}{\underline{|3}}P_3 + \dots . \tag{43}$$

This is an important identity. The argument of the P functions is z. When $x = 0$, we reduce this identity to the old one, (35). We may also write it thus

$$\epsilon^{\sigma t} = \{1 + \alpha + \beta + \alpha^2 + \beta^2 + \dots\}P_0 = (\Sigma\alpha^n)P_0, \tag{44}$$

where n has to receive all integral values from $-\infty$ to ∞ See also (29) and (29A) above, which make, combined with (43)

$$P_0 + \epsilon^{\sigma t} = \frac{\epsilon^u + \epsilon^w}{1 - (\alpha\beta)^{-1}}. \tag{45}$$

In connection with (33) we may take note of the effect of alternating signs. Thus, given

$$V\epsilon^{\rho t} = e\left(P_0 - 2wP_1 + 2\frac{w^2}{\underline{|2}}P_2 - \dots\right), \tag{46}$$

to represent the V wave. It makes $V_0\epsilon^{\rho t} = e\epsilon^{-\sigma t}$ at the origin, that is, $V_0 = e\epsilon^{-Rt/L}$ is the impressed voltage. The corresponding current wave may be obtained by the second of (31), which makes

$$\begin{aligned} Lv C\epsilon^{\rho t} &= e(1 - 2a + 2a^2 + \dots)(1 - 2a + 2a^2 + \dots)P_0(z) \\ &= e(1 - 4a + 8a^2 - 12a^3 + 16a^4 - \dots)P_0(z) \\ &= e(P_0 - 4wP_1 + 8\frac{w^2}{\lfloor 2}P_2 - \dots)(z). \end{aligned} \tag{47}$$

Also note that we are not restricted to integral values of m in P_m solutions. Thus, if

$$V_0\epsilon^{\rho t} = \frac{\epsilon^{\sigma t}}{(\frac{1}{2}\pi\sigma t)^{\frac{1}{2}}} = I_{\frac{1}{2}}(\sigma t) + I_{-\frac{1}{2}}(\sigma t) \tag{48}$$

shows the impressed voltage V_0, then the V wave is

$$V\epsilon^{\rho t} = \frac{w^{\frac{1}{2}}}{\lfloor \frac{1}{2}}P_{\frac{1}{2}}(z) + \frac{w^{-\frac{1}{2}}}{\lfloor -\frac{1}{2}}P_{-\frac{1}{2}}(z) = \frac{\text{shin}\, z}{(\pi u)^{\frac{1}{2}}} + \frac{\cosh z}{(\pi w)^{\frac{1}{2}}}. \tag{49}$$

This is done by simply generalising (48) so as to harmonise with it and contain powers of w as initial factors.

Deduction of the V and C Waves when V_0 is Constant from the Case $V_0 = e\epsilon^{-Kt/S}$. Expansion of $\epsilon^{\rho t}$ in I Functions. Construction of the Wave of V due to any Impressed Voltage.

§ 382. Now pass on to a further extension. We know that when $V_0\epsilon^{\rho t} = e\epsilon^{\sigma t}$, the current wave is $C\epsilon^{\rho t} = eP_0(z)$. From this deduce the current wave when $V_0 = e$, constant. Referring to § 368, equations (106), (107), we see that the operand $\epsilon^{\sigma t}e$ there used has to be altered to $\epsilon^{\rho t}e$. This is the same as changing the operator $p/(p-\sigma)$ in (107) to $p/(p-\rho)$; or the same as operating on the old solution by $(p-\sigma)/p-\rho$. That is, when the impressed voltage is constant, the current wave is

$$C\epsilon^{\rho t} = \frac{p-\sigma}{p-\rho}\, eP_0(z). \tag{50}$$

First put p in terms of α, β. Then

$$\frac{p-\sigma}{p-\rho} = 1 - \frac{1 - \sigma/\rho}{1 - (\sigma/2\rho)(\alpha+\beta)}. \tag{51}$$

But if we expand this in rising powers of $\alpha+\beta$, the result must be a series of P_m functions with u^m as well as w^m factors. This will not do. Besides, we found before that we should employ a differentiations or β integrations. Do so here, therefore, substituting a^{-1} for β in (51). We then get

$$\frac{p-\sigma}{p-\rho}=1+\frac{2(\rho/\sigma-1)a}{1-(2\rho a/\sigma-a^2)} \tag{52}$$

$$=1+\left(1-\frac{\sigma}{\rho}\right)\left\{\frac{2\rho}{\sigma}a+\frac{2\rho}{\sigma}a\left(\frac{2\rho}{\sigma}a-a^2\right)+\ldots\right\}, \tag{53}$$

that is, a series in rising powers of $(2\rho a/\sigma - a^2)$. The result must now be a proper $w^m P_m$ series, namely, if $c=2\rho/\sigma$ for brevity,

$$\begin{aligned}C\epsilon^{\rho t} = eP_0+e\left(1-\frac{\sigma}{\rho}\right)\Big\{&cwP_1+c^2\frac{w^2}{\underline{|2}}P_2-c\frac{w^3}{\underline{|3}}P_3\\ &+c^3\frac{w^3}{\underline{|3}}P_3-2c^2\frac{w^4}{\underline{|4}}P_4+c\frac{w^5}{\underline{|5}}P_5\\ &+c^4\frac{w^4}{\underline{|4}}P_4-3c^3\frac{w^5}{\underline{|5}}P_5+3c^2\frac{w^6}{\underline{|6}}P_6-c\frac{w^7}{\underline{|7}}P_7+\&c.\Big\}.\end{aligned} \tag{54}$$

Though complicated, the law is obvious, since the coefficients follow the rule of $(c-a)^n$. There is only one thing needed now, to verify that it is correct at the origin. This we can do by comparison with (77), (76), § 364, where the value of C at the origin is expressed in a special form. Put $x=0$ in (54). Then

$$\begin{aligned}C_0\epsilon^{\rho t} = eI_0+e\left(1-\frac{\sigma}{\rho}\right)c\{&I_1+cI_2-I_3+c^2I_3-2cI_4+I_5\\ &+c^3I_4-3c^2I_5+3cI_6-I_7+\ldots\},\end{aligned} \tag{55}$$

the argument of the I functions being σt. In (54) it is z. If now we write out the first few terms of the I functions, we shall obtain the previous result (77), (76), exactly. Therefore (54) is correct.

Continuing the above, find the V wave due to steady V_0 at the origin, by deriving it from the case $V_0\epsilon^{\rho t}=e\epsilon^{\sigma t}$. In the latter case we have the solution (33) above. Operate on it by the operator (52) expanded as in (53). The result is

$$\begin{aligned}V\epsilon^{\rho t} &= \frac{p-\sigma}{p-\rho}\,e\,\frac{1+a}{1-a}P_0\\ &= e(1+2a+2a^2+2a^3+\ldots)P_0\\ &\quad+\left(1-\frac{\sigma}{\rho}\right)c\,\{a+2a^2+2a^3+2a^4+\ldots+c(a^2+2a^3+2a^4+\ldots)\\ &\quad-(a^3+2a^4+2a^5+\ldots)+c^2(a^3+2a^4+2a^5+\ldots)\\ &\quad-2c(a^4+2a^5+\ldots)+(a^5+2a^6+\ldots)+\ldots\}P_0.\end{aligned} \tag{56}$$

Arrange in powers of a, and then turn to $w^m P_m$ functions. We get

$$V\epsilon^{\rho t} = e\left\{P_0 + \frac{2\rho}{\sigma} w P_1 + 2\left(\frac{2\rho^2}{\sigma^2} - 1\right)\frac{w^2}{\underline{|2}} P_2 + 2\left(\frac{4\rho^3}{\sigma^3} - \frac{3\rho}{\sigma}\right)\frac{w^3}{\underline{|3}} P_3 + \dots\right\} \tag{57}$$

This is the complete wave. To verify, at $x=0$ it makes

$$V_0\epsilon^{\rho t} = e\left\{I_0 + 2\frac{\rho}{\sigma} I_1 + 2\left(\frac{2\rho^2}{\sigma^2} - 1\right)I_2 + \dots\right\}(\sigma t). \tag{58}$$

Here $V_0 = e$, so we have an expansion of $\epsilon^{\rho t}$ in I functions. It is an identity.

Furthermore, in the light of the preceding, this is the process whereby the complete V wave can be developed when V_0 is given as any function of the time. Expand $V_0\epsilon^{\rho t}$ in the form $\Sigma\, A_n I_n(\sigma t)$. Then

$$V\epsilon^{\rho t} = \Sigma\, A_n \frac{w^n}{\underline{|n}} P_n(z) \tag{59}$$

is the required result. The process of expansion in the Bessel series is merely the equation of coefficients of powers of σt, and can be done specially in special cases. But it will be more satisfactory now to give the general formula.

Identical Expansions of Functions in I_n Functions. Formula for $(\frac{1}{2}x)^n$. Electromagnetic Applications.

§ 383. It is a commonplace in mathematical physics to require the expansion of an arbitrary function in a series of functions of a given type. The dynamics indicates that a certain kind of function represents a normal vibration, for instance; and also shows how to obtain every possible variety of this normal vibration so as to harmonise with the boundary conditions. Then, since in the real motion there is no limitation to any particular motion, provided it be consistent with the conditions imposed, a physically-minded man can at once conclude that every possible sort of motion is included in the special normal motions, and therefore that an arbitrary function can be expanded in a series of normal functions. And so it always is, of course.

But it is not with these expansions in normal functions that we have to deal at present, but with something quite different. If we have $f(x) = \Sigma\, A_n \sin nx$, a function of x being expanded in sine functions, to hold good between certain limits, and to satisfy some conditions at the limits; or if we have $f(x) = \Sigma\, A_n J_0(nx)$, under similar restrictions; although these equations are loosely called identities, they are not real identities.

An absolute identity is such that the expression upon one side of the equation expressing it can be converted to that on the other side by mere rearrangement of parts. After cancellations, it means no more than $0 = 0$, or $1 = 1$, or $x = x$, &c. Expansions in normal functions are not of this sort; say that $f(x) = \Sigma\, A_n \sin nx$, and $f(x)$ is expressed by a power series, then it is not usually the case that the coefficient of x on the left side is the same as on the right side, or similarly for any other power of x. If it were so, we should have an absolute identity. This may sometimes occur, *e.g.*, when $f(x) = \sin nx$; but, in general, the coefficients of powers of x in the normal series assume infinite values. So we do not have identity, but merely equivalence, under limitations; for example, the form of $f(x)$ must be varied in different ranges of the variable x in order to preserve the equivalence.

But when we say that $f(x) = \Sigma\, A_n I_n(x)$, where $I_n(x)$ is not a normal function (or is one only in a changed sense), it is with an absolute or true identity that we are concerned. By mere rearrangement of parts, aided by mutually destructive additional terms, the function $f(x)$ has to be turned to the form

$$f(x) = A_r I_r(x) + A_s I_s(x) + A_t I_t(x) + \dots, \qquad (60)$$

where r, s, t, &c., and the coefficients A are determined by the nature of $f(x)$ alone, without reference to boundary conditions.

Our physical problem indicates plainly enough that the expansion is either possible in general, or else is the form from which to derive other series in cases of primary failure. And so it works out, without difficulty. As an easy preliminary example of this sort of expansion, consider the function

$$e_n = \frac{x^n}{\lfloor n} + \frac{x^{n+1}}{\lfloor n+1} + \frac{x^{n+2}}{\lfloor n+2} + \dots, \qquad (61)$$

where n has any value. We have $x^n / \lfloor n = e_n - e_{n+1}$. It follows obviously that any power series can be identically expanded in a series of e_n functions, whether the n's be integral or fractional, save for failures when n is a negative integer.

Now the Bessel expansions are by no means so simple as the last, but the same principle is concerned precisely. Say it is $(\frac{1}{2}x)^n/\lfloor\underline{n}$ that is to be expanded. This function is the first term in $I_n(x)$. So if we equate $f(x)$ to $I_n(x)$, the right side is redundant in the other terms of $I_n(x)$, which involve x^{n+2}, x^{n+4}, and so on. Therefore introduce I_{n+2}. This will make, with a proper factor, the coefficient of x^{n+2} be zero. Then the addition of I_{n+4} with a proper factor will make the coefficient of x^{n+4} be zero. And so on. Thus,

$$\frac{(\frac{1}{2}x)^n}{\lfloor\underline{n}} = A_n\left\{\frac{(\frac{1}{2}x)^n}{\lfloor\underline{n}} + \frac{(\frac{1}{2}x)^{n+2}}{\lfloor\underline{1}\,\lfloor\underline{n+1}} + \frac{(\frac{1}{2}x)^{n+4}}{\lfloor\underline{2}\,\lfloor\underline{n+2}} + \dots\right\}$$
$$+ A_{n+2}\left\{\frac{(\frac{1}{2}x)^{n+2}}{\lfloor\underline{n+2}} + \frac{(\frac{1}{2}x)^{n+4}}{\lfloor\underline{1}\,\lfloor\underline{n+3}} + \dots\right\}$$
$$+ A_{n+4}\left\{\frac{(\frac{1}{2}x)^{n+4}}{\lfloor\underline{n+4}} + \dots\right\} + \dots \tag{62}$$

makes

$$A_n = 1, \quad \frac{A_{n+2}}{\lfloor\underline{n+2}} + \frac{A_n}{\lfloor\underline{1}\,\lfloor\underline{n+1}} = 0, \quad \frac{A_{n+4}}{\lfloor\underline{n+4}} + \frac{A_{n+2}}{\lfloor\underline{1}\,\lfloor\underline{n+3}} + \frac{A_n}{\lfloor\underline{2}\,\lfloor\underline{n+2}} = 0,$$

&c., so all the coefficients become known. The result is

$$\frac{(\frac{1}{2}x)^n}{\lfloor\underline{n}} = \left\{I_n - \frac{n+2}{\lfloor\underline{1}} I_{n+2} + \frac{n+1}{\lfloor\underline{2}}(n+4) I_{n+4} - \frac{(n+1)(n+2)}{\lfloor\underline{3}} \times \right.$$
$$\left.(n+6) I_{n+6} + \frac{(n+1)(n+2)(n+3)}{\lfloor\underline{4}}(n+8) I_{n+8} + \dots\right\}(x). \tag{63}$$

By the mode of construction, this formula is valid for any value of n from $-\infty$ to $+\infty$. But it fails, or gives a useless identity, like $I_n = I_{-n}$, when n is a negative integer. This case, and its application to electromagnetic waves, will be done separately, as will the case of the logarithm. Examples:—

$$1 = (I_0 - 2\,I_2 + 2\,I_4 - 2\,I_6 + 2\,I_8 - \dots)(x), \tag{64}$$
$$\tfrac{1}{2}x = (I_1 - 3\,I_3 + 5\,I_5 - 7\,I_7 + 9\,I_9 - \dots)(x), \tag{64A}$$
$$(\tfrac{1}{2}x)^2/\lfloor\underline{2} = (I_2 - 4\,I_4 + 9\,I_6 - 16\,I_8 + 25\,I_{10} - \dots)(x), \tag{64B}$$
$$(\tfrac{1}{2}x)^3/\lfloor\underline{3} = (I_3 - 5\,I_5 + 14\,I_7 - 30\,I_9 + 55\,I_{11} - \dots)(x), \tag{64C}$$
$$(\tfrac{1}{2}x)^{\frac{1}{3}}/\lfloor\underline{\tfrac{1}{3}} = (I_{\frac{1}{3}} - 2\tfrac{1}{3}\,I_{2\frac{1}{3}} + 2\tfrac{8}{9}\,I_{4\frac{1}{3}} - 3\tfrac{23}{81}\,I_{6\frac{1}{3}} + \dots)(x). \tag{64D}$$

To make no mistake about the application, put σt for x, then the last formula shows that if the impressed voltage at the origin is expressed by $V_0\epsilon^{\rho t} = e(\frac{1}{2}\sigma t)^{\frac{1}{3}}/|\underline{\tfrac{1}{3}}$, then the wave of voltage generated is

$$V\epsilon^{\rho t} = e\left\{\frac{w^{\frac{1}{3}}}{|\underline{\tfrac{1}{3}}}P_{\frac{1}{3}} - 2\tfrac{1}{3}\frac{w^{2\frac{1}{3}}}{|\underline{2\tfrac{1}{3}}}P_{2\frac{1}{3}} + 2\tfrac{8}{9}\frac{w^{4\frac{1}{3}}}{|\underline{4\tfrac{1}{3}}}P_{4\frac{1}{3}} - \dots\right\}(z). \qquad (65)$$

the function $I_n(\sigma t)$ being turned to $(w^n/|\underline{n})\, P_n(z)$.

Since (63) is an identity, it does not matter what x^n, x^{n+2}, &c., mean, provided $(a \pm b)x^m = ax^m \pm bx^m$, when a and b are numbers, and m is any index concerned. For in any case the cancellations bring us to $x^n = x^n$. The formula may be written more symmetrically, to show the structure. Thus,

$$(\tfrac{1}{2}x)^n = \left\{\frac{|\underline{n-1}}{|\underline{0}}nI_n - \frac{|\underline{n}}{|\underline{1}}(n+2)I_{n+2} + \frac{|\underline{n+1}}{|\underline{2}}(n+4)I_{n+4} - \dots\right\}(x). \qquad (66)$$

It is now perfectly regular from the beginning, and we see that it is a complete series, because carrying it backward will only introduce zero terms, $|\underline{-1}$, $|\underline{-2}$, &c., being infinite.

Expansion of any Power Series in I_n Functions. Examples.

§384. Having got the expansion of x^n, we are in full possession of the expansion of the series $\Sigma B_n x^n$. There may be any number of terms in this series, and the indices need have no connection with one another. But should there be no connection given, the expansion in I functions will consist merely of (63) or (66) repeated again and again, with various values given to n, and with initial coefficients B, and we cannot simplify further. But practically the indices will follow some law, as unit step or step 2 from one to the next. Then we can collect terms and get useful formulæ. The most important case is step 1, with 0 for first index. Thus, let

$$B_0 + B_1(\tfrac{1}{2}x) + B_2\frac{(\frac{1}{2}x)^2}{|\underline{2}} + \dots = (A_0I_0 + A_1I_1 + A_2I_2 + \dots)(x). \quad (67)$$

Given the B's, find the A's. There are two ways. First equate the coefficients of the different powers of x. This will give, first A_0, then A_1, then A_2, and so on. The other way is to use the formula (63), and rearrange terms. Either way we come to the result

$$\Sigma B_n \frac{(\frac{1}{2}x)^n}{\underline{|n}} = B_0 I_0 + B_1 I_1 + (B_2 - 2B_0) I_2 + (B_3 - 3B_1) I_3$$
$$+ (B_4 - 4B_3 + 2B_0) I_4 + (B_5 - 5B_3 + 5B_1) I_5 + \ldots, \qquad (68)$$

the argument of the I functions being x throughout. But this does not go far enough to exhibit the law of the A coefficients. Nevertheless, it is very simple, and is exhibited in the following table of the numerical coefficients in the A's, from A_0 up to A_{13} :—

0	1						
1	1						
2	1	2					
3	1	3					
4	1	4	2				
5	1	5	5				
6	1	6	9	2			
7	1	7	14	7			
8	1	8	20	16	2		
9	1	9	27	30	9		
10	1	10	35	50	25	2	
11	1	11	44	77	55	11	
12	1	12	54	112	105	36	2
13	1	13	65	156	182	91	13

The numbers to the left of the vertical line show which A is referred to. In the corresponding row are the numerical coefficients for that A. Equation (68) gives them up to A_5. Then, by the table,

$$A_6 = B_6 - 6B_4 + 9B_2 - 2B_0, \quad A_7 = B_7 - 7B_5 + 14B_3 - 7B_1,$$

and so on. The first column of 1's explains itself. So does the second column. All the rest is constructed by this rule :—The sum of the first m numbers in any column after the first is the mth number in the next column. Thus, $2+3+4$ in the second column $= 9$, the third number in the next column. Similarly, 105 in the fifth column is the sum of the first five numbers in the previous column. Practically, $50+55=105$, and $77+105=182$, as shown by the sloping lines. It is easier to work with such a table than refer to a general formula. Here are some examples.

$$\epsilon^x = (I_0 + 2I_1 + 2I_2 + 2I_3 + \ldots)(x), \qquad (69)$$

from which come ϵ^{-x}, shin x and cosh x. Compare with (33), (35), § 381, where the special electromagnetic problem is concerned.

$$\tfrac{1}{2}\sin x = (\mathrm{I}_1 + \mathrm{I}_3 + \mathrm{I}_5 + \mathrm{I}_7 + \dots)(x), \qquad (70)$$

$$\sin \tfrac{1}{2}x = (\mathrm{I}_1 - 4\mathrm{I}_3 + 11\mathrm{I}_5 - 29\mathrm{I}_7 + 76\mathrm{I}_9 - 199\mathrm{I}_{11} + \dots)(x), \qquad (71)$$

$$\epsilon^{\frac{1}{2}x} = \mathrm{I}_0 + (\mathrm{I}_1 - \mathrm{I}_2 - 2\mathrm{I}_3 - \mathrm{I}_4 + \mathrm{I}_5 + 2\mathrm{I}_6) + (\mathrm{I}_7 - \mathrm{I}_8 - 2\mathrm{I}_9 - \mathrm{I}_{10} + \mathrm{I}_{11} + 2\mathrm{I}_{12}) + \dots . \qquad (72)$$

These are constructed by the table. As an example of application, if the impressed voltage is given by $\frac{1}{2}\sin \sigma t = \mathrm{V}_0\epsilon^{\rho t}/e$, where e is constant, then the wave of voltage, according to (70), is given by

$$\mathrm{V}\epsilon^{\rho t} = e\left\{ w_1\mathrm{P}_1 + \frac{w^3}{\underline{|3}}\mathrm{P}_3 + \frac{w^5}{\underline{|5}}\mathrm{P}_5 + \dots \right\}(z). \qquad (73)$$

As another example, the formula (58), § 382, relating to steady impressed voltage, may be referred to. The table will give the development to any extent required. Another example is

$$\mathrm{V}_0\epsilon^{\rho t} = e(\tfrac{1}{2}\sigma t)(\mathrm{I}_0 + \mathrm{I}_1)(\sigma t) = e(\mathrm{I}_1 + 2\mathrm{I}_2 + 3\mathrm{I}_3 + 4\mathrm{I}_4 + \dots)(\sigma t) . \qquad (74)$$

This will make

$$\mathrm{V}\epsilon^{\rho t} = e\left\{ w\mathrm{P}_1 + 2\frac{w^2}{\underline{|2}}\mathrm{P}_2 + 3\frac{w^3}{\underline{|3}}\mathrm{P}_3 + \dots \right\}(z), \qquad (75)$$

because it is right at the origin, and at the wave front, and satisfies the characteristic everywhere between.

In the above the step in the index was 1. This includes step 2, and other cases, by the vanishing of certain of the B's. But it is now necessary to give the formula for a power series when the indices are not integers, but still so that there is unit step from one to the next. For this purpose use (66). Thus,

$$\Sigma\,\mathrm{B}_n\frac{(\frac{1}{2}x)^n}{\underline{|n}} = \dots\dots\dots\dots\dots\dots$$

$$+\frac{\mathrm{B}_{n-1}}{\underline{|n-1}}\left\{\frac{\underline{|n-2}}{\underline{|0}}(n-1)\mathrm{I}_{n-1} - \frac{\underline{|n-1}}{\underline{|1}}(n+1)\mathrm{I}_{n+1} + \frac{\underline{|n}}{\underline{|2}}(n+3)\mathrm{I}_{n+3} - \dots\right\}$$

$$+\frac{\mathrm{B}_{n}}{\underline{|n}}\left\{\frac{\underline{|n-1}}{\underline{|0}}n\mathrm{I}_{n} - \frac{\underline{|n}}{\underline{|1}}(n+2)\mathrm{I}_{n+2} + \frac{\underline{|n+1}}{\underline{|2}}(n+4)\mathrm{I}_{n+4} - \dots\right\}$$

$$+\frac{\mathrm{B}_{n+1}}{\underline{|n+1}}\left\{\frac{\underline{|n}}{\underline{|0}}(n+1)\mathrm{I}_{n+1} - \frac{\underline{|n+1}}{\underline{|1}}(n+3)\mathrm{I}_{n+3} + \frac{\underline{|n+2}}{\underline{|2}}(n+5)\mathrm{I}_{n+5} - \right\}$$

$$+ \dots\dots\dots\dots\dots\dots \qquad (76)$$

Now collect the I_n terms together, and the I_{n+2} terms, and so on. The result is that the coefficient of I_n is

$$A_n = n\left\{\frac{B_n}{\underline{|n}}\frac{\underline{|n-1}}{\underline{|0}} - \frac{B_{n-2}}{\underline{|n-2}}\frac{\underline{|n-2}}{\underline{|1}} + \frac{B_{n-4}}{\underline{|n-4}}\frac{\underline{|n-3}}{\underline{|2}} - \frac{B_{n-6}}{\underline{|n-6}}\frac{\underline{|n-4}}{\underline{|3}} + \dots\right\} \tag{77}$$

This is written symmetrically to show the structure. But it may be simplified to

$$A_n = B_n - \frac{n}{\underline{|1}}B_{n-2} + \frac{n}{\underline{|2}}(n-3)B_{n-4} - \frac{n}{\underline{|3}}(n-4)(n-5)B_{n-6}$$
$$+ \frac{n}{\underline{|4}}(n-5)(n-6)(n-7)B_{n-8}$$
$$- \frac{n}{\underline{|5}}(n-6)(n-7)(n-8)(n-9)B_{n-10} + \dots, \tag{78}$$

which also shows the structure, save near the beginning. This formula applies to all the A's. That is, changing n to $n+m$ produces A_{n+m}. When n is integral, and the first B is B_0, we obtain the previous results, in (68) and the table.

The series for A_n must be continued until it stops, by exhaustion of the B's in descending order. But if the B series in

$$\Sigma\, B_n \frac{(\frac{1}{2}x)^n}{\underline{|n}} = \Sigma\, A_n I_n(x), \tag{79}$$

does not stop, going backward, or with decreasing n, then the series for A_n does not stop either. This occurs when the power series is endless in the negative direction. It is not assumed that the B series begins with any particular n. Thus the power series

$$f(x) = ax^{-\frac{3}{2}} + bx^{-\frac{1}{2}} + cx^{\frac{1}{2}} + dx^{\frac{3}{2}} + \dots, \tag{80}$$

which begins with $x^{-\frac{3}{2}}$, will give a Bessel series expansion involving $I_{-\frac{3}{2}}$, $I_{-\frac{1}{2}}$, $I_{\frac{1}{2}}$, &c., on indefinitely, whether the series $f(x)$ stops or not in the positive direction, and the coefficients A will in either case be finite series. But the series

$$f(x) = \dots\dots + ax^{-\frac{3}{2}} + bx^{-\frac{1}{2}} + cx^{\frac{1}{2}} + dx^{\frac{3}{2}}, \tag{81}$$

which stops at the term $x^{\frac{3}{2}}$, will make the A's be infinite series when $f(x)$ is endless in the negative direction. And should $f(x)$ be endless in both directions, we may expect that the series for the A's may be divergent. But (79) will be an identity for all that.

This matter of the endlessness of the A_n series, when the original power series is endless backwards, turns up in a curious form when integral powers are concerned. One example will be sufficient. Let $f(x) = 1$. We know the proper result already. *See* (64) above. Or, in (78), let all the B's be zero except $B_0 = 1$, and then give n all values 0, 1, 2, &c., to find the A's. But in using (79), we may regard the B series as being endless in the negative direction, and stopping at the term $B_0 x^0$. For $\underline{|-1}$, $\underline{|-2}$, &c., are infinite, and therefore B_{-1}, B_{-2}, &c., may have any finite values without interfering with the value of the power series being 1. But if we do this, we shall obtain endless series for the A's. Will the results be wrong, then? Test this.

The formula (78) makes $A_0 = 1$ as before, and then

$$A_1 = -B_{-1} - B_{-3} - 2B_{-5} - 5B_{-7} - \dots,$$
$$A_2 = -2 - B_{-2} - 2B_{-4} - 5B_{-6} - \dots.$$

These are the coefficients of $I_1(x)$ and $I_2(x)$. The first may have any value, the second also, although the first term is correct.

The explanation is that in virtue of the inclusion of the arbitrary B's below B_0, we require to include $I_{-1}(x)$, $I_{-2}(x)$, &c. Doing this, we shall find that $A_{-1} = -A_1$, and $A_{-2} = -2 - A_2$. But I_{-1} is the same as I_1, and I_{-2} the same as I_2. So, joining them together, all the arbitrary B's are eliminated, and the true result already obtained is arrived at. Our formula rejects all the redundancies. Practically, of course, (63) should be used, with $n = 0$, and similarly in other cases.

Expansion of a Power Series in J_n Functions. Examples.

§ 385. In another application, to be made shortly, we shall require the expansion of a function of x, not in a series of I_n functions, but of J_n functions. The connection is $I_n(xi) = i^n J_n(x)$, and this makes $J_n(x)$ be an oscillating function. The expansions are still of an identical nature, and the question now is, how to modify the preceding formulæ to suit the changed circumstances. Say that

$$\Sigma B_n \frac{(\tfrac{1}{2}x)^n}{\underline{|n}} = \Sigma A_n I_n(x) = \Sigma E_n J_n(x). \qquad (82)$$

We know the A's in terms of the B's. Now find the E's in terms of the B's. Put xi for x, then the first equation becomes

$$\Sigma\, B_n i^n \frac{(\frac{1}{2}x)^n}{\lfloor n} = \Sigma\, A_n i^n J_n(x). \tag{83}$$

In this use (77) or (78). The right side becomes

$$\Sigma\left\{B_n i^n + \frac{n}{\lfloor 1} B_{n-2} i^{n-2} + \frac{n}{\lfloor 2}(n-3) B_{n-4} i^{n-4} + \dots\right\} J_n(x), \tag{84}$$

since $i^2 = -1$, $i^4 = 1$, &c. Now for $B_n i^n$ write B_n, for $B_{n-2} i^{n-2}$ write B_{n-2}, &c. The result is

$$\Sigma B_n \frac{(\frac{1}{2}x)^n}{\lfloor n} = \Sigma\left\{B_n + \frac{n}{\lfloor 1} B_{n-2} + \frac{n}{\lfloor 2}(n-3) B_{n-4} + \dots\right\} J_n(x), \tag{85}$$

the required result. That is,

$$E_n = B_n + \frac{n}{\lfloor 1} B_{n-2} + \frac{n}{\lfloor 2}(n-3) B_{n-4} + \frac{n}{\lfloor 3}(n-4)(n-5) B_{n-6} + \dots\,. \tag{86}$$

To expand the function in J_n series, therefore, we have merely to alter all the − signs in the formula for A_n to + signs, to produce E_n.

Thus, when the indices are integers beginning with 0, (68) above becomes

$$\Sigma\, B_n \frac{(\frac{1}{2}x)^n}{\lfloor n} = B_0 J_0 + B_1 J_1 + (B_2 + 2B_0) J_2 + (B_3 + 3B_1) J_3$$
$$+ (B_4 + 4B_2 + 2B_0) J_4 + (B_5 + 5B_3 + 5B_1) J_5 + \dots . \tag{87}$$

The same table is to be used, with all signs taken positively. Examples :—

$$\cos x = (J_0 - 2J_2 + 2J_4 - 2J_6 - \dots)(x) \tag{88}$$
$$1 = (J_0 + 2J_2 + 2J_4 + 2J_6 + \dots)(x) \tag{89}$$
$$\epsilon^x = (J_0 + 2J_1 + 6J_2 + 14J_3 + 34J_4 + \dots)(x) \tag{90}$$

These are done by the table. *E.g.*, the 34 in the last formula is $2^4 \times 1 + 2^2 \times 4 + 2$, by the fourth row of figures.

Sometimes the expansions in I functions and J functions are quite similar, except in the signs. But this is not general. Thus,

$$\operatorname{shin} x = 2(I_1 + I_3 + I_5 + I_7 + \dots)(x), \tag{91}$$
$$= 2(J_1 + 7J_3 + 41J_5 + \dots)(x), \tag{92}$$
$$\sin x = 2(J_1 - J_3 + J_5 - J_7 + \dots)(x), \tag{93}$$
$$= 2(I_1 - 7I_3 + 41I_5 - \dots)(x). \tag{94}$$

In treatises on Bessel functions will be found various expansions of the identical kind in J_n functions. They are usually derived from definite integrals by trigonometrical processes, and by the properties of Bessel functions. Perhaps this may be the best way sometimes; but I think, in general, the above method has the advantage of simplicity of reasoning and of working, besides being comprehensive. The fractional cases, for example, are done by the same formulæ, and these might sometimes be very difficult by integrals and trigonometry.

The Waves of V and C due to any V_0 developed in $w^m P_m(z)$ Functions from the Operational Form of $V_0\epsilon^{pt}$.

§ 386. There is another way of obtaining the waves of voltage and current generated by a source at the origin in the form of a series of waves of the type $\epsilon^{-\rho t}P_m(z)w^m/\underline{|m}$. Instead of, as already explained, expanding $V_0\epsilon^{pt}$ in I functions, and then generalising, we may operate upon the known fundamental solution $P_0(z)$ in a suitable manner directly. Of this process a special example was given in § 382, relating to V_0 being constant. It can be generalised thus. We know that

$$V=\epsilon^{-qx}V_0, \quad \text{or} \quad V\epsilon^{pt}=\epsilon^{-rx}V_0\epsilon^{pt}, \tag{95}$$

where r stands for $(p^2-\sigma^2)^{\frac{1}{2}}/v$. Now let the operator which generates a function out of 1 be denoted by enclosing the function in square brackets; for example, $V_0\epsilon^{pt}=[V_0\epsilon^{pt}]1$. Then

$$V\epsilon^{pt}=\epsilon^{-rx}[V_0\epsilon^{pt}]. \tag{96}$$

Now we know that

$$I_0(\sigma t)=\frac{p}{(p^2-\sigma^2)^{\frac{1}{2}}}, \quad \text{or,} \quad 1=\frac{(p^2-\sigma^2)^{\frac{1}{2}}}{p}I_0(\sigma t). \tag{97}$$

Put this at the end of (96), then

$$V\epsilon^{pt}=\epsilon^{-rx}[V_0\epsilon^{pt}]\frac{(p^2-\sigma^2)^{\frac{1}{2}}}{p}I_0(\sigma t). \tag{98}$$

But we also know that

$$\epsilon^{-rx}I_0(\sigma t)=P_0(z); \tag{99}$$

therefore (98) reduces to

$$V\epsilon^{pt}=[V_0\epsilon^{pt}]\frac{(p^2-\sigma^2)^{\frac{1}{2}}}{p}P_0(z), \tag{100}$$

the required result.

Correspondingly,

$$\mathrm{L}v\mathrm{C}\epsilon^{pt} = \left(\frac{p-\sigma}{p+\sigma}\right)^{\frac{1}{2}}\mathrm{V}\epsilon^{pt} = [\mathrm{V}_0\epsilon^{pt}]\,\frac{p-\sigma}{p}\,\mathrm{P}_0(z), \qquad (101)$$

by (100).

The advantage of conversion of functions to operational form is that whereas we cannot shift the order of conjoined functions and operators, we can do so with operators, if we work properly. Thus, in passing from (98) to (100), we shifted ϵ^{-rx} two steps forward; and in passing to (101), we shifted the fractional operator one step forward.

The structure of (100) is worth notice. It may be simply shown thus. Let F_0 and G_0 be two functions of t, and $[\mathrm{F}_0]$ and $[\mathrm{G}_0]$ their generating operators. Then

$$\mathrm{F}_0 = [\mathrm{F}_0]\,[\mathrm{G}_0]^{-1}\,\mathrm{G}_0. \qquad (102)$$

This is obvious, being mere mathematical jugglery. But if we change G_0 to G, a function of x as well as of t, reducing to G_0 when $x=0$, then F_0 will also become changed to a function of x and t. Thus,

$$\mathrm{F} = [\mathrm{F}_0]\,[\mathrm{G}_0]^{-1}\,\mathrm{G}. \qquad (103)$$

Now, if G satisfies a certain characteristic partial (variables x and t), so does F, by its construction. Therefore F is the x, t solution produced by F_0 just as G is produced by G_0.

In our application, equation (100), F_0 is $\mathrm{V}_0\epsilon^{pt}$, which is any function of t, and G is $\mathrm{P}_0(z)$, the special function of x and t, reducing to $\mathrm{I}_0(\sigma t)$ at the origin. Put $x=0$ in (100), and it makes $\mathrm{V}_0\epsilon^{pt} = [\mathrm{V}_0\epsilon^{pt}]\,1$. When x is not 0, (100) represents the complete wave.

But time differentiations on $\mathrm{P}_0(z)$ are complicated. The proper simply working differentiators are α and β. Use them. Put $\frac{1}{2}\sigma(\alpha+\beta)$ for p. We can at once eliminate β, because β^{-1} and α are equivalent with operand $\mathrm{P}_0(z)$. So put $\frac{1}{2}\sigma(\alpha+\alpha^{-1})$ for p in (100). This makes

$$\frac{p-\sigma}{p} = \frac{(1-\alpha)^2}{1+\alpha^2}, \qquad \left(\frac{p-\sigma}{p+\sigma}\right)^{\frac{1}{2}} = \frac{1-\alpha}{1+\alpha}, \qquad \frac{(p^2-\sigma^2)^{\frac{1}{2}}}{p} = \frac{1-\alpha^2}{1+\alpha^2}. \qquad (104)$$

Therefore (100), (101) become

$$\mathrm{V}\epsilon^{pt} = [\mathrm{V}_0\epsilon^{pt}]\frac{1-\alpha^2}{1+\alpha^2}\mathrm{P}_0(z), \qquad (105)$$

$$\mathrm{L}v\mathrm{C}\epsilon^{pt} = [\mathrm{V}_0\epsilon^{pt}]\frac{(1-\alpha)^2}{1+\alpha^2}\mathrm{P}_0(z), \qquad (106)$$

where, of course, the first operator is to be expressed in terms of a. When this has been done, if the combined operator acting on $P_0(z)$ makes a power series, say,

$$V\epsilon^{pt} = (\Sigma A_n a^n) P_0(z), \tag{107}$$

then we at once get the full development

$$V\epsilon^{pt} = \Sigma A_n \frac{w^n}{\underline{|n}} P_n(z). \tag{108}$$

Remarkable Formula for the Expansion of a Function in I_n Functions, and Examples. Modification of § 386, and Example.

§ **387.** Incidentally, we get this interesting application of physical to pure mathematics. Take $x=0$ in (107) or (105). Then we expand $V_0\epsilon^{pt}$ in I functions. That is, F_0 being any function of the time,

$$F_0 = [F_0]\frac{1-a^2}{1+a^2}, \tag{109}$$

is its expansion in I functions, provided a^n is turned to $I_n(\sigma t)$ in the result. First turn F_0 to $[F_0]1$, $[F_0]$ being a function of p, then put $\frac{1}{2}\sigma(a+a^{-1})$ for p. Then multiply by $(1-a^2) \times (1+a^2)^{-1}$. The result is $\Sigma A_n a^n$, and is the expansion of F_0 in I functions, if a^n means $I_n(\sigma t)$.

The theorem is very striking. It is obviously true by the method employed, based upon a special use of (103), a simple property; but by pure rigorous mathematics there seems nothing whatever about (109) even suggestive of Bessel functions, let alone the result.

By using (109), all the previous results of the kind may be got. It is only necessary to give one or two examples for the sake of explicit illustration. Say

$$F_0 = \frac{(\frac{1}{2}\sigma t)^n}{\underline{|n}} = \left(\frac{\sigma}{2p}\right)^n = \left(\frac{a}{1+a^2}\right)^n; \tag{110}$$

then (109) makes

$$\frac{(\frac{1}{2}\sigma t)^n}{\underline{|n}} = \frac{(1-a^2)a^n}{(1+a^2)^{n+1}}. \tag{111}$$

Expand in rising powers of a by division, or otherwise. The result is the expansion (63), with the variable σt here instead of x there.

Again,

$$F_0 = \sin(\tfrac{1}{2}\sigma t) = (\tfrac{1}{2}\sigma t) - \frac{(\frac{1}{2}\sigma t)^3}{\lfloor 3} + \dots$$

$$= \frac{\sigma}{2p} - \left(\frac{\sigma}{2p}\right)^3 + \dots = \frac{\sigma/2p}{1+(\sigma/2p)^2} = \frac{\alpha(1+\alpha^2)}{\alpha^2+(1+\alpha^2)^2}. \quad (112)$$

Therefore, by (109),

$$\sin\tfrac{1}{2}\sigma t = \frac{\alpha(1-\alpha^2)}{1+\alpha^2(3+\alpha^2)}. \quad (113)$$

Expand by division. The result is equation (71) before got.

A modification of (100) is just worth mentioning. Though not wanted, it will serve as a good example of the treatment of operators. We have

$$\epsilon^{\rho t}V_0 = \epsilon^{\rho t}[V_0] = [V_0(p-\rho)]\epsilon^{\rho t} = [V_0(p-\rho)]\frac{p}{p-\rho}, \quad (114)$$

so we may put the last result for $[V_0\epsilon^{\rho t}]$ in (100). $[V_0(p-\rho)]$ means $[V_0]$ with $p-\rho$ put for p. And, using α instead, (100) becomes

$$V\epsilon^{\rho t} = [V_0(p-\rho)]\frac{1-\alpha^2}{1-\alpha(2\rho/\sigma-\alpha)}P_0(z). \quad (115)$$

One example, different from the former, will be enough. Say

$$V_0 = \sin st = \frac{sp}{p^2+s^2}. \quad (116)$$

Now turn p to $p-\rho$, and then put $\frac{1}{2}\sigma(\alpha+\alpha^{-1})$ for p. The result is

$$\frac{2s}{\sigma}\alpha\frac{1-(2\rho/\sigma)\alpha+\alpha^2}{\{1-(2\rho/\sigma)\alpha+\alpha^2\}^2+s^2(2\alpha/\sigma)^2}. \quad (117)$$

Using this in (115), we get

$$V\epsilon^{\rho t} = \frac{2s}{\sigma}\alpha\frac{1-\alpha^2}{\{1-(2\rho\alpha/\sigma-\alpha^2)\}^2+s^2(2\alpha/\sigma)^2}P_0(z); \quad (118)$$

or, expanding by division just the beginning part,

$$V\epsilon^{\rho t} = \frac{2s}{\sigma}\alpha\left\{1+\frac{4\rho}{\sigma}\alpha - \alpha^2\left(3-\frac{12\rho^2}{\sigma^2}+\frac{4s^2}{\sigma^2}\right)+\dots\right\}P_0(z). \quad (119)$$

This is correct as far as α^3. For, with $x=0$, we get the expansion of $\epsilon^{\rho t}\sin st$ in I functions, as can be verified by the formula (68). It will be seen that this is not so good a way as the other. That is, it is simpler to make $V_0\epsilon^{\rho t}$ instead of V_0 be the time function under treatment, as in the former method, equation (100).

Impulsive Impressed Voltages, and the Impulsive Waves and their Tails generated.

§ 388. Although the development of t^n in I functions fails to give immediately, by the proper generalisation, the complete wave of voltage generated when the impressed voltage V_0 varies as $\epsilon^{-\rho t}t^n$ and n is any negative integer, but requires a particular treatment which will be given; yet the formula may still be employed without change in these exceptional cases to obtain the waves generated by impulsive voltages, simple or multiple. Thus, let

$$\frac{V_0}{e_0}\epsilon^{\rho t} = p1 = \tfrac{1}{2}\sigma\frac{(\frac{1}{2}\sigma t)^{-1}}{\lfloor\underline{-1}}. \tag{120}$$

This means that V_0 is an impulsive function, the impulse being e_0, at the moment $t = 0$, because the variation of $\epsilon^{\rho t}$ does not count in the no time of an impulse. Expand in I functions by the formula (63). The result is

$$p1 = \tfrac{1}{2}\sigma(I_{-1} - I_1). \tag{121}$$

Now I_1 and I_{-1} are identical when t is finite, but they are not absolutely identical. The difference is most important here, being the impulsive function itself. Generalising (121) in the usual way, we get

$$\frac{V}{e_0}\epsilon^{\rho t} = \tfrac{1}{2}\sigma\left(\frac{w^{-1}}{\lfloor\underline{-1}}P_{-1} - \frac{w}{\lfloor\underline{1}}P_1\right), \tag{122}$$

the argument of the P functions being z as usual, instead of the σt of the I functions. By inspecting the formula for u^nP_n, we see that

$$\frac{w^{-1}}{\lfloor\underline{-1}} = \frac{u^{-1}}{\lfloor\underline{-1}}P_{-1} - \frac{u}{\lfloor\underline{1}}P_1. \tag{123}$$

This is the generalisation of (121). It expresses $0 = 0$ for all finite values of u and w. In fact, we have

$$\frac{u^n}{\lfloor\underline{n}}P_n = \frac{w^{-n}}{\lfloor\underline{-n}}P_{-n}, \tag{124}$$

for any integral value of n, positive or negative, provided u and w are finite. The difference is merely impulsive, and may be quite negligible in general. But not when impulses are actually in question, as now. Using (123) in (122,, we get

$$\frac{V}{e_0}\epsilon^{\rho t} = \tfrac{1}{2}\sigma\left(\frac{w^{-1}}{\lfloor\underline{-1}} + \frac{\sigma x}{v}P_1\right), \tag{125}$$

because $u - w = \sigma x/v$. This expresses the complete wave, including the head, which may, in fact, be the most important part. The first term, which is the same as $\frac{1}{2}\sigma\beta 1$, only exists at the place $w = 0$, that is, at $x = vt$, and its time integral is 1. The rest expresses the tail, continuous and finite from the origin to the wave front.

The meaning may be made plainer thus. Put on V_0 steady at the origin. The result is a finite wave of V, falling from V_0 at the origin to $V_0\epsilon^{-\rho t}$ at the wave front, and then dropping suddenly to zero. But if we let V_0 remain on only the very short time τ, and be followed by $V_0 = 0$, the effect is the same as keeping on the steady V_0, but followed by a second impressed voltage $-V_0$, starting at the moment $t = \tau$. The real V is then the difference of the two waves, and consists of a head, of depth $v\tau$, at the wave front, consisting of the uncancelled part of the first wave, and of a tail, resulting from the difference of the two waves. Shortening t indefinitely, and increasing V_0, we finally come to an impulsive impressed voltage. It generates an impulsive pure electromagnetic wave, in which $V = LvC = \infty$, but with finite time totals, viz., the same as the impulsive voltage at the origin, attenuated by the factor $\epsilon^{-\rho t}$ expressing the effect of absorption in transit. Behind this impulsive wave is the continuous tail. Both are represented in (125) and previous formula for V. The corresponding C formula is got by

$$(2/\sigma e_0)LvC\epsilon^{\rho t} = (2/\sigma e_0)\frac{1-a}{1+a}V\epsilon^{\rho t} = \frac{1-a}{1+a}\left(\frac{1}{a} - a\right)P_0,$$

$$= (a^{-1} - 2 + a)P_0 = \frac{w^{-1}}{\lfloor\underline{-1}}P_{-1} + \frac{w}{\lfloor\underline{1}}P_1 - 2P_0. \qquad (126)$$

Using (123), this makes

$$\frac{LvC}{e_0}\epsilon^{\rho t} = \frac{1}{2}\sigma\left\{\frac{w^{-1}}{\lfloor\underline{-1}} + \sigma t P_1 - 2P_0\right\}. \qquad (127)$$

expressing the impulsive C at the wave front, and the continuous tail of C following it.

Without using the preliminary expansion in I functions, the same results come out of the operational solution (105). We have

$$\frac{V_0}{e_0}\epsilon^{\rho t} = p1 = \frac{1}{2}\sigma(a + a^{-1}), \qquad (128)$$

which makes, by (105),

$$\frac{\mathrm{V}\epsilon^{pt}}{e_0} = \tfrac{1}{2}\sigma\frac{1-a^2}{a}\mathrm{P}_0 = \tfrac{1}{2}\sigma(\beta - a)\mathrm{P}_0, \qquad (129)$$

expressing the same as (125).

After this detailed notice of a simple impulse, the peculiarities connected with multiple impulses will be readily understood. They are to be done in the same way as continuous functions, only the terms which vanish in treating the latter must be carefully retained. Thus,

$$(\tfrac{1}{2}\sigma)^{-2}p^2 1 = \frac{(\tfrac{1}{2}\sigma t)^{-2}}{\lfloor -2} = \mathrm{I}_{-2} - \mathrm{I}_2, \qquad (130)$$

by (63). If this is $\mathrm{V}_0\epsilon^{pt}$ (constant omitted), then

$$\mathrm{V}\epsilon^{pt} = \frac{w^{-2}}{\lfloor -2}\mathrm{P}_{-2} - \frac{w^2}{\lfloor 2}\mathrm{P}_2 = \frac{w^{-2}}{\lfloor -2} + \frac{w^{-1}u}{\lfloor -1} + \frac{u^2 - w^2}{\lfloor 2}\mathrm{P}_2. \qquad (131)$$

This shows the multiple impulsive wave at the front, followed by the continuous tail.

Similarly,

$$(\tfrac{1}{2}\sigma)^{-3}p^3 1 = \frac{(\tfrac{1}{2}\sigma t)^{-3}}{\lfloor -3} = \mathrm{I}_{-3} + \mathrm{I}_{-1} - \mathrm{I}_1 - \mathrm{I}_3; \qquad (132)$$

which, if representing $\mathrm{V}_0\epsilon^{pt}$, produces

$$\mathrm{V}\epsilon^{pt} = \frac{w^{-3}}{\lfloor -3}\mathrm{P}_{-3} + \frac{w^{-1}}{\lfloor -1}\mathrm{P}_{-1} - \frac{w}{\lfloor 1}\mathrm{P}_1 - \frac{w^3}{\lfloor 3}\mathrm{P}_3$$

$$= \left(\frac{w^{-3}}{\lfloor -3} + \frac{uw^{-2}}{\lfloor -2} + \frac{u^2w^{-1}}{\lfloor 2\lfloor -1} + \frac{w^{-1}}{\lfloor -1}\right) + \frac{u^3 - w^3}{\lfloor 3}\mathrm{P}_3 + \frac{u - w}{\lfloor 1}\mathrm{P}_1, \qquad (133)$$

the multiply impulsive wave at the front being shown in the brackets.*

The next case is

$$\frac{(\tfrac{1}{2}\sigma t)^{-4}}{\lfloor -4} = (\mathrm{I}_{-4} - \mathrm{I}_4) + 2(\mathrm{I}_{-2} - \mathrm{I}_2). \qquad (134)$$

---

* Since the terms in the brackets are of different orders of infinities, we may apparently ignore all except the highest, and consider that $w^{-3}/\lfloor -3$ represents the impulsive wave multiplied by ϵ^{pt}. It is, in fact, the impulse at the origin shifted in position. The other terms, however, are needed in order that $\mathrm{V}\epsilon^{pt}$ and $\mathrm{L}v\mathrm{C}\epsilon^{pt}$ may satisfy their mutual connections. Construct the waves U and W, for example, and see that $a\mathrm{U} = \mathrm{W}$, and $\beta\mathrm{W} = \mathrm{U}$. Similarly with the other impulses in the text which have more than one term in the expression of the impulsive wave.

After that we require a condensed notation. Say (n) stands for $I_{-n} - I_n$. Then the expansions of $(\frac{1}{2}\sigma t)^{-n}/\lfloor -n$ are

$$n=5, \qquad 1(5)+3(3)+2(1)$$
$$n=6, \qquad 1(6)+4(4)+5(2)$$
$$n=7, \qquad 1(7)+5(5)+9(3)+5(1)$$
$$n=8, \qquad 1(8)+6(6)+14(4)+14(2)$$

&c. To develop these to $V\epsilon^{pt}$ waves, substitute $w^{-n}P_{-n}/\lfloor -n - w^n P_n/\lfloor n$ for (n), and retain the terms which vanish when w is finite, in order to represent the impulsive wave at the front.

Tendency of Distortion to vanish in rapid Fluctuations. Effect of increased Resistance in rounding off corners and distorting.

§ 389. The object of using impulses, involving infinite forces acting for infinitely small periods of time, is to be able to represent with comparative simplicity effects which, considered finitely, might be nearly the same in character, but vastly more complicated in expression. Considered finitely, the effect of a multiple impulse may be thus stated. Let us operate on the beginning of a cable by an impressed voltage V_0. Let us first send a single impulse. It generates an impulsive wave followed by a tail. In time, as the head decreases according to $\epsilon^{-\rho t}$, caused by absorption in transit, it is the tail that is the significant phenomenon, the head having practically vanished. But instead of a single impulse, send a multiple one. Let V_0 vary anyhow with extreme rapidity in the small interval τ, and then cease. The result is a nearly pure electromagnetic wave of depth $v\tau$, travelling at speed v, in which all the variations in V_0 are (nearly) faithfully copied, but attenuated in transit according to $\epsilon^{-\rho t}$. At any distance, therefore, if sufficiently sensitive means existed of registering rapid variations of voltage or current, we could faithfully receive the complicated "message" sent by V_0 in its variations, provided they were fast enough. As for the tail, that would depend upon the time total of V_0 principally, and might therefore be serious. But if V_0 consists of fluctuations about zero, the tail need not be of any importance compared with the head, the different parts of the head producing cancelling tails. Then we see that the message need not be a short multiple impulse, but may be continuous, and still travel nearly without distortion. The important point here is extreme rapidity of fluctuation. There is no distortion at the very front, and little near the front of the continuous wave due to V_0 steady, so there is little distortion if we cut it short. By sufficiently rapid fluctuations we use the heads and destroy the tails. But the attenuator $\epsilon^{-\rho t}$ is a serious factor in a long cable. Self-induction is salvation.

The tendency for all fluctuations, if sufficiently rapid, to be transmitted without distortion, naturally includes the simply periodic train of waves. The variation of the attenuation with the frequency tends to vanish, and likewise the variation of phase; so that, as the frequency increases, the distortion tends to vanish, provided we keep only to rapid fluctuations. Say that V_0 is $f(t)$, then

$$V = f(t - x/v)\epsilon^{-\rho x/v}, \qquad (135)$$

which is exactly true in a distortionless circuit, all along the wave from the origin to the wave front, may still be approximately true in a circuit in which σ is not zero, under the circumstances stated as regards the variations of V_0. And it is to be borne in mind that (135) is always true, in the theory propounded, at the wave front itself; *i.e.*, at the place momentarily fixed by $x = vt$, provided V_0 commences at the moment $t = 0$, the value of V is $f(0)\epsilon^{-\rho t}$.

But there is a distinction to be drawn between the ideal perfection of a theory and the reality of which it professes to be an approximate representation, by the usual process of ignoration. Theories generally fail in application when pushed to extremes, owing to the ignored circumstances assuming importance. It is always desirable to develop a theory exactly when it can be done profitably (this has no reference to professedly rigorous methods of working), in order to know what to expect under given circumstances, without confusion between an approximation to a definite theory, and the approximation of that theory to the reality, which is another matter. This has a distinct application here. When the constants of a circuit are truly constant, as imagined, the impression of V_0 steadily at the origin produces a wave in which V falls from V_0 at the origin to $V_0\epsilon^{-\rho x/v}$ at the wave front (at distance x), and then drops abruptly to zero. It is this abruptness that is impossible in reality. There are probably no real jumps like this in any physical phenomenon of wave character, perhaps not even in the ether itself. What we should expect is the rapid, but not ins antaneous assumption of the full value, or of most of it, as the wave front passes a place. This means the rounding off of the two sharp corners in the abrupt jump. First of all will come a slow rise, then a very rapid one, nearly to full value; and, lastly, a relatively slow assumption of the rest of the full value. The theory of constancy of ρ and σ substitutes square corners for rounded ones. To get the rounded curve we must enlarge the theory, and allow for the variation of the constants of the circuit. Now, K and S do not need to have any important change made in their values at different frequencies, because the conductivity involved in K is so small; but L has a sensible variation, and R may have a large one, because the conductivity involved in R is (relatively) so great. This is important precisely in this question of the state of things at the wave front. The resistance increases as we pass from the origin to the wave front in the case of the continued wave due to steady V_0, and may be largely increased at and near the wave front. The reduction in L is relatively a small matter, unless iron wires are in question. But they are no good. The penetration of the wires by the magnetic force not being instantaneous, and augmenting the resistance in the way described, will have the effect of removing the abruptness at the wave front. The imperfect penetration is a cause of such a substantial character that other causes having a similar effect need not be considered.

In the simply periodic case we can fully trace the effects of increased R and reduced L, taking count of the wire and external conducting boundary. As regards the state of things tended to at a very high frequency, it is represented by equation (135), but with a changed value of ρ, which increases with the frequency, being the same as $R/2L + K/2S$. Consequently, there is a cause of distortion left uncompensated, unless ρ should tend to ultimate constancy. Now R, when calculated by the magnetic theory, does not tend to constancy, since it varies as the square root of the frequency. But it is not likely to follow this law for ever. If we allow for the permittivity of the conductors it does not. It tends then to ultimate constancy, after first following the law just mentioned. The frequency needed, however, is excessive, trillions per second, or thereabouts, according to an old calculation of mine. This does not necessarily put the question entirely on the shelf, however, because in the production of a wave when V_0 is discontinuous, *i.e.*, containing a jump, harmonic analysis involves the consideration of infinite frequency; consequently, by the harmonic integrational method of calculation, we are bound to take count of the extreme tendencies, so far as they can be determined.

There is no end to elaborations when we introduce minor effects, involving other considerations than the four constants of a circuit, and their variation with the frequency. Keeping to the latter entirely, it may be inquired whether it is feasible to tackle the problem of finding the real shape of the V wave due to V_0 steady (a fundamental case) as modified by imperfect penetration. It is feasible, but very complicated. It is far more difficult than the simply periodic case. It is true that the expressions for the effective R and L at a given frequency are very complicated themselves in that case. But then the type of the formula for a train of waves remains the same when R and L vary with the frequency as when they are constants; and since there is no necessity to write out their full expressions in the wave formula, there is an effective reduction to simplicity. But the case is different in calculating a continued wave like V due to steady V_0. The harmonic integrational method is impracticable. In the operational method, on the other hand, we have $V = \epsilon^{-qx}V_0$, where q is $(YZ)^{\frac{1}{2}}$, and Y is $K + Sp$, whilst Z is $R + Lp$ in the theory of this chapter. There is no need to alter Y, but for Z we must use the full resistance operator of unit length of the wires, derived from magnetic considerations, to allow for imperfect penetration. It is a known function of p, and we may either treat the wires as mere conductors, or allow for their permittivity as well. Consequently $V = \epsilon^{-qx}V_0$ becomes a definite solution in operational form, and may be directly algebrised by similar ways to those already employed in the practical case of constancy of R and L. But the work is so complicated that I have not been able to bring it to a manageable form, so I say nothing more about it now, but return to the main question.

Wave Due to Impressed Voltage varying as $\epsilon^{-pt}t^{-1}$.

§ 390. Let us now tackle the postponed case of $V_0\epsilon^{pt}$ varying inversely as an integral power of the time. Say $V_0\epsilon^{pt} \propto t^{-1}$, to begin with. A difficulty arises at the very beginning. We know already that when $V_0\epsilon^{pt}$ varies as $p1$, or $t^{-1}/\underline{|-1}$, it is impulsive, and generates an impulsive head wave followed by a finite and continuous tail. Now t^{-1} is an infinite multiple of $t^{-1}/\underline{|-1}$. Therefore, if $V_0\epsilon^{pt}$ varies as t^{-1}, with a finite factor, the result must be $V=\infty$ from the origin up to the wave front, and zero beyond. Put in this way, the problem is not physically realisable.

Why is V infinite wherever it exists, however? The mere fact that V_0 is initially infinite does not account for it. At the wave front, of course, V must be infinite, but this is not necessarily of any importance, The practical value of an infinity may be zero. The real reason why V is infinite all along, although V_0 is only momentarily infinite, is that the momentary infiniteness is infinitely impulsive. That is, the time integral of V_0 from just before to just after $t=0$ is infinite. So there is generated an infinitely impulsive electromagnetic wave as head, and the mere dregs of it make a tail of infinite intensity.

We may notice here three kinds of infinity, of different significance. If V_0 is always finite and varies in any way, the V wave is always finite. But if V_0 contains momentary infinities, so does the V wave at corresponding points travelling at speed v. Now if $\int V_0 dt$ is zero for the momentary infinity, it has no importance. The infinity has nothing in it. But if $\int V_0 dt$ is finite, it means a finite impulse, generating a finite tail. Finally, if $\int V_0 dt$ is infinite, the impulsive wave is infinite, and so is the tail. This will continue for ever, if V_0 is kept on, though finite save initially. To restore finiteness, we may send an infinite negative impulse. It can never catch up the first impulse, of course, so, between the two impulses, V must be infinite, but it may be finite in the rear of the second impulse, if of the right sort.

There are also infinite impulses of higher orders, and multiple, but one is enough now. In the case of $V_0\epsilon^{\rho' t} \propto t^{-1}$, to have finite results we must cut off the first part of V_0. Say by having $V_0\epsilon^{\rho t} \propto (t+\tau)^{-1}$, where τ is small and positive. The wave is finite now. But a fresh formula will be wanted. There is another way which is neater, and which allows us to use the old formula. Cutting off the beginning is the same as superposing the beginning infinite part negatively without cutting off the beginning positively. In the limit this means sending an infinite negative impulse initially to cancel the infinite initial impulsiveness of t^{-1}. The result is finite now, except at the head.

If
$$V\epsilon^{\rho t} = \Sigma A_n \frac{w^n}{\underline{|n}} P_n(z), \tag{136}$$

and the series on the right be differentiated with respect to n, the result is another wave solution having a different generating V_0. Similarly, by further differentiations with respect to n, we generate other solutions. These derived solutions differ from the original, inasmuch as they are due to impressed voltages which cannot be expanded in the form $\Sigma A_n I_n(\sigma t)$, but can in series involving I_n and its differential coefficients with respect to n.

Thus, differentiate $t^n/\underline{|n}$ with respect to n. The result is

$$\frac{t^n}{\underline{|n}} \log t + t^n g'(n), \tag{137}$$

if $g(n)$ means the reciprocal of $\underline{|n}$, and $g'(n)$ is its derivative. When n is a negative integer $g(n)=0$, and $g'(n)$ is finite. Then (137) may be taken to represent a function which is finite, and varies inversely as a negatively integral power of t, together with an initial impulse. The function, for our purposes, only begins at $t=0$. Say $n=-1$, then we get

$$\frac{t^{-1}}{\underline{|-1}} \log t + t^{-1}. \tag{138}$$

The initial impulse is infinite, and negative, so it may destroy the infiniteness of the wave generated by t^{-1} alone. It does so, as we may see by developing the wave.

Use formula (63), or

$$\frac{(\frac{1}{2}\sigma t)^n}{\underline{|n}} = \mathrm{I}_n - (n+2)\,\mathrm{I}_{n+2} + \frac{n+1}{\underline{|2}}(n+4)\,\mathrm{I}_{n+4} - \dots \qquad (139)$$

Differentiate to n, and then put $n = -1$. We get

$$\mathrm{X} + (\tfrac{1}{2}\sigma t)^{-1} = \mathrm{I}'_{-1} - \mathrm{I}'_1 - \mathrm{I}_1 + \tfrac{3}{2}\mathrm{I}_3 - \tfrac{5}{6}\mathrm{I}_5 + \tfrac{7}{12}\mathrm{I}_7 - \dots, \qquad (140)$$

where X is the initial auxiliary impulse given by

$$\mathrm{X} = \frac{(\frac{1}{2}\sigma t)^{-1}}{\underline{|-1}} \log(\tfrac{1}{2}\sigma t), \qquad (141)$$

and the accent means differentiation to n. Thus, I'_{-1} means $d\mathrm{I}_n/dn$, with $n = -1$ after the differentiation.

Consequently, if the left side of (140) represents $\mathrm{V}_0\epsilon^{\rho t}$, the wave it generates is

$$\mathrm{V}\epsilon^{\rho t} = \frac{d}{dn}\left(\frac{w^n}{\underline{|n}}\mathrm{P}_n - \frac{w^{n+2}}{\underline{|n+2}}\mathrm{P}_{n+2}\right)_{(n=-1)}$$

$$- w\mathrm{P}_1 + \frac{3}{2}\frac{w^3}{\underline{|3}}\mathrm{P}_3 - \frac{5}{6}\frac{w^5}{\underline{|5}}\mathrm{P}_5 + \frac{7}{12}\frac{w^7}{\underline{|7}}\mathrm{P}_7 - \dots, \qquad (142)$$

by generalising the I functions in (140) in the usual way.

Like (139), the derived equation (140) must express an absolute identity. The terms on the left side must be repeated on the right side, and the rest on the right side must come to 0, like life. To verify this, differentiate I_n to n. Thus,

$$\mathrm{I}'_n = \mathrm{I}_n \log \tfrac{1}{2}\sigma t + (\tfrac{1}{2}\sigma t)^n g'(n) + (\tfrac{1}{2}\sigma t)^{n+2}\frac{g'(n+1)}{\underline{|1}} + \dots, \qquad (143)$$

from which

$$\mathrm{I}'_{-1} = \mathrm{I}_{-1}\log\tfrac{1}{2}\sigma t + (\tfrac{1}{2}\sigma t)^{-1} + (\tfrac{1}{2}\sigma t)\frac{g'(0)}{\underline{|1}} + (\tfrac{1}{2}\sigma t)^3\frac{g'(1)}{\underline{|2}} + \dots \qquad (144)$$

$$\mathrm{I}'_1 = \mathrm{I}_1\log\tfrac{1}{2}\sigma t + \tfrac{1}{2}\sigma t g'(1) + (\tfrac{1}{2}\sigma t)^3\frac{g'(2)}{\underline{|1}} + (\tfrac{1}{2}\sigma t)^5\frac{g'(3)}{\underline{|2}} + \dots. \qquad (145)$$

Taking the difference, we obtain

$$\mathrm{I}'_{-1} - \mathrm{I}'_1 = \mathrm{X} + (\tfrac{1}{2}\sigma t)^{-1} + (\tfrac{1}{2}\sigma t) + \tfrac{1}{2}\frac{(\frac{1}{2}\sigma t)^3}{\underline{|1}\,\underline{|2}} + \tfrac{1}{3}\frac{(\frac{1}{2}\sigma t)^5}{\underline{|2}\,\underline{|3}} + \dots. \qquad (146)$$

Comparing with (140) we see that

$$\tfrac{1}{2}\sigma t + \tfrac{1}{2}\frac{(\frac{1}{2}\sigma t)^3}{\underline{|1}\,\underline{|2}} + \tfrac{1}{3}\frac{(\frac{1}{2}\sigma t)^5}{\underline{|2}\,\underline{|3}} + \ldots = \mathrm{I}_1 - \tfrac{3}{2}\mathrm{I}_3 + \tfrac{5}{6}\mathrm{I}_5 - \ldots . \qquad (147)$$

Here we have an identity of the original kind, a power series expanded in I functions. Its truth may be now verified by the same formula (63) from which we started, or, more easily, by (68).

But (147) by itself is of no use. The terms omitted from (140) are vital in the electromagnetic wave problem we started with. Equation (147) may, of course, be generalised to a wave, if the left side be taken to represent $\mathrm{V}_0\epsilon^{\rho t}$; but that is another question. It is (142) that is the generalisation of (140), and requires development. The differentiated part is

$$\left(\frac{w^{-1}}{\underline{|-1}}\mathrm{P}_{-1} - \frac{w}{\underline{|1}}\mathrm{P}_1\right)\log w + w^{-1}g'(-1) + ug'(0) + \frac{u^2 w}{\underline{|2}}g'(1)$$

$$+ \frac{w^3 u^2}{\underline{|3}}g'(2) + \ldots - wg'(1) - uw^2 g'(2) - \frac{u^2 w^3}{\underline{|2}}g'(3) - \ldots, \qquad (148)$$

so the full wave is

$$\frac{\mathrm{V}\epsilon^{\rho t}}{e} = \frac{w^{-1}}{\underline{|-1}}\log w + \frac{1}{w} + \frac{\sigma w}{v}\mathrm{P}_1 \log w$$

$$+ ug'(0) + \frac{u^2 w}{\underline{|2}}g'(1) + \frac{u^3 w^2}{\underline{|3}}g'(2) + \ldots$$

$$- wg'(1) - uw^2 g'(2) - \frac{u^2 w^3}{\underline{|2}}g'(3) - \ldots$$

$$- w\mathrm{P}_1 + \frac{3}{2}\frac{w^3}{\underline{|3}}\mathrm{P}_3 - \frac{5}{6}\frac{w^5}{\underline{|5}}\mathrm{P}_5 + \frac{7}{12}\frac{w^7}{\underline{|7}}\mathrm{P}_7 - \ldots . \qquad (149)$$

This is complete, and is finite everywhere save at the wave front. As regards the values of $g'(n)$, they are

$$g'(0) = \gamma = \cdot 5772, \qquad g'(1) = \gamma - 1, \qquad g'(2) = \frac{\gamma - 1 - \frac{1}{2}}{\underline{|2}},$$

$$g'(n) = \frac{\gamma - 1 - \frac{1}{2} - \frac{1}{3} - \ldots - n^{-1}}{\underline{|n}}, \qquad (150)$$

when n is a positive integer. Also, on the negative side, it is easier still, thus,

$$g'(-1) = 1, \quad g'(-2) = -\lfloor 1, \quad g'(-3) = +2, \quad g'(-4) = -\lfloor 3,$$
$$g'(-n) = (-1)^{n-1}\lfloor n-1. \tag{151}$$

Not being able to do everything at once, I must defer the proof of these side results. They may be found in treatises, truly, but the treatment of the subject in high mathematical works is not of a nature to encourage physical students. What is necessary for physical purposes can be more simply given. Taking the constants for granted at present, we may collect terms involving γ, and produce

$$\frac{V\epsilon^{\rho t}}{e} = \frac{w^{-1}}{\lfloor -1} \log w + \frac{1}{w} + \frac{\sigma x}{v} P_1(z)(\log w + \gamma)$$
$$- wP_1 + \frac{3}{2}\frac{w^3}{\lfloor 3}P_3 - \frac{5}{6}\frac{w^5}{\lfloor 5}P_5 + \dots$$
$$- \frac{u^2 w}{\lfloor 2\lfloor 1} - \frac{w^3 w^2}{\lfloor 3\lfloor 2}(1+\tfrac{1}{2}) - \frac{u^4 w^3}{\lfloor 4\lfloor 3}(1+\tfrac{1}{2}+\tfrac{1}{3}) - \dots$$
$$+ w + \frac{uw^2}{\lfloor 1\lfloor 2}(1+\tfrac{1}{2}) + \frac{u^2 w^3}{\lfloor 2\lfloor 3}(1+\tfrac{1}{2}+\tfrac{1}{3}) + \dots. \tag{152}$$

I have given a good deal of detail in this case, in order to show explicitly how the cases $n = -2, -3$, &c., may be similarly fully developed. Particular attention must be paid to the impulsive terms if real electromagnetic waves are wanted, otherwise the mathematics is useless. Nothing would be more natural at first than to omit the impulsive term from both sides of the equation, and so have apparently a finite wave (except at the front) arising from $V_0\epsilon^{\rho t}$ varying as t^{-1}. All conditions would seem to be satisfied, including an apparently identical expansion of t^{-1} in I and I′ functions; but the initial momentary failure of this expansion would be fatal to the electromagnetic vitality.

Wave Due to V_0 varying as $\epsilon^{-\rho t}\log t$.

§391. Another case of "transcendental" character occurs when the impressed voltage varies as $\epsilon^{-\rho t}\log t$. The logarithm of t cannot be expanded in a convergent power series, and therefore also not in a series of I functions. But $\log t$ is the value of the differential coefficient of t^n with respect to n when $n=0$. So if $V_0\epsilon^{\rho t} = e \log \frac{1}{2}\sigma t$, the wave generated is

$$V\epsilon^{\rho t} = \frac{d}{dn}\left\{\lfloor n\left(\frac{w^n}{\lfloor n}P_n - \frac{n+2}{\lfloor 1}\frac{w_{n+2}}{\lfloor n+2}P_{n+2} + \dots\right)\right\}_{(n=0)} \qquad (153)$$

where n is to be put $=0$ after the differentiation. Now develop, using the values of $g'(n)$ already given, and we shall come to a numerically calculable formula.

Or thus. By (137),

$$\frac{d}{dn}\frac{t^n}{\lfloor n} = \log t + \gamma, \qquad \text{when } n = 0, \qquad (154)$$

from which we see that $\log t + \gamma$ is expansible in I and I′ functions. But γ is a constant, and expansible in I functions itself. So we get an expansion of $\log t$ in I and I′ functions, which can be generalised to make an electromagnetic wave.

The fundamental expansion is

$$\begin{aligned}\log \tfrac{1}{2}\sigma t = & -\gamma(I_0 - 2I_2 + 2I_4 - 2I_6 + \dots) \\ & + (I'_0 - 2I'_2 + 2I'_4 - 2I'_6 + \dots) \\ & - I_2 + \frac{5}{\lfloor 2}I_4 - \frac{20}{\lfloor 3}I_6 + \frac{94}{\lfloor 4}I_8 - \dots, \end{aligned} \qquad (155)$$

where the coefficient of $-\gamma$ is equivalent to 1. If then, we have $V_0\epsilon^{\rho t} = e \log \frac{1}{2}\sigma t$, where e is constant, the wave generated is

$$\begin{aligned}\frac{V\epsilon^{\rho t}}{e} = & -\gamma\left\{P_0 - 2\frac{w^2}{\lfloor 2}P_2 + 2\frac{w^4}{\lfloor 4}P_4 - 2\frac{w^6}{\lfloor 6}P_6 - \dots\right\} \\ & - \left\{\frac{w^2}{\lfloor 2}P_2 - \frac{5}{\lfloor 2}\frac{w^4}{\lfloor 4}P_4 + \frac{20}{\lfloor 3}\frac{w^6}{\lfloor 6}P_6 - \frac{94}{\lfloor 4}\frac{w^8}{\lfloor 8}P_8 + \dots\right\} \\ & + P'_0 - 2\left(\frac{w^2}{\lfloor 2}P_2\right)' + 2\left(\frac{w^4}{\lfloor 4}P_4\right)' - 2\left(\frac{w^6}{\lfloor 6}P_6\right)' + \dots, \end{aligned} \qquad (156)$$

where in the last line, for example, $(w^6P_6)'$ means $(w^nP_n)'$ with $n=6$. To develop the last line, we have

$$\begin{aligned}\left(\frac{w^n}{\lfloor n}P_n\right)' = & \frac{w^n}{\lfloor n}P_n \log w \\ & + w^n\left\{\frac{g'(n)}{\lfloor 0} + \frac{uw}{\lfloor 1}g'(n+1) + \frac{u^2w^2}{\lfloor 2}g'(n+2) + \dots\right\}. \end{aligned} \qquad (157)$$

Consequently (156) becomes

$$\frac{V\epsilon^{pt}}{e} = (\log w - \gamma)\left\{P_0 - 2\frac{w^2}{\lfloor 2}P_2 + 2\frac{w^4}{\lfloor 4}P_4 - \dots\right\}$$

$$- \left\{\frac{w^2}{\lfloor 2}P_2 - \frac{5\,w^4}{\lfloor 2\,\lfloor 4}P_4 + \frac{20}{\lfloor 3}\frac{w^6}{\lfloor 6}P_6 - \dots\right\}$$

$$+ [g'(0) + uwg(1)g'(1) + u^2w^2g(2)g'(2) + \dots]$$

$$- 2w^2[g'(2) + uwg(1)g'(3) + u^2w^2g(2)g'(4) + \dots]$$

$$+ 2w^4[g'(4) + uwg(1)g'(5) + u^2w^2g(2)g'(6) + \dots] - \text{\&c.}, \quad (158)$$

in which everything is known.

It is to be observed that there is nothing impulsive about the logarithm of t. Its initial momentary time integral is zero, so no impulsive terms occur in the wave, although the value of V at the wave front is infinite.

Also observe that by differentiating the V wave with respect to t we can obtain the wave due to the impressed voltage similarly differentiated. At first sight this looks as though a finite wave arose when $V_0\epsilon^{pt}$ varied as t^{-1}, since $p \log t = t^{-1}$ by the differential calculus. But that is not true here. The function $\log \frac{1}{2}\sigma t$ is zero before and existent after $t = 0$. So

$$p(\log \tfrac{1}{2}\sigma t \,.\, 1) = \log \tfrac{1}{2}\sigma t \,.\, p1 + t^{-1}. \qquad (159)$$

This is the true differential coefficient, since it includes the initial impulse. Therefore we shall by time differentiation be led to the wave generated by $V_0\epsilon^{pt}$ varying as t^{-1} *and* the initial impulse.

Effect of a Terminal Resistance as expected in 1887 and as found in 1896.

§ 392. Now for a little change, to break the monotony inseparably connected with regular developments. It is grievous that they should be so dry, but it is necessary for some one to do the work; though, of course, there are cynics who may say they do not see the necessity; and, in fact, it is easy to become cynical oneself after, say, an attack of influenza, which is a demoralising disease, itself unnecessary without question.

We need not depart far from the preceding environment to obtain the change of air and scene desirable for reinvigoration.

Leaving the subject of the development and progress of waves, let us consider in some degree what happens to them when they arrive at the end of the guiding circuit. The proper treatment will differ materially according as the waves are long or short. If they are really long without question, as in telegraphic and telephonic applications, and further still, we may sum up the action of terminal "apparatus" in the form of a terminal condition, say, $V = ZC$, about which a good deal was said in the last chapter on diffusion. The treatment is quite similar when self-induction and leakage are included, with extended meanings of the symbols and operations. But we need not go into that now, except to remark that long waves are essential, because such an equation as $V = ZC$, when constructed in the usual way to represent the action of a set of condensers, resistances and inductive coils, only does so on the hypothesis of instantaneous action and reaction between the different parts of the apparatus.

But it may well happen that this procedure is insufficient. It must certainly become insufficient when the waves are shortened sufficiently. As an extreme case we may imagine the wave length to be only a fractional part of the size of the apparatus, when clearly there is no opportunity for anything like an equilibrium theory (in a certain sense) to be established by mutual actions, which are really the resultant of waves transmitted to and fro at finite speed between all parts of the apparatus. A resultant terminal condition of the form $V = ZC$ is still obtainable; but the form of Z will be quite different, and much more complicated.

Take the very simplest case of all for initial illustration. I showed in 1887 that electromagnetic waves sent along a circuit were under certain circumstances completely absorbed by a terminal resistance. If the circuit is distortionless, and the amount of the terminal resistance is Lv, where L is the inductance per unit length of the circuit and v the speed of light, then the absorption is complete, or there is no terminal reflection. The reason is very simple. The relation between V and C in the circuit in a wave of any sort going in the positive direction is $V = LvC$. The relation at the terminal is $V = RC$, if R is the terminal resistance. So, if R and Lv are equal, the resistance behaves to the circuit itself in the same way as a continuation of the circuit—that is, there is no back effect at all. The same thing may be approximately true when the circuit is not distortionless, since distortion takes time, and is cumulative. With very high frequency, for instance, the reflection may be nearly destroyed.

But although in the truly distortionless circuit itself the consideration of the frequency is unnecessary, this is not altogether true as regards the terminal resistance, for the reason given above. The waves must evidently be rather long waves as regards the terminal resistance. If they are shortened sufficiently, it is clear that the kind of terminal resistance requires consideration. There will be a different theory for every different arrangement of resistance, even though in the long-wave theory they would be all alike—viz., $V = RC$, where R is a constant. In another form, R requires to be generalised to a Z of complicated structure in order to represent the course of events. Similar remarks must apply to the numerous experiments with waves along wires after Hertz and Lodge, when terminal effects are in question. It is difficult to make more than a rough guess as to how short waves may be allowed to be before an assumed mere terminal resistance needs to be studied in detail as regards its reaction upon arriving disturbances.

It is very interesting, however, to observe that the experiments of Dr. E. H. Barton and Mr. G. B. Bryan (*Phil. Mag.*, January, 1897) with waves $8\frac{1}{2}$ metres long, on a circuit of a pair of parallel copper wires 1·5 mm. in diameter and 8 cm. apart (length of line 116 m.), showed that a fair approximation to complete extinction of the reflected waves could be obtained. A small coil wound non-inductively was unsatisfactory by failure of insulation, and its use was abandoned. It is questionable whether it would behave like a mere resistance apart from the question of insulation. But when the terminal resistance was constructed by pencil markings on glass, the results were satisfactory. A terminal resistance of 261 ohms produced large reflection of one kind. Another of 1,336 to 1,355 ohms produced still larger reflection of a different kind. But a resistance of intermediate value, of 549 to 560 ohms, about the value of Lv, produced a quite small reflection in com parison with either. At the same time there was far from perfect extinction of the reflected waves, though quite as much, if not more, than I could have ventured to expect with certainty. The waves of $8\frac{1}{2}$ metres are certainly fairly long as regards the transverse dimensions of the line itself; and the experiments show that they are also fairly long, though probably not to the same degree, as regards the particular sort of terminal resistance concerned, viz., pencil markings, forming an extremely thin sheet. The area covered is not stated, but it could not have been large.

Having a sort of kindly paternal interest in Dr. Barton's experiments, or at any rate in the results expected and obtained, I have sought for and found the exact solution of a case of waves along a straight wire circuit terminated by a resistance. It casts some light on the subject, and is fortunately of a kind admitting of easy description. In the first place it should be remembered that in order that the V and C theory of waves along wires may be an exact theory in plane waves, we have to regard the wires as mere guides, and distribute their resistance uniformly in the ether outside them, not as electrical resistivity, however, but as magnetic conductivity, as I have explained in Chapter IV., Vol. I. Having got truly plane waves in this manner, if there are any intermediate or terminal influences, they too must be transferred to planes at the proper places, so as to act evenly on the plane waves. For instance, an intermediate conducting bridge across the practical circuit must be transformed to an infinitely extended and infinitely thin plane sheet of uniform conductance, its effective conductance from wire to wire being made equal to that of the practical bridge. Similarly a terminal wire bridge must be replaced by a transverse plane sheet of the same effective conductance.

Now, in the case of an intermediate bridge, the reflection coefficient, or ratio of the reflected V_2 to an incident disturbance V_1, is

$$\frac{V_2}{V_1} = -\frac{Lv}{2R + Lv} = -\frac{\mu v}{2r + \mu v}. \tag{1}$$

See "Electrical Papers," Vol. II., p. 142, for the first form, in which R is written here for the bridge resistance. The second form occurs in the true plane wave theory, L becoming μ, and R becoming , which is the resistance of unit area of the bridge sheet. In the second form we consider a tube of energy flux of unit cross-section. We cannot make V_2 vanish except by $R = \infty$ or $r = \infty$. This is equivalent to no bridge. The wave goes right through.

Now, it is vital to the success of formula (1) that the bridge should be really intermediate—*i.e.*, the main circuit must be continued, however short a distance, on the other side of the bridge. The formula does not fully apply at a real terminal, when there is mere ether on the other side of the plane bridge resistance, because the transmitted wave cannot go on entirely as a plane wave when it has lost its linear guide. We must consider the change of type that occurs. It does not look probable that a cancellation of the reflected wave is even possible. I shall, however, show that, with a certain value of r, there is a complete annihilation of the reflected wave along the wire itself. The reflected wave runs along the plane instead, and is virtually lost from the wire.

Reflection at the Free Ends of a Wire. A Series of Spherical Waves.

§ 393. In the first volume of this work, § 53 to 61, I have described several cases in which the simplest kind of spherical electromagnetic wave, published by me in 1888 ("Elec. Papers," Vol. II., p. 403), occurs in an instructive manner. For instance, the sudden stoppage of a charge moving in a straight line at the speed of light, accompanied by a plane electromagnetic wave, generates an expanding spherical wave joined on to the plane wave. It is with this sort of phenomena that we are concerned when reflection occurs at the free end of a wire. In fact, it is easy to see that there must be something of the kind, because the electrification is transferred along a wire at the speed of light, and is then suddenly stopped and sent back when it comes to a free end. As regards the wire itself, it may be regarded as a cone of infinitely small angle, and is therefore included in the investigations just referred to. An easy modification enables us to include the effect produced by a transverse terminal resisting plane plate.

Let there be a fine straight wire of no resistance. To generate a spherical wave upon it, it is merely necessary to produce a voltaic (or electromotive) impulse in the wire at any point, say A. If the impulse acts from left to right, then the state of things at time t later is a positive charge at B and an equal negative charge at C, the poles of a spherical surface of radius $vt = AB = AC$. The displacement is joined from B to C in the simplest way possible in the sheet. By symmetry, no other way is possible than by following the lines of longitude evenly. The arrow-heads show this symbolically. In the sheet of displacement put magnetic induction following the lines of latitude, and the complete wave is represented; $E = \mu v H$, or $D = cvB$, is the relation between the intensities of **E** and **H** or the densities of **D** and **B**.

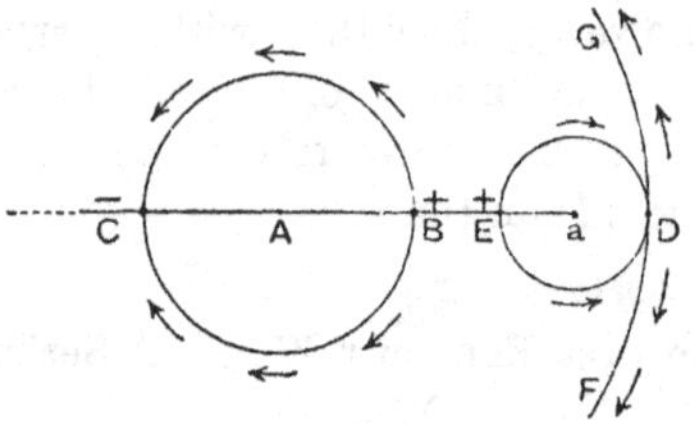

Now, let a be the free end of the wire. Then, a little while after the wave reaches the free end, the state of things is represented by the *two* spherical waves FDG and DE. The first is nothing more than the expanded original wave (a portion only, in the diagram), with its core removed. The electrification, on arriving at the end a, at once reversed its motion. It has gone back to E. The displacement in the original spherical wave is joined on to the charge at E by the secondary wave. That is, the displacement goes along the small spherical surface from E to D, and then diverges into the big one. The magnetic induction must be put in as before, taking care to have its direction right. It is up through the paper at the top in the big wave, but down in the small one. The flux of energy **VEH** settles the direction of **H**.

If, instead of an impulsive wave, we generate one of finite but not great depth, the superposition in the neighbourhood of D of the original and secondary waves has the result of practically cancelling the sharp corner at D. The displacement does not go as far as D, save insensibly, but bends round in curves from one sphere into the other at a little distance from D. The deeper the original wave, the further off does the bending occur. Too many diagrams would be needed to show the various cases, but there is no difficulty in the general idea.

Keeping to the unaltered diagram, the further course of events is got by expanding the spheres centred at A and *a*. If the wire is infinitely long on the left side, nothing more happens. But if it is of finite length, then a third spherical wave is generated directly the negative charge (at C in the first case illustrated) reaches it. A fourth spherical wave will be generated when the positive charge (at E in the second case illustrated) reaches the left end of the wire. Two more waves are generated by the reflection of the last two at the right end of the wire; and so on. It will be seen that any schoolboy, with a pair of compasses, can follow up the subsequent history to any extent. (In fact, there seems no reason why instruction in electromagnetic waves should not become an elementary subject in the "Board" schools, as they are absurdly called.)

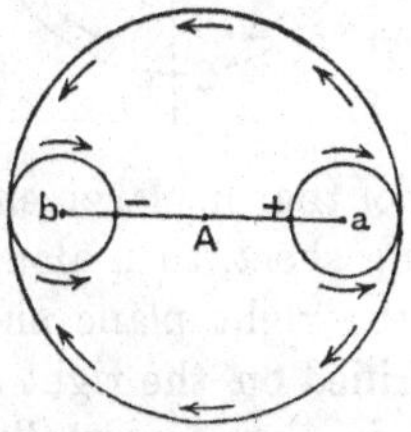

This diagram, for instance, shows the state of things due to an impulsive voltage at A, in the middle of the wire *ba*, a little while after the first wave (the big one) reached the ends. The arrow-heads show the course of the displacement at the moment in question. It is unnecessary to multiply diagrams of this sort.

Reflection at the End of a Circuit terminated by a Plane Resisting Sheet.

§ 394. So far there has been nothing that is not virtually included in the previous work referred to above. It is, however, necessary for preliminary purposes to make the following relating to a terminal resistance easier to understand. Put a uniformly resisting plane sheet at the end of the wire, and let it be struck flush by an incident plane sheet wave. If the source was, as before, at a point, that point should be a long way from the end of the wire.

In the diagram, *ba* is the wire terminating at *a*, at right angles to which is the resisting plane sheet. Two parallel planes are also represented. The one on the right is the transmitted plane wave sheet (or part of very large spherical wave); that on the left is the reflected plane wave sheet, at the moment of time when the charge has gone back on the wire to the point marked +. These two sheets of displacements are joined together by a spherical sheet, and the arrow-

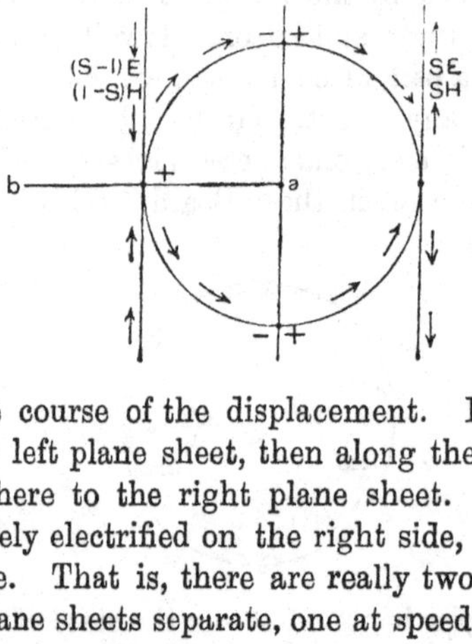

heads show the course of the displacement. It is inward tc the wire in the left plane sheet, then along the lines of longitude of the sphere to the right plane sheet. The resisting plane is positively electrified on the right side, and negatively on the left side. That is, there are really two hemispherical waves. The plane sheets separate, one at speed v to the right, the other similarly to the left, whilst the connecting spherical sheet expands so as to keep up with them. It only remains to specify the intensities in different parts.

If s is the transmission coefficient at the resisting plane, that is, the ratio V_3/V_1 or E_3/E_1 of the electric force E_3 transmitted to that incident, then we have

$$\frac{E_3}{E_1} = s = \frac{2r}{2r + \mu v}, \tag{2}$$

if r is, as before, the resistance of unit area of the plane.

The size of s is anything between 0 and 1, being 0 when the plane is a perfect conductor, and 1 when it is of zero conductance. If, then, E and H are the intensities of electric and magnetic force in an incident very thin electromagnetic sheet, sE and sH are the values in the transmitted sheet on the right, and $(s-1)$ E and $(1-s)$ H are the values in the reflected sheet. Observe that **H** is reflected positively, and there is persistence of the total induction. On the other hand, **E** is reflected negatively, and there is loss of displacement. If q is the charge in the initial incident wave, the charge reflected is the sum of the displacements leaving the wire for the spherical and plane waves at the point +, that is,

$$sq + (s-1)q = (2s-1)q = \frac{2r - \mu v}{2r + \mu v} q. \qquad (3)$$

The displacement in the right plane wave is continuous with that in the spherical wave. So sq is the total charge on the right side of the resisting plane (in the complete ring), and $-sq$ is the total on the left side.

If the initial wave is not impulsive, but is of finite depth, then the displacement in the plane wave on the left will turn round into the spherical wave without going right up to the wire, save insensibly. Similarly on the right side. But there is not usually continuity on the left side, on account of the charge on the wire. It may be positive or negative. When $2r = \mu v$, there is complete annihilation of the reflected charge. Then $s = \frac{1}{2}$, the two plane waves are alike (except as regards the directions of the displacement and motion), and there is perfect continuity at the point +. Practically, then, with a sheet of finite depth, the displacement runs away from the wire altogether on the left side, as well as on the right, for the sphere expands, and its practical junction with the left plane wave moves away from the wire. That is, the complicated wave is lost altogether, so far as the wire is concerned.

Instead of $r = \mu v$ (or $\mathrm{R} = \mathrm{L}v$), which is the condition of complete absorption by a terminal resistance in the long wave theory applied to a condensed resistance, we find that the condition of no reflection along the wire is $r = \frac{1}{2}\mu v$, when the terminal resistance is spread over the complete plane wave. In the former case all the energy is wasted in the terminal resistance. In the present case, only one half is thus wasted, and this occurs in the act of transmission through the plate. The other half is wasted in another way. It remains (in our somewhat ideal state of things) in the spherical and two plane waves.

Practically, then, from the above considerations, we should expect the value of the terminal resistance which annulled the reflection most completely to lie between Lv and $\frac{1}{2}Lv$, according to its arrangement, these being the extreme values.

But with a pair of wires terminating upon the transverse resisting terminal plane, the process of development of the reflected wave is more complicated. Just at the beginning, it is true, the process is similar, but duplicated, as shown in the next diagram. A spherical wave of the above kind is generated at each wire terminal. The course of the displacement is fully shown by the arrow-heads, assuming that there is a

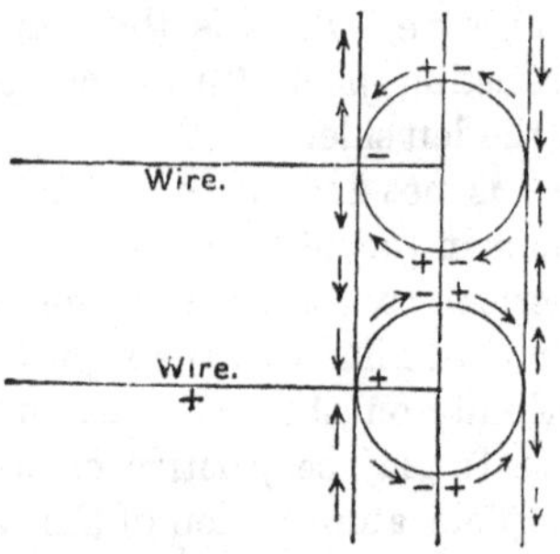

positive reflected charge on the lower and a negative on the upper wire. But this state of things only lasts until the spherical waves reach the opposite wires. A fresh kind of disturbance then begins by reflection from the wires themselves. It is, of course, not so important as the spherical and plane waves. These disturbances, originating on the wires, lead to an infinite series of minor disturbances, because any disturbance from one wire comes into collision with the other. Whether the resultant of this complicated process of terminal reflection can be a fully effective and complete annihilation of the reflection along the wires at a distance from the ends, I am not prepared to say. There is nothing peculiar to electromagnetics in complications of this kind. It is the same in all mechanics when we go into detail.

Long Wave Formulæ for Terminal Reflection.

§ 395. The only way to make a plane or spherical wave run clean off the end of a wire without alteration appears to be to carry forward the charge. For instance, when a thin plane sheet of displacement reaches the end, let the particle it is then centred upon be dislodged, and carried forward at the speed of light. Then it will keep up with the plane wave, and no auxiliary spherical wave will be generated. Similarly, the plane wave with carried core can be slipped on to another wire without disturbance, by simply letting its core impinge on the free end. The wire will then serve as guide.

In the Hertz-Lodge experiments the reflection at the transmitting end is a much more difficult matter, and can scarcely be attacked at all in detail. But if the waves can be really treated as long waves, then the regular procedure for long waves may be applied, at least to some extent. Say, the terminal arrangement is a condenser and induction coil.

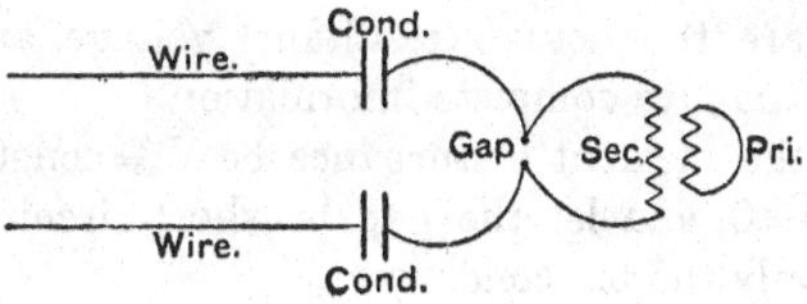

Then the resistance operator of the arrangement is

$$Z = \frac{V}{C} = \frac{1}{S_0 p} + (R_2 + L_2 p) - \frac{M^2 p^2}{R_1 + L_1 p}. \tag{4}$$

If the gap is short-circuited, we have only the condenser term, or $Z = (S_0 p)^{-1}$, where S_0 is the effective permittance of the two condensers. This first term in Z should be generalised to $(S_0 p)^{-1} + r + lp$, to allow for the resistance and inductance of the part between the condensers and the gap. If the gap is open and non-conductive, but the primary is not closed, then use the second part as well, $R_2 + L_2 p$ being the resistance operator of the secondary. If, in addition, the primary is closed, add on the third part, involving M, the mutual inductance, and $R_1 + L_1 p$ the resistance operator of the primary coil by itself. If the gap is conductive to a definite amount, then it acts as a shunt to the induction coil. Say its resistance is r, then $(r^{-1}+s^{-1})^{-1}$ is the resistance operator to be added to that of the condenser, s being what is added in equation (4). This r may be generalised to $(r^{-1}+s_0p)^{-1}$, if the gap be regarded as a condenser of permittance s_0 and resistance r.

All cases of this kind can be readily developed to show the reflected wave train in the case of simply periodic waves. The reflection coefficient for V is

$$\rho=\frac{R-Lv}{R+Lv}, \quad \text{or} \quad \rho=\frac{Z-Lv}{Z+Lv}=\frac{V_2}{V_1}, \tag{5}$$

the first being with a terminal resistance R, the second with R turned to Z, the resistance operator of the terminal arrangement, whatever it may be; for instance, the Z in equation (4). V_1 is an incident, and V_2 the corresponding reflected disturbance (transverse voltage). We have

$$V_1+V_2=V, \quad \text{and} \quad C_1+C_2=C=\frac{V_1-V_2}{Lv}, \tag{6}$$

if V and C are the actual (resultant) voltage and current. These equations give complete information.

Thus, let the incident disturbance be $V_1=$ constant, beginning when $t=0$, and let the gap be short-circuited. Then, considering only the big condensers,

$$\frac{V_2}{V_1}=\frac{(S_0p)^{-1}-Lv}{(S_0p)^{-1}+Lv}=1-2\epsilon^{-t/S_0Lv}, \tag{7}$$

and
$$V=\frac{2V_1}{1+S_0Lvp}=2V_1(1-\epsilon^{-t/LvS_0}), \tag{8}$$

i.e., V rises from 0 to $2V_1$, not instantly, but very quickly.

In the simply periodic case, the reflected disturbance in terms of the incident is got by putting $p=ni$, where $n=2\pi\times$ frequency; thus

$$V_2=\frac{1-S_0pLv}{1+S_0pLv}V_1=\frac{(1-S_0pLv)^2}{1+S_0{}^2n^2L^2v^2}V_1$$

$$=\frac{(1-S_0{}^2L^2v^2n^2)-2S_0Lvp}{1+S_0{}^2L^2v^2n^2}V \tag{9}$$

showing the relation of V_2 to V_1 explicitly.

As regards the resultant V at the terminal, that is much simpler. Thus

$$V = \frac{2V_1}{1 + S_0 Lvp} = \frac{2(1 - S_0 Lvp)V_1}{1 + (S_0 Lvn)^2}, \tag{10}$$

explicitly. In general

$$V = \frac{2V_1}{1 + LvZ^{-1}}. \tag{11}$$

The other cases can be done in the same way; but I fear that the inclusion of the induction coil and shunt at the gap will lead to results of very questionable validity when the short waves usual in this kind of experiment are employed, *i.e.*, only a few metres long.

Reflection of Long Waves in General.

§ 396. Passing now to long-wave reflection, without limitation to approximately distortionless transmission on the line itself, it is to be remarked that there are a few simple cases where we can develop the complete solutions out of the initial wave generated, without further investigation. This occurs when the reflection of an impulse at a terminal is also impulsive. Then an incident wave of any sort generates a reflected wave of the same type. But in general we must allow for a change of type. This may be done by means of the reflection coefficients, say, ρ_0 and ρ_1, at $x = 0$ and l respectively.

Thus, let V_1, V_2 be corresponding elements in an incident and the reflected wave at the terminal $x = l$. Then

$$V_1 + V_2 = V, \qquad C_1 + C_2 = C = \frac{V_1 - V_2}{Z} = \frac{V_1 + V_2}{Z_1}, \tag{1}$$

if V and C are the resultant voltage and current at the terminal. Z is the resistance operator of the line when infinitely long, and Z_1 that of the terminal arrangement. So

$$\rho_1 = \frac{V_2}{V_1} = -\frac{C_2}{C_1} = \frac{Z_1 - Z}{Z_1 + Z}, \qquad \rho_0 = \frac{Z_0 - Z}{Z_0 + Z}, \tag{2}$$

are the reflection coefficients for the voltage. Knowing V_1, the structure of ρ_1 or ρ_0 enables us to find V_2. All the successive waves are therefore developable.

In the distortionless case, Z is the constant Lv. The reflected waves are then of the same type as the incident if Z_0 and Z_1 are also constant—that is, mere resistances in the obvious case, or else equivalent distortionless circuits. But in the theory of this chapter,

$$Z = \left(\frac{R + Lp}{K + Sp}\right)^{\frac{1}{2}} = Lv\left(\frac{p + \rho + \sigma}{p + \rho - \sigma}\right)^{\frac{1}{2}}, \tag{3}$$

so we cannot get clean reflection without change of type unless the terminal resistance operator is a constant multiple of Z, which means that Z_0 and Z_1 may be other lines with suitably chosen constants, or else practical imitations. In all cases of this kind, it is sufficient to find the formula for the initial wave. All the following waves may be obtained from it by changing the origin and argument suitably.

Let V_0 be voltage impressed at $x = 0$, and V_1 the first wave. Then

$$V_1 = \epsilon^{-qx} V_0. \tag{4}$$

On arrival at $x = l$, reflection begins. To produce the reflected wave, put $x = l$ in V_1, multiply by ρ_1, and then introduce the attenuator $\epsilon^{-q(l-x)}$. Thus

$$V_2 = \rho_1 \epsilon^{-q(2l-x)} V_0 \tag{5}$$

is the second wave. On its arrival at $x = 0$, the third wave begins. Put $x = 0$ in V_2, multiply by ρ_0, and then by ϵ^{-qx}, producing

$$V_3 = \rho_0 \rho_1 \epsilon^{-q(2l+x)} V_0. \tag{6}$$

Observe that to get V_3 from V_1 we multiply by $\rho_0\rho_1\epsilon^{-2ql}$. In the same way V_4 is got from V_2, and any V_n from V_{n-2}. So the total V is

$$V = (1 + \rho_0\rho_1\epsilon^{-2ql} + \rho_0^2\rho_1^2\epsilon^{-4ql} + \dots)(V_1 + V_2), \tag{7}$$

or, which means the same,

$$V = \frac{\epsilon^{-qx} + \rho_1\epsilon^{-q(2l-x)}}{1 - \rho_0\rho_1\epsilon^{-2ql}} V_0, \tag{8}$$

which is the condensed form of solution for V in terms of V_0. The latter is not the real voltage at $x = 0$, except when the line is infinitely long, when, however, (4) above is sufficient. Also the relation between V_0 and e, when e is impressed force producing V_0, is

$$V_0 = Z\frac{e}{Z_0 + Z}. \tag{9}$$

Operational solutions of the condensed kind, as (8), and its modifications, may be developed to Fourier series by means of the expansion theorem, § 282 and after. It is unfortunate that the results, save in relatively simple cases, should be unmanageable for numerical calculation. Even to prove their identity with the wave series by rigorous methods involves the evaluation of definite integrals of a formidable nature. But that is hardly the right way to work. When in mathematical physics we find, by following Fourier's methods, say, that our solutions involve certain series and integrals, not recognisable as known forms, the practical way to evaluate them is to find another way of solving the same problems, and then equate the results to the former ones. This may not be rigorous. But it may be better than that Whole families of new results may "tumble out" of themselves, altogether beyond rigorous treatment. But this is an episode.

If we write

$$V = V_1 + V_2 + V_3 + V_4 + \dots, \tag{10}$$

although we know that only a limited number of terms on the right side may be in existence, we may do so without ambiguity, whether the individual waves are represented by algebraical formulæ or by operational formulæ. The latter are (4), (5), (6), &c., the former are their algebrisations. In (8) all the waves are added, though they may not be in existence. But there is no metaphysics here. If non-existent they do nothing. The function V_0 is zero before and begins when $t=0$. The derived wave V_1 has the same property, modified by the operator ϵ^{-qx}. It only exists for positive values of t which are not less than x/v. Or, more shortly, it begins when t reaches the value x/v. Similarly V_2 begins when t reaches $(2l - x)/v$; V_3 begins when t reaches $(2l + x)/v$, and so on. Before these epochs they are zero. So, considering the complete formula (10), whether it be algebraical or operational, we see that the terms come into existence one after another at the proper moments indicated by their structure. Thus, from $t=0$ up to $t=l/v$, we have $V=V_1$ only, and its range is only from $x=0$ to vt. From $t=l/v$ up to $2l/v$, we have $V=V_1$ only, between $x=0$ and $2l-x$, but $V=V_1+V_2$ beyond, up to $x=l$. And so on to the rest of the waves. It is only necessary to follow the wave front running to and fro at speed v to see the extent of operation of the waves.

In the algebrised formulæ themselves, the positivity of the arguments limits the existence of special terms. Thus, if the first wave involves $w_1{}^n\mathrm{P}_n(z_1)$, where

$$w_1=\tfrac{1}{2}\sigma(t-x/v), \qquad z_1=\sigma(t^2-x^2/v^2)^{\frac{1}{2}}; \tag{11}$$

then the second wave will involve $w_2{}^n\mathrm{P}_n(z_2)$, where

$$w_2=\tfrac{1}{2}\sigma\{t-(2l-x)/v\}, \qquad z_2=\sigma\{t^2-(2l-x)^2/v^2\}^{\frac{1}{2}}; \tag{12}$$

and the third will involve $w_3{}^n\mathrm{P}_n(z_3)$, where

$$w_3=\tfrac{1}{2}\sigma\{t-(2l+x)/v\}, \qquad z_3=\sigma\{t^2-(2l+x)^2/v^2\}^{\frac{1}{2}}; \tag{13}$$

and so on. In all cases the w's must be positive, or at least zero, and the same as regards the quantities under the radical. This consideration makes the formula (10) explicitly correct all along.

Terminal Reflection without Loss. Wave Solutions.

§ 397. Now, let V_0 be known. Let it be expanded in the form

$$\mathrm{V}_0=\epsilon^{-\rho t}\,\Sigma\,\mathrm{A}_n\mathrm{I}_n(\sigma t), \tag{14}$$

of which I have given numerous examples. Then, as before explained, the first wave is expressed by

$$\mathrm{V}_1=\epsilon^{-\rho t}\,\Sigma\,\mathrm{A}_n\frac{w_1{}^n}{\lfloor n}\mathrm{P}_n(z_1), \tag{15}$$

and the second by

$$\mathrm{V}_2=\rho_1\epsilon^{-\rho t}\,\Sigma\,\mathrm{A}_n\frac{w_2{}^n}{\lfloor n}\mathrm{P}_n(z_2), \tag{16}$$

and the third by

$$\mathrm{V}_3=\rho_0\rho_1\epsilon^{-\rho t}\,\Sigma\,\mathrm{A}_n\frac{w_3{}^n}{\lfloor n}\mathrm{P}_n(z_3), \tag{17}$$

and so on, introducing the factors ρ_1 and ρ_0 alternately, and changing x to $2l-x$, $2l+x$, $4l-x$, $4l+x$, &c.

Thus, the solution is fully completed when ρ_0 and ρ_1 are constants. When not constants, but operators, then there is change of type. Equations (15), (16), (17) are still true, but require the performance of the ρ_0 and ρ_1 operations to fully algebrise them. How to do this will come later. At present note that there are four practical cases in which the reflection coefficients are constants (besides the general case of terminal arrangements having resistance operators constant multiples of that of the line), namely terminal short-circuit or insulation at either or both ends. There is complete negative reflection of voltage at a short-circuit, that is, ρ_0 or $\rho_1 = -1$; and there is complete positive reflection of voltage at a disconnection, that is, ρ_0 or $\rho_1 = +1$. The current is also completely reflected, but in the opposite sense to the voltage. So, by letting ρ_0 or ρ_1 be $+1$ or -1, we can construct the full solutions for a line of finite length. [In this connection the case of a circuit closed upon itself should be noticed. Here an impressed force at any part of the circuit generates two waves, to right and left respectively. They are similar as regards C, but opposite as regards V. They are also the same as if they entered infinitely long cables. But they travel round and round and overlap each other and one another; so the resultant V and C are represented by infinite series of waves just as in cases of terminal reflection. The above description is sufficient to build up the complete formulæ.]

It usually happens that the initial wave is expressed by an infinite series of the form (15). Then, of course, all the derived waves are similar infinite series. There is only one practical case where the waves are expressed by a single term. If either there be no leakage, and V_0 be steady, which is thoroughly practical; or else, next best, if there be leakage, but V_0 impressed be made to subside according to the leakage law, *i.e.*, $V_0 = e_0\epsilon^{-Kt/S}$ where e_0 is steady; then, as before seen,

$$C_1 = \frac{e_0}{Lv} P_0(z_1)\epsilon^{-\rho t} \tag{18}$$

is the first current wave. It follows that if there be short circuit at both terminals, the complete result due to the impressed force V_0 at $x = 0$ is

$$C = \frac{e_0}{Lv}\epsilon^{-\rho t}\left\{P_0(z_1) + P_0(z_2) + P_0(z_3) + P_0(z_4) + \dots\right\}, \tag{19}$$

because there is complete positive reflection of the current at the terminals.

On the other hand, under the same circumstances, the first voltage wave is, by § 381, equation (33),

$$V_1 = e_0\epsilon^{-\rho t}\left\{P_0(z_1) + 2w_1P_1(z_1) + 2\frac{w_1^2}{\lfloor 2}P_2(z_1) + \dots\right\}. \qquad (20)$$

It follows that the second wave is

$$V_2 = -e_0\epsilon^{-\rho t}\left\{P_0(z_2) + 2w_2P_1(z_2) + 2\frac{w_2^2}{\lfloor 2}P_2(z_2) + \dots\right\}, \qquad (21)$$

because there is negative reflection of the voltage. And the third wave is

$$V_3 = +e_0\epsilon^{-\rho t}\left\{P_0(z_3) + 2w_3P_1(z_3) + 2\frac{w_3^2}{\lfloor 2}P_2(z_3) + \dots\right\}, \qquad (22)$$

and so on.

When a steady V_0 is applied to a distortionless circuit the curve of V (and C also) is shaped according to $\epsilon^{-Kx/Sv}$, or $\epsilon^{-Kt/S}$, if $x = vt$. If, therefore, we let V_0 decrease with the time according to the same law, the resulting curve of V will be a straight line, and $V = \epsilon^{-Kt/S} \times$ constant will represent the wave, extending from $x = 0$ to vt. No doubt this simplicity is the ultimate reason why the primary wave of C just treated is represented by one term only.

If the line is insulated at $x = l$, whilst still short-circuited at $x = 0$, we have $\rho_0 = -1$, $\rho_1 = +1$. The complete C wave is therefore

$$C = \frac{e_0}{Lv}\epsilon^{-\rho t}\left\{P_0(z_1) - P_0(z_2) - P_0(z_3) + P_0(z_4) + \dots\right\}. \qquad (23)$$

If insulated at $x = 0$ there is negative reflection of the current. An impressed e produces no C, and therefore no V either. We must shift e away from the origin, or make some other change before e can work.

Comparison with Fourier Series. Solution of Definite Integrals.

§ **398.** Now, to show the connection with the Fourier series form of solution, take the fundamental case of (19), and do it by the expansion theorem for normal solutions. The first and second waves are

$$C_1 = \epsilon^{-qx}\frac{V_0}{Z}, \quad C_2 = -\rho_1\epsilon^{-q(2l-x)}\frac{V_0}{Z}, \tag{24}$$

therefore the complete C is

$$C = (1 + \rho_0\rho_1\epsilon^{-2ql} + \dots)(C_1 + C_2) = \frac{\epsilon^{-qx} - \rho_1\epsilon^{-q(2l-x)}}{(1 - \rho_0\rho_1\epsilon^{-2ql})}\cdot\frac{V_0}{Z}. \tag{25}$$

Here we must put $\rho_0 = -1 = \rho_1$, and $V_0 = e_0\epsilon^{-Kt/S} = e_0\epsilon^{(\sigma-\rho)t}$. The result is

$$C = \frac{\cosh q(l-x)}{\text{shin}\, ql}\,\frac{q}{R + Lp}\,\epsilon^{(\sigma-\rho)t}e_0, \tag{26}$$

and the corresponding V formula is

$$V = \frac{\text{shin}\, q(l-x)}{\text{shin}\, ql}\epsilon^{(\sigma-\rho)t}e_0. \tag{27}$$

We see that the determinantal equation gives $ql = n\pi i$ in the case of V, excepting the zero root, but that in the case of C there is an additional root $p = -R/L$. But in the expansion theorem the operand is steady. Here it is a function of the time. This is easily put right. Shift the exponential time function to the left, whilst increasing p to $p + \sigma - \rho$ at the same time. Then (26) becomes

$$C = \epsilon^{(\sigma-\rho)t}\frac{\cosh(l-x)(p^2+2p\sigma)^{\frac{1}{2}}/v}{\text{shin}\, l(p^2+2p\sigma)^{\frac{1}{2}}/v}\,\frac{(v^2+2p\sigma)^{\frac{1}{2}}/v}{L(p+2\sigma)}e_0, \tag{28}$$

where the operand is now steady. The expansion theorem turns it (by the usual work, presenting nothing special) to

$$C = e_0\epsilon^{(\sigma-\rho)t}\left\{\frac{1-\epsilon^{-2\sigma t}}{2Ll\sigma} + \frac{2\epsilon^{-\sigma t}}{Ll}\Sigma\cos\frac{n\pi x}{l}\frac{\sin\lambda t}{\lambda}\right\}, \tag{29}$$

where
$$\lambda = \{(n\pi v/l)^2 - \sigma^2\}^{\frac{1}{2}} = (m^2v^2 - \sigma^2)^{\frac{1}{2}}, \tag{30}$$

and in the summation n is integral, ranging from 1 to ∞. This solution (29) is therefore equivalent to (19) above.

In a similar manner the expansion theorem turns (27) to

$$V = e_0\epsilon^{(\sigma-\rho)t}\left\{\left(1 - \frac{x}{l}\right) - \frac{2\epsilon^{-\sigma t}}{\pi}\Sigma\frac{1}{n}\sin\frac{n\pi x}{l}\left(\cos + \frac{\sigma}{\lambda}\sin\right)\lambda t\right\}, \tag{31}$$

which is equivalent to the sum of (20), (21), (22), &c.

If the line is infinitely long, the summations become definite integrals. Put $n\pi/l = m$, then the step dm is π/l, and (29) becomes

$$C = e_0\epsilon^{-\rho t}\frac{2}{L\pi}\int_0^\infty \cos mx \frac{\sin \lambda t}{\lambda}\,dm, \tag{32}$$

where λ is as in the second form of (30). At the same time there is only one wave, viz., (18) above, the rest having no chance to make a start. So, equating (18) and (32), we obtain

$$P_0(z_1) = \frac{2v}{\pi}\int_0^\infty \cos mx \frac{\sin \lambda t}{\lambda}\,dm. \tag{33}$$

Again, equate the C's in (19) and (29). Multiply both sides by $\cos(n\pi x/l)$ and integrate to x from 0 to l. We obtain

$$\int_0^l \{P_0(z_1) + P_0(z_2) + \dots\} \cos mx\,dx = v\frac{\sin \lambda t}{\lambda}. \tag{34}$$

On the left side the function integrated ranges in the way before explained. When the line is infinitely long (34) becomes

$$\int_0^{vt} P_0(z_1)dx = v\,\frac{\text{shin}\,\sigma t}{\sigma}. \tag{35}$$

Use this in (33). Then follows

$$\frac{\text{shin}\,\sigma t}{\sigma} = \frac{2}{\pi}\int_0^\infty \frac{\sin mvt}{m}\frac{\sin \lambda t}{\lambda}\,dm. \tag{36}$$

The other formula (31), when $l = \infty$, reduces to

$$V = e_0\epsilon^{(\sigma-\rho)t}\left\{1 - \epsilon^{-\sigma t}\frac{2}{\pi}\int_0^\infty \frac{\sin mx}{m}\left(\cos + \frac{\sigma}{\lambda}\sin\right)\lambda t.dm\right\}, \tag{37}$$

and this is equivalent to the first wave, equation (20). So we obtain

$$P_0(z_1) + 2w_1P_1(z_1) + 2\frac{w_1^2}{\underline{|2}}P_2(z_1) + 2\frac{w_1^3}{\underline{|3}}P_3(z_1) + \dots$$
$$= \epsilon^{\sigma t} - \frac{2}{\pi}\int_0^\infty \frac{\sin mx}{m}\left(\cos + \frac{\sigma}{\lambda}\sin\right)\lambda t\,.dm. \tag{38}$$

But, with l finite, equating the sum of the V waves to (31), multiplying both sides by $\sin(n\pi x/l)$, and integrating to x from 0 to l, we obtain

$$\int_0^l \{F(x) - F(2l - x) + F(2l + x) - \dots\} \sin mx\,dx$$
$$= \frac{1}{m}\left(\epsilon^{\sigma t} - \cos \lambda t - \frac{\sigma}{\lambda}\sin \lambda t\right), \tag{39}$$

where $$F(x) = V_1\epsilon^{\rho t}/e_0 = P_0(z) + 2nP_1(z) + \dots, \tag{40}$$
as in (20), the change of argument from x to $(2l-x)$, &c., being to introduce the second, third and following waves as they come on. If, for instance, t is not greater than l/v, the limits in (39) may be 0 and vt, and then only the first F function occurs, the rest being zero.

If, in (38), x is greater than vt, all the left side is zero, as no disturbance has reached x. So we get

$$\epsilon^{\sigma t} = \frac{2}{\pi}\int_0^\infty \frac{\sin mx}{m}\left(\cos + \frac{\sigma}{\lambda}\sin\right)\lambda t.dm. \tag{41}$$

This is true when x is the least bit greater than vt. But it is not true when x is actually $= vt$, because if it is the least bit less, the left side of (41) should be $\epsilon^{\sigma t} - 1$, for the left side of (38) is $P_0(0)$, or 1, when $x = vt$. So, taking the mean, we obtain

$$\epsilon^{\sigma t} - \tfrac{1}{2} = \frac{2}{\pi}\int_0^\infty \frac{\sin mvt}{m}\left(\cos + \frac{\sigma}{\lambda}\sin\right)\lambda t\,.\,dm. \tag{42}$$

Change the sign of σ to obtain another formula. This peculiar behaviour of definite integrals at places of discontinuity has to be very carefully remembered, or it may be a dangerous source of error. In any case, it is very annoying. In the operational and wave formulæ, on the other hand, there is no puzzling change of formula involved at the wave front.

From (42) the formula (36) may be derived. Similarly, it may be noted that in (33) the value of the right member is zero when $x > vt$. This is plain by the manner of construction. But at the wave front itself we must take the mean value just before and behind the wave front to evaluate the integral. Thus,

$$\tfrac{1}{2} = \frac{2v}{\pi}\int_0^\infty \cos mvt\,\frac{\sin \lambda t}{\lambda}dm. \tag{43}$$

Perhaps, on the whole, it is as well to keep away from the definite integrals, if we can get formulæ clear of them.

When the impressed force is steady, say e_0 itself, under the same circumstances terminally, *i.e.*, short circuits, then the expansion theorem applied to (27) above, without the exponential operand, gives at once

$$V = e_0 \frac{\text{shin}\,(l-x)(\text{RK})^{\frac{1}{2}}}{\text{shin}\,l(\text{RK})^{\frac{1}{2}}}$$

$$- \frac{2e_0}{l}\epsilon^{-\rho t}\Sigma \frac{m \sin mx}{m^2 + \text{RK}}\left(\cos \lambda t + \frac{\rho}{\lambda}\sin \lambda t\right), \quad (44)$$

and
$$C = e_0\left(\frac{\text{K}}{\text{R}}\right)^{\frac{1}{2}} \frac{\cosh\,(l-x)\,(\text{RK})^{\frac{1}{2}}}{\text{shin}\,l(\text{RK})^{\frac{1}{2}}} - \frac{e_0}{\text{R}l}\epsilon^{-\text{R}t/\text{L}}$$

$$- \frac{2e_0}{\text{R}l}\epsilon^{-\rho t}\Sigma \frac{\cos mx}{m^2 + \text{RK}}\left\{\text{RK}\left(\cos - \frac{\sigma}{\lambda}\sin\right)\lambda t - \frac{\text{R}m^2}{\text{L}}\frac{\sin \lambda t}{\lambda}\right\}, \quad (45)$$

where m is $n\pi/l$ as before.

Now, in § 382, equations (54) (57), I have given the V and C solutions in $w^n \text{P}_n$ functions for the first wave, and have shown how to carry on the formulæ to any extent in § 384 in another way. The succeeding waves are obtainable by change of argument, as explained above. Therefore we may equate the wave solutions to (44) and (45) respectively, and derive a fresh batch of solved definite integrals if required. But, as they are all included in the equivalence of (44), (45) to the wave formulæ, we need not go far in elaborating them. Considering only the first wave, $l = \infty$ in (44) makes

$$V = e_0\epsilon^{-x\sqrt{\text{RK}}} - \frac{2e_0}{\pi}\epsilon^{-\rho t}\int_0^\infty \frac{m \sin mx}{m^2 + \text{RK}}\left(\cos \lambda t + \frac{\rho}{\lambda}\sin \lambda t\right) dm \quad (45\text{A})$$

equivalent to (57) § 382, with $e = e_0$.

General Way of Finding Second and Following Waves due to Terminal Reflection.

§ **399**. The first wave sent along a circuit when voltage is applied at its beginning is independent of the nature of the arrangement at the far end. This follows from the property of propagation of disturbances at finite speed. It follows, again, that if we construct, by the expansion theorem for normal functions generalised, the series which represents the complete solution due to impressed force at the origin, so as to suit given terminal arrangements, that series will, for a short period of time, have a known meaning—viz., representing the first wave only. The series will contain, in every term, symbols standing for the electrical properties of the terminal arrangements, but in such a manner that only those in Z_0 have any effect upon the numerical meaning in the first wave—that is, in the region from $x=0$ up to vt, provided $vt<l$. But as soon as vt reaches the value l, then the symbols in Z_1 become operative, but only in the part of the circuit occupied by the second wave. In the remainder, next the origin, the Z_1 symbols are still inoperative. But after $t=2l/v$, both the Z_0 and Z_1 symbols are operative, and we cannot change their values without affecting the numerical meaning of the formula. Though far-reaching, these conclusions are perfectly safe and sound, because the fundamental reason is sound, and has no exceptions in its application. Similar conclusions, of a more comprehensive nature, obtain in all electrical problems involving waves when done in terms of normal functions.

If, as in a few examples lately given, we know the nature of the second and succeeding waves, then we have the full interpretation of the Fourier series. But, in more general cases, from the hidden complex roots involved, it may be of a practically unmanageable nature. I shall now explain how to obtain the formulæ for the series of waves, one after another, and illustrate by relatively easy examples. There is this to be said for the solutions in the Fourier series. They are readily obtainable, and are comprehensive. Against, they may be excessively difficult in the interpretation. On the other hand, the wave formulæ are harder to obtain, but when got are easily calculable. Although the British Association blundered sadly about the electrical units, I am grateful to that body for its tables of the $I_n(x)$ functions. If tables of the K_n function (continuous) and the G_n function (oscillating) are in preparation, about which I know nothing, I venture to strongly recommend that they be standardised as done in this work, § 336 and later. Much trouble will be saved in the future, even though it be not, like 4π, a question of practical importance.

That the calculation of the second and succeeding waves is feasible may be seen thus. The first and second waves being

$$V_1=\epsilon^{-qx}V_0, \qquad V_2=\rho_1\epsilon^{-q(2l-x)}V_0, \tag{46}$$

the second wave differs from the first in the change from x to $(2l-x)$ and in the ρ_1 factor. Put $2l-x=y$, then we may write

$$V_2 = \epsilon^{-qy}\rho_1 V_0. \tag{47}$$

Comparing now with the first wave, we see that V_2 is the wave sent into an unlimited circuit by the impressed voltage $\rho_1 V_0$. Similarly

$$V_3 = \epsilon^{-qz}\rho_0\rho_1 V_0, \tag{48}$$

if $z = 2l + x$. So V_3 is the wave sent in by the impressed voltage $\rho_0\rho_1 V_0$. And so on. This is not a bad way of looking at the matter, because it shows that all the waves may be obtained like the first, every one from its proper impressed voltage. Thus,

$$V_0 = \epsilon^{-\rho t}\sum A_n I_n(\sigma t) \qquad \text{makes} \qquad V_1 = \epsilon^{-\rho t}\sum A_n \frac{w_1^n}{\underline{|n}} P_n(z_1)$$

Therefore, if we write the voltage $\rho_1 V_0$ thus:—

$$\rho_1 V_0 = \epsilon^{-\rho t}\sum B_n I_n(\sigma t), \tag{49}$$

the second wave will be

$$V_2 = \epsilon^{-\rho t}\sum B_n \frac{w_2^n}{\underline{|n}} P_n(z_2); \tag{50}$$

and so on. But, though easy to follow in principle, this way is not convenient in practice, if it be understood to mean the prior determination of $\rho_1 V_0$, $\rho_0\rho_1 V_0$, &c., as functions of the time, and their subsequent expansion in I_n functions, and then the generalisation to waves. There is a better way of carrying out the work.

We found that a great simplification arose by the use of the differentiators α and β, instead of d/dx and d/dt, and practically by using only one of them, with the operand $P_0(z)$ throughout. This can now be followed up.

Knowing that $$V_1\epsilon^{\rho t} = f_1(\alpha)P_0(z), \tag{51}$$
where, by (105), §386,

$$f_1(\alpha) = [V_0\epsilon^{\rho t}]\frac{1-\alpha^2}{1+\alpha^2}, \tag{52}$$

we shall now have

$$V_2\epsilon^{\rho t} = f_2(\alpha)P_0(z), \tag{53}$$

where $f_2(\alpha)$ is the modification in $f_1(\alpha)$ required corresponding to the passage from the first of (46) to (47). To find it, note that

$$V_2 = \rho_1\{\epsilon^{-\rho t} f_1(a) P_0(z)\}. \tag{54}$$

So, ρ_1 being a function of p, turn p to $p - \rho$ (allowing us to shift ρ_1 forward to the other side of $\epsilon^{-\rho t}$), and then put $\frac{1}{2}\sigma(a + a^{-1})$ for p. Let the result be σ_1. Then

$$f_2(a) = \sigma_1 f_1(a). \tag{55}$$

Similarly $$V_3 \epsilon^{\rho t} = f_3(a) P_0(z), \tag{56}$$

where $$f_3(a) = \sigma_0 \sigma_1 f_1(a), \tag{57}$$

σ_0 being the function of a obtained from ρ_0 by turning p to $\frac{1}{2}\sigma(a + a^{-1}) - \rho$.

The matter is now reduced to plain algebra. Thus, as regards V_2. Expand $f_2(a)$ in powers of a; then, by (53),

$$V_2 \epsilon^{\rho t} = \Sigma B_n a^n P_0(z) = \Sigma B_n \frac{w_2^n}{\lfloor n} P_n(z_2) \tag{58}$$

is the fully developed formula. It is only just at the last that we need think about the proper arguments, as w_2 and z_2 in the second wave, w_3 and z_3 in the third, and so on.

The successive waves of current may be done in the same way. See (106), § 386 for the first wave. Then introduce $-\sigma_1$ and $-\sigma_0$ as factors, instead of $+\sigma_1$ and $+\sigma_0$ above, to obtain the following waves: Say,

$$LvC_1 \epsilon^{\rho t} = g_1(a) P_0(z), \qquad g_1 = [V_0 \epsilon^{\rho t}] \frac{(1-a)^2}{1+a^2}, \tag{59}$$

$$LvC_2 \epsilon^{\rho t} = g_2(a) P_0(z), \qquad g_2 = -\sigma_1 g_1, \tag{60}$$

and so on.

Application to Terminal Resistances. Full Solutions with the Critical Resistances. Second Wave with any Resistance.

§ **400.** The next thing is to show how the above plan works out. The simplest case is that of a terminal mere resistance, say R_1, at $x = l$. Then

$$-\rho_1 = \frac{Z - Z_1}{Z + Z_1} = \frac{Lv\left(\dfrac{p+\rho+\sigma}{p+\rho-\sigma}\right)^{\frac{1}{2}} - R_1}{\dots\dots\dots\dots\dots\dots + \dots}. \tag{61}$$

Make the above-described changes in p (or use the middle formula in (104), § 386). We get

$$- \sigma_1 = \frac{Lv(1+a)(1-a)^{-1} - R_1}{\ldots\ldots\ldots\ldots\ldots + \ldots}. \tag{62}$$

Or $\qquad \sigma_1 = -\dfrac{c+a}{1+ca}$ if $c = \dfrac{Lv - R_1}{Lv + R_1}.$ (63)

The quantity c varies from -1 to $+1$, vanishing when $R_1 = Lv$. We may then expect a great simplification. This terminal condition destroys reflection in the distortionless case. At present it will destroy a lot of terms. We have $\sigma = -a$. So when the first wave is

$$V_1 = \epsilon^{-\rho t} \sum A_n a^n P_0(z), \tag{64}$$

the second wave is

$$V_2 = -\epsilon^{-\rho t} \sum A_n a^{n+1} P_0(z) ; \tag{65}$$

and if $R_0 = Lv$ is the similar terminal condition at $x = 0$, we have $\sigma_0 = -a$ also; so that the third wave is

$$V_3 = \epsilon^{-\rho t} \sum A_n a^{n+2} P_0(z). \tag{66}$$

The complete result is therefore

$$V = \epsilon^{-\rho t} \sum A_n \left\{ \frac{w_1{}^n}{\lfloor n} P_n(z_1) - \frac{w_2{}^{n+1}}{\lfloor n+1} P_{n+1}(z_2) + \frac{w_3{}^{n+2}}{\lfloor n+2} P_{n+2}(z_3) - \ldots \right\} \tag{67}$$

arising from any voltage impressed at $x = 0$ producing

$$V_0 = \epsilon^{-\rho t} \sum A_n I_n(\sigma t), \tag{68}$$

with terminal resistances of the critical amount.

Similarly as regards the C waves. Thus, when $V_0 = e_0 \epsilon^{-Kt/S}$, where e_0 is constant, the simple case so often used before, we get

$$e_0^{-1} Lv C_1 = \epsilon^{-\rho t} P_0(z_1), \tag{69}$$

$$e_0^{-1} Lv C_2 = \epsilon^{-\rho t} w_2 P_1(z_2), \tag{70}$$

$$e_0^{-1} Lv C_3 = \epsilon^{-\rho t} \frac{w_3{}^2}{\lfloor 2} P_2(z_3), \tag{71}$$

and so on, when there is the critical resistance Lv at both ends of the line. The reflection coefficients are now $-\rho_1$ and $-\rho_0$, so the differentiator $+\alpha$ derives any wave from the preceding one

But if there be a short circuit at $x=0$, so that $-\rho_0=1$, and therefore $-\sigma_0=1$, the operators of derivation are alternately α and 1, and we obtain

$$e_0^{-1}\text{L}v\text{C}\epsilon^{pt} = \text{P}_0(z_1) + \{w_2\text{P}_1(z_2) + w_3\text{P}_1(z_3)\} + \left\{\frac{w_4^{\,2}}{\underline{|2}}\text{P}_2(z_4) + \frac{w_5^{\,2}}{\underline{|2}}\text{P}_2(z_5)\right\} + \dots \quad (72)$$

Notice in the above that all waves after the first are zero at their fronts. This is because the front of the first wave travels without distortion, and since there is no reflection at all of a pure plane wave when $\text{R}_1=\text{L}v$ in the distortionless case, there is no reflection of wave front now. Any departure of R_1 from the value $\text{L}v$ must produce a reflected wave which is finite at its front. This will also appear in the formulæ to follow.

When R_1 has any value, then

$$\frac{c+\alpha}{1+c\alpha} = c + \frac{(1-c^2)\alpha}{1+c\alpha} = c + (1-c^2)\alpha(1-\alpha c+\alpha^2c^2-\dots). \quad (73)$$

This is the expression for $-\sigma_1$, in rising powers of α. So, when

$$e_0^{-1}\text{L}v\text{C}_1\epsilon^{pt} = \text{P}_0(z_1),$$

as in (69), is the first wave of current, the second is

$$e_0^{-1}\text{L}v\text{C}_2\epsilon^{pt} = c\text{P}_0 + (1-c^2)\left(w\text{P}_1 - c\frac{w^2}{\underline{|2}}\text{P}_2 + c^2\frac{w^3}{\underline{|3}}\text{P}_3 - \dots\right). \quad (74)$$

For distinctness the arguments are omitted. Use w_2 and z_2.

When $c=\pm1$ we reduce to the cases of terminal earth or insulation. When $c=0$ we fall back on the case already considered. When c is $+$, then $\text{L}v>\text{R}_1$, and the second wave of current is $+$ at its front, the terminal resistance being insufficient to destroy the initial reflection. When c is $-$, $\text{R}_1>\text{L}v$, and the reflected wave of current is negative.

Similarly, if the first wave is of a complicated type, say

$$\text{L}v\text{C}_1\epsilon^{pt} = \Sigma\,\text{A}_n\alpha^n\text{P}_0(z), \quad (75)$$

this being a sufficient expression in general, then

$$\text{L}v\text{C}_2\epsilon^{pt} = \frac{c+\alpha}{1+c\alpha}\Sigma\,\text{A}_n\alpha^n\text{P}_0(z). \quad (76)$$

Here we may use the development (73), and apply it to every term in the summation by itself, one after another. This in, in fact, what we must do unless we can bring $\sum A_n a^n$ to a simpler form with advantage, which may be done by specialising it. The part belonging to A_n is

$$A_n\{ca^n + (1-c^2)(a^{n+1} - ca^{n+2} + c^2a^{n+3} - \dots)\}P_0(z), \tag{77}$$

the meaning of which is known.

Inversion of Operations. Derivation of First Wave from the Second.

§ **401.** As an illustration, we may reverse operations, and ask this question: Given the second wave, what was the first one? To answer this, operate on the second wave by $-\sigma_1^{-1}$, the reciprocal operator. Thus, if

$$e_0^{-1}LvC_2\epsilon^{pt} = P_0(z_2), \tag{78}$$

or the second wave is of the fundamental simplest type, then the first is such that

$$\sum A_n a^n = \frac{1+ca}{c+a} = \frac{1}{c} + \left(1 - \frac{1}{c^2}\right)\left(a - \frac{a^2}{c} + \frac{a^3}{c^2} - \dots\right); \tag{79}$$

which makes

$$e_0^{-1}LvC_1\epsilon^{pt} = \frac{P_0}{c} + \left(1 - \frac{1}{c^2}\right)\left(a - \frac{a^2}{c} + \frac{a^3}{c^2} - \dots\right)P_0, \tag{80}$$

in which w_1 and z_1 must be used. The C_0 impressed at $x=0$ is now obtainable. It is

$$e_0^{-1}LvC_0\epsilon^{pt} = \frac{I_0}{c} + \left(1 - \frac{1}{c^2}\right)\left(I_1 - \frac{I_2}{c} + \frac{I_3}{c^2} - \dots\right)(\sigma t). \tag{81}$$

That this is the proper answer may be verified by taking (80) as a given primary wave and calculating the second by the operator $-\sigma_1$, *i.e.*, as exhibited in (76). The result is the second wave (78). But it does not follow that the first wave is a physically possible one. It might be that the given second wave is such that no finite first wave could produce it. Now (81) looks suspiciously divergent when c is < 1, but I think a further examination (which, this being an episode of curiosity, need not be entered upon) will show that the primary wave obtained is usually quite a fair one. There is, however, one case of evident failure. When $c=0$ we obtain an infinite first wave. But now the reason is also evident. We know already that when $R_1 = Lv$ the disturbance must be zero at the front of the second wave when the first wave is finite. But our proposed second wave is not zero but finite at the wave front. So the first wave must needs be infinitely intense. Save in this case, I see no reason for failure. Naturally, when c differs little from 1, the first wave must be very intense.

Derivation of Third and Later Waves from the Second.

§ 402. In a similar manner we can work back from any wave to any preceding wave, and determine the possibility by examining the result, when the possibility or impossibility is not evident beforehand. Returning to the main question, the second wave being exhibited generally by (76), (77), take the case $n=0$. Then

$$LvC_1\epsilon^{pt} = A_0P_0(z_1), \qquad (82)$$

$$LvC_2\epsilon^{p} = A_0\{c + (1-c^2)(a - ca^2 + c^2a^3 - \ldots)\}P_0, \qquad (83)$$

with z_2 and w_2 in the second wave. Find the third wave, when the terminal arrangement at the origin is a mere resistance. We shall have

$$-\sigma_0 = \frac{c_0 + a}{1 + c_0 a}, \qquad c_0 = \frac{Lv - R_0}{Lv + R_0}. \qquad (84)$$

Here σ_0 and c_0 only differ from σ_1 and c in the substitution of R_0 for R_1. The third wave comes from the second, therefore, in the same manner precisely as the second came from the first, with the change mentioned. Thus if the second wave is of the type (78) or (82), the third is of the type (83), with C_3, w_3, z_3, and c_0. But taking matters literally as in (82), (83), namely, that these equations represent the first and second waves, then, to find the third, we require to expand $\sigma_0\sigma_1$ in powers of a. Here σ_0 is like σ_1, and is given by

$$-\sigma_0 = c_0 + (1 - c_0^2)a(1 - c_0 a + c_0^2 a^2 - \ldots). \qquad (85)$$

This may be applied to (83) directly. The result is, as far as P_3,

$$LvC_3\epsilon^{\rho t} = A_0[cc_0 + a(c+c_0)(1-cc_0)$$
$$+ a^2\{cc_0(c^2+c_0^2-2) + (1-c^2)(1-c_0^2)\}$$
$$+ a^3\{cc_0^2(1-c_0^2) + c_0c^2(1-c^2) - (c+c_0)(1-c^2)(1-c_0^2)\} + \dots]P_0. \quad (86)$$

This, using w_3 and z_3, represents the third wave. Remember that a^nP_0 means $(w^n/\underline{|n})P_n(z)$ throughout, with the proper w and z.

Similarly, the fourth wave may be obtained from (86) by multiplying the right member by $-\sigma_1$; and the result multiplied by $-\sigma_0$ will give the fifth wave; and so on to any extent. The work is so entirely mechanical and simple in principle, being Algebra, Chapter II., or thereabouts, on the formation of products by multiplication, that it is unnecessary to elaborate the developments unless they are actually wanted. It will be seen that the full expressions for the successive waves get more and more complicated, just as, in fact, the algebraical expansions of σ_1, $\sigma_0\sigma_1$, $\sigma_0\sigma_1^2$, &c., grow complicated. So it is best to use the condensed forms, and say $\sigma_0^m\sigma_1^nP_0(z)$, after n reflections at $x=l$, and m reflections at $x=0$, understanding by σ_0, σ_1 certain known functions of a to be expressed in power series.

Summarised Complete Solutions.

§ 403. From the above it will be seen that the following is a convenient way of writing or describing the complete solution for V :—

$$V\epsilon^{\rho t} = (1+\sigma_1+\sigma_0\sigma_1+\sigma_0\sigma_1^2+\sigma_0^2\sigma_1^2+\dots)V_1\epsilon^{\rho t}$$
$$= (1+\sigma_1+\sigma_0\sigma_1+\sigma_0\sigma_1^2+\dots)\Sigma A_n a^n P_0(z), \quad (87)$$

when $V_1\epsilon^{\rho t}$ is the first wave, and the summation in the second line expresses its equivalent to be obtained, in the way already explained, out of $V_0\epsilon^{\rho t}$. After the development of the σ products for any wave the proper z's and w's are to be inserted. The last equation does not strictly represent the operational solution in wave form. That contains exponential terms, along with the ρ_0, ρ_1 coefficients. But their effect is allowed for, along with the change from the ρ's to the σ's, in the final correct choice of w and z for any wave.

Similarly,

$$\begin{aligned} C\epsilon^{pt} &= (1 - \sigma_1 + \sigma_0\sigma_1 - \sigma_0\sigma_1^2 + \ldots)C_1\epsilon^{pt} \\ &= (1 - \sigma_1 + \sigma_0\sigma_1 - \sigma_0\sigma_1^2 + \ldots)\Sigma B_n a^n P_0(z) \end{aligned} \tag{88}$$

expresses the complete C in terms of the first wave. But it is not necessary to go through the whole work twice over. The C waves can be obtained from the V waves, or conversely. For we know that

$$LvC_1\epsilon^{pt} = \frac{1-a}{1+a}V_1\epsilon^{pt}, \tag{89}$$

by (101) and (104), § 386. This is for the first wave. So the same operator, taken positively or negatively, as the case may be, derives any C wave from the corresponding V wave; and conversely with the inverse operator.

The just described summary does not apply merely when ρ_0 and ρ_1, or σ_0 and σ_1 after them, have the particular forms belonging to terminal resistances, but has general application to any terminal arrangements. One or two examples of other arrangements will now be examined.

Reflection by a Condenser.

§ **404.** When the reflector is a non-conductive condenser, its resistance operator is $(S_1p)^{-1}$, if S_1 is the permittance. Therefore we must put

$$Z_1 = \frac{1}{S_1p}, \qquad -\rho_1 = \frac{Z - Z_1}{Z + Z_1}, \tag{90}$$

in the previous results. Putting $\frac{1}{2}\sigma(a + a^{-1}) - \rho$ for p, we obtain σ out of ρ, thus

$$-\sigma_1 = \frac{Lv\dfrac{1+a}{1-a} - \dfrac{2a}{S_1\{\sigma(1+a^2) - 2a\rho\}}}{\ldots + \ldots\ldots}. \tag{91}$$

Clearing of fractions, and arranging in powers of a, makes

$$-\sigma_1 = \frac{1 + a\left(1 - \dfrac{2\rho}{\sigma} - \dfrac{2}{\theta\sigma}\right) + a^2\left(1 - \dfrac{2\rho}{\sigma} + \dfrac{2}{\theta\sigma}\right) + a^3}{1 + a\left(1 - \dfrac{2\rho}{\sigma} + \dfrac{2}{\theta\sigma}\right) + a^2\left(1 - \dfrac{2\rho}{\sigma} - \dfrac{2}{\theta\sigma}\right) + a^3}, \tag{92}$$

where θ is the time period S_1Lv. This σ_1 being the reflection coefficient for the voltage $\times\,\epsilon^{pt}$, only needs development in rising powers of a to show its effect upon a given incident voltage.

The easiest way is by long division, and the result, as far as a^3, which is sufficient for illustration, is

$$-\sigma_1 = 1 - \frac{4a}{\theta\sigma} + \frac{8a^2}{\theta\sigma}\left(1 - \frac{\rho}{\sigma} + \frac{1}{\theta\sigma}\right)$$
$$- \frac{4a^3}{\theta\sigma}\left(1 - \frac{4\rho}{\sigma} + \frac{4\rho^2}{\sigma^2} + \frac{8}{\theta\sigma} - \frac{8\rho}{\theta\sigma^2} + \frac{4}{\theta^2\sigma^2}\right) + \dots \quad (93)$$

So, if the first wave is represented by $V_1\epsilon^{\rho t} = eP_0(z_1)$, e being constant, the second wave is

$$V_2\epsilon^{\rho t} = -e\left\{P_0 - \frac{4}{\theta\sigma}w_2P_1 + \frac{8}{\theta\sigma}\left(1 - \frac{\rho}{\sigma} + \frac{1}{\theta\sigma}\right)\frac{w_2^2}{\underline{|2}}P_2 + \dots\right\}, \quad (94)$$

following (93) on to P_3 if wanted, and further still to any extent by carrying on the long division. Use z_2 with the P functions, of course.

We may readily verify that the last solution reduces correctly when $\sigma = 0$. It is then the case of reflection at a terminal condenser in a distortionless circuit. The voltage impressed on the condenser by the first wave is $e\epsilon^{-\rho t}$. So at the condenser itself

$$V_2 = -\frac{1 - Z_1/Z}{1 + Z_1/Z}V_1 = -\frac{1 - (\theta p)^{-1}}{1 + (\theta p)^{-1}}e\epsilon^{-\rho t}, \quad (95)$$

or

$$V_2\epsilon^{\rho t} = -e\frac{1 - \{\theta(p - \rho)\}^{-1}}{1 + \{\theta(p - \rho)\}^{-1}}$$
$$= -e\left\{1 - \frac{2}{\theta(p - \rho)} + \frac{2}{\theta^2(p - \rho)^2} - \dots\right\}$$
$$= -e\left\{1 - \frac{2}{\theta p}\left(1 + \frac{\rho}{p} + \frac{\rho^2}{p^2} + \dots\right)\right.$$
$$\left. + \frac{2}{\theta^2p^2}\left(1 + \frac{\rho}{p} + \dots\right)^2 - \frac{2}{\theta^3p^3}(1 + \dots)^2 + \dots\right\}. \quad (96)$$

The work is developed in this particular way because we only want the result as far as p^{-3}, and so far as that all is shown. Or, collecting terms,

$$V_2\epsilon^{\rho t} = -e\left\{1 - \frac{2}{\theta p} + \frac{2}{\theta p^2}\left(\frac{1}{\theta} - \rho\right) - \frac{2}{\theta p^3}\left(\frac{1}{\theta^2} - \frac{2\rho}{\theta} + \rho^2\right) + \dots\right\} \quad (97)$$

Integrating this, we turn p^{-n} to $t^n/\underline{|n}$. But the initial moment is l/v after the incident wave left the beginning of the line. So, instead of t write $t - l/v$, and we obtain the voltage of the reflected wave at the condenser.

To compare with (94), remember that $\sigma=0$ reduces all the P functions to unity, and that w^n contains the factor σ^n. The result is that all terms disappear save those in which w is divided by σ, w^2 by σ^2, and so on. What is left is identical with (97), save that it is more general, since it gives the voltage all along the reflected wave, as well as its terminal value. Put $x=l$ in w_2 to produce exact agreement. The complete solution in the reduced case may be readily obtained by remembering that the distortionless circuit behaves towards disturbances coming from the condenser, just as if it were a resistance Lv.

Reflection by an Inductance Coil.

§ **405.** When the reflector is a coil of inductance L_1 and no resistance, its resistance operator is $L_1 p$, and therefore

$$\rho = -\frac{Z - L_1 p}{Z + L_1 p} \tag{98}$$

is the reflection coefficient for the voltage. Turn p to $p-\rho$ to obtain the reflection coefficient for the voltage multiplied by $\epsilon^{\rho t}$. Lastly, put p in terms of a to obtain the expression for σ_1. It is

$$-\sigma_1 = \frac{Lv\dfrac{1+a}{1-a} - L_1\left\{\sigma\dfrac{1+a^2}{2a} - \rho\right\}}{\quad . \quad . \quad . \quad + \quad . \quad . \quad . \quad . \quad . \quad . \quad . \quad}, \tag{99}$$

and, clearing of fractions and re-arranging, this becomes

$$\sigma_1 = \frac{1 - a\left(1 + \dfrac{2\rho}{\sigma} + \dfrac{2}{\theta\sigma}\right) + a^2\left(1 + \dfrac{2\rho}{\sigma} - \dfrac{2}{\theta\sigma}\right) - a^3}{1 - a\left(1 + \dfrac{2\rho}{\sigma} - \dfrac{2}{\theta\sigma}\right) + a^2\left(1 + \dfrac{2\rho}{\sigma} + \dfrac{2}{\theta\sigma}\right) - a^3}, \tag{100}$$

where θ is the time period L_1/Lv. This somewhat resembles (92), but is essentially different in detail. By division we get

$$\sigma_1 = 1 - \frac{4a}{\theta\sigma} - \frac{8a^2}{\theta\sigma}\left(1 + \frac{\rho}{\sigma} - \frac{1}{\theta\sigma}\right) + \ldots \tag{101}$$

So, the first wave being

$$V_1\epsilon^{\rho t} = e_0 a^n P_0 = e_0 \frac{w_1^n}{\underline{|n}} P_n(z_1)\,; \tag{102}$$

the second is, by (101),

$$\sigma_2 = e_0\left\{\frac{w_2^n}{\underline{|n}}P_n - \frac{4}{\theta\sigma}\frac{w_2^{n+1}}{\underline{|n+1}}P_{n+1} - \frac{8}{\theta\sigma}\left(1 + \frac{\rho}{\sigma} - \frac{1}{\theta\sigma}\right)\frac{w_2^{n+2}}{\underline{|n+2}}P_{n+2} + \ldots\right\} \tag{103}$$

with z_2 in the P functions instead of z_1. I omit fuller development. A whole book would be wanted to carry it out in full for various practical terminal arrangements. It is sufficient to give enough to show how the method works out.

A comprehensive case is that of a conducting condenser in sequence with a resisting coil, making

$$Z_1 = R_1 + L_1 p + \frac{1}{K_1 + S_1 p}.$$

Here the fraction corresponding to (92) and (100) goes as far as a^5 in the numerator and denominator, and is too lengthy to print in one line. Other combinations may be made up readily by attending to the fundamental property of resistance operators, that they combine like resistances.

The working of the human mind is slow. I set myself the above problems 22 years ago, when I first recognised as a consequence of Maxwell's theory of self-induction, combined with W. Thomson's theory of the electric telegraph, that all disturbances travelled at finite speed, and, therefore, that the Fourier solutions could be broken up into an infinite series of distinct solutions. But it was not until 10 years later that I managed to carry out this analysis in the simple case of terminal earth or insulation. Another 10 years later, by following up the previous work, I have extended the method to any terminal arrangement. And it is remarkable how simply it goes, excepting, of course, in the complication of the development of products of the powers of σ_0 and σ_1 (*see* (87), (88), § 403.) Perhaps I may have done the work wrongly. If so, I shall be glad to be corrected.

Initial States. Expression of Results by Definite Integrals.

§ **406**. It is now desirable to say something about initial states. To show that a given initial state is transformed when left to itself, at any time later on, to some other state, may be regarded as the "classical" way of expressing results. When, however, it is considered that the formula must involve an integration applied to the initial state coupled with some other function, and that the execution of the integration may be impracticable, it is clear that the only essential part of the matter is to know how an initial state confined to a single spot behaves when left to itself. As regards the effects due to special initial states, they may, I think, usually be determined better by other ways than through the integral.

I have shown (in § 371, equations (121), (122)) that the V and C at x due to the charge Q initially at the origin are

$$V = \frac{Q}{2Sv}\epsilon^{-\rho t}\left(\frac{d}{dt}+\sigma\right)P_0(z), \qquad (1)$$

$$C = \frac{1}{2}Qv\epsilon^{-\rho t}\left(-\frac{d}{dx}\right)P_0(z), \qquad (2)$$

where z has the usual meaning $\sigma(t^2-x^2/v^2)^{\frac{1}{2}}$. If Q is at the point y, then put $(x-y)$ for x in z. Now $Q = SV_0dy$, when we have a distribution of Q, or of V_0, the initial transverse voltage. Given, then, V_0 as a function of y, the state at time t is

$$V = \frac{\epsilon^{-\rho t}}{2v}\int_{-\infty}^{\infty} V_0\left(\frac{d}{dt}+\sigma\right)P_0(z')dy, \qquad (3)$$

$$C = \tfrac{1}{2}Sv\epsilon^{-\rho t}\int_{-\infty}^{\infty} V_0\left(-\frac{d}{dx}\right)P_0(z')dy, \qquad (4)$$

where $\qquad z' = \sigma\{t^2-(x-y)^2/v^2\}^{\frac{1}{2}}$.

But here $P_0(z')$ is a discontinuous function, and requires special attention because it is differentiated. Moreover, the finite speed of propagation makes the practical limits be $x+vt$ and $x-vt$, that is, at the distance vt on each side of the point x where V and C are to be found. If y is outside this range, the V_0 there has had no effect at x.

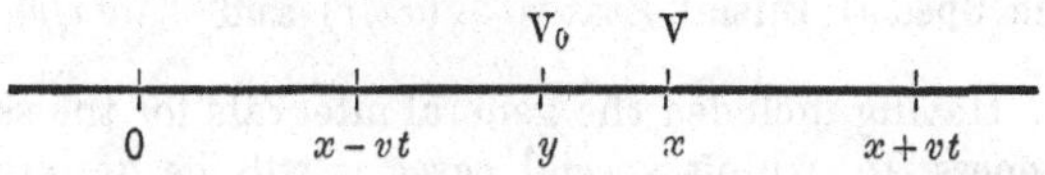

It is just at the new limits that $z'=0$, and $P_0(z')$ drops from 1 to 0. It results that when y is between the new limits we do not need any change, but at the limits we want extra terms, showing the effect of the discontinuity in $P_0(z')$. Considering the space and time variations separately at the wave front of $P_0(z')$, the operators $d/d(vt)$ and $-d/dx$ with unit operands both represent unit impulses at $y=x+vt$, whilst $d/d(vt)$ and $+d/dx$ represent unit impulses at $x-vt$. The result is that if we use the new limits we transform (3), (4) to

$$V = \epsilon^{-\rho t}\left\{\frac{V_{01}+V_{02}}{2}+\frac{1}{2v}\int_{x-vt}^{x+vt} V_0\left(\frac{d}{dt}+\sigma\right)P_0(z')dy\right\}, \qquad (5)$$

$$LvC = \epsilon^{-\rho t}\left\{\frac{V_{01}-V_{02}}{2}+\frac{1}{2}\int_{x-vt}^{x+vt} V_0\left(-\frac{d}{dx}\right)P_0(z')dy\right\}, \qquad (6)$$

where V_{01} is the value of V_0 at the lower, and V_{02} its value at the upper limit.

Similarly, the effect of an initial distribution of C_0 is to be obtained by interchanging V and C, V_0 and C_0, L and S, R and K, producing

$$C = \epsilon^{-\rho t}\left\{\frac{C_{01}+C_{02}}{2} + \frac{1}{2v}\int_{x-vt}^{x+vt} C_0\left(\frac{d}{dt} - \sigma\right)P_0(z')dy\right\}, \qquad (7)$$

$$SvV = \epsilon^{-\rho t}\left\{\frac{C_{01}-C_{02}}{2} + \frac{1}{2}\int_{x-vt}^{x+vt} C_0\left(-\frac{d}{dx}\right)P_0(z')dy\right\}. \qquad (8)$$

As the reasoning about the limits is, from the purely mathematical point of view, troublesome, it may be as well to repeat that the fundamental formulæ (1), (2) not only represent continuous distributions of V and C between the limits $\pm vt$, but also impulsive waves at the limits; and of course the impulses must not be overlooked. At $x = +vt$, (1) indicates that V is impulsive to the amount $(\frac{1}{2}Q/S)\epsilon^{-\rho t}$, and (2) that C is impulsive to the amount $(\frac{1}{2}Qv)\epsilon^{-\rho t}$. The ratio of the first to the second is Lv, showing that there is a pure electromagnetic wave at the front, $x = vt$. At the back, where $x = -vt$, V has the same sign and value, but C is reversed. The impulsive wave goes the other way.

The Special Initial States $J_0(\sigma x/v)$ and $J_n(\sigma x/v)$.

§ 407. Having included the general integrals for the sake of completeness, to exhibit special cases it will be desirable to leave them alone, and make use of previously-obtained results. In all the preceding, when generating waves, the $P_0(z)$ or $P_n(z)$ functions have been discontinuous, or only existent within certain limits. We shall now remove this restriction.

Let

$$V_0 = e\, J_0\left(\frac{\sigma x}{v}\right), \qquad (9)$$

where e is constant, represent an initial state. It is an oscillating function, like the cosine, of amplitude e at the origin, but decreasing in amplitude as we pass away in either direction, according to a law which tends ultimately to that of variation inversely as the square root of the distance. The biggest hump is in the middle ($x = 0$), and it is of extra length also.

As time passes on, two things happen. First, there is the attenuation all over according to the time factor $\epsilon^{-\rho t}$. Besides that, the function $J_0(\sigma x/v)$ is converted to $J_0\{\sigma(x^2/v^2 - t^2)^{\frac{1}{2}}\}$, which is the same as $P_0(z)$ or $I_0(z)$. That is,

$$V = e\,\epsilon^{-\rho t}P_0(z), \tag{10}$$

is what arises from the given initial state under certain circumstances to be considered presently.

There are two regions to be considered. If $x < vt$ (on either side) we have the original middle hump spread out, without any oscillation. It represents the $P_0(z)$ solution, z^2 being positive. But if $x^2 > v^2t^2$, z^2 is negative, and the function is oscillatory as at first. As time goes on, all the nodes and humps and hollows move out to the right on the right side, and to the left on the left side, leaving behind only the middle hump attenuated and changed in shape, widely spread and flattened. This is a rough general description of what occurs. The proof is that (10) is known to satisfy the characteristic, that it does so independently of the sign of z^2, and that it, when $t = 0$, expresses the initial state. What has been omitted is the consideration of the initial and subsequent current.

As regards the speed of motion of the nodes, let y_m be a root of $J_0(y) = 0$. Then

$$y_m^2 = \sigma^2\left(\frac{x^2}{v^2} - t^2\right), \quad \text{or} \quad \frac{\sigma x}{v} = (y_m^2 + \sigma^2t^2)^{\frac{1}{2}}, \tag{11}$$

and

$$\frac{dx}{dt} = \frac{\sigma v t}{(y_m^2 + \sigma^2t^2)^{\frac{1}{2}}}. \tag{12}$$

The initial speed is zero. The ultimate speed is v. The greater y_m, the less the speed. The values of the first three y's are

$$y_1 = 2{\cdot}404, \quad y_2 = 5{\cdot}520, \quad y_3 = 8{\cdot}653.$$

It is, no doubt, the big hump in the middle that drives away the smaller ones.

In a similar manner if the state at time t is

$$V = e\,\epsilon^{-\rho t}\frac{w^n}{\lfloor n}P_n(z), \tag{13}$$

or, in full, as regards w and z,

$$V = e\,\epsilon^{-\rho t}\frac{\sigma^n(t - x/v)^n}{2^n\lfloor n}P_n\{\sigma(t^2 - x^2/v^2)^{\frac{1}{2}}\}, \tag{14}$$

holding good for all values of x, the initial state may be

$$V_0 = e\ J_n(\sigma x/v) \times (-1)^n. \qquad (15)$$

But for a better understanding, it is necessary to consider the state of current as well as that of voltage.

The States of V and C resulting from any Initial States, V_0 and C_0, expansible in J_n functions.

§ **408.** For this purpose the functions U and W, expressing the momentary pure waves, are convenient (§ 378). Suppose that

$$U = B_0P_0 + (A_1u + B_1w)P_1 + (A_2u^2 + B_2w^2)\frac{P_2}{\underline{|2}} + \dots . \qquad (16)$$

Then, since $W = \alpha U$, we have

$$W = A_1P_0 + (A_2u + B_0w)P_1 + (A_3u^2 + B_1w^2)\frac{P_2}{\underline{|2}} + \dots, \qquad (17)$$

with argument z. In these put $t = 0$, then

$$U_0 = B_0J_0 + (A_1 - B_1)J_1 + (A_2 + B_2)J_2 + (A_3 - B_3)J_3 + \dots, \quad (18)$$

$$W_0 = A_1J_0 + (A_2 - B_0)J_1 + (A_3 + B_1)J_2 + (A_4 - B_2)J_3 + \dots . \quad (19)$$

Now, suppose that $W_0 = 0$ (initial state of W). Then all the A's become known in terms of the B's, and equations (16, 17) are reduced to

$$U = [B_0(1+\beta^2) + B_1(\alpha - \beta^3) + B_2(\alpha^2 + \beta^4) + B_3(\alpha^3 - \beta^5) + \dots]P_0, \qquad (20)$$

$$W = [B_0(\alpha+\beta) + B_1(\alpha^2 - \beta^2) + B_2(\alpha^3 + \beta^3) + B_4(\alpha^4 - \beta^4) + \dots]P_0. \qquad (21)$$

As a further test, note that $\beta W = U$. We now have a system of U and W such that $W_0 = 0$ and

$$U_0 = B_0J_0 + B_1J_1 + (B_0 + B_2)J_2 + (B_1 + B_3)J_3 + \dots . \qquad (22)$$

The argument of the J functions is always $\sigma x/v$.

Now, I have shown how to expand any power function in Bessel functions, § 385; do this for U_0; say

$$U_0 = \Sigma\ C_nJ_n(\sigma x/v) = C_0J_0 + C_1J_1 + \dots . \qquad (23)$$

Comparison with (22) finds the B's in terms of C's, and reduces U to

$$U = [C_0(1+\beta^2) + C_1(\alpha - \beta^3) + (C_2 - C_0)(\alpha^2 + \beta^4) + (C_3 - C_1)(\alpha^3 - \beta^5)$$
$$+ (C_4 - C_2)(\alpha^4 + \beta^6) + (C_5 - C_3)(\alpha^5 - \beta^7) + \dots]P_0. \qquad (24)$$

We have now the complete solution for the initial state U_0, coupled with the initial state W_0, for we may derive W by $\alpha U = W$. I use the convenient notation $\alpha^n P_0$ and $\beta^n P_0$ as previously, because it simplifies the work. *See* (27), (28), § 380. We may also rearrange (24) thus:

$$U = [C_0(1 + \beta^2 - \beta^4 + \ldots - \alpha^2 + \alpha^4 - \alpha^6 + \ldots)$$
$$- C_1(\beta^3 - \beta^5 + \beta^7 - \ldots - \alpha + \alpha^3 - \alpha^5 + \ldots)$$
$$+ C_2(\alpha^2 - \alpha^4 + \alpha^6 - \ldots + \beta^4 - \beta^6 + \beta^8 - \ldots) - \ldots]P_0. \quad (25)$$

Or

$$U = \left[\left(\frac{1}{1+\alpha^2} + \frac{\beta^2}{1+\beta^2}\right)C_0 + \left(\frac{\alpha}{1+\alpha^2} - \frac{\beta^3}{1+\beta^2}\right)C_1 + \ldots\right]P_0; \quad (26)$$

or

$$U = \left(C_0 + \alpha C_1 + \alpha^2 C_2 + \alpha^3 C_3 + \ldots\right)\frac{P_0}{1+\alpha^2}$$
$$+ \left(C_0 - \beta C_1 + \beta^2 C_2 - \beta^3 C_3 + \ldots\right)\frac{\beta^2 P_0}{1+\beta^2}; \quad (27)$$

of which the full expansion is

$$U = C_0\left\{ P_0 - \frac{w^2 - u^2}{\lfloor 2}P_2 + \frac{w^4 - u^4}{\lfloor 4}P_4 - \frac{w^6 - u^6}{\lfloor 6}P_6 + \ldots\right\}$$
$$+ C_1\left\{ w_1 P_1 - \frac{w^3 + u^3}{\lfloor 3}P_3 + \frac{w^5 + u^5}{\lfloor 5}P_5 - \frac{w^7 + u^7}{\lfloor 7}P_7 + \ldots\right\}$$
$$+ C_2\left\{\frac{w^2}{\lfloor 2}P_2 - \frac{w^4 - u^4}{\lfloor 4}P_4 + \frac{w^6 - u^6}{\lfloor 6}P_6 - \frac{w^8 - u^8}{\lfloor 8}P_8 + \ldots\right\}$$
$$+ \quad \ldots\ldots\ldots\ldots \quad (28)$$

To be paired with this is the solution

$$W = \left[\left(\frac{\alpha}{1+\alpha^2} + \frac{\beta}{1+\beta^2}\right)C_0 + \left(\frac{\alpha^2}{1+\alpha^2} - \frac{\beta^2}{1+\beta^2}\right)C_1 + \ldots\right]P_0, \quad (29)$$

which expands to

$$W = C_0\left\{(u + w)P_1 - \frac{u^3 + w^3}{\lfloor 3}P_3 + \frac{u^5 + w^5}{\lfloor 5}P_5 - \ldots\right\}$$
$$+ C_1\left\{\frac{w^2 - u^2}{\lfloor 2}P_2 - \frac{w^4 - u^4}{\lfloor 4}P_4 + \frac{w^6 - u^6}{\lfloor 6}P_6 - \ldots\right\}$$
$$+ \quad \&c., \quad \&c. \quad (30)$$

It is now obvious at a glance from the meaning of u and w that W vanishes initially. So, taking this form (30) as known (which it was not at the beginning), we can at once derive $U = \beta W$. We can at any rate utilise this knowledge to derive the similar solutions for U and W so that $U_0=0$, without the troublesome work. Thus

$$U=(C_0+C_1\beta+C_2\beta^2+C_3\beta^3+\dots)\frac{\beta}{1+\beta^2}P_0$$

$$+(C_0-C_1\alpha+C_2\alpha^2-C_3\alpha^3+\dots)\frac{\alpha}{1+\alpha^2}P_0 \qquad (31)$$

makes $U_0=0$. The corresponding W is αU, or

$$W=(C_0+C_1\beta+C_2\beta^2+C_3\beta^3+\dots)\frac{P_0}{1+\beta^2}$$

$$+(C_0-C_1\alpha+C_2\alpha^2-C_3\alpha^3+\dots)\frac{\alpha^2P_0}{1+\alpha^2}. \qquad (32)$$

Remember that $\alpha\beta=1$ in these workings. The initial states are

$$U_0=0, \qquad W_0=\Sigma\, C_nJ_n(\sigma x/v). \qquad (33)$$

We have now completed the solution of the problem of the states of V and C resulting from any initial distributions which admit of expansion in Bessel integral functions. The fractional cases need not be entered into at present.

Some Fundamental Examples.

§ **409.** Here are some illustrative examples :

$$U_0=0, \qquad W_0=J_0(\sigma x/v), \qquad (34)$$

$$U=(u+w)P_1-\frac{u^3+w^3}{\lfloor 3}P_3+\frac{u^5+w^5}{\lfloor 5}P_5-\dots, \qquad (35)$$

$$\alpha U=W=P_0-\frac{u^2-w^2}{\lfloor 2}P_2+\frac{u^4-w^4}{\lfloor 4}P_4-\dots. \qquad (36)$$

In a similar manner we have

$$U_0=J_0(\sigma x/v), \qquad W_0=0. \qquad (37)$$

$$U=P_0-\frac{w^2-u^2}{\lfloor 2}P_2+\frac{w^4-u^4}{\lfloor 4}P_4-\dots, \qquad (38)$$

$$W=(u+w)P_1-\frac{u^3+w^3}{\lfloor 3}P_3+\frac{u^5+w^5}{\lfloor 5}P_5-\dots. \qquad (39)$$

In the latter case $V_0=LvC_0$ initially; *i.e*, a momentary pure positive wave of the form $J_0(\sigma x/v)$ gives rise to (38) (39). At time t, $V\epsilon^{\rho t}=U+W$, and $LvC\epsilon^{\rho t}=U-W$ find V and C. In the former case the initial state was a pure negative wave, resulting in (35) (36).

Again, we know that

$$\epsilon^{\sigma t} = P_0 + (u+w)P_1 + \frac{u^2+w^2}{\lfloor 2}P_2 + \ldots, \tag{40}$$

by (43), § 381. If we change σ to σi, u becomes ui, w becomes wi, and uw becomes $-uw$; so

$$\epsilon^{\sigma ti} = P_0 + i(u+w)P_1 - \frac{u^2+w^2}{\lfloor 2}P_2 - i\frac{u^3+w^3}{\lfloor 3}P_3 + \ldots, \tag{41}$$

with argument zi, or $\sigma(x^2/v^2-t^2)^{\frac{1}{2}}$. Now interchange t and x/v, then we get

$$U = \sin(\sigma x/v) = (u-w)P_1 - \frac{u^3+w^3}{\lfloor 3}P_3 + \ldots, \tag{42}$$

$$aU = W = \cos(\sigma x/v) = P_0 - \frac{u^2+w^2}{\lfloor 2}P_2 + \frac{u^4+w^4}{\lfloor 4}P_4 - \ldots, \tag{43}$$

with argument z as usual.

Here $\sigma x/v = u - w$. We see that U, W forms a system of V, C subsiding according to $\epsilon^{-\rho t}$ without any shifting of the nodes. Of this an easier proof was given before, § 379; the present is an interesting way of exhibiting the independence of the expansions of $\sin(\sigma x/v)$ and $\cos(\sigma x/v)$ of the time, although, of course, t occurs in every term.

The General Solution for any Initial State, and some Simple Examples.

§ 410. Another way of getting solutions of the characteristic equation of V or C should not be passed over, having interesting connections with the preceding methods. The characteristic of $V\epsilon^{\rho t}$ being

$$v^2\Delta^2(V\epsilon^{\rho t}) = (p^2-\sigma^2)(V\epsilon^{\rho t}), \tag{1}$$

by (5), (6), § 378, we formerly used the form of solution

$$V\epsilon^{\rho t} = \epsilon^{sx}A + \epsilon^{-sx}B, \tag{2}$$

where A and B are time functions, and s is the operator $v^{-1}(p^2-\sigma^2)^{\frac{1}{2}}$. But write (1) in the form

$$(v^2\Delta^2+\sigma^2)(V\epsilon^{\rho t}) = p^2(V\epsilon^{\rho t}), \tag{3}$$

and we see that another form of solution is

$$V\epsilon^{\rho t} = \cosh \nu t \,.\, A + \frac{\text{shin}\, \nu t}{\nu} B, \tag{4}$$

where A and B are now x functions, and

$$\nu = (v^2\Delta^2 + \sigma^2)^{\frac{1}{2}}. \tag{5}$$

The form (4) is suitable for the derivation of the results due to given initial states. Thus, given that $V = V_0$ and $C = C_0$ at the moment $t = 0$. These data serve to determine A and B, and therefore V and C, with the assistance of (5), (6), § 378. The results are

$$V = \epsilon^{-\rho t}\left\{\left(\cosh + \frac{\sigma}{\nu}\,\text{shin}\right)\nu t \,.\, V_0 - \frac{\text{shin}\,\nu t}{\nu}\frac{\Delta}{S}C_0\right\}, \tag{6}$$

$$C = \epsilon^{-\rho t}\left\{\left(\cosh - \frac{\sigma}{\nu}\,\text{shin}\right)\nu t \,.\, C_0 - \frac{\text{shin}\,\nu t}{\nu}\frac{\Delta}{L}V_0\right\}. \tag{7}$$

Notice that the second of these is derivable from the first by interchanging V and C, V_0 and C_0, R and K, L and S. If one solution has been worked out, another having a different physical meaning is immediately derivable by the interchanges mentioned. Also notice that the functions of νt are even functions; that is, by (5), they are functions of Δ^2. So the V due to V_0 is derivable by even differentiations with respect to x, whilst the V due to C_0 is derivable by odd differentiations. We have here the means of constructing any number of solutions, provided V_0 and C_0 are given continuous functions of x, so that the operations can be carried out by ordinary rules.

For example, if $V_0 = Ax$. Then $\Delta^2 x = 0$, so it is the same as taking $\nu = \sigma$. The result is

$$V = \epsilon^{-\rho t}\epsilon^{\sigma t}Ax = Ax\,\epsilon^{-Kt/S}. \tag{8}$$

Here V subsides by leakage, but without change of type. The corresponding current is

$$C = -\frac{A}{L\sigma}\epsilon^{-\rho t}\text{shin}\,\sigma t = \frac{A}{2L\sigma}\left(\epsilon^{-Rt/L} - \epsilon^{-Kt/S}\right). \tag{9}$$

The current is the same everywhere, and is negative, whether σ be positive or negative.

Similarly, if $C_0 = Bx$, we obtain

$$C = Bx\,\epsilon^{-Rt/L}, \qquad V = \frac{B}{2S\sigma}\left(\epsilon^{-Rt/L} - \epsilon^{-Kt/S}\right). \tag{10}$$

It is now V that is negative for any value of σ.

If we combine the two initial states, and make $B = A/Lv$, then $V_0 = LvC_0$ initially, a positive wave, and

$$V = Ax\,\epsilon^{-Kt/S} + \frac{Av}{2\sigma}\left(\epsilon^{-Rt/L} - \epsilon^{-Kt/S}\right) \tag{11}$$

is the resulting V. When $\sigma = 0$, we reduce to

$$V = LvC = A(x - vt)\epsilon^{-\rho t}, \tag{12}$$

representing a positive wave.

Similarly, the results due to $V_0 = Ax^2$, Ax^3, &c., may be developed. They are much more complicated. Passing on to a complete power series, the simplest is $V_0 = A\epsilon^{hx}$. Here the potence of Δ is h, so the results are

$$V = \epsilon^{-\rho t}\left(\cosh \nu t + \frac{\sigma}{\nu}\,\text{shin}\,\nu t\right)A\epsilon^{hx}, \tag{13}$$

$$L\nu C = -A\epsilon^{-\rho t}\,\text{shin}\,\nu t\, h\epsilon^{hx}, \tag{14}$$

where ν means the positive constant $(v^2h^2 + \sigma^2)^{\frac{1}{2}}$, by (5).

More important than the last case is the simply periodic. If V_0 and C_0 are simply periodic functions of x, say of $\sin mx$ and $\cos mx$, the potence of Δ^2 is $-m^2$, and (6), (7) represent the solutions explicitly, provided ν has the constant value $(\sigma^2 - m^2v^2)^{\frac{1}{2}}$. Or, if $\lambda = (m^2v^2 - \sigma^2)^{\frac{1}{2}}$, which may be more convenient, then

$$V = \epsilon^{-\rho t}\left\{\left(\cos + \frac{\sigma}{\lambda}\sin\right)\lambda t.\ V_0 - \frac{\sin\lambda t}{\lambda}\frac{\Delta}{S}C_0\right\}, \tag{15}$$

$$C = \epsilon^{-\rho t}\left\{\left(\cos - \frac{\sigma}{\lambda}\sin\right)\lambda t.\ C_0 - \frac{\sin\lambda t}{\lambda}\frac{\Delta}{L}V_0\right\} \tag{16}$$

are the simply periodic solutions. They may also be regarded as the general solutions, provided $-\lambda^2 = v^2\Delta^2 + \sigma^2$ defines λ.

A curious case is $\lambda = 0$, or $m = \sigma/v$. This we had before, § 379. There is no change of type as the time advances,

Conversion to Definite Integrals. Short Cut to Fourier's Theorem.

§ 411. All the above solutions labour under the defect that they are not practical in this respect, that the initial states extend continuously over the whole range of x. Moreover, if V_0 is represented by x, x^2, x^3, &c., it increases infinitely as we pass away from the origin. Even $V_0 = \epsilon^{hx}$ becomes infinite on one side. And although the simply periodic solution has no infinities it does not vanish at infinity. A more practical sort of initial state would be $V_0 = A\epsilon^{-hx^2}$, vanishing at infinity on both sides, and by an expedient admitting of concentration at the origin. But the working out of this case is rather elaborate.

Now since (7), (8) are the general operational solutions they are true for any V_0 and C_0, discontinuous as well as continuous. It is by the consideration of states which are initially discontinuous that we should obtain practical solutions showing the genesis and progression of waves. So we have to find how to algebrise (7), (8) for discontinuous V_0 or C_0. There are two ways. One is to convert the general solutions to Fourier integrals, and then evaluate them. The other way is to find a transformation of the general operational solutions which will allow of their direct algebrisation. The direct way was the first way I attacked this problem. My second way was a similar process applied to an impressed force. The third way was the indirect process of going viâ Fourier integrals, corroborating the results got by the previous ways.

Here first do the integrals, to get it over. It is readily proved in works on plane trigonometry that

$$\pm 1 = \frac{4}{\pi}\left(\sin\frac{\pi x}{l} + \tfrac{1}{3}\sin\frac{3\pi x}{l} + \tfrac{1}{5}\sin\frac{5\pi x}{l} + \dots\right), \tag{17}$$

being $+1$ when x is between 0 and l and -1 when between 0 and $-l$. So there is a jump at the origin. This is important. Make l infinite. Denote the coefficient of x in any sine term by m, then the step dm is $2\pi/l$, so (17) is converted to

$$\pm 1 = \frac{2}{\pi}\int_0^\infty \frac{\sin mx}{m}\,dm, \tag{18}$$

being $+1$ when x is $+$ and -1 when x is $-$. The jump is 2 at the origin. Next differentiate to x. The result is (dividing by 2),

$$\Delta 1 = \frac{1}{\pi}\int_0^\infty \cos mx\, dm, \tag{19}$$

i.e., an impulsive function of total 1, condensed at the origin. This is the effective meaning for physical purposes. So

$$\frac{f(y)dy}{\pi}\int_0^\infty \cos m(x-y)\, dm, \tag{20}$$

represents zero, except at $x=y$, where the total is $f(y)dy$. Finally,

$$f(x) = \frac{1}{\pi}\int_{-\infty}^{\infty} f(y)dy \int_0^\infty \cos m(x-y)\, dm, \tag{21}$$

by integrating (20) from $y = -\infty$ to $+\infty$.

The above is a short cut to Fourier's theorem. But it is (19) or (20) we want here. Go back to (6), (7), and let the initial state be $V_0 dy$ at the point $x=y$. Then

$$V = \epsilon^{-\rho t}(p+\sigma)\frac{\text{shin}\,\nu t}{\nu}\,\frac{V_0 dy}{\pi}\int_0^\infty \cos m(x-y)\, dm. \tag{22}$$

In this form the ν operator can act on the discontinuous operand by direct differentiations applied to the cosine function. The potence of Δ^2 is $-m^2$, and the result is

$$V = \epsilon^{-\rho t}(p+\sigma)\,\frac{V_0 dy}{\pi}\int_0^\infty \frac{\sin \lambda t}{\lambda}\cos m(x-y)dm. \tag{23}$$

The corresponding C is

$$C = \epsilon^{-\rho t}\frac{V_0 dy}{\pi}\int_0^\infty \frac{m}{L}\frac{\sin\lambda t}{\lambda}\sin m(x-y)\, dm, \tag{24}$$

by doing the same work for the part of (7) depending on V_0. These being the V and C arising from $V_0 dy$ at the point $y=x$, integration from $y = -\infty$ to $+\infty$ will give V and C due to any initial distributions V_0, C_0. So the general solutions (6), (7) are equivalently expressed by

$$V = \frac{\epsilon^{-\rho t}}{\pi}\int_0^\infty\int_{-\infty}^{\infty}\left\{V_0(p+\sigma) - C_0\frac{\Delta}{S}\right\}\frac{\text{shin}\,\nu t}{\nu}\cos m(x-y)\,dm\,dy, \tag{25}$$

$$C = \frac{\epsilon^{-\rho t}}{\pi}\int_0^\infty\int_{-\infty}^{\infty}\left\{C_0(p-\sigma) - V_0\frac{\Delta}{L}\right\}\frac{\text{shin}\,\nu t}{\nu}\cos m(x-y)\,dm\,dy, \tag{26}$$

where $\nu = (\sigma^2 - m^2 v^2)^{\frac{1}{2}}$, and V_0, C_0 are expressed as functions of y.

The Space Integrals of V and C, due to Elements at the Origin.

§ 412. We can evaluate these integrals in any case in which V and C have been determined, say by the operational solutions (6), (7). For example, we know the results due to V_0 = constant, or x, ϵ^{hx}, &c. But these special evaluations do not help us to evaluate for a discontinuity. Say V_0 is constant on the left, and zero on the right side of the origin. The full realisation then depends upon evaluating this integral, which is the fundamental one,

$$u = \frac{2}{\pi}\int_0^\infty \frac{\sin \lambda t}{\lambda} \cos mx\, dm, \tag{27}$$

where $\lambda = (m^2v^2 - \sigma^2)^{\frac{1}{2}}$. When this is done we can easily derive the V and C due to elements of V_0 and C_0 anywhere. Before solving (27), however, and without knowing the result, we can get some information out of (23), (24).

Thus, let us find the space-integral of V due to $V_0 dy$ at the origin. We have

$$\frac{V}{V_0 dy} = \frac{\epsilon^{-\rho t}}{\pi}\int_0^\infty \cos mx \left(\cosh + \frac{\sigma}{\nu}\text{shin}\right)\nu t . dm, \tag{28}$$

and therefore

$$\frac{\int_0^x V dx}{V_0 dy} = \frac{\epsilon^{-\rho t}}{\pi}\int_0^\infty \frac{\sin mx}{m}\left(\cosh + \frac{\sigma}{\nu}\text{shin}\right)\nu t . dm. \tag{29}$$

This can be evaluated when $x = \infty$. Look back to equation (18). When x is made infinitely great, the effective meaning is that the 1 is concentrated at the origin of m. It is the only place where the infinitely rapid oscillations do not effectively cancel. So put $m = 0$ in the t function. We make $\nu = \sigma$, and get

$$\frac{\int_0^\infty V dx}{V_0 dy} = \frac{\epsilon^{-\rho t}}{\pi}\frac{\pi}{2}\epsilon^{\sigma t} = \tfrac{1}{2}\epsilon^{-Kt/S}. \tag{30}$$

That is, a charge initially at one spot subsides in amount according to $\epsilon^{-Kt/S}$, however it may be redistributed, which we are not supposed to know. The $\frac{1}{2}$ factor shows that half the charge only goes to the right.

Exactly similarly, if the initial state is $C_0 dy$, we have

$$\frac{\int_0^\infty C dx}{C_0 dy} = \frac{\epsilon^{-\rho t}}{\pi}\frac{\pi}{2}\epsilon^{-\sigma t} = \tfrac{1}{2}\epsilon^{-Rt/L}. \tag{31}$$

That the total charge subsides according to $\epsilon^{-Kt/S}$, and the total momentum according to $\epsilon^{-Rt/L}$, may be proved in an elementary manner through the properties of the distortionless circuit.

Since there is no V beyond $x = vt$, we may give x in (29) any value exceeding vt, and still have the result (30), and therefore

$$\epsilon^{\sigma t} = \frac{2}{\pi}\int_0^\infty \frac{\sin mx}{m}\left(\cosh + \frac{\sigma}{\nu}\,\text{shin}\right)vt\,.\,dm, \qquad (32)$$

if $x > vt$. But if x is less than vt, the value of this integral is a function of x in general. Except when x is just under vt. We then exclude the head of the disturbance, namely, the wave going to the right, of amount $\frac{1}{2}V_0 dy\,\epsilon^{-\rho t}$. Deduct this from $\int_0^\infty V dx$ in (30), therefore. This means that the value of the right number of (32) is $\epsilon^{\sigma t} - 1$, when x is just under vt. When x is actually vt, take the mean, and say $\epsilon^{\sigma t} - \frac{1}{2}$. See (41), (42), § 398, for another arrival at these results.

The Time Integral of C, due to Elements at the Origin.

§ 413. Next we may find the time-integral of C at any point. First, we may prove by ordinary integrations that

$$\int_0^t \epsilon^{-\rho t}\frac{\sin \lambda t}{\lambda}\,dt = \left[-\epsilon^{-\rho t}\frac{\cos \lambda t + (\rho/\lambda)\sin \lambda t}{v^2(m^2 + RK)}\right]_0^t, \qquad (33)$$

$$\int_0^t \epsilon^{-\rho t}\left\{\cos \lambda t - \frac{\sigma}{\lambda}\sin \lambda t\right\}dt$$

$$= \left[\epsilon^{-\rho t}\frac{-2b\cos \lambda t + (m^2v^2 + 2b\sigma)\lambda^{-1}\sin \lambda t}{v^2(m^2 + RK)}\right]_0^t, \qquad (34)$$

where $b = K/2S$. Therefore, if initially we have $V_0 dy$ and $C_0 dy$ at the origin, we obtain, by (24) for the result due to V_0, and by the companion to (23) for the result due to C_0,

$$\int_0^\infty C dt = \frac{dy}{\pi}\int_0^\infty \frac{KS^{-1}LC_0\cos mx + V_0 m \sin mx}{Lv^2(m^2 + RK)}\,dm. \qquad (35)$$

This is a well-known form of integral. It makes

$$\int_0^\infty C dt = dy\left\{\tfrac{1}{2}LC_0\left(\frac{K}{R}\right)^{\frac{1}{2}} + \tfrac{1}{2}SV_0\right\}\epsilon^{-x\sqrt{RK}}, \qquad (36)$$

showing separately how much is due to V_0 and how much to C_0. The latter vanishes when there is no leakage, and which is more striking, increases indefinitely with K at the origin; whereas the former is greatest when $K=0$, and least (zero) when $K=\infty$.

If $V_0 = LvC_0$, so that we start with a positive wave to the right and $Q = SV_0 dy$ is the initial charge, then the last formula makes

$$\int_0^\infty C dt = \tfrac{1}{2}Q\left\{1+\left(\frac{KL}{RS}\right)^{\frac{1}{2}}\right\}\epsilon^{-x\sqrt{RK}}. \tag{37}$$

All the charge goes to the right when $L/R = K/S$, and never returns. But we get only $\frac{1}{2}Q$ when there is no leakage. Though all of Q goes to the right at first, $\frac{1}{2}Q$ comes back again. It is the resistance that causes this. When K/S exceeds R/L, the time integral of the current exceeds Q. This is abnormal, because R/L is always greater than K/S practically. But in the application to electromagnetic waves in general, both cases may have to be considered.

Evaluation of the Fundamental Integral.

§414. Next, we have to evaluate the integral u, equation (27), before progress can be made to complete knowledge by the Fourier integral method. We may expand the function $(\sin\lambda t)/\lambda$ in a series in various ways. For example, the expansion (166), §375, may be employed, with special meanings of the symbols, to find $\epsilon^{\nu t}/\nu$, and $\epsilon^{-\nu t}/\nu$, and therefore $(\sin\lambda t)/\lambda$. We may then, if we can, effect the integration concerned in u term by term. Or other expansions may be employed in the same manner. But such work is very complicated, and, without guidance, might not be satisfactory. There is, however, a short cut whereby the complications due to using a full algebraical expansion of $(\sin\lambda t)/\lambda$ may be avoided.

Thus, λ is a function of σ^2. So, by Taylor's Theorem,

$$f(\sigma^2) = f(0) + \sigma^2 f'(0) + \frac{\sigma^4}{\underline{|2}} f''(0) + \dots . \tag{38}$$

But it may easily be proved by differentiating $\epsilon^{\nu t}/\nu$ that

$$\frac{d^2}{dt\,d\sigma^2}\frac{\epsilon^{\nu t}}{\nu} = \tfrac{1}{2}t\frac{\epsilon^{\nu t}}{\nu}. \tag{39}$$

Therefore
$$\frac{d}{d\sigma^2}\frac{\epsilon^{\nu t}}{\nu}=\int \tfrac{1}{2}t\frac{\epsilon^{\nu t}}{\nu}dt. \tag{40}$$

The operators $d/d\sigma^2$ and $\frac{1}{2}p^{-1}t$ are therefore equivalent, and we may employ the changed operator to obtain the expansion of $\epsilon^{\nu t}/\nu$ by integrations to t instead of differentiations to σ^2. The same is true of the function $\lambda^{-1}\sin\lambda t$. Thus,

$$\frac{\sin\lambda t}{\lambda}=\left(1+\tfrac{1}{2}\sigma^2p^{-1}t+\frac{(\frac{1}{2}\sigma^2p^{-1}t)^2}{\underline{|2}}+\dots\right)\frac{\sin mvt}{mv}, \tag{41}$$

because $\sigma=0$ reduces λ to mv. Or, to avoid misunderstanding of the meaning of $p^{-1}t$,

$$\frac{\sin\lambda t}{\lambda}=\frac{\sin mvt}{mv}+\tfrac{1}{2}\sigma^2\int_0^t t\frac{\sin mvt}{mv}dt+\frac{(\frac{1}{2}\sigma^2)^2}{2}\int_0^t tdt\int_0^t t\frac{\sin mvt}{mv}dt+\dots. \tag{42}$$

That this gives a correct expansion of the function concerned may be verified by comparison with the formula (166), § 375, already referred to. The process here indicated, in fact, provides an alternative method of finding the expansion of $\epsilon^{\nu t}/\nu$, and the connected functions. But the point at present is *not* to employ the developed expansion, but the operator which develops if out of $(\sin mvt)/mv$. Thus, by (27) and (41),

$$u=\left(1+\tfrac{1}{2}\sigma^2p^{-1}t+\frac{(\frac{1}{2}\sigma^2p^{-1}t)^2}{\underline{|2}}+\dots\right)\frac{2}{\pi}\int_0^\infty \cos mx\frac{\sin mvt}{mv}dm, \tag{43}$$

by shifting the operator outside the sign of integration to m.

Now, since
$$\cos mx\sin mvt=\tfrac{1}{2}\sin m(vt+x)+\tfrac{1}{2}\sin m(vt-x),$$
we can see, by comparison with (18) above, that the value of the integral in the last equation multiplied by $2/\pi$ is v^{-1} when $vt>x$, and zero when $vt<x$. Therefore

$$u=\frac{1}{v}\left(1+\tfrac{1}{2}\sigma^2p^{-1}t+\frac{(\frac{1}{2}\sigma^2p^{-1}t)^2}{\underline{|2}}+\dots\right)(0\text{ or }1). \tag{44}$$

That the operand is 0 when $t<x/v$ has the result that if we use the operand 1 only, the limits of integration must be from x/v to t instead of from 0 to t. So.

$$u=\frac{1}{v}\left\{1+\tfrac{1}{2}\sigma^2\int_{x/v}^t t\,dt+\frac{(\frac{1}{2}\sigma^2)^2}{\underline{|2}}\int_{x/v}^t tdt\int_{x/v}^t t\,dt+\dots\right\}. \tag{45}$$

The first term in the brackets is 1; the second is

$$\tfrac{1}{2}\sigma^2\left[\tfrac{1}{2}t^2\right]_{x/v}^{t} = \tfrac{1}{2}\,\tfrac{1}{2}\sigma^2(t^2 - x^2/v^2)\,; \tag{46}$$

the third is

$$\tfrac{1}{2}\frac{(\tfrac{1}{2}\sigma^2)^2}{\underline{|2}}\int_{x/v}^{t} t(t^2 - x^2v^2)dt = \tfrac{1}{8}\frac{(\tfrac{1}{2}\sigma^2)^2}{\underline{|2}}(t^2 - x^2/v^2)^2\,; \tag{47}$$

the fourth

$$\tfrac{1}{8}\frac{(\tfrac{1}{2}\sigma^2)^3}{\underline{|3}}\int_{x/v}^{t} t(t^2 - x^2/v^2)^2dt = \tfrac{1}{48}\frac{(\tfrac{1}{2}\sigma^2)^3}{\underline{|3}}(t^2 - x^2/v^2)^3\,: \tag{48}$$

and so on, the process being to multiply any term by t and then integrate to obtain the next term. We can see at a glance now what the complete formula is. It is

$$u = \frac{2}{\pi}\int_0^\infty \cos mx\,\frac{\sin\lambda t}{\lambda}\,dm = \frac{1}{v}\,\mathrm{I}_0[\sigma(t^2 - x^2/v^2)^{\frac{1}{2}}]\,. \tag{49}$$

provided $t > x/v$. And $u = 0$, if $t < x/v$.

Compare with (33), § 398, where this result was arrived at in another way. Having got it out of the Fourier integral, we can now if we like proceed to obtain the formulæ for the waves of V and C due to $\mathrm{V}_0 dy$ and $\mathrm{C}_0 dy$. They agree fully with the previous results, as in § 371 for instance, and so need not be repeated here.

Generalisation of the Integral. Both kinds of Bessel Functions.

§ **415.** The following generalisation of the result is of importance in some other electromagnetic problems, and may therefore be conveniently put here for future reference. Using the notation for Bessel functions explained towards the end of the last chapter, we have

$$\mathrm{J}_0[(n^2/v^2 - m^2)^{\frac{1}{2}}r] = \frac{2}{\pi}\int_0^\infty \frac{\cos m\lambda\,\sin nv^{-1}(r^2 + \lambda^2)^{\frac{1}{2}}}{(r^2 + \lambda^2)^{\frac{1}{2}}}\,d\lambda, \tag{50}$$

$$\mathrm{G}_0[(n^2/v^2 - m^2)^{\frac{1}{2}}r] = \frac{2}{\pi}\int_0^\infty \frac{\cos m\lambda\,\cos nv^{-1}(r^2 + \lambda^2)^{\frac{1}{2}}}{(r^2 + \lambda^2)^{\frac{1}{2}}}\,d\lambda, \tag{51}$$

provided $n/v > m$. The first of these is equivalent to (49). The second is its companion, concerning the other Bessel function. But if $n/v < m$, then the value of the integral in

(50) is zero instead of $J_0(\ldots)$; and instead of $G_0(\ldots)$ on the left side of (51) we must write

$$K_0[(m^2 - n^2/v^2)^{\frac{1}{2}}r]. \qquad (52)$$

Also, whether m be greater or less than n/v, we have this result:—

$$K_0[(m^2 + p^2/v^2)^{\frac{1}{2}}r]\sin nt$$
$$= \frac{1}{\pi}\int_{-\infty}^{\infty} \cos m\lambda \frac{\sin n\,[t - v^{-1}(r^2 + \lambda^2)^{\frac{1}{2}}]}{(r^2 + \lambda^2)^{\frac{1}{2}}} d\lambda. \qquad (53)$$

This formula includes the preceding. [Here p is the time time differentiator as usual, and the special previous results depend upon (56), § 342].

The C due to initial V_0. Operational Method, and Modification.

§ 416. Returning to the investigation in § 414, there are some things about it worth attention in view of understanding the inner workings of the mechanism of analysis. We converted the original operational solution to a Fourier integral because we assumed that we did not see how to algebrise it directly. But then again, not seeing how to evaluate the integral simply by commonly-employed orthodox methods, we employed an operational process to avoid the complication of a full development of the functions concerned. Now, consideration will show that the Fourier integral plays an unnecessary part in the work. It may be skipped over without any loss. On the other hand, there will be a gain in simplicity by making the whole of the work operational.

Thus, given initially $V = V_0$, constant on the left side, and zero on the right side of the origin, the current that results, by (7), is

$$C = V_0 \epsilon^{-\rho t} \frac{\text{shin}\, \nu t}{L\nu} \Delta 1, \qquad (54)$$

because $-\Delta V_0 = V_0 \Delta 1$ here. Now use the operational expansion (41), or, which is the same,

$$\frac{\text{shin}\, \nu t}{\nu} = \epsilon^{\frac{1}{2}\sigma^2 p^{-1} t}\, \frac{\text{shin}\, vt\Delta}{v\Delta}. \qquad (55)$$

Then we get

$$C = \frac{V_0 \epsilon^{-\rho t}}{Lv} \epsilon^{\frac{1}{2}\sigma^2 p^{-1} t}\, \text{shin}\, vt\Delta \,.\, 1. \qquad (56)$$

Now $\epsilon^{vt\Delta}1$ is 1 from $x = -vt$ to ∞, and $\epsilon^{-vt\Delta}1$ is 1 from $x = +vt$ to ∞, by Taylor's theorem, § 276. So their difference represents 1 from $x = -vt$ to $+vt$. Therefore

$$C = \frac{V_0\epsilon^{-\rho t}}{2Lv}\,\epsilon^{\frac{1}{2}\sigma^2 p^{-1}t}[1]_{-vt}^{+vt}, \tag{57}$$

where the 1 means a function of x existent between the limits indicated. Now the limits for the time integration are 0 and t as usual, but since the operand is 0 when $x > vt$ or $< -vt$, the effective limits are x/v and t on the right side of the origin, and $-x/v$ and t on the left side. The rest of the work is as in § 414, for the right side, making

$$C = \frac{V_0\epsilon^{-\rho t}}{2Lv}I_0\{\sigma(t^2 - x^2/v^2)^{\frac{1}{2}}\}, \tag{58}$$

as before found, § 367. It is true on both sides of the origin.

It will be observed that the only troublesome part of this investigation is the initial recognition of the expansion (55). Granting that, the rest is plain enough.

The result (58) suggests a modification which did not present itself naturally previously. If two operators are strictly identical, however different in appearance, one may be substituted for the other in general, perhaps universally, if worked rightly. But the substitution may also be made when the operators are not identical, if their effects are the same under the circumstances considered. Thus, as above, 2 shin $vt\Delta.1$ means 1 between $x = +vt$ and $-vt$. The original operand 1 is a positively existent function of x, and the operator turns it to a function of x and t. (They are not functions at all in the limited sense of function theorists, but that does not matter. Physics is above mathematics, and the slave must be trained to work to suit the master's convenience.) Now compare it with $\epsilon^{-(x^2/v^2)D}1$, where D means the t^2 differentiator. The 1 is now a positively existent function of t^2, and the operator destroys the part between $t^2 = 0$ and x^2/v^2. The resulting function is therefore 1 provided t^2 is greater than x^2/v^2. So it is the same function of x and t as before. Instead of (56), therefore, we may write

$$C = \frac{V_0\epsilon^{-\rho t}}{2Lv}\,\epsilon^{\frac{1}{2}\sigma^2 p^{-1}t}\epsilon^{-(x^2/v^2)D}1. \tag{59}$$

The significance of the change will now appear. For since $D = \frac{1}{2}t^{-1}p$, therefore

$$D^{-1} = 2p^{-1}t,$$

and (59) is the same as

$$C = \frac{V_0\epsilon^{-\rho t}}{2Lv}\epsilon^{\frac{1}{4}\sigma^2 D^{-1}}\epsilon^{-(x^2/v^2)D}1. \qquad (60)$$

There is now only one differentiator in question, and the corresponding variable does not appear in the operators. They may therefore be interchanged, making

$$C = \frac{V_0\epsilon^{-\rho t}}{2Lv}\epsilon^{-(x^2/v^2)D}\epsilon^{\frac{1}{4}\sigma^2 D^{-1}}1. \qquad (61)$$

Now we know that

$$\epsilon^{\frac{1}{4}\sigma^2 D^{-1}}1 = I_0(\sigma t), \qquad (62)$$

as in (19), § 377. Or we may verify (62) on the spot. Then the other operator in (61) turns t^2 to $t^2 - x^2/v^2$, and brings us to the result (58) again. It is not necessary to interchange the operators as above done; but it has the result of considerably simplifying the work. We could not interchange in the form (56), because the variable t is present in the second operator, and is acted upon by the corresponding differentiator in the first.

The V due to initial V_0.

§ **417.** The corresponding treatment of the V due to initial V_0 is more difficult. We have, by (6)

$$V = V_0\epsilon^{-\rho t}(p+\sigma)\frac{\text{shin}\,\nu t}{\nu}1, \qquad (63)$$

if initially $V = V_0$, constant on the + side of the origin. Now use (55). We get

$$V = \frac{V_0\epsilon^{-\rho t}}{2v}(p+\sigma)\epsilon^{\frac{1}{4}\sigma^2 p^{-1}t}(\epsilon^{vt\Delta} - \epsilon^{-vt\Delta})x, \qquad (64)$$

where the x stands for $\Delta^{-1}1$. It is only positively existent.

Or,
$$V = \frac{V_0\epsilon^{-\rho t}}{2v}(p+\sigma)\epsilon^{\frac{1}{2}\sigma^2 p^{-}t}\{[x+vt] - [x-vt]\}, \qquad (65)$$

where the functions in the [ ]'s are only positively existent. That is, when x is less than $-vt$, (x being on the negative side), the operand is zero, and V is zero. And when x between $-vt$ and $+vt$, the operand is $x+vt$. Finally, when x is greater than vt, the operand is $2vt$. This triple function of x and t is fully represented either in (64) or in (65) under the limitation to positivity mentioned.

Now, when x is greater than vt, we know the result by elementary reasoning. The disturbance has not reached x, so V there is the original V_0 attenuated by the leakance. It is $V = V_0\epsilon^{-Kt/S}$. We should verify this. The operand is $2vt$, and

$$(1+\tfrac{1}{2}\sigma^2p^{-1}t+\ldots)t = t+\tfrac{1}{2}\sigma^2\frac{t^3}{3}+\frac{(\frac{1}{2}\sigma^2)^2}{\underline{|2}}\frac{t^5}{3\,.\,5}+\ldots = \sigma^{-1}\operatorname{shin}\sigma t. \quad (66)$$

This makes

$$V = V_0\epsilon^{-\rho t}(p+\sigma)\frac{\operatorname{shin}\sigma t}{\sigma} = V_0\epsilon^{-\rho t}\epsilon^{\sigma t}, \quad (67)$$

which is the required result.

But when x lies between $\pm vt$, and the operand is $x+vt$, the work is considerably complicated. We found before, by another method, that an infinite series of Bessel functions was needed. When x is +, or in the region of V_0, as the time goes from 0 to t, the operand is $2vt$ from $t=0$ to x/v, and is then $x+vt$ from $t=x/v$ to the full t. When x is −, the operand is 0 until the time reaches $-x/v$, and is then $x+vt$ till the full t is reached. It is not necessary to do both sides. The side away from V_0, the negative side here, is the easier to manage. We have to find U, given by

$$U = (1+\tfrac{1}{2}\sigma^2p^{-1}t+\ldots)\,(x+vt), \quad (68)$$

the limits in the integrations being $-x/v$ and t. The beginning of the resulting series going only as far as p^{-2}, is

$$U = x+vt+\tfrac{1}{2}\sigma^2\left\{\frac{x}{2}\left(t^2-\frac{x^2}{v^2}\right)+\frac{v}{3}\left(t^3+\frac{x^3}{v^3}\right)\right\}$$
$$+\frac{(\frac{1}{2}\sigma^2)^2}{\underline{|2}}\left\{\frac{x}{2}\left[\frac{1}{4}\left(t^4-\frac{x^4}{v^4}\right)-\frac{x^2}{2v^2}\left(t^2-\frac{x^2}{v^2}\right)\right]+\frac{v}{3}\left[\frac{1}{5}\left(t^5+\frac{x^5}{v^5}\right)\right.\right.$$
$$\left.\left.+\frac{x^3}{2v^3}\left(t^2-\frac{x^2}{v^2}\right)\right]\right\} \quad (69)$$

This is horrid, and need not be pursued further, although doubtless, by extension and proper arrangement, it can be more conveniently exhibited. If the initial V_0 is on the left side, the corresponding U on the right side is obtained by changing the sign of x.

We have already found a concise and convenient way of developing the C wave when the V wave is known, and conversely, §§ 382, 386. In the present case we have (58) for the C wave, and therefore

$$V = V_0 \epsilon^{-\rho t} \frac{p+\sigma}{2v\Delta} I_0(z), \tag{70}$$

for the V wave. For distinctness let V_0 be on the negative side, then we calculate V on the positive side, so as not to be troubled with the subsiding V_0. Now use the α, β differentiators, and we have

$$\frac{p+\sigma}{v\Delta} = \frac{1+\frac{1}{2}(\alpha+\beta)}{\frac{1}{2}(\alpha-\beta)} = \frac{1+\alpha}{1-\alpha}, \tag{71}$$

as explained in § 382. This makes

$$V = \tfrac{1}{2} V_0 \epsilon^{-\rho t}(1 + 2\alpha + 2\alpha^2 + \ldots)P_0(z) \tag{72}$$

$$= \tfrac{1}{2} V_0 \epsilon^{-\rho t}\left(P_0 + 2wP_1 + 2\frac{w^2}{\underline{|2}} P_2 + \ldots\right)(z). \tag{73}$$

See also § 381. The investigations are practically the same on the right side of the origin; but on the left side there is a distinction. If the source is an impressed force at the origin, then C at x and at $-x$ are identical, whilst V at $-x$ is the negative of V at x, assuming that x is positive. But when the source is the distribution V_0 above assumed, then we must add on $V_0 \epsilon^{-Kt/S}$ to the negative of the V in (73) to obtain V at $-x$.

Final Investigation of the V and C due to initial V_0 and C_0.

§ **418.** We now come to the very last way I shall give of algebrising the operational solutions for initial states. It has some historical interest, and had especial value in being the first way in which I was able to obtain full mathematical confirmation of the results relating to the generation of positive and negative tails obtained by a consideration of the distortionless circuit.

For explicitness, say that $V = V_0$ initially, without any C. Then, by (6), (7), the resulting states of V and C are

$$V = \epsilon^{-\rho t}(p+\sigma)\frac{\text{shin}\,\nu t}{\nu}V_0, \qquad (74)$$

$$C = -\epsilon^{-\rho t}\frac{\text{shin}\,\nu t}{L\nu}\Delta V_0. \qquad (75)$$

Now in the distortionless case $\epsilon^{\nu t}$ reduces to $\epsilon^{vt\Delta}$, which is merely a translational operator, and enables the algebrisation to be immediately effected. But since when there is distortion the property of propagation of elementary disturbances at speed v remains true, if we express $\nu^{-1}\epsilon^{\nu t}$ in the form $\epsilon^{vt\Delta}f(v\Delta)$, then we separate the distortion from the translation. Whether this is useful depends upon the form assumed by $f(v\Delta)$. If $f(v\Delta)V_0$ can be algebrised, then the work is done.

The required expansion may be effected by § 375, equation (166). Turn y to t, s^2 to σ^2, and p to $v\Delta$.

The result is

$$\frac{\epsilon^{\nu t}}{\nu} = \frac{\epsilon^{vt\Delta}}{v\Delta}\left\{1 + \frac{s\sigma t}{2}r_1 + \left(\frac{s\sigma t}{2}\right)^2\frac{r_2}{\lfloor 2} + \left(\frac{s\sigma t}{2}\right)^3\frac{r_3}{\lfloor 3} + \dots\right\}, \qquad (76)$$

where $s = \sigma/v\Delta$, and the r's are functions of $(v\Delta)^{-1}$, thus,

$$r_1 = 1 - \frac{1}{vt\Delta}, \qquad r_2 = 1 - \frac{3}{vt\Delta} + \frac{3}{(vt\Delta)^2}, \quad \text{\&c.}, \qquad (77)$$

as described in § 375, only with yp there changed to $vt\Delta$ here.

Since changing the sign of t gives us $\epsilon^{-\nu t}/\nu$, it is clear that the solutions (74), (75) are expanded as required, and since $f(v\Delta)$ contains Δ inversely in s and in the r's, the expansion is suitable for the treatment of discontinuities.

In the practical execution it is perhaps as well, or better, to rearrange (76) in powers of $(v\Delta)^{-1}$. This is to be done by inspection of the r functions. The result is

$$\frac{\epsilon^{\nu t}}{\nu} = \frac{\epsilon^{vt\Delta}}{v\Delta}\{1 + sg_1 + s^2g_2 + s^3g_3 + \dots\}, \qquad (78)$$

where the g's are time functions, namely

$$g_1 = \frac{1}{2}\sigma t, \qquad g_2 = -\frac{1}{2} + \frac{\sigma^2t^2}{2.4}, \qquad g_3 = -\frac{3\sigma t}{8} + \frac{\sigma^3t^3}{2.4.6}, \qquad (79)$$

being the same as in § 376, equation (179), where they turned up in another way. The next three are

$$g_4=\frac{3}{8}-\frac{\sigma^2t^2}{8}+\frac{\sigma^4t^4}{2.4.6.8},\qquad g_5=\frac{5\sigma t}{16}-\frac{5}{4}\frac{\sigma^3t^3}{2.4.6}+\frac{\sigma^5t^5}{2.4.6.8.10},$$

$$g_6=-\frac{5}{16}+\frac{15}{16}\frac{\sigma^2t^2}{2.4}-\frac{3}{2}\frac{\sigma^4t^4}{2.4.6.8}+\frac{\sigma^6t^6}{2.4.6.8.10.12}. \tag{80}$$

Similarly we can develop $(p+\sigma)\epsilon^{\nu t}/\nu$, either by differentiating (78), or by using the A series in § 375. We get

$$(p+\sigma)\frac{\epsilon^{\nu t}}{\nu}=\epsilon^{\nu t\Delta}(1+sf_1+s^2f_2+s^3f_3+\dots), \tag{81}$$

where the f's are also time functions, of which the first six are

$$f_1=1+\frac{\sigma t}{2},\quad f_2=\frac{\sigma t}{2}\Big(1+\frac{\sigma t}{4}\Big),\quad f_3=-\frac{1}{2}\Big(1+\frac{\sigma t}{4}\Big)+\frac{\sigma^2t^2}{2.4}\Big(1+\frac{\sigma t}{6}\Big),$$

$$f_4=-\frac{3\sigma t}{8}\Big(1+\frac{\sigma t}{6}\Big)+\frac{\sigma^3t^3}{2.4.6}\Big(1+\frac{\sigma t}{8}\Big),$$

$$f_5=\frac{3}{8}\Big(1+\frac{\sigma t}{6}\Big)-\frac{\sigma^2t^2}{2.4}\Big(1+\frac{\sigma t}{8}\Big)+\frac{\sigma^4t^4}{2.4.6.8}\Big(1+\frac{\sigma t}{10}\Big),$$

$$f_6=\frac{5\sigma t}{16}\Big(1+\frac{\sigma t}{8}\Big)-\frac{5}{4}\frac{\sigma^3t^3}{2.4.6}\Big(1+\frac{\sigma t}{10}\Big)+\frac{\sigma^5t^5}{2.4.6\,8.10}\Big(1+\frac{\sigma t}{12}\Big). \tag{82}$$

By changing the sign of σ, and also that of t, we get

$$(p-\sigma)\frac{\epsilon^{-\nu t}}{\nu}=\epsilon^{-\nu t\Delta}(1-sf_1+s^2f_2-s^3f_3+\dots)\,; \tag{83}$$

and so, by taking half the sum of (81) and (83), (74) becomes expanded.

Our original equations (74), (75) are therefore brought to the forms

$$\begin{aligned}\mathrm{V}&=\tfrac{1}{2}\epsilon^{-\rho t}\,\epsilon^{\nu t\Delta}(1+sf_1+s^2f_2+s^3f_3+\dots)\mathrm{V}_0\\&+\tfrac{1}{2}\epsilon^{-\rho t}\,\epsilon^{-\nu t\Delta}(1-sf_1+s^2f_2-s^3f_3+\dots)\mathrm{V}_0,\end{aligned} \tag{84}$$

$$\begin{aligned}\mathrm{L}\nu\mathrm{C}&=-\tfrac{1}{2}\epsilon^{-\rho t}\epsilon^{\nu t\Delta}(1+sg_1+s^2g_2+s^3g_3+\dots)\mathrm{V}_0,\\&+\tfrac{1}{2}\epsilon^{-\rho t}\epsilon^{-\nu t\Delta}(1-sg_1+s^2g_2-s^3g_3+\dots)\mathrm{V}_0.\end{aligned} \tag{85}$$

The only operations to be performed now, except the translations to right and left respectively, are the integrations $s\mathrm{V}_0$, $s^2\mathrm{V}_0$, &c. Suppose, for example, V_0 is constant, but exists only from the point y to $+\infty$, and we want V outside the V_0

region. Then the second line in (84) is zero, because V_0 is zero; and in the first line, $\Delta^{-n}V_0$ is $V_0(x-y)^n/\underline{|n}$, when $x>y$, and zero on the other side. That is, all these integrals exist from y to ∞. Then $\epsilon^{vt\Delta}$ turns x to $x+vt$, and makes (84) become

$$V=\tfrac{1}{2}\epsilon^{-\rho t}V_0\left\{1+\frac{\sigma y_1}{v}f_1+\left(\frac{\sigma y_1}{v}\right)^2\frac{f_2}{\underline{|2}}+\left(\frac{\sigma y_1}{v}\right)^3\frac{f_3}{\underline{|3}}+\dots\right\}, \quad (86)$$

where $y_1=x+vt-y$. Here y_1 must be positive. If it be negative, V is zero. This diagram will illustrate :—

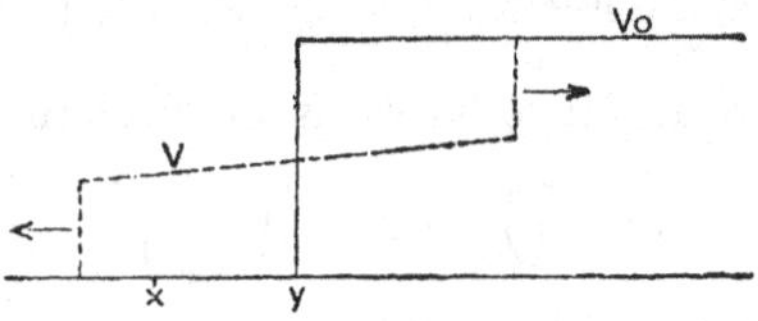

The dotted line shows the beginning of the V wave before it has attenuated much at its front, without any allowance for the attenuation of V_0, if there be any. The curve of V is not a straight line, as mistakenly represented. But later diagrams will show the approximate shape under different circumstances.

Similarly, the other formula (85) makes

$$LvC=-\tfrac{1}{2}\epsilon^{-\rho t}V_0\left\{1+\frac{\sigma y_1}{v}g_1+\left(\frac{\sigma y_1}{v}\right)^2\frac{g_2}{\underline{|2}}+\left(\frac{\sigma y_1}{v}\right)^3\frac{g_3}{\underline{|3}}+\dots\right\}. \quad (87)$$

It gives the current between the limits $x=y+vt$ and $y-vt$. It is pretty easy to calculate these formulæ when t is not too big. In fact, to show how the wave begins, they are easier than the corresponding Bessel formulæ.

If the initial V_0 is on the left side of the origin, that is, from $-\infty$ to $x=0$, then the proper change in the meaning of y_1 makes

$$V=\tfrac{1}{2}V_0\epsilon^{-\rho t}\left[1+\sigma t f_1\left(1-\frac{x}{vt}\right)+\frac{\sigma^2t^2}{\underline{|2}}f_2\left(1-\frac{x}{vt}\right)^2+\dots\right] \quad (88)$$

express V at x on the positive side, from $x=0$ to vt; beyond which, V is zero. And the corresponding C is given by

$$LvC=\tfrac{1}{2}V_0\epsilon^{-\rho t}\left[1+\sigma t g_1\left(1-\frac{x}{vt}\right)+\frac{\sigma^2t^2}{\underline{|2}}g_2\left(1-\frac{x}{vt}\right)^2+\dots\right]. \quad (89)$$

To obtain the formula for the C due to C_0 change the sign of σ in (88), explicit and in the f's, of course also writing C and C_0 instead of V and V_0. To obtain the formula for the V due to C_0, change LvC to SvV, and V_0 to C_0, in (89), without other change, because reversing the sign of σ leaves $\sigma t g_1$, &c., the same.

Comparing different ways of working, we may collect some of the results thus :—

$$\frac{\epsilon^{-t(v^2\Delta^2+\sigma^2)^{\frac{1}{2}}}}{(v^2\Delta^2+\sigma^2)^{\frac{1}{2}}} v\Delta 1 = \frac{\epsilon^{-xv^{-1}(p^2-\sigma^2)^{\frac{1}{2}}}}{(p^2-\sigma^2)^{\frac{1}{2}}} p 1 \qquad (90)$$

$$= \left(\frac{p-\sigma}{p+\sigma}\right)^{\frac{1}{2}} \epsilon^{-xv^{-1}(p^2-\sigma^2)^{\frac{1}{2}}} \epsilon^{\sigma t} = I_0\{\sigma(t^2 - x^2/v^2)^{\frac{1}{2}}\} \qquad (91)$$

$$= \frac{2v}{\pi}\int_0^\infty \cos mx \frac{\sin t(m^2v^2-\sigma^2)^{\frac{1}{2}}}{(m^2v^2-\sigma^2)^{\frac{1}{2}}} dm \qquad (92)$$

$$= 1 + \sigma t g_1\left(1 - \frac{x}{vt}\right) + \frac{\sigma^2 t^2}{\lfloor 2} g_2\left(1 - \frac{x}{vt}\right)^2 + \dots. \qquad (93)$$

There are several other forms, but these are the principal. There are also combination formulæ, when there is initially existent both V_0 and C_0; for example, to show how an initial pure wave casts its tail behind. But they do not involve fresh investigation, and as space is running short, I now bring the development of these formulæ to a conclusion. I have gone over pretty well all the ground. Another way of attacking the mathematics of the subject will be found in Webster's lately published "Electricity and Magnetism," § 255, due to Boussinesq, and methods by Poincaré and Picard are also referred to there. In these the reader may revel if he wants more. I have had enough of it for the present. Naturally, I prefer my own ways, because I can understand them better. And best of all I like the distortionless circuit, because everything essential to a general understanding can be foreseen by its means without any of the complicated mathematics, which however is of course necessary to complete the matter, and permit particular people to admit the truth of the conclusions, and to allow of close calculations being made if desired ("El. Pa.," Vol. 2, pp. 119 to 155, and pp. 381, 427, 475. &c.).

Undistorted Waves without and with attenuation.

§ 419. Having in Chap. IV., Vol. I., described pretty fully the meaning of the theory, in its application to conducting circuits as well as to plane waves in a uniform medium with two conductivities, it is unnecessary to repeat the information here. What follows is supplementary, particularly for the assistance of those readers who may desire to go into numerical detail, and draw the curves according to the preceding formulæ.

First take the case of no resistance and no leakage. Represent V by the ordinate erected on the base line which stands for the circuit. Then let there be two equal waves travelling towards one another. The one on the left is the positive wave. In it $V = LvC$. On the right is the negative wave, in which $V = -LvC$. This being the state at a certain moment, the subsequent history is got by moving the positive wave to the right and the negative to the left, at the same

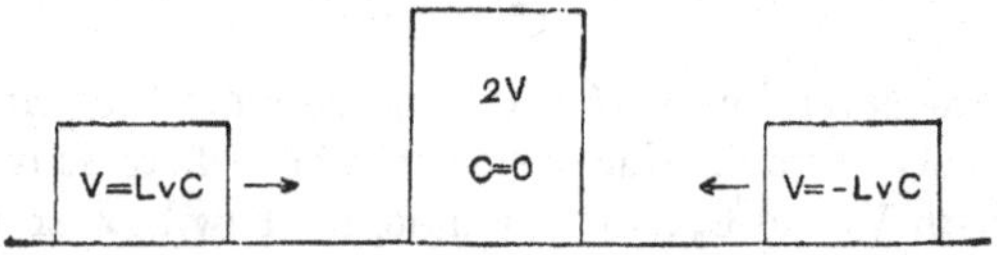

speed v. When they overlap the real V and C are got by superposition. It follows that when they coincide in position there is no C, whilst V is doubled. After that the negative wave emerges on the left, and the positive on the right.

This also shows how an initial state of transverse voltage only, without magnetic force, behaves. Say it is 2V, as in the middle. Then it splits immediately into the positive and negative waves, not as shown in the diagram, but as they would be after emerging from coincidence. Similarly as regards an initial state of current only. Say it is 2C. Then it splits into a positive wave in which $V = LvC$ going to the right, and a negative wave in which $-V = LvC$ going to the left.

After this rudimentary preliminary, pass to the distortionless circuit with resistance. Besides the undistorted transmission as above, there is now attenuation in transit. The time factor of attenuation is $\epsilon^{-Rt/L}$. The curve of attenuation along the circuit is therefore $y = \epsilon^{-\mathrm{R}x\ \mathrm{L}v}$. Draw it, and then let an elementary plane wave slide along it at speed v, and it will correctly represent the transit of the wave as modified by the resistance and the leakage.

All the elements of a given arbitrary wave, say a positive wave, travel at the same speed, and attenuate at the same rate.

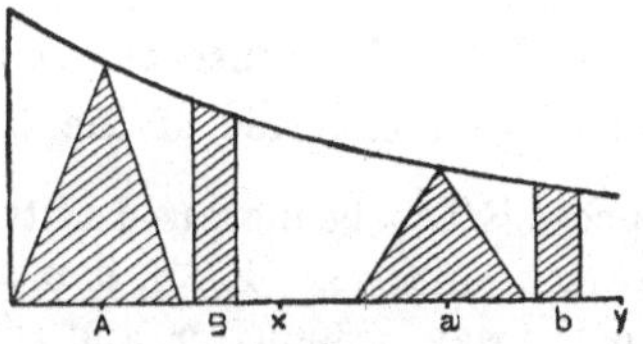

FIG. 1.

To illustrate, if A, B shows the curve of voltage at a certain moment, then a, b will show what it becomes a little later, the time interval being the distance Aa (or Bb) divided by v. Similarly a circle will shrink to an ellipse. Note that this shrinkage does not count as distortion at all. The variations of voltage (and of current, since $\mathrm{V} = \mathrm{L}v\mathrm{C}$) as the wave passes any point x, are identically repeated when it passes y further on, only reduced in size. If the voltage is impressed at the left end, according to $f(t)$, then it makes $f(t - x/v) \times \epsilon^{-\mathrm{R}x/\mathrm{L}v}$ at any later point x.

Now consider the initial state $\mathrm{V} = 2\mathrm{V}_0$ constant on the left side and zero on the right side of the origin, without any C.

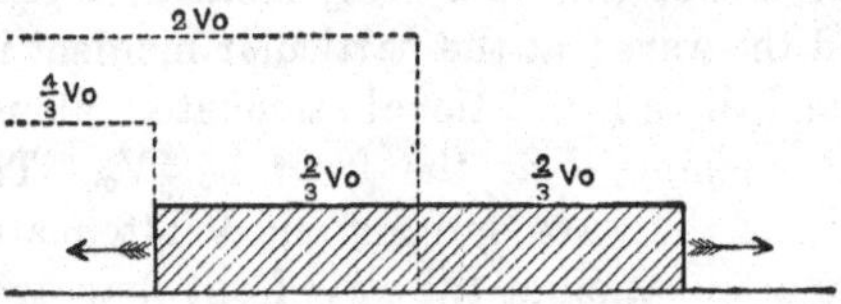

FIG. 2

What results is to be found by dividing the initial state into two equal parts, one with positive current, making $\mathrm{V}_0 = \mathrm{L}v\mathrm{C}_0$, the other with equal negative current making $\mathrm{V}_0 = -\mathrm{L}v\mathrm{C}_0$, and then moving the first to the right and the second to the left, at speed v. In the figure, the upper dotted line shows the initial state, and the lower dotted line what it attenuates to at the later time making $\epsilon^{-Kt/S} = \frac{2}{3}$. The shaded region shows the electromagnetic wave $V = LvC$ spreading both ways. The dotted part of the curve of voltage has no magnetic force associated with it. The voltage is there attenuating by leakage only, but at the same rate as the V and C in the wave.

Effects of Resistance and Leakance on an Initial State of Constant V_0 on one side of the Origin.

§ 420. Next suppose R/L to be increased a little, K/S remaining the same. Then ρ, or $R/2L + K/2S$, is made greater than K/S. New $\epsilon^{-\rho t}$ is the wave attenuator, and $\epsilon^{-Kt/S}$ the leakage attenuator. It follows that, with the same initial circumstances as in the last case, the attenuation at the wave front is greater than at the origin of the wave. So, instead of the

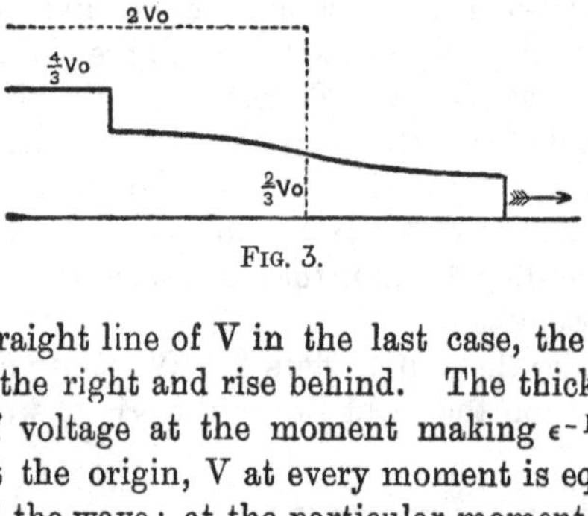

FIG. 3.

horizontal straight line of V in the last case, the curve of V must fall to the right and rise behind. The thick line shows the curve of voltage at the moment making $\epsilon^{-Kt/S} = \frac{2}{3}$, and $\epsilon^{-\rho t} = \frac{4}{9}$. At the origin, V at every moment is equal to half the V behind the wave; at the particular moment in question this is $\frac{2}{3}V_0$, and there is additional attenuation along the wave to the right, making V at the front be $\frac{4}{9}V_0$. The general characteristic as the wave spreads and V attenuates all over is the greater attenuation at the wave front than at the origin.

The case is quite altered if, instead of increasing R/L, we reduce it, taking the distortionless case as initial standard. For we shall now make ρ less than K/S. The remarkable result is that the attenuation at the forward wave front, and in the forward wave generally, is less than at the origin, and in the region behind the wave occupied by V without any C. In the diagram, Fig. 4, the same attenuation by leakage alone is shown, and the same value at the origin, but at the forward wave front V is $\frac{8}{9}V_0$.

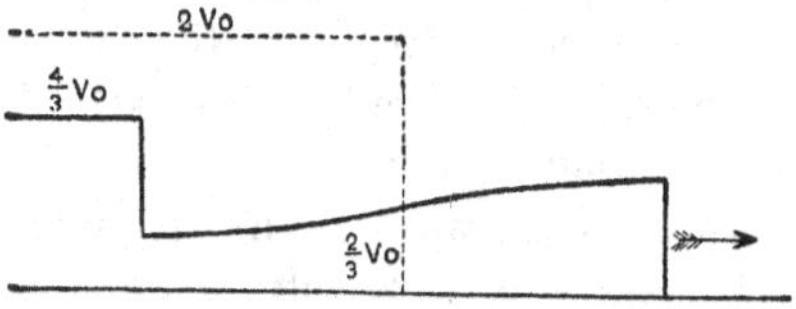

FIG. 4.

Next pass to the two extremes partially indicated. One is $K=0$, the other $R=0$. If $K=0$, we have $\rho=R/2L=\sigma$ as well. With the same initial state there is no attenuation behind the wave region, and the voltage at the origin

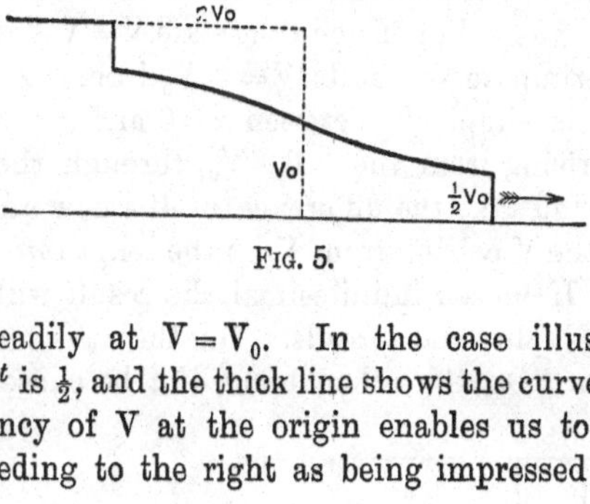

FIG. 5.

remains steadily at $V=V_0$. In the case illustrated, the value of $\epsilon^{-\rho t}$ is $\frac{1}{2}$, and the thick line shows the curve of voltage. The constancy of V at the origin enables us to regard the wave proceeding to the right as being impressed by a steady voltage V_0.

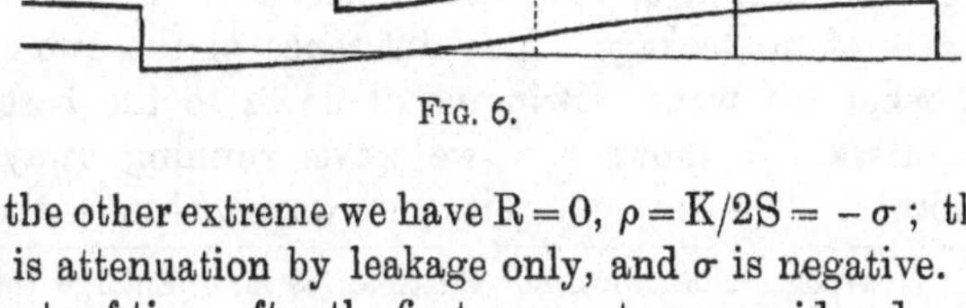

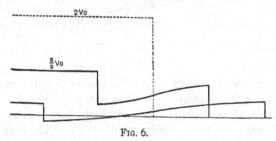

FIG. 6.

At the other extreme we have $R=0$, $\rho=K/2S=-\sigma$; that is, there is attenuation by leakage only, and σ is negative. Two moments of time after the first moment are considered, and the corresponding two stages of the voltage curve are shown. When $2V_0$ has fallen to $\frac{8}{9}V_0$, the voltage at the origin is $\frac{4}{9}V_0$, and at the forward wave front $\frac{2}{3}V_0$. And at the later moment, when this wave front has gone twice* as far, the initial $2V_0$ has attenuated to $\frac{2}{9}V_0$, and at the origin we have $\frac{1}{9}V_0$, and at the forward wave front $\frac{1}{3}V_0$. On the left side of the origin V has become negative in a portion of the wave. The ultimate limit of this is easily seen; the place of zero voltage moves towards the origin as the wave spreads and attenuates.

Division of Charge Initially at the Origin into Two Waves with Positive or Negative Charge between them.

§ 421. From the previous curves may readily be derived those representing how a charge initially confined to a limited region splits and spreads. For if upon the state $V = V_0$ from $x = -\infty$ to 0, we superimpose the state $V = -V_0$ from $-\infty$ to $x = +a$, the resultant is simply V_0 between $x = 0$ and a. So shift the curve of V arising from the full $+V_0$, through the distance a to the right, and take the difference of the new and old curve to represent the V arising from V_0 in the length a. It may be any length. If chosen infinitesimal the result will show how a point charge splits and spreads. But then the curve requires magnification. On the other hand, if a be chosen large, we

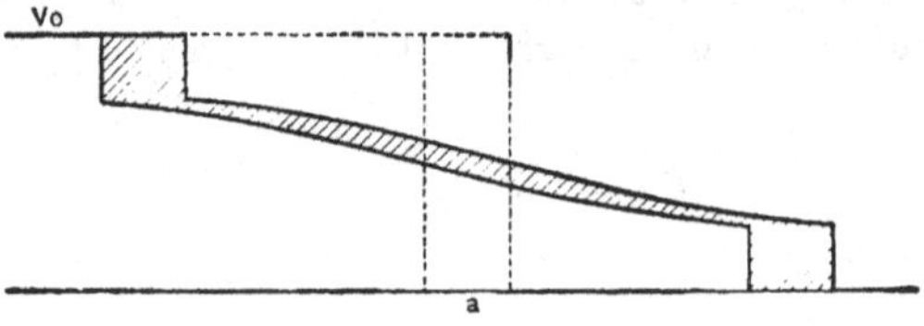

FIG. 7.

do not get clear representation of the waves, owing to overlapping. So take a of a moderate size, as in Fig. 7. This is the case of no leakage. The difference of the two curves shows what we want. Bringing it down to the base line, Fig. 8 arises. It shows the two waves running away from one another. They are not quite pure waves, because they are

* This is wrong. Not twice, but 2·7 times as far. Enlarge the time scale to suit in the second curve.

of finite depth. Between them is the diffused electrification cast behind by the two heads as they travel; that is, it is the double tail formed by the superposition of the tails of the two waves. The complete area of two heads and the double tail is the same as the area of the original rectangle, shown dotted. At any rate, it ought to be so, and by the look of it, is not

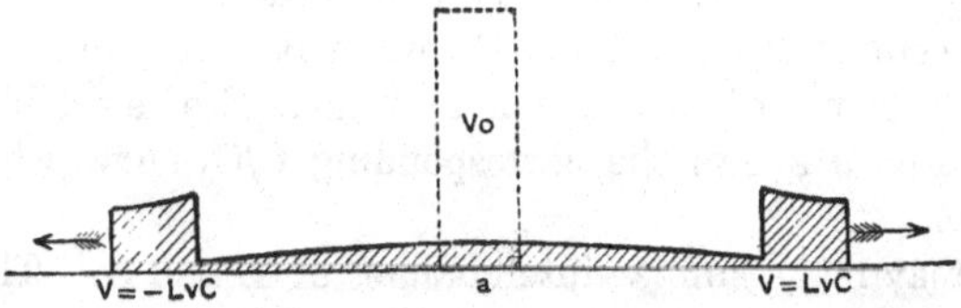

FIG. 8.

far wrong. As the heads attenuate to zero, the curve of V approximates towards the curve of diffusion in Fourier's theory.

When σ is negative, the intermediate V is negative instead of positive. In Fig. 9 the case $R=0$ is illustrated, or $\sigma = -K/2S = -\rho$. The moment of time is such that $\sigma t = -1$. I have roughly calculated that the value of ϵ^{-1} being ·3678,

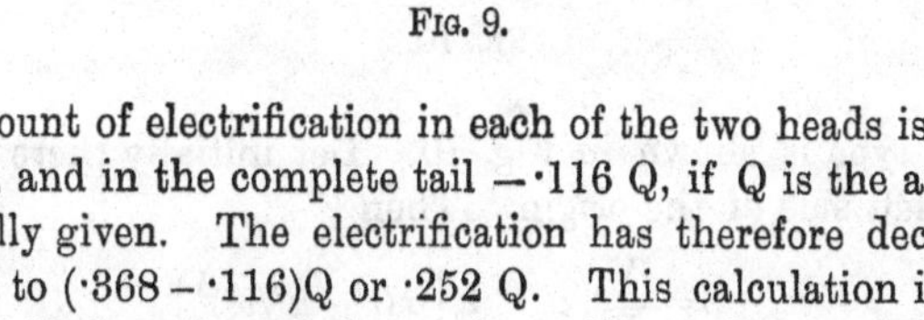

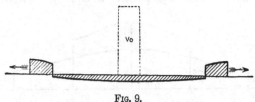

FIG. 9.

the amount of electrification in each of the two heads is about ·184 Q, and in the complete tail −·116 Q, if Q is the amount originally given. The electrification has therefore decreased from Q to (·368 − ·116)Q or ·252 Q. This calculation is done for an initial condensed charge, but will not be greatly wrong under the circumstances.

So far relating to the V-waves arising from initial V_0, the C-waves due to initial C_0 may be readily deduced. In brief, the C due to C_0 is the same function as the V due to V_0 with the sign of σ changed. This means that if a first case concerned a given R and K, and a second one is made by interchanging their values, then V/V_0 in the first case is the same as C/C_0 in the second, if the distributions V_0 and C_0 are similar and equal.

So the V/V_0 curve in Fig. 3, where R is in excess, or σ is positive, also represents the C/C_0 curve when K is in excess, making σ negative. Similarly Fig. 4 shows the C/C_0 curve when R is in excess. And Fig. 8 shows the C/C_0 curve arising from a central distribution of C_0, when R is zero (σ negative), with persistence of mementum now instead of electrification, whilst Fig. 9 shows the corresponding C/C_0 curve when σ is positive.

We may make similar interchanges of L and S. The $\epsilon^{-\rho t}$ factor remains the same when R/L and K/S are transposed, whereas σ is reversed.

The Current due to Initial Charge on one side of the Origin.

§ 422. There remain for consideration the curves of C due to V_0 and of V due to C_0. These are alike. Not only that, but they are similar when σ is positive or equally negative. The

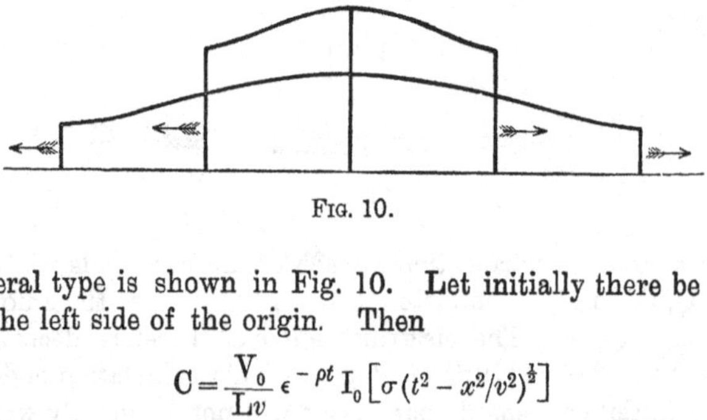

FIG. 10.

general type is shown in Fig. 10. Let initially there be $2V_0$ on the left side of the origin. Then

$$C = \frac{V_0}{Lv}\epsilon^{-\rho t} I_0\left[\sigma(t^2 - x^2/v^2)^{\frac{1}{2}}\right]$$

is the formula for C. Here $\rho = \sigma(1 + K/S\sigma)$, so, taking σ positive, we require to give ρ a greater value, unless K is zero. Say $\sigma t = 1$, then $\epsilon^{-\rho t}$ must not be greater than ϵ^{-1}. So $\frac{1}{3}$ is a good value to take, and in the figure are shown the curves of C when $\sigma t = 1$ and 2, the corresponding values of $\epsilon^{-\rho t}$ being $\frac{1}{3}$ and $\frac{1}{9}$. The voltage at the origin is $V_0\epsilon^{-Kt/S}$.

From these curves, by taking differences as before, we can derive the curves showing the current due to initial V_0 through a certain distance. Two stages are shown in Fig. 11. The current is positive in the wave going to the right, and negative

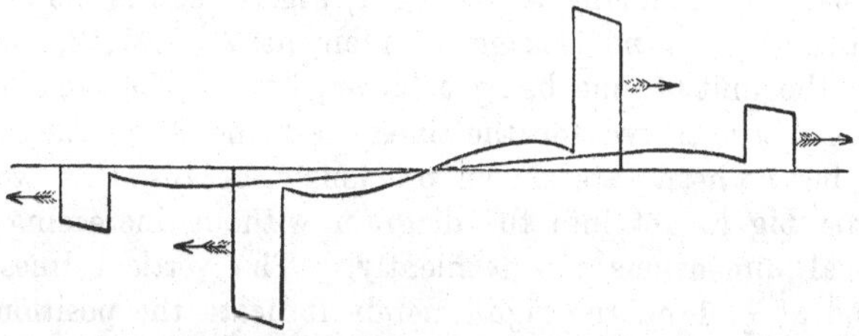

FIG. 11.

in the one going to the left, and the diffused current between them is positive and negative also. The positive wave appears to have a positive tail; the negative a negative tail. This appears anomalous, but it is a mixed up case. It is the negative part of the intermediate current that represents (in a great measure) the tail of the positive wave, whilst the positive part represents that of the negative wave, as will be more plainly seen presently.

The After-effects of an Initially Pure Wave. Positive and Negative Tails.

§ **423.** To obtain the formulæ showing the tail cast behind by an initially pure wave in which $V_0 = LvC_0$, we have merely to add together the V resulting from V_0 to that resulting from LvC_0 to find the V due to both, and similarly as regards the C due to both causes. The expansions in powers of $(1 - x/vt)$ are convenient for immediate rough numerical calculation. But the following Fig. 12 has been obtained by an entirely different process, namely, by making numerical interchanges. It would take up too much of the rapidly-decreasing space remaining in this volume to enter into detail concerning this method. The reader who will study the short account given elsewhere ("Elec. Pa.," Vol. II., p. 318), and will take it in, will easily be able to construct tail curves, as well as a variety of others. The results are rough, but show all the main characteristics in size and shape. In this method you go behind the scenes, as it were, and see the inner workings of the formulæ when time and space vary. Both the voltage and the current curves are obtained from the same series of numbers.

Supposing there is initially a pure wave of small depth at the place O, advancing to the right, Fig. 12 shows the tail of V when there is no leakage, at moments $t = \frac{1}{2}$, 1, 2, and 3 later, the unit of time being arbitrary here. The remains of the head are shown for the times $t = 2$ and 3, on the right side, but no heads are shown previously, because they would be too big to get into the diagram without increasing its vertical dimensions inconveniently. The vertical lines at 0, and at $\frac{1}{2}$, 1 on the right merely indicate the position of the middle of the head; on the other hand, its depth as well as height are shown at 2 and 3.

There are signs of the maximum having been reached in the curve 1. In the later curve 2 nearly all the electrification is in the tail, and the maximum is a long way from the head. In the later curve 3 the head does not count for much, the maximum has moved back a little, the curve has become more symmetrical with respect to the middle point, and the change of curvature shows itself markedly on the left side, and perhaps a little on the right side as well. Later on, when the head disappears, the tail (widely spread) continues to tend to become symmetrical with respect to O.

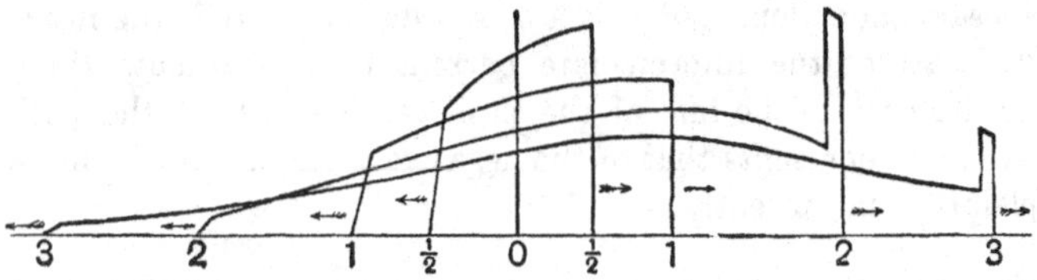

Fig. 12.

In the next diagram, Fig. 13, is represented the development of the tail of current, on the same scale, and at the same moments of time. The heads 2 and 3 are shown the same as in Fig. 12. This means that the curve is that of LvC, not C. The heads for $t = 0$, $\frac{1}{2}$, and 1 are left out as before. At first the tail is entirely negative, thickest at the tip, and decreases towards the head. Note in passing that the sloping lines at the tips in Figs. 12 and 13 are merely consequences of the finite depth chosen for the head, which causes overlapping. These sloping lines are vertical when the head is of infinitesimal thickness. Going further, at the moment $t=1$, the

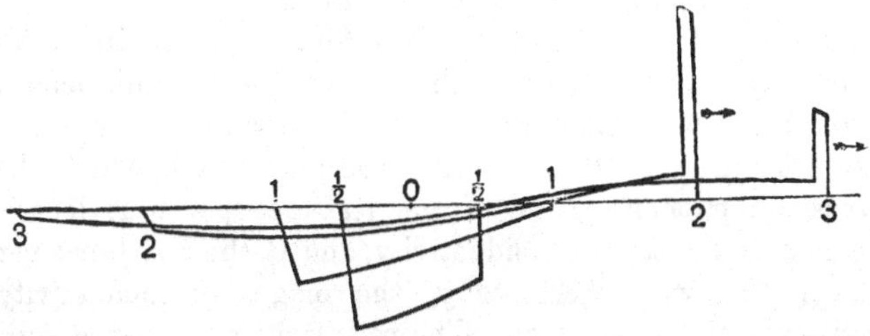

FIG. 13.

negativity of the tail has nearly ceased at its junction with the head. At the moment $t=2$, it is positive next the head, and a long way behind. Remember that the head is greatly reduced from the original head at O. At the moment $t=3$, further progress has been made towards symmetry, and the positive part of the tail has a maximum. After that, there is a continuation of the progress towards symmetry.*

Figs. 1 to 13 described in terms of Electromagnetic Waves in a doubly Conducting Medium.

§ 424. In the preceding, nearly all equations and results have been expressed in terms of V and C (voltage and current), instead of E and H (the electric and magnetic forces). The main reason for this is on account of the direct application that can be made to telegraphic, telephonic and Hertzian waves along wires. But for plane electromagnetic waves in general in a conducting medium, it is, of course, more natural to consider E and H, and think of what goes on within and along

* I am unable to explain by my theory why the waves at 2 and 3 should lean backwards. We must have recourse to a scienticulistic explanation. The spy who drives the cart informed me that the retardation of the top of the wave with respect to the bottom was indicative of incipient action of "the KR law," making the speed of the current be less at the top than at the bottom. Bob the tanner confirmed this, with an oath. Both are scienticulistically impregnated.

a single tube of flux of energy of unit section. It is sufficient here merely to restate how to translate solutions from the concrete to the specific. It would scarcely be necessary were it not for the singular part played by R when we pass from real wire waves to pure electromagnetic waves.

Turn V and C to E and H. The former are the line integrals of the latter. Then L, the inductance per unit length becomes μ the inductivity, and S the permittance per u.l. becomes c the permittivity of the medium through which the waves are passing. Also, K the leakance per unit length becomes k, the electric conductivity, and R the resistance per u.l. of the wires becomes g, the magnetic conductivity (fictitious) of the medium. The peculiarities connected with the last should be noted in the following, since if they are not understood the subject is rather puzzling sometimes.

Referring to Figs. 1 to 13, Fig. 1 shows how plane waves of E and H of any type would be transmitted in a medium possessing both conductivities of such amounts as to make $g/\mu = k/c$. Similarly, in Fig. 2, where the initial state is a uniform field E_0 on the left side of the plane $x=0$. The wave arising may stand for either E or H, since $E = \mu v H$.

Passing to Fig. 3, here it is the magnetic conductivity that is in excess, and the wave of E arising from the initial state E_0 is represented. It equally well represents the wave of H, arising from a similar initial state H_0, when the electric conductivity is in excess.

In Fig. 4 the curve shows the wave of E arising from E_0 when k is in excess. Or else, the wave of H arising from H_0 when g is in excess.

In Fig. 5 we have the wave of E due to E_0 when the conductivity is wholly magnetic. Or else, the wave of H due to H_0 when the conductivity is wholly electric.

Fig. 6 shows the wave of E due to E_0 when the conductivity is wholly electric. Or else, the wave of H due to H_0 when it is wholly magnetic. In these examples, if one case is fictitious owing to the non-existence of g, it may be turned to a real case by considering E instead of H, or H instead of E.

Fig. 8 shows how an initial state of E_0 in a plane slab divides into two equal plane slabs, which separate and attenuate, and leave part of their contents behind and between them, when $k=0$, and g is finite. There is persistence of total displacement. Or else it shows the division of initial H_0, when $g=0$, and k is finite. It is the induction that now persists.

Fig. 9 shows how the initial state of E_0 divides and spreads when $g=0$, and k is finite. Or else how a state of H_0 divides when $k=0$, and g is finite.

Fig. 10 shows the wave of H due to initial uniform E_0 on the left side of the origin. Or else the wave of E due to similar H_0.

Fig. 11 shows the wave of H due to a slab of E_0 at the origin. Or else the wave of E due to a similar slab of H_0.

Fig. 12 shows the tail of E in different stages of growth arising from the initial pure wave $E_0=\mu v H_0$, when the conductivity is wholly magnetic. Or else the tail of H due to the same initial state when the conductivity is wholly electric.

Finally, Fig. 13 shows the tail of H in different stages due to the same initial state as last when there is magnetic conductivity alone. Or else the tail of E due to the same when there is electric conductivity alone.

I am aware of the deficiencies of the above diagrams. Perhaps some electrical student who possesses the patient laboriousness sometimes found associated with early manhood may find it worth his while to calculate the waves thoroughly and give tables of results, and several curves in every case.* It should be a labour of love, of course; for although if done thoroughly there would be enough to make a book, it would not pay, and the most eminent publishers will not keep a book in stock if it does not pay, even though it be a book that is well recognised to be a valuable work, and perhaps to a great extent the maker of other works of a more sellable nature. Storage room is too valuable.

* Webster's "Electricity and Magnetism" contains a few curves, but not enough for what I refer to. Gray's work, in Vol. II., may perhaps go further.

CHAPTER VIII.

GENERALISED DIFFERENTIATION AND DIVERGENT SERIES.

A Formula for $\lfloor n$ obtained by Harmonic Analysis.

§ 425. I am reminded that this volume is growing fat. Therefore I come to the last matter to be considered in it. Omitting some other investigations concerning plane waves, I must say a few words on the subject of generalised differentiation and divergent series. I have put it off to the last for several reasons. It is not the main subject for one thing, and only turns up incidentally. Again, I have asked myself, if investigations of this matter are not good enough for Another Place, how can they be good enough for Electromagnetic Theory? I really cannot answer such a poser. Then again, it is not easy to get up any enthusiasm after it has been artificially cooled by the wet blankets of rigorists. Nevertheless, I have been informed that I have been the means of stimulating some interest in the subject in certain places. Perhaps not in England to any extent worth speaking of, but certainly in Paris it is a fact that a big prize has been offered lately on the subject of the part played by divergent series in analysis. Well, that is better than the wet blanket, and sets a lesson that may or may not be followed in other quarters. I hope the prize-winner will have something substantial to say. I wish him success, whoever he may be.

The easiest way out of my difficulty would be to just shove in here my three papers "On Operators in Physical Mathematics," Royal Soc. *Proceedings* (Part I., Dec. 15, 1892; Part II., June 8, 1893, and Part III., down for reading June 21, 1894, and not read). But this would take about 120 pages, which is out of the question. An account of certain parts of the papers will be quite sufficient for the present work, and can be given in a limited space, especially because a good deal has been given already in one form or another.

So, passing over §§ 1 to 14, Part I., O.P.M., already given with large extensions, come at once to the $\lfloor n$ function in its relation to fractional differentiations and integrations. Let Δ be a differentiator, say with respect to x, then we have had to consider not only $\Delta^{-n}1$ when n is integral, but also when it is fractional, say $\frac{1}{2}$ or $\frac{1}{3}$. We are obliged to consider such cases, because our operational solutions involve them. The case $n=\frac{1}{4}$ also occurs in cylindrical problems, and no doubt other values turn up in other problems. We want to find what $\lfloor n$ is in the equation

$$\Delta^{-n}1=\frac{x^n}{\lfloor n}, \tag{1}$$

subject to

$$\lfloor n = n\lfloor n-1, \qquad \lfloor 1 = 1, \tag{2}$$

the last datum being necessary to fix the size of the function, or to standardise it. We know that there is such a function as is defined by (2) given in all treatises on the Integral Calculus (equivalently) under the name of Euler's Gamma function, though I think most books only treat of it from $n=-1$ to $+\infty$, and have nothing to say about the region from $-\infty$ to -1, and still less about the application to generalised differentiation. With a different zero it is also called Gauss's II. function. But put aside from memory for the present what may be said about the Gamma function in books, and consider it in the light of practical work in operational mathematics.

A formula for $\lfloor n$, based upon (1) may be readily found. Use harmonic analysis. We have

$$\Delta 1=\frac{1}{\pi}\int_0^\infty \cos mx\, dm, \tag{3}$$

a remarkable formula already used many times. So

$$\Delta^{\;n}1=\Delta^{-(n+1)}\Delta 1=\frac{\Delta^{-(n+1)}}{\pi}\int_0^\infty \cos mx\, dm. \tag{4}$$

Now with the simply periodic operand $\cos mx$, the potence

of Δ^2 is $-m^2$. That is, $\Delta = mi$, the sign of equality merely meaning equivalence under the circumstances, not identity in general. Or, the algebraic imaginary i is equivalent to Δ/m. So

$$\Delta^{-(n+1)} = \frac{i^{-(n+1)}}{m^{n+1}} = \frac{\epsilon^{-\frac{\pi}{2}(n+1)i}}{m^{n+1}} = \frac{\epsilon^{-\frac{\pi}{2}(n+1)\frac{\Delta}{m}}}{m^{n+1}}. \tag{5}$$

That is, we assume that the equivalence of Δ to mi persists when n is fractional. On this understanding, the final form in (5) involves only the operator of Taylor's theorem, so (4) becomes

$$\Delta^{-n}1 = \int \frac{\epsilon^{-\frac{\pi}{2}(n+1)\frac{\Delta}{m}}}{\pi m^{n+1}} \cos mx \, dm$$

$$= \frac{1}{\pi}\int_0^\infty \frac{\cos\left[mx - \frac{1}{2}\pi\, n+1)\right]}{m^{n+1}}\, dm. \tag{6}$$

This, observe, is a formula evaluating $\Delta^{-n}1$, obtained without any reference to the $\underline{|n}$ function. According to (1), then, we have

$$\frac{x^n}{\underline{|n}} = \frac{1}{\pi}\int_0^\infty \frac{\cos\left[mx - \frac{1}{2}\pi(n+1)\right]}{m^{n+1}}\, dm, \tag{7}$$

and so, putting $x = 1$, and denoting the reciprocal of $\underline{|n}$ by $g(n)$, we have

$$g(n) = \frac{1}{\pi}\int_0^\infty \frac{\cos\left[m - \frac{1}{2}\pi(n+1)\right]}{m^{n+1}}\, dm. \tag{8}$$

Thus it is not $\underline{|n}$, but its reciprocal, that should be investigated for our purposes.

Formula (8) is quite correct and intelligible when n lies between -1 and 0. The integral is then convergent. But just at the limit $n = 0$, it fails, giving only half the real value. It may be readily checked that $\underline{|-\frac{1}{2}} = \pi^{\frac{1}{2}}$, as before obtained in another way. The interpretation outside the limits named need not be considered, because equation (2) finds n all over when it is known between -1 and 0.

Algebraical Construction of $g(n)$. Value of $g(n)g(-n)$.

§ 426. Although the above way (not given in O.P.M.) of getting a formula for $g(n)$ is very direct and short, the result is a definite integral, which must be worried to display its effective meaning. The curve of $g(n)$ can be drawn approximately at once by plotting its known values when n is integral, viz., 0 for every negative integer, and 1, 1, $\frac{1}{2}$, $\frac{1}{6}$, $\frac{1}{24}$, &c., at $n=0, 1, 2, 3, 4$, &c. By the look of it, we see that the curve has a maximum between 0 and 1, and is oscillatory on the negative side, whilst continuously positive on the positive side. This is confirmed by plotting the midway values, also known, for $n=\frac{1}{2}, \frac{3}{2}$, &c. The range of the oscillations is now approximately indicated. As n increases negatively, the range increases.

O.P.M., §§ 17, 18, 19. If we turn right to left, we shall obtain the similar curve of $g(-n)$, oscillatory on the right side of the origin, and continuously positive on the left side, with a maximum between 0 and -1.

The two curves are remarkably connected. Form the curve of the product $g(n)\,g(-n)$. It is oscillatory on both sides. It has a zero for every integral value of n, positive and negative, and has the value 1 at the origin. This defines the function $(\sin n\pi)/n\pi$. That is,*

$$g(n)\,g(-n)=\frac{\sin n\pi}{n\pi}, \tag{9}$$

for all values of n. The values of $g(\frac{1}{2})$, &c., follow at once. Now (9) is the same as

$$g(n)\,g(-n)=(1-n^2)(1-\tfrac{1}{4}n^2)(1-\tfrac{1}{9}n^2)\ \ldots\ldots, \tag{10}$$

therefore, since the zeros of $g(n)$ are known,

$$g(n)=\mathrm{X}(1+n)(1+\tfrac{1}{2}n)(1+\tfrac{1}{3}n)\ \ldots\ldots, \tag{11}$$

$$g(-n)=\mathrm{X}^{-1}(1-n)(1-\tfrac{1}{2}n)(1-\tfrac{1}{3}n)\ \ldots\ldots, \tag{12}$$

where X is some function which does not introduce fresh zeros. Since $n=0$ makes $g(0)=\mathrm{X}$ apparently in (11), it might be thought that $\mathrm{X}=1$. But it is ∞ instead when n is $-$, and 0 when n is $+$, as may be easily seen by drawing the curve

$$y=(1+n)(1+\tfrac{1}{2}n)\ldots\left(1+\frac{n}{r}\right) \tag{13}$$

with a limited number of products. This is an oscillatory curve crossing the n-axis at the points -1, -2, &c., up to $-r$.

* There might be an extra factor. But there is not, to harmonise $g(n)$ and $g(-n)$. Or we may regard the value of $g(-\frac{1}{2})$ as known, to determine the factor to be 1.

Now as r is increased, the range of the oscillations gets smaller, and tends to become infinitely small when $r=\infty$. This is confirmed by the fact that (13) is identically the same as

$$y=1+n+\frac{n(n+1)}{\lfloor 2}+\ldots+\frac{n(n+1)(n+2)\ldots(n+r-1)}{\lfloor r}, \quad (14)$$

viz., the first $r+1$ terms in the expansion of $(1-1)^{-n}$ by the binomial theorem. When r is ∞, y is zero for all negative and ∞ for all positive values of n.

The X multiplier must therefore be ∞ when n is negative. There is a radical distinction between the vanishing of y at $n=-1, -2$, &c., and the vanishing between these values. In the first case the zeros are absolute, and cannot be magnified. The curve crosses the axis at the points named. But the vanishing of y between can be cancelled by infinite magnification, so as to lift up the curve of y from the zero line to the $g(n)$ curve. Thus, choose X so that equation (11) shall be true when there are r products and n is an integer. This makes

$$\frac{1}{\lfloor n}=\mathrm{X}\frac{\lfloor n+r}{\lfloor n\,\lfloor r}, \quad \text{or} \quad \mathrm{X}=\frac{\lfloor r}{\lfloor n+r}.$$

For example, if n is 10 and r is 100, then

$$\mathrm{X}=\frac{1\,.\,2\,.\,3\,.\,4\,\ldots\,.\,99\,.\,100}{1\,.\,2\,.\,3\,.\,4\,\ldots\,.\,109\,.\,110}=\frac{1}{101\ldots110}.$$

As r is indefinitely increased this becomes r^{-n}, so

$$g(n)=r^{-n}(1+n)(1+\tfrac{1}{2}n)\ldots.(1+n/r)=r^{-n}(1-1)^{-n}. \quad (15)$$

The limit that is tended to when r is made infinite is the value of $g(n)$.

By logarising (15) we get

$$\log\lfloor n=-n\mathrm{C}+\frac{n^2}{2}\mathrm{S}_2-\frac{n^3}{3}\mathrm{S}_3+\frac{n^4}{4}\mathrm{S}_4-\ldots, \quad (16)$$

where $\mathrm{S}_m=1+\frac{1}{2^m}+\frac{1}{3^m}+\ldots, \qquad \mathrm{C}=\mathrm{S}_1-\log r=0\cdot5772. \quad (17)$

From (16) may be derived a formula for $g(n)$ in rising powers of n, namely

$$g(n)=1+\mathrm{C}n+(\mathrm{C}^2-\mathrm{S}_2)\frac{n^2}{2}+(\mathrm{C}^3-3\mathrm{CS}_2+2\mathrm{S}_3)\frac{n^3}{\lfloor 3}$$
$$+(\mathrm{C}^4-6\mathrm{C}^2\mathrm{S}_2+8\mathrm{CS}_3+3\mathrm{S}_2{}^2-6\mathrm{S}_4)\frac{n^4}{\lfloor 4}+\ldots. \quad (18)$$

Like π, the value of C has been calculated to a large number of places of decimals, though I do not know that anything is gained by it. Something wonderful might however turn up, for instance, a long succession of 0's.

The cosine function may similarly be split into two, making

$$f(n)f(-n) = \cos \pi n, \quad \text{and} \quad (n-\tfrac{1}{2})f(n) = f(n-1). \qquad (19)$$

But this $f(n)$ is not useful like the $g(n)$ function.

Generalisation of Exponential Function.

§ 427. Passing on to O.P.M., § 20, 21, the usual formula defining $\lfloor n$ and expressing Euler's Gamma integral is equivalent to

$$1 = \int_0^\infty \frac{\epsilon^{-x} x^n}{\lfloor n} dx, \qquad (20)$$

provided n is over -1. There are two ways of developing the indefinite integral "by parts," say

$$w_1 = \epsilon^{-x}\left(\frac{x^{n+1}}{\lfloor n+1} + \frac{x^{n+2}}{\lfloor n+2} + \frac{x^{n+3}}{\lfloor n+3} + \dots\right), \qquad (21)$$

$$-w_2 = -\epsilon^{-x}\left(\frac{x^n}{\lfloor n} + \frac{x^{n-1}}{\lfloor n-1} + \frac{x^{n-2}}{\lfloor 2-2} + \dots\right). \qquad (22)$$

So $[w_1]$ between any two limits is equal to $[-w_2]$ between the same limits. That is, $[w_1 + w_2]$ between any two limits, is zero. Therefore $w_1 + w_2$ does not change its value with x. But does its value depend upon n? No, because if n is > -1, we may divide the integral (20) into two, one going from 0 to x, the other from x to ∞. For the first use (21), since $w_1 = 0$ when $x = 0$; and for the second use (22), since $w_2 = 0$ when $x = \infty$. We get

$$1 = w_1 + w_2, \qquad (23)$$

or, which is the same,

$$\epsilon^x = \dots + \frac{x^{n-1}}{\lfloor n-1} + \frac{x^n}{\lfloor n} + \frac{x^{n+1}}{\lfloor n+1} + \dots = \Sigma \frac{x^n}{\lfloor n}, \qquad (24)$$

the series to be continued both ways. The assumption was that $n > -1$. But the change of n to $n-1$ in (24) makes no difference. So (24) is true for all values of n. The series on the right is a periodic function of n in appearance, since the change of n to $n+1$ reproduces the series. But as regards the value, the periodicity is only apparent. It is always ϵ^x. In numerical calculation only the initial convergent part of the series on its divergent side is to be counted, though all on the convergent side, as usual.

Note that x is a positive quantity all through. When negative another formula is needed, to be given later on.

The function w_1 may be developed in a rising power series operationally thus,

$$w_1 = \Delta^{-1}\epsilon^{-x}\Delta^{-n}1 = \Delta^{-1}(\Delta+1)^{-n}\epsilon^{-x} = \Delta^{-1}(\Delta+1)^{-n}\frac{\Delta 1}{\Delta+1}$$

$$= (\Delta+1)^{-(n+1)} = \Delta^{-(n+1)} - (n+1)\Delta^{-(n+2)}$$

$$+ \frac{(n+1)(n+2)}{\lfloor 2}\Delta^{-(n+3)} - \ldots, \quad (25)$$

expanding by the binomial theorem; and now algebrising, we get

$$w_1 = \frac{x^{n+1}}{\lfloor n+1} - (n+1)\frac{x^{n+2}}{\lfloor n+2} + \frac{(n+1)(n+2)}{\lfloor 2}\frac{x^{n+3}}{\lfloor n+3} - \ldots. \quad (26)$$

This may be confirmed by (21). But a similar operational treatment of w_2 fails. So expand w_2 in (22) in a power series by direct multiplication, using the usual formula for ϵ^{-x}. The result is

$$w_2 = \ldots \frac{x^{n-1}}{\lfloor n-1}(1-1)^{n-1} + \frac{x^n}{\lfloor n}(1-1)^n + \frac{x^{n+1}}{\lfloor n+1}(1-1)^{n+1} + \ldots - w_1. \quad (27)$$

But $w_1 + w_2 = 1$, so

$$1 = \Sigma\frac{(1-1)^n}{\lfloor n}x^n. \quad (28)$$

This is the result we shall get by developing (24) to a power series, multiplying it by the usual ϵ^{-x} expansion. A more general formula including (28) will occur later. The apparent numerical unintelligibility is no necessary bar to the use of (28) as a working and transforming formula. It often turns up. By $(1-1)^n$ is to be understood the expansion according to the binomial theorem, in powers of the second 1.

Application of the Generalised Exponential to a Bessel Function, to the Binomial Theorem, and to Taylor's Theorem.

§ 428. It is easy to generalise a common stopping series in which n is integral to another in which n is fractional, but the result is not to be expected in general to be equivalent to the stopping series, which is a special case of the generalised series. Yet there may be apparent equivalence. Thus (O.P.M., § 22) the solution of

$$(\Delta^2 + x^{-1}\Delta)u = u$$

in rising powers of x is

$$u = \mathrm{I}_0(x) = 1 + \frac{x^2}{2^2} + \frac{x^4}{2^2 4^2} + \frac{x^6}{2^2 4^2 6^2} + \ldots = \epsilon^{\mathrm{D}^{-1}} 1, \qquad (29)$$

by putting $y = \frac{1}{4}x^2$, and denoting d/dy by D. Substitute the generalised exponential, and see what we get. The result is

$$u = \Sigma \frac{\mathrm{D}^{-n}}{\lfloor n} 1 = \Sigma \frac{y^n}{(\lfloor n)^2}. \qquad (30)$$

Now is u the same as $\mathrm{I}_0(x)$? Take $n = \frac{1}{2}$, then

$$u = \frac{2}{\pi}\left(x + \frac{x^3}{3^2} + \frac{x^5}{3^2 5^2} + \ldots + \frac{1}{x} + \frac{1^2}{x^3} + \frac{1^2 3^2}{x^5} + \ldots\right), \qquad (31)$$

By numerical calculation this is $\mathrm{I}_0(x)$, without any fault to find. When n is not $\frac{1}{2}$, I also found fair agreement at first, with only rough values of $\lfloor n$. Later on I found that the assumed equivalence led to wrong results, and in Part 3 found the explanation. This will come later; at present it is enough to say that u is equivalent to $\mathrm{I}_0(x)$ for integral n and midway between, but is only apparently equivalent otherwise.

Another application of the generalised exponential is to the binomial theorem. Thus (O.P.M. § 23)

$$\frac{(1+x)^n}{\lfloor n} = \epsilon^{\Delta} \frac{x^n}{\lfloor n} = \epsilon^{\Delta}\Delta^{-n} 1 = \Sigma \frac{\Delta^{r-n}}{\lfloor r}, \qquad (32)$$

where we use the generalised exponential, and r is the general exponent, not n, which is constant. So, algebrising, we get

$$\frac{(1+x)^n}{\lfloor n} = \Sigma \frac{x^{n-r}}{\lfloor r \lfloor n-r}. \qquad (33)$$

If $r=0$ or any integer we have the form

$$\frac{(1+x)^n}{\lfloor\underline{n}} = \frac{x^n}{\lfloor\underline{n}} + \frac{x^{n-1}}{\lfloor\underline{n-1}} + \frac{x^{n-2}}{\lfloor\underline{2}\,\lfloor\underline{n-2}} + \dots \tag{34}$$

If $r=n$ we have the form

$$\frac{(1+x)^n}{\lfloor\underline{n}} = \frac{1}{\lfloor\underline{n}} + \frac{x}{\lfloor\underline{1}\,\lfloor\underline{n-1}} + \frac{x^2}{\lfloor\underline{2}\,\lfloor\underline{n-2}} + \dots\,. \tag{35}$$

These are doubly-stopping series when n is integral, and become identical. When n is fractional they stop only one way. In general the series (33) does not stop either way.

In the case $r=\frac{1}{2}$ we get

$$\frac{(1+x)^n}{\lfloor\underline{n}} = \dots + \frac{x^{n-\frac{3}{2}}}{\lfloor\underline{\tfrac{3}{2}}\,\lfloor\underline{n-\tfrac{3}{2}}} + \frac{x^{n-\frac{1}{2}}}{\lfloor\underline{\tfrac{1}{2}}\,\lfloor\underline{n-\tfrac{1}{2}}} + \frac{x^{n+\frac{1}{2}}}{\lfloor\underline{-\tfrac{1}{2}}\,\lfloor\underline{n+\tfrac{1}{2}}} + \dots\,. \tag{36}$$

If $n=1$, this makes

$$\frac{\pi}{4}(1+x) = x^{\frac{1}{2}}\left(1 + \frac{x+x^{-1}}{1\,.\,3} - \frac{x^2+x^{-2}}{3\,.\,5} + \frac{x^3+x^{-3}}{5\,.\,7} - \dots\right). \tag{37}$$

In special cases (O.P.M., § 25) we can get fully convergent formulæ for $\lfloor\underline{n}$. Thus, take $r=\frac{1}{4}$, $n=\frac{1}{2}$, $x=1$ in (33). Then

$$\left(\frac{8}{\pi}\right)^{\frac{1}{2}}\left(\lfloor\underline{\tfrac{1}{4}}\right)^2 = 1 + \tfrac{2}{5}(1 - \tfrac{3}{9}(1 - \tfrac{7}{13}(1 - \tfrac{11}{17}(1 - \dots\,. \tag{38}$$

Similarly, $r=\frac{2}{3}$, $n=\frac{1}{3}$, $x=1$ in (33) makes

$$\frac{8\pi^2}{81\,.\,4^{\frac{1}{3}}\left(\lfloor\underline{\tfrac{1}{3}}\right)^3} = 1 - \tfrac{1}{5}(1 - \tfrac{4}{8}(1 - \tfrac{7}{11}(1 - \tfrac{10}{14}(1 - \tfrac{13}{17}(1 - \dots\,. \tag{39}$$

So we know $g(\frac{1}{2})$, $g(\frac{1}{3})$, and $g(\frac{1}{4})$, without going to the general formula or to tables.

Applying the generalised exponential to Taylor's theorem,

$$f(x+h) = \epsilon^{h\Delta} f(x) = \Sigma\, \frac{h^r\Delta^r}{\lfloor\underline{r}} f(x), \tag{40}$$

if $f(x)$ is a power series, we should expect the generalised exponential to apply as well as to the special function $x^n/\lfloor\underline{n}$. We get (O.P.M., § 24)

$$f(x+h) = \left\{\frac{h^r\Delta^r}{\lfloor\underline{r}} + \frac{h^{r+1}}{\lfloor\underline{r+1}}\Delta^{r+1} + \dots + \frac{h^{r-1}}{\lfloor\underline{r-1}}\Delta^{r-1} + \dots\right\} f(x). \tag{41}$$

The extent of application may be left open.

If $f(x)=1$, then we get

$$1=\frac{h^r x^{-r}}{\underline{|r}\,\underline{|-r}}+\frac{h^{r-1}x^{1-r}}{\underline{|r-1}\,\underline{|1-r}}+\ldots+\frac{h^{r+1}x^{-r-1}}{\underline{|r+1}\,\underline{|-r-1}}+\ldots\,. \qquad (42)$$

Put $h/x=c$ and use (9) above in every term, then

$$1=\Sigma\,\frac{c^r \sin r\pi}{r\pi}, \qquad (43)$$

where c is to be positive. This is another formula of the nature of (28) above. In fact (28) may be derived from (43).

Algebraical Connection of the Convergent and Divergent Series for the Zeroth Bessel Function.

§ 429. Another application (O.P.M., §§ 27, 28) of the generalised exponential is to show the algebraical connection between the zeroth Bessel function in the usual ascending series, and the equivalent descending series.

Let $$A=1+\frac{x^2}{2^2}+\frac{x^4}{2^2 4^2}+\frac{x^6}{2^2 4^2 6^2}+\ldots, \qquad (44)$$

$$B=\frac{2}{\pi}\left(x+\frac{x^3}{3^2}+\frac{x^5}{3^2 5^2}+\ldots+\frac{1}{x}+\frac{1^2}{x^3}+\frac{1^2 3^2}{x^5}+\ldots\right), \qquad (45)$$

$$C=\frac{\epsilon^x}{(2\pi x)^{\frac{1}{2}}}\left(1+\frac{1^2}{8x}+\frac{1^2 3^2}{\underline{|2}(8x)^2}+\frac{1^2 3^2 5^2}{\underline{|3}(8x)^3}+\ldots\right). \qquad (46)$$

Here A is $I_0(x)$, and C is an equivalent form. Both come out of the same operator, as in § 353 above. They satisfy the same differential equation, and are numerically equivalent as well. They also act equivalently as operators, certainly sometimes. But aside from the extent of equivalence, what is the algebraic connection between A and C? How turn one to the other by mere algebra? It cannot be done with the usual form of ϵ^x because then we have integral even powers in A and nothing but fractional powers in C. But use the generalised ϵ^x, that is

$$\epsilon^x=\Sigma\,\frac{x^r}{\underline{|r}}, \qquad (47)$$

with the special value $r=\frac{1}{2}$, that is, in the representative term, the others being $1\frac{1}{2}$, $2\frac{1}{2}$, &c., $-\frac{1}{2}$, $-1\frac{1}{2}$, &c. Then C becomes an integral power series. But it does not agree with A, for it contains all integral powers of x, positive and negative. Evaluating coefficients, we find that

$$C = \tfrac{1}{2}(A+B). \tag{48}$$

Thus C is, on the understanding mentioned about ϵ^x, half the sum of the usual $I_0(x)$, and of another form already found to be equivalent to the same, in § 428 above. Later on this process of algebraical conversion will be extended to any Bessel function. Remember, however, that x is to be positive, in general. The common theory of i does not hold in general in these transformations.

It is noteworthy that if in the differential equation

$$(\Delta^2 + x^{-1}\Delta)u = u, \tag{49}$$

which is the characteristic of A, B and C, we substitute

$$u = a_0 + a_1 x + a_2 x^2 + \dots + b_1 x^{-1} + b_2 x^{-2} + \dots, \tag{50}$$

we shall obtain the result

$$u = aA + bB, \tag{51}$$

where a and b are independent constants. But A and B are not independent solutions of the characteristic, but are equivalent forms of one solution only. There are only two independent convergent solutions of (49), but any number of divergent.

The result (O.P.M., §§ 29 and 30)

$$I_0(x) = \epsilon^x (1 + 2\Delta^{-1})^{-\frac{1}{2}} 1, \tag{52}$$

by using the generalised binomial expansion, generalises to

$$u = \epsilon^x \underline{\lfloor -\tfrac{1}{2}} \Sigma \frac{(2x)^r}{(\underline{\lfloor r})^2 \underline{\lfloor -\tfrac{1}{2} - r}}. \tag{53}$$

This is equivalent to $I_0(x)$ when $r=0$ and $-\frac{1}{2}$, but its full meaning is considered later.

Limiting Form of Generalised Binomial Expansion when Index is -1.

§ 430. (O.P.M., §§ 31, 32.) When the exponent is a negative integer the generalised binomial expansion becomes ambiguous by the existence of vanishing factors. In

$$\frac{(1+x)^n}{\lfloor n} = \Sigma \frac{x^r}{\lfloor r \lfloor n-r}, \tag{54}$$

put $n=-1$. We get

$$\frac{(1+x)^{-1}}{\lfloor -1} = \frac{x^r}{\lfloor r \lfloor -1-r}\left\{(1-x+x^2-x^3+\ldots) - (x^{-1}-x^{-2}+x^{-3}-x^{-4}+\ldots)\right\}. \tag{55}$$

On the left side there is the vanishing factor $g(-1)$. So, on right side we should have

$$1-x+x^2-\ldots = x^{-1}-x^{-2}+x^{-3}-\ldots \quad . \tag{56}$$

This asserts the equivalence of the two forms of expansion of $(1+x)^{-1}$ in integral power series.

But if $r=0$ as well, (55) takes the form $0\times(1+x)^{-1}=0\times 0$. To make plainer, first put $n=-1+s$ in (54), and then let both r and s be infinitely small, but without any connection. Using

$$g(-1+s)=s, \qquad g(-1+s-r)=s-r, \tag{57}$$

true when r and s are infinitesimal, (43) becomes

$$s(1+x)^{-1+s} = \frac{x_r}{\lfloor r}(s-r)\left\{1+\frac{-1+s-r}{r+1}x+\ldots + \frac{r}{s-r}x^{-1}+\frac{r(r-1)}{(s-r)(1+s-r)}x^{-2}+\ldots\right\} \tag{58}$$

Dividing by s, and then putting $r=0$, $s=0$, $r/s=c$, we have

$$(1+x)^{-1} = (1-c)(1-x+x^2-\ldots)+c(x^{-1}-x^{-2}+x^{-3}-\ldots). \tag{59}$$

That is, the limiting form of (54) when n is -1 and $r=0$ is indeterminate, as it consists of the two forms in (56) combined in any ratio, the quantity c having any value.

Remarks on the Operator $(1+\Delta^{-1})^n$.

§ 431. By using the two extreme forms of expansion of $(1+\Delta^{-1})^{-\frac{1}{2}}$, we are led to two equivalent formulæ for the function $\epsilon^{-\frac{1}{2}x}I_0(\frac{1}{2}x)$, as done before, § 353. This suggested the examination of the operator $(1+\Delta^{-1})^n$, partially carried out in O.P.M., §§ 33 to 42. Say

$$u=\left(1+n\Delta^{-1}+\frac{n(n-1)}{\lfloor 2}\Delta^{-2}+\ldots\right)1, \tag{60}$$

$$v=\Delta^{-n}\left(1+n\Delta+\frac{n(n-1)}{\lfloor 2}\Delta^{2}+\ldots\right)1, \tag{61}$$

with unit operand, making

$$u=1+nx+\frac{n(n-1)}{\lfloor 2\,\lfloor 2}x^2+\frac{n(n-1)(n-2)}{\lfloor 3\,\lfloor 3}x^3+\ldots, \tag{62}$$

$$v=\frac{x^n}{\lfloor n}\left(1+\frac{n^2}{x}+\frac{n^2(n-1)^2}{x^2\lfloor 2}+\frac{n^2(n-1)^2(n-2)^2}{x^3\lfloor 3}+\ldots\right). \tag{63}$$

The suggestion is that u and v may be equivalent. If they are, and we put Δ^{-1} for x in the last equations and use unit operand again, we shall obtain two new functions which might be equivalent. This process could be carried on indefinitely. It is like equating the two forms of the expression

$$\Sigma\frac{x^r\lfloor n}{(\lfloor r)^m\lfloor n-r} \tag{64}$$

which result by putting $r=0$ and n in turn.

The results of numerical comparison of (62), (63), are briefly these. The functions u and v are equivalent when n is $-\frac{1}{2}$, as already considered. Now the case $+\frac{1}{2}$ differs by a whole differentiation, and may be expected to give excellent results. It does, with $x=1, 2, 3$. Increasing n to $\frac{3}{4}$ and $\frac{9}{10}$ also produces apparent equivalence. There is identity when $n=1$. The cases of $n=\frac{1}{4}$ and $-\frac{1}{4}$ also show an apparent equivalence. So far, then, it is suggested that there is real equivalence all along, and not merely at $n=1, \frac{1}{2}, 0, -\frac{1}{2}$. But this is not made probable by further examination. For if we go down to $n=-\frac{3}{4}$, the function v for $x=1$ and 2 is rather too small, and a part of the term following the l. c. t. (or last convergent term) must be added to make the value of u. We may expect $n=-\frac{9}{10}$ to be worse. It is much worse. Going from $n=-\frac{9}{10}$ to -1 accentuates the difference; and finally, when $n=-1$, we have

$$u=\epsilon^{-x}, \qquad v=\frac{x^{-1}}{\lfloor -1}\left(1+\frac{1}{x}+\frac{\lfloor 2}{x^2}+\ldots\right), \tag{65}$$

which show no sort of agreement, v being zero.

Yet even in this case of extreme apparent failure there is a way of reconciling the two forms of operator. Thus

$$u = (1 - \Delta^{-1} + \Delta^{-2} - \ldots)1, \tag{66}$$

$$v = (\Delta - \Delta^2 + \Delta^3 - \ldots)1. \tag{67}$$

That u is ϵ^{-x} is obvious. Now if we apply harmonic analysis to the other form, we shall get ϵ^{-x}. Thus,

$$v = \frac{\Delta}{1+\Delta}1 = \frac{1-\Delta}{1-\Delta^2}\frac{1}{\pi}\int_0^\infty \cos mx\,dm$$

$$= (1-\Delta)\frac{1}{\pi}\int_0^\infty \frac{\cos mx}{1+m^2}dm = (1-\Delta)\tfrac{1}{2}\epsilon^{-\sqrt{x^2}} = \epsilon^{-x}, \tag{68}$$

when x is $+$, and 0 when x is $-$. This change from 0 to ϵ^{-x} is also true of u, since the operand 1 only begins when $x = 0$.

Two suggestions arise: First, that as the generalised binomial theorem becomes indeterminate in form when n is -1, so we may expect anomalous results, requiring special interpretation. The other is that, except with special values of n, u and v are not really equivalent, but that an auxiliary series is wanted, say $u = v + w$, where w is quite zero for the special values of n; but that when n has other values producing apparent equivalence, the value of w is too small (with the values of x tested) to influence the result; whilst finally, with the values of n from $-\frac{3}{4}$ to -1 the function w becomes important. At any rate I know that this is what happens in some other cases of apparent equivalence.

As regards the extension to (64), there will be some notes later.

Remarks on the Use of Divergent Series.

§ 432. In O.P.M., §§ 43 to 48, I have stated the growth of my views about divergent series up to that time:—I have avoided defining the meaning of equivalence. That has to be found out by experience and experiment. The definitions will make themselves in time. Starting from complete ignorance, my first notion of a series was that to have a finite value it must be convergent, of course. A divergent series also, of course, has an infinite value. Solutions of physical problems should always be in finite terms or in convergent series, otherwise nonsense is made.

Then came a partial removal of ignorant blindness. In some physical problems divergent series are actually used, notably by Stokes, referring to the divergent formula for the oscillating function $J_n(x)$. He showed that the error was less than the last term included. Now here the terms are alternately + and −. This seems to give a clue. After initial convergence, the terms do get bigger and bigger, but the alternation of sign is significant. It is possible to imagine a *finite* quantity divided into parts in this way. It is a bad arrangement of parts, but as the initial convergence guides one to the value, it may be practical.

But by the same reasoning, a continuously divergent series, with all its terms +, is infinite in value, of course. It cannot represent the solution of a physical problem involving finite values. This seems to be what Boole maintained in his "Differential Equations" (3rd Ed., p. 475):—"It is known that in the employment of divergent series an important distinction exists between the cases in which the terms of the series are ultimately all positive, and alternately positive and negative. In the latter case we are, according to a known law, permitted to employ that portion of the series which is convergent for the calculation of the entire value." He illustrated this by integrals ascribed to Petzval. It is equivalent to this. The equation

$$(\Delta^2 + x^{-1}\Delta)u = u \tag{69}$$

is satisfied by

$$A = \frac{\epsilon^x}{x^{\frac{1}{2}}}\left(1 + \frac{1^2}{8x} + \frac{1^2 3^2}{(8x)^2 \lfloor 2} + \dots\right), \tag{70}$$

$$B = \frac{\epsilon^{-x}}{x^{\frac{1}{2}}}\left(1 - \frac{1^2}{8x} + \frac{1^2 3^2}{(8x)^2 \lfloor 2} - \dots\right), \tag{71}$$

that is, multiples of $H_0(x)$ and $K_0(x)$. According to Boole, we may use B, but not A, when x is +. But if x is −, use A, but not B. So there is only *one* solution of the characteristic obtained in this way. The plausibility of this argument is evident, as evident as that A is ∞ when x is +.

But, later on, divergent series presented themselves to me in a different way—viz., as differentiating operators. The operators may be the same functions of Δ, or d/dx, as would make convergent series, or else series which would be either alternatingly or continuously divergent if x took the place of Δ. Introducing an operand and algebrising, the solution of a problem arises, and in a convergent form. Here, then, is the secret of the continuously divergent series. They are numerically meaningless, using x; the proper use is as differentiating operators to obtain convergent solutions. So the series A and B above are really independent solutions, and neither should be rejected when used as operators.

But this view is soon found to be imperfect. For an operator may lead to a convergent solution by one way, and to a divergent by another. This and other considerations show that divergent series, even when continuously divergent, must be considered numerically, as well as algebraically and analytically. In the analytical use, every term must be used, if the result is a convergent series. But we cannot use all terms in the numerical case, because there is no limit. And numerical examination shows that the initial convergence determines the value of a continuously divergent series in the same sense as an alternatingly divergent series. The supposed distinction between the two cases asserted by Boole disappears, and we seem to have something like a distinct theory.

Examination into the reason why two series, one convergent, the other divergent, are equivalent, leads us to consider generalised differentiation, and the connected generalised series. There are certainly three kinds of equivalence. The first use I made of equivalent series, one of which is continuously divergent, was analytical only. The second one was numerical. The third was algebraical, connecting the series concerned by means of generalised series. Equivalence does not mean identity. The investigation in § 429 above illustrates several points in this connection. The identical connection is given by $C = \frac{1}{2}(A+B)$, using a particular formula for ϵ^x. This further explains some other things, viz., the different behaviour of A and C (which are numerically equivalent when x is +) on making x imaginary, as will be shown later on. The series B effects the reconciliation.

But the numerical meaning of divergent series still remains obscure. The property of estimation of value by the initial convergence is a very valuable one, and is true in a large number of cases. But it seems to fail in a marked manner sometimes. (See the examples above and suggested explanation.) The use of divergent series, however, is not merely for numerical purposes. In the analytical use, as operators, the question of value does not arise. In any case we must not be misled by apparent unintelligibility to ignore the subject. The error fallen into by Boole was striking There was a time, too, when mathematicians of the highest repute could not see the validity of investigations involving the algebraic imaginary. The results reached, they considered, were only suggestive, and required independent corroboration. But there is now a theory of the imaginary. There will have to be a theory of divergent series, or, say, a larger theory of functions than the present, including convergent and divergent series in one harmonious whole.

Logarithmic Formula derived from Binomial.

§ 433. The limit of $(d/dn)x^n$ when $n=0$ is $\log x$. Using

$$(1+x)^n = \Sigma \frac{x^r \lfloor n}{\lfloor r \lfloor n-r}, \tag{72}$$

where r has any leading value, and the step is unity, we obtain (O.P.M. §§ 49, 50)

$$\log(1+x) = \Sigma \frac{x^r}{\lfloor r} \frac{d}{dn} \frac{g(n-r)}{g(n)}, \quad (n=0), \tag{73}$$

$$= \Sigma \frac{x^r}{\lfloor r \lfloor -r} \left(\frac{g'(-n)}{g(-n)} - \frac{g'(0)}{g(0)} \right), \tag{74}$$

where the accent indicates the derivative with respect to n. That is,

$$\log(1+x) = \Sigma\, x^r \frac{\sin r\pi}{r\pi} \left(\frac{g'(-n)}{g(-n)} - \frac{g'(0)}{g(0)} \right), \tag{75}$$

by equation (9). But here

$$g(0) = 1, \qquad g'(0) = C = 0{\cdot}5772, \qquad \Sigma\, x^r \frac{\sin r\pi}{r\pi} = 1, \tag{76}$$

by (17) and (28); so we reduce to

$$\log(1+x) = -C + \Sigma\, x^r g(r) g'(-r). \tag{77}$$

The common formula is the case $r=0$. To obtain it, we must use

$$g'(-1)=1,\ g'(-2)=-1,\ g'(-3)=\lfloor 2,\ g'(-4)=-\lfloor 3,\ \&c. \quad (78)$$

If we differentiate (75) to x, we get (O.P.M. § 51),

$$(1+x)^{-1}=\Sigma\, x^{r-1}g(r-1)g'(-r); \quad (79)$$

or, increasing r by unity,

$$(1+x)^{-1}=\Sigma\, x^{r}g(r)g'(-r-1). \quad (80)$$

This reduces to $1-x+x^2-\dots$ when $r=0$, and it may be readily tested to be numerically right when $r=\frac{1}{2}$.

If the last result (80) can be regarded (O.P.M., § 52) as true when Δ^{-1} is put for x, and we introduce unit operand, the left side will make ϵ^{-x}, and so the right side should be a generalised formula for ϵ^{-x}, wanted in the theory of Bessel functions. But this formula turned out to be incorrect later, so need not be given. Its failure led to finding the proper one, to be given later. Similarly, O.P.M., §§ 53, 54, may be skipped, being based on the erroneous formula, and only partially valid.

Logarithmic Formulæ derived from Generalised Exponential.

§ 434. From the generalised exponential we may derive (O.P.M., §§ 56 to 62) some formulæ involving the logarithm. Thus, differentiate

$$\epsilon^{x}=\Sigma\, x^{r}g(r), \quad (81)$$

with respect to r. We get

$$0=\epsilon^{x}\log x+\Sigma\, x^{r}g'(r). \quad (82)$$

A second and a third differentiation give

$$0=-\epsilon^{x}(\log x)^2+\Sigma\, x^{r}g''(r), \quad (83)$$

$$0=+\epsilon^{x}(\log x)^3+\Sigma\, x^{r}g'''(r), \quad (84)$$

and so on. In another form, we may write

$$\log x=-\frac{\Sigma\, x^{r}g'(r)}{\Sigma\, x^{r}g(r)}=-\frac{\Sigma\, x^{r}g''(r)}{\Sigma\, x^{r}g'(r)}=\&c. \quad (85)$$

These various formulæ may be combined to make others, to show a fit. Thus

$$x\epsilon^x = \epsilon^x\{1 + \log x + \frac{(\log x)^2}{\underline{|2}} + \ldots\}, \tag{86}$$

by the use of (85) becomes

$$= \Sigma\, x^r(g - g' + \frac{g''}{\underline{|2}} - \ldots)(r) \tag{87}$$

$$= \Sigma\, x^r g(r-1) = x\Sigma\, x^r g(r) = x\epsilon^x, \tag{88}$$

by using Taylor's theorem.

If we use (82) thus,

$$-\log x = \Sigma x^r\left(1 - x + \frac{x^2}{\underline{|2}} + \frac{x^3}{\underline{|3}} - \ldots\right) g'(r), \tag{89}$$

to obtain a formula for $-\log x$ in powers of x, by rearrangement of terms we are led to

$$\Sigma x^r \frac{d}{dr}\,\frac{1 - r + \frac{1}{2}r(r-1) - \ldots}{\underline{|r}} = \Sigma x^r \frac{d}{dr}\,\frac{(1-1)^r}{\underline{|r}}, \tag{90}$$

which is striking, if not usable. A power series for x^{-1} also fails. The equation (43) comes in, and brings us back to x^{-1}.

Now take $r = 0$ in (82) to specialise. We get

$$-\log x = \epsilon^{-x}[g'(0) + xg'(1) + x^2g'(2) + \ldots$$
$$+ x^{-1}g'(-1) + x^{-2}g'(-2) + \ldots]. \tag{91}$$

The values of g' for negative integers were given before, (78). For positive integers we have

$$g'(0) = \mathrm{C}, \qquad g'(1) = \mathrm{C} - 1, \qquad g'(2) = \tfrac{1}{2}(\mathrm{C} - 1 - \tfrac{1}{2})$$
$$g'(n) = \frac{1}{\underline{|n}}\left\{\mathrm{C} - (1 + \tfrac{1}{2} + \tfrac{1}{3} + \ldots + \frac{1}{n})\right\}. \tag{92}$$

These bring (91) to

$$-(\log x + \mathrm{C}) = \epsilon^{-x}\left\{\frac{\underline{|0}}{x} - \frac{\underline{|1}}{x^2} + \frac{\underline{|2}}{x^3} - \frac{\underline{|3}}{x^4} + \ldots\right\}$$
$$-\epsilon^{-x}\left\{x + \frac{x^2}{\underline{|2}}(1 + \tfrac{1}{2}) + \frac{x^3}{\underline{|3}}(1 + \tfrac{1}{2} + \tfrac{1}{3}) + \ldots\right\}. \tag{93}$$

There is only poor initial convergency for calculation with small values of x. The second line of (93) may also be written

$$-\left\{x - \tfrac{1}{2}\frac{x^2}{\underline{|2}} + \tfrac{1}{3}\frac{x^3}{\underline{|3}} - \tfrac{1}{4}\frac{x^4}{\underline{|4}} + \ldots\right\}. \tag{94}$$

A different way of reaching (93) is thus :—Integrate the identity

$$\frac{1}{x}=\frac{\epsilon^{-x}}{x}+1-\frac{x}{|\underline{2}}+\frac{x^2}{|\underline{3}}-\frac{x^4}{|\underline{4}}+\dots \tag{95}$$

with respect to x. We get

$$\log x+\mathrm{C}=-\epsilon^{-x}\left(\frac{1}{x}-\frac{|\underline{1}}{x^2}+\frac{|\underline{2}}{x^3}+\dots\right)+x-\tfrac{1}{2}\frac{x^2}{|\underline{2}}+\dots, \tag{96}$$

where C is the constant of integration, given by the limit of

$$\mathrm{C}=x-\tfrac{1}{2}\frac{x^2}{|\underline{2}}+\tfrac{1}{3}\frac{x^3}{|\underline{3}}-\dots-\log x, \tag{97}$$

when $x=\infty$.

By giving x the values 1, 2, 3, &c., in turn we approximate pretty quickly to the value of C, much faster than by

$$\mathrm{C}=1+\tfrac{1}{2}+\tfrac{1}{3}+\dots+\frac{1}{r}-\log r, \tag{98}$$

with r increased to ∞.

Connections of the Zeroth Bessel Functions.

§ 435. In a similar way, by differentiating the function

$$u=\Sigma\, y^r[g(r)]^2 \tag{99}$$

with respect to r, we obtain various formulæ involving the zeroth Bessel function $\mathrm{I}_0(x)$, and its companion with the logarithm, if, as in § 428, y stands for $\frac{1}{4}x^2$. But the results (O.P.M., §§ 64 to 67) were considered on the idea that u represents $\mathrm{I}_0(x)$, which is only true specially, though, as before mentioned, there is apparent numerical equivalence beyond the special cases. So this is superseded by the proper investigation later.

The companion zeroth Bessel functions in divergent series are got by algebrising

$$\mathrm{H}_0(qx)=\frac{2\Delta}{(\Delta^2-q^2)^{\frac{1}{2}}}1,\qquad \mathrm{K}_0(qx)=\frac{2\Delta}{(q^2-\Delta^2)^{\frac{1}{2}}})1. \tag{100}$$

The first was done in § 353, equivalently. Introduce ϵ^{qx} as prefactor, and ϵ^{-qx} as postfactor, at the same time turning Δ to $\Delta-q$. Then put $\Delta\,(\Delta+q)^{-1}$ for the final ϵ^{-qx}, and algebrise. The result is

$$\mathrm{H}_0(qx)=\left(\frac{2}{\pi qx}\right)^{\frac{1}{2}}\left\{1+\frac{1}{8qx}+\frac{1^2 3^2}{(8qx)\,|\underline{2}}+\dots\right\}. \tag{101}$$

Now it is striking that the second operator in (100) similarly treated leads to the companion function, by introducing the prefactor ϵ^{-qx}. Thus, for distinctness, and to show how (101) was got (O.P.M. §§ 68 to 71),

$$K_0(qx) = \epsilon^{-qx}\epsilon^{qx}\frac{2\Delta}{(q^2-\Delta^2)^{\frac{1}{2}}}1 = 2\epsilon^{-qx}\frac{\Delta - q}{(2q\Delta-\Delta^2)^{\frac{1}{2}}}\frac{\Delta}{\Delta - q}1$$

$$= 2\epsilon^{-qx}\left(\frac{\Delta}{2q-\Delta}\right)^{\frac{1}{2}}1 = \epsilon^{-qx}\left\{1+\frac{\Delta}{4q}+\frac{1\cdot 3}{\lfloor 2}\left(\frac{\Delta}{4q}\right)^2+\dots\right\}\left(\frac{2}{\pi qx}\right)^{\frac{1}{2}}$$

$$= \epsilon^{-qx}\left(\frac{2}{\pi qx}\right)^{\frac{1}{2}}\left\{1-\frac{1}{8qx}+\frac{1^2 3^2}{(8qx)^2\lfloor 2}-\dots\right\}. \quad (102)$$

The function (101) only differs from the last in all signs being positive. The equivalent convergent series is

$$K_0(qx) = \frac{2}{\pi}\left\{-I_0(qx)[\log(\tfrac{1}{2}qx)+C]\right.$$
$$\left.+\frac{q^2x^2}{2^2}+\frac{q^4x^4}{2^24^2}(1+\tfrac{1}{2})+\frac{q^6x^6}{2^24^26^2}(1+\tfrac{1}{2}+\tfrac{1}{3})+\dots\right\}, \quad (103)$$

in which $I_0(qx)$ is given by

$$I_0(qx) = 1+\frac{q^2x^2}{2^2}+\frac{q^4x^4}{2^24^2}+\frac{q^6x^6}{2^24^26^2}+\dots, \quad (104)$$

as usual. The functions (104) and (103) are, being convergent, the companion solutions of rigorous mathematicians, save in the standardisation of (103) by the two constants $2/\pi$ and C, in which respects there is no settled practice.

Now the oscillating functions come out by putting $q = si$ and letting s be real. Thus

$$I_0(qx) = J_0(sx) = 1-\frac{s^2x^2}{2^2}+\frac{s^4x^4}{2^24^2}-\frac{s^6x^6}{2^24^26^2}+\dots \quad (105)$$

is the original Fourier-Bessel function. But $q = si$ in (103) makes

$$K_0(qx) = G_0(sx) - iJ_0(sx), \quad (106)$$

where the new function $G_0(sx)$ is given by

$$G_0(sx) = \frac{2}{\pi}\left\{-J_0(sx)\,[\log(\tfrac{1}{2}sx)+C]\right.$$
$$\left.-\frac{s^2x^2}{2^2}+\frac{s^4x^4}{2^24^2}(1+\tfrac{1}{2})-\frac{s^6x^6}{2^24^26^2}(1+\tfrac{1}{2}+\tfrac{1}{3})+\dots\right\} \quad (107)$$

It is G_0 that is the proper companion (oscillating) to J_0.

Similarly $q=si$ in H_0 and K_0 makes

$$\overline{H}_0(qx)=\bar{J}_0(sx)-i\bar{G}_0(sx), \tag{108}$$

$$K_0(qx)=\bar{G}_0(sx)-i\bar{J}_0(sx), \tag{109}$$

the bars indicating that divergent series are used, and

$$\bar{J}_0(sx)=\left(\frac{2}{\pi sx}\right)^{\frac{1}{2}}(R\cos+S'\sin)(sx-\tfrac{1}{4}\pi), \tag{110}$$

$$\bar{G}_0(sx)=\left(\frac{2}{\pi sx}\right)^{\frac{1}{2}}(S'\cos-R\sin)(sx-\tfrac{1}{4}\pi), \tag{111}$$

where R and S′ are the real functions of sx given by

$$R=1-\frac{1^2 3^2}{(8sx)^2\lfloor 2}+\frac{1^2 3^2 5^2 7^2}{(8sx)^4\lfloor 4}-\dots, \tag{112}$$

$$S'=\frac{1}{8sx}-\frac{1^2 3^2 5^2}{(8sx)^3\lfloor 3}+\frac{1^2 3^2 5^2 7^2 9^2}{(8sx)^5\lfloor 5}-\dots. \tag{113}$$

Equations (110) and (105) are equivalent, and so are (111) and (107).

Here we come to another matter explained by the generalised exponential series. The functions $H_0(qx)$ and $2I_0(qx)$ are equivalent. But the transformation $q=si$ makes them discrepant. We get $(\bar{J}_0-i\bar{G}_0)(sx)$ from the first and $2J_0(sx)$ from the second. But I have shown that the identical connection is

$$\overline{H}_0(qx)=I_0(qx)+\frac{2}{\pi}\left\{\frac{1}{qx}+\frac{1^2}{q^3x^3}+\dots+qx+\frac{q^3x^3}{1^2 3^2}+\dots\right\}. \tag{114}$$

In this put $q=si$, and there results

$$\overline{H}_0(qx)=J_0(sx)-i\bar{\bar{G}}_0(sx),$$

where

$$\bar{\bar{G}}_0(sx)=\frac{2}{\pi}\left\{\frac{1}{sx}-\frac{1^2}{s^3x^3}+\frac{1^2 3^2}{s^5x^5}-\dots-sx+\frac{s^3x^3}{1^2 3^2}-\frac{s^5x^5}{1^2 3^2 5^2}+\dots\right\}. \tag{115}$$

So we produce harmony by the function $\bar{\bar{G}}_0(sx)$, which is an equivalent form of $G_0(sx)$ and $\bar{G}_0(sx)$. The series (115) occurs in Lord Rayleigh's "Sound" (Vol. I., p. 154, 1st Ed) in a split form, and not identified, or rather equivalised with $G_0(sx)$.

We have

$$J_0(sx)\Delta G_0(sx)-G_0(sx)\Delta J_0(sx)=-2/\pi x, \tag{116}$$

when using the pair (105), (107), or else the pair (110), (111). And similarly

$$H_0(qx)\Delta K_0(qx)-K_0(qx)\Delta H_0(qx)=-4/\pi x, \tag{117}$$

using the pair (101), (102). But in the transition from (117) to (116) by $q = si$ it is indifferent whether we substitute $2J_0(sx)$ or $J_0(sx) - iG_0(sx)$ for $H_0(qx)$.

Operational Properties of the Zeroth Bessel Functions.

§ 436. The following (O.P.M., §§ 72 to 76) brings together compactly the principal mutual relations of H_0 and K_0. These formulæ occur naturally in the treatment of columnar elastic waves, and should be studied in immediate connection therewith, for which there is no space here. Let there be two variables, say, r and vt, whose differentiators are Δ and q. Then we have

$$\left.\begin{aligned}
[\mathrm{P}] = \frac{\Delta q}{(\Delta^2 - q^2)^{\frac{1}{2}}}1 &= \mathrm{I}_0(qr)q1, && [a] \\
= \epsilon^{qr}\left(\frac{\Delta}{2q+\Delta}\right)^{\frac{1}{2}} q1 &= \tfrac{1}{2}\mathrm{H}_0(qr)q1, && [b] \\
= \epsilon^{-vt\Delta}\left(\frac{q}{2\Delta - q}\right)^{\frac{1}{2}}\Delta 1 &= \tfrac{1}{2}\mathrm{K}_0(vt\Delta)\Delta 1, && [c] \\
&= \frac{1}{\pi}\frac{1}{(r^2 - v^2t^2)^{\frac{1}{2}}}, && [d] \\
&= \frac{1}{\pi}\mathrm{I}_0(vt\Delta)\frac{1}{r}, && [e]
\end{aligned}\right\} \qquad (118)$$

This makes one set. Another set is obtained by interchanging r and vt, and Δ and q throughout.

To get $[a]$ from $[\mathrm{P}]$, expand $[\mathrm{P}]$ in descending powers of Δ, and algebrise.

To get $[b]$, introduce the prefactor ϵ^{qr} to $[\mathrm{P}]$, and expand the properly transformed operator in descending powers of q, and algebrise.

To get $[c]$, introduce the prefactor $\epsilon^{-vt\Delta}$ to $[\mathrm{P}]$, and expand the properly transformed operator in descending powers of Δ, and algebrise.

The result $[d]$ may be derived from $[a]$, or $[b]$ or $[c]$. Thus, from $[b]$ to $[d]$,

$$\begin{aligned}
\tfrac{1}{2}q\mathrm{H}_0(qr)1 &= \frac{\epsilon^{qr}}{(2\pi r)^{\frac{1}{2}}}\left\{1 + \frac{1}{8qr} + \frac{1^2 3^2}{\lfloor\underline{2}\,(8qr)^2} + \dots\right\}\frac{1}{(\pi vt)^{\frac{1}{2}}} \\
&= \epsilon^{qr}\left\{1 + \tfrac{1}{2}\frac{vt}{2r} + \frac{1.3}{2^2\lfloor\underline{2}}\left(\frac{vt}{2r}\right)^2 + \dots\right\}\frac{1}{\pi(2vtr)^{\frac{1}{2}}} \\
&= \frac{\epsilon^{qr}}{\pi}\frac{1}{(vt)^{\frac{1}{2}}(2r - vt)^{\frac{1}{2}}} = \frac{1}{\pi(r^2 - v^2t^2)^{\frac{1}{2}}}.
\end{aligned} \qquad (119)$$

Similarly from [c] to [d],

$$\tfrac{1}{2}\Delta K_0(vt\Delta)1 = \frac{\epsilon^{-vt\Delta}}{(2\pi vt)^{\frac{1}{2}}}\left\{1 - \frac{1}{8vt\Delta} + \frac{1^2 3^2}{\lfloor 2(8vt\Delta)^2} - \dots\right\}\Delta^{\frac{1}{2}}1$$

$$= \frac{\epsilon^{-vt\Delta}}{\pi(2rvt)^{\frac{1}{2}}}\left\{1 - \tfrac{1}{2}\left(\frac{r}{2vt}\right) + \frac{1.3}{2^2\lfloor 2}\left(\frac{r}{2vt}\right)^2 - \dots\right\}$$

$$= \frac{\epsilon^{-vt\Delta}}{\pi}\frac{1}{r^{\frac{1}{2}}(2vt+r)^{\frac{1}{2}}} = \frac{1}{\pi(r^2 - v^2t^2)^{\frac{1}{2}}}. \qquad (120)$$

To get [d] from [a], use harmonic analysis, thus,

$$I_0(qr)q1 = I_0(qr)\frac{1}{\pi}\int_0^\infty \cos svt\, ds = \frac{1}{\pi}\int_0^\infty J_0(sr)\cos svt\, ds, \qquad (121)$$

a known integral, equivalent to [d].

To get [d] from [e] is obvious by executing the differentiations. Similar remarks apply to the second set above referred to, the change from one to the other set being usually related to the change from inward- to outward-going waves.

In the theory of pure diffusion there are analogous results. One of the simplest is this. Change the meaning of q from $d/d(vt)$ to $\{d/d(vt)\}^{\frac{1}{2}}$. On this understanding we shall have

$$\frac{\pi}{2}K_0(qr)q1 = I_0(qr)\frac{1}{2vt} = \frac{\epsilon^{-r^2/4vt}}{2vt}. \qquad (122)$$

As regards the proper use, sometimes of one method sometimes of another, that cannot be understood save in concrete applications.

Remarks on Common and Generalized Mathematics.

§ 437. Coming to O.P.M., Part. 3, 60 pp. of *Proc.* R.S. = about 80 or 90 here, which must be boiled down to 20 or 30 by omission of details.

When algebra reached a certain stage of development, the imaginary turned up. It was exceptional, however, and unintelligible, and therefore to be evaded, if possible. But it would not submit to be ignored. It demanded consideration, and has since received it. The algebra of real quantity is now a specialisation of the algebra of the complex quantity, say $a + bi$, and great extensions of mathematical knowledge have arisen out of the investigation of this once impossible and non-existent quantity. It may be questioned whether it is entitled to be called a quantity, but there is no question as to its usefulness, and the algebra of real quantity would be imperfect without it.

It has no essential connection with vectors or with quaternions or with circular functions, though it may be used illustratively. It turns up by itself in these subjects just as in scalar algebra, and in the same way. It is *sui generis*, and is *the* imaginary.

Some writers on double algebra appear to think that when they multiply together complexes they are multiplying vectors together. In an illustrative sense, vectorial work is being done. But it is versorial rather than vectorial. Say

$$(a+bi) \times (c+di) = (ac-bd)+(ad+bc)i.$$

This is strictly scalar algebra. But even illustratively, it is not the multiplication of the vector $c+di$ by the vector $a+bi$ that is done, when these are represented by lines in a plane, to form a new vector. Really i is a quadrantal versor, and $a+bi$ and $c+di$ are also versors with a stretching faculty as well, and their product is another operator of the same sort. Thus, if $\mathbf{x}$ is a real vector in the plane, then

$$(a+bi)(c+di)\mathbf{x} = [(ac-bd)+(ad+bc)i]\,\mathbf{x}$$

is a proper vectorial equation. The $\mathbf{x}$ may be a unit vector, and may be omitted altogether. The point is that the multiplication does not refer to vectors at all. The idea is too prevalent (though of late years it has been disappearing fast) that vectors ought to possess the associative property in multiplication. It is not in their nature to possess it. Versors possess this property naturally.

Again, in Quaternionics, all directions in space were made "equally imaginary, and therefore equally real." But this imaginary foundation gave way long ago. The real imaginary enters into quaternions just as usual. Yet quaternionists profess to be doing vector work, and go on confounding versors with vectors. It is not a new subject by any means. I think there is much room for improvement in the way of expounding the science of quaternions to those who are hardy enough to attack what is, as at present expounded, a puzzling subject.

It is the more desirable because there is no doubt that Quaternionics has been unduly neglected, and should have a useful field of its own. This is quite independent of the question of the suitability of the present Quaternionics for the purposes of physical inquirers. For that I prefer to discard the quaternionic idea completely, for the usual purposes, and use a simple algebra based upon vectorial ideas.

Another use of i is in the treatment of physical differential equations whose solutions are required to be simply periodic functions. There are two ways. One is to assume a complex form of solution at the beginning. It comes out complex at the end. Then either of its two parts may be selected for a real solution. The algebra is that of the real imaginary. But in the other way, if i be used at all, it is only a spurious imaginary. Say that $C = Ye$ determines C from e, a given function of the time, through the operator Y, containing p, the time differentiator. Then if we specialise e to be simply periodic, and also C, the power of p^2 in Y is $-n^2$, if the frequency of e be $n/2\pi$. This reduces C to the form $C = (Y_1 + Y_2 p)e$, when Y_1 and Y_2 are real functions of n^2. The solution is full and explicit. But if we say $p = ni$, we come to the result $C = (Y_1 + Y_2 ni)e$, which looks imaginary. But it is not, for the i means p/n, a differentiator. Either of these methods, the algebraical complex method or the differential one, may be done illustratively by lines in a plane as before alluded to.

Now just as the imaginary first presented itself in algebra as an unintelligible anomaly, so does fractional differentiation turn up in physical mathematics. It seems meaningless, and that suggests its avoidance in favour of more roundabout but understandable methods. But it refuses to be ignored. Starting from the ideas associated with complete differentiations, we come in practice quite naturally to fractional ones and combinations. This occurs when we know unique solutions to exist, and asserts the necessity of a proper development of the subject. Besides, as the imaginary was the source of a large branch of mathematics, so I think it must be with generalised analysis and series. Ordinary analysis is a specialised form of it. There is a universe of mathematics lying in between the complete differentiations and integrations.

The bulk of it may not be useful, when found, to a physical mathematician. The same can be said of the imaginary lore.

But some of it I have found to be very useful, and to furnish the most ready way of getting results simply. Compare, for instance, the simplicity of the processes used in §§353,4, with the complication of the subject when done by ordinary analysis. As regards some of the problems worked out later, I do not know any way of doing them by ordinary analysis. But it is unnecessary to appeal to utility. As the subject may be developed, what is useful will find its way out.

There is another analogy to be drawn. It is true that we cannot fully understand the usual algebra of convergent series without the imaginary. It is equally true that we cannot fully understand algebra, whether real or imaginary, without generalised analysis. I do not say that it is fully understandable with it, without more light. But a little light is better than darkness. In illustration of this I may refer to the equivalences treated of in § 429. Ordinary algebra furnishes no reason whatever why the series A and C should be equivalent. The generalised analysis does. This is not an isolated example; great extension follows.

The question of physical application raises another. Generalisations are to some extent arbitrary, according to the direction they take and the nature of the controlling ideas. It may be possible to elaborate generalised analysis in different ways. But to be useful in physical applications, it should be developed to suit them. In physical mathematics the quantities concerned are not arbitrary, but are controlled by the special relations involved in certain laws, involving, for instance, the necessary positiveness and singleness of certain quantities, simple themselves, or it may be complicated functions of other quantities, and their continuity of existence in time or space, or both, and their variation in time and space according to definite laws. So we have a definite march of events from one state to another, without that complicated multiplicity so common in pure mathematics. It is these general characteristics that seem to give reality to the mathematics, and serve to guide one along safe paths to useful results. Every one who has gone seriously into the mathematical theory of a physical subject (though it may be professedly only an ideal theory) knows how important it is not to look upon the symbols as standing for mere quantities (which might have any meaning), but to bear in mind the physics in a broad way, and obtain the important assistance of physical guidance in the actual work of getting solutions. This being the case generally, when the mathematics is well known, it is clear that when one is led to ideas and processes which are not understood, and when one has to find ways of attack, the physical guidance becomes more important still. If it be wanting, we are left nearly in the dark. The Euclidean logical way of development is out of the question. That would mean to stand still. First get on, in any way possible, and let the logic be left for later work.

These remarks are caused by certain experiences in the interval between Parts 2 and 3, O.P.M., when performing some very complicated and laborious calculations. Prof. Klein distinguishes three main classes of mathematicians—the intuitionists, the formalists or algorithmists, and the logicians. Now it is intuition that is most useful in physical mathematics, for that means taking a broad view of a question, apart from the narrowness of special mathematics. For what a physicist wants is a good view of the physics itself in its mathematical relations, and it is quite a secondary matter to have logical demonstrations. The mutual consistency of results is more satisfying, and exceptional peculiarities are ignored. It is more useful than exact mathematics.

But when intuition breaks down, something more rudimentary must take its place. This is groping, and it is experimental work, with of course some induction and deduction going along with it. Now, having started on a physical foundation in the treatment of irrational operators, which was successful, in seeking for explanation of some results, I got beyond the physics altogether, and was left without any guidance save that of untrustworthy intuition in the region of pure quantity. But success may come by the study of failures. So I made a detailed and close examination of some of the obscurities before alluded to, beginning with numerical groping. The result was to clear up most of the obscurities, correct the errors involved, and by their revision to obtain correct formulæ and extend the results considerably. Leaving out the details of the groping, some account of the results will follow.

It is of some importance to distinguish between a function in the physical sense, and its mode of expression in symbols standing for numbers. A physically-minded man need have no difficulty in conceiving the existence of a function of position and time, for instance, varying according to certain laws, and with any number of discontinuities in it, without any power to find formulæ which will have the necessary properties. Perhaps, if found, they would be much too comprehensive, the physical application requiring various limitations and reservations. It is characteristic of rigorous mathematicians, I think, that they think too much of the formula, and consider that it is the function. No, it is only the dress, and need not be a convenient fit. It is generally too large; or, may be, several dresses are needed, disconnected, for what would be simply a single function in its physical meaning. One form of expression of a function is a divergent series, and I take the view that the whole series is significant functionally, and not merely the few terms that may be utilised numerically. In a convergent series, though we cannot reach the value of the function by adding on terms at a uniform rate, we may go as near as we like, and it is easy to imagine practically taking the later terms in larger and larger groups, so on to include the whole in a finite time. But in a divergent series we cannot do this, though the initial convergence may guide us to the value approximately. How far does this property extend? It can be demonstrated to be true in certain cases, but something more general is wanted. Yet it can hardly be considered to be generally true, for we might make up series arbitrarily, having no particular characteristics. Nevertheless, the principle has a very wide application, to continuously divergent as well as alternatingly divergent series. I have employed it in the examination of a large number of divergent series, to test their equivalence to other convergent series, and have found it very useful. It has enabled me to distinguish between true and apparent equivalences, even when showing very small difference. But I do not think that the size of the smallest term does always govern the size of the error. May be, the terms preceding and following the smallest influence the value.

This simple illustration will serve to illustrate the nature of convergence. This series $1+\frac{1}{2}+\frac{1}{3}+\frac{1}{4}+\ldots$ *ad inf.* is called a divergent series because it sums up to infinity. Yet, at first glance, it might be thought to be convergent, because the terms get smaller and smaller continuously. We may say that it is convergent, only that it converges to infinity instead of to a finite value. Now suppose we slightly alter the law of the successive terms in a suitable manner so that the convergence ceases at a finite distance, after which the series becomes divergent. Then the point of convergence finds the value, as near as the smallest term will allow. But I do not say that all formulæ admit of a continuity of calculation if they pass from a state of finiteness through infinity to divergency. The meaning of the formula may change at the same time.

The Generalised Zeroth Bessel Function Analysed.

§ 438. Consider the generalised function

$$\mathrm{U}=\Sigma\frac{y^r}{(\underline{|r})^2}=\Sigma\, y^r[g(r)]^2. \qquad (123)$$

Let $y=\frac{1}{4}x^2$. Then, if r is zero or any integer, u is $\mathrm{I}_0(x)$. In any case it satisfies the characteristic of $\mathrm{I}_0(x)$. But that has two solutions, $\mathrm{I}_0(x)$ and $\mathrm{K}_0(x)$. Is then u a function of both when r is not integral?

On first experience it would seem not, but that U remains $\mathrm{I}_0(x)$ as r varies. This is closely true when $r=\frac{1}{2}$ by calculation. This includes $1\frac{1}{2}$, $2\frac{1}{2}$, &c. Besides, in this case we have had separate verification by algebraical use of the $r=\frac{1}{2}$ case. I also tested u for $r=\frac{1}{4}$, $\frac{3}{4}$, and $\frac{1}{10}$, though only in a rapid way, having no tables of $g(r)$. There was apparently equivalence, though not minutely verified.

But an anomaly presented itself. Differentiate U to r. Thus,

$$\mathrm{U}'=\Sigma\, y^r\Big([g(r)]^2\log y+2g(r)g'(r)\Big) \qquad (124)$$

$$=\Sigma\frac{y^r}{(\underline{|r})^2}\Big(\log y+2\mathrm{G}(r)\Big), \qquad (125)$$

if $\mathrm{G}(r)=g'(r)/g(r)$, a rather important function.

Now if U is really $I_0(x)$, U′ is zero. But it is not zero according to (125), in the case $r=0$. We get

$$(r=0) \qquad U' = -\pi K_0(x), \tag{126}$$

where $K_0(x)$ is the other solution, as in (103) above. So U must involve $K_0(x)$, in spite of its apparent equivalence to $I_0(x)$. Say

$$U = I_0 + bK_0, \tag{127}$$

where b is independent of x. It must be a periodic function of r, with period 1, and must vanish when r is integral, and when r is $\frac{1}{2}$, and must satisfy (126) when $r=0$. These considerations suggest a sine function, viz.,

$$U = I_0 - \tfrac{1}{2}\sin 2\pi r\, K_0, \tag{128}$$

$$U' = -\pi \cos 2\pi r . K_0. \tag{129}$$

It is now easy to see why, if (128) is true, U seemed to be I_0 always. For, with the not small values of x which are needed to produce marked initial convergence in U, the extra term is like a small satellite. The function I_0 is 1 at the origin, 2·279 when $y=1$, 4·252 when $y=2$, and so on, rapidly increasing to ∞. On the other hand, K_0 is ∞ at the origin, but decreases so fast that it is only 0·072 when $y=1$, 0·027 when $y=2$, 0·009 when $y=3$, and so on. So, unless specially sought for, the distinction between U and I_0 may be invisible numerically.

The next thing is to see whether the satellite really shows itself in the numerical results. We require

$$(r=0) \quad U' = -\pi K_0, \quad \text{and } U' = +\pi K_0, \ (r=\tfrac{1}{2}), \tag{130}$$

$$(r=\tfrac{1}{4}) \quad U = I_0 - \tfrac{1}{2}K_0, \text{ and } U' = 0, \tag{131}$$

$$(r=\tfrac{3}{4}) \quad U = I_0 + \tfrac{1}{2}K_0, \tag{132}$$

$$U = I_0 - \pi r K_0, \text{ when } r \text{ is very small.} \tag{133}$$

I have tested all these, and find them true when carefully calculated, and with values of x and r suitable for allowing the satellite to show itself. It is not necessary to give details, so I will quote one or two results.

Test (133). Say, $r=\frac{1}{100}$. Then

$$g(\tfrac{1}{100}) = 1{\cdot}005706, \qquad g^2(\tfrac{1}{100}) = 1{\cdot}011443,$$

and when $y=1$, these make $U=2{\cdot}27728$. But the value of I_0 is 2·27958. So

$$I_0 - U = 0{\cdot}00220. \tag{134}$$

Now this difference though itself small is far larger than is permissible according to the size of the l.c.t. But

$$\frac{\pi}{100}K_0 = 0{\cdot}00227, \tag{135}$$

which agrees with (134) up to the fourth figure. This makes matters right, for the error is now brought to be smaller than the l.c.t., and that is all that we can do.

With $y=\frac{1}{10}$, the results are

$$I_0 = 1{\cdot}1025, \qquad I_0 - U = 0{\cdot}0139, \tag{136}$$
$$U = 1{\cdot}0886, \qquad \frac{\pi}{100}K_0 = 0{\cdot}0147,$$

The smallness of r allows small values of y to be used here, but in other cases, $r=\frac{1}{4}$, for instance, larger values of y are required for clearly showing that the satellite serves to bring the error within the limits permitted by the size of the l.c.t. I usually count $\frac{1}{2}$ the l.c.t. in the series U, and make note of the size of this half-term; and usually employ smallish values of y to avoid the very lengthy calculations involved when y is large.

The other cases may be skipped. Nearly the whole of the evidence supports the truth of the formulæ (128), (129), and the little that does not is of a dubious nature, perhaps arising from errors in calculation. I only refer to these tests because the case is a typical one. We establish not merely that U *may* be the function of I_0 and K_0 given above, but also that the error may be regarded as limited by the size of the smallest term in the divergent series U. If there had been distinct failure, the assumed equivalence would have been rejected at once. Algebraical proof will follow.

Expression of the Divergent $H_0(x)$ and $K_0(x)$ in Terms of Two Generalised Bessel Functions. Generalisation of ϵ^{-x}.

§ 439. Putting $r=0$ and $\frac{1}{2}$ in turn in (129) and subtracting, using (125), we get the result

$$\pi K_0(x) = \Big(\sum_{r=\frac{1}{2}} - \sum_{r=0}\Big) y^r g(r) g'(r) \tag{137}$$

This expresses the function K_0 without the logarithm. Now we had before a remarkable connection between the usual descending series for H_0 and the $r=0$ and $\frac{1}{2}$ cases of the present function U. Is there a similar connection between the usual descending K_0 and (137)? If so, and (137) expressed an identity, not a mere equivalence, we should obtain the two oscillating functions J_0 and G_0 out of the right side of (137) by putting $x=zi$. But this fails on trial.

What then is the function

$$K_0(x)=\left(\frac{2}{\pi x}\right)^{\frac{1}{2}}\epsilon^{-x}\left\{1-\frac{1}{8x}+\frac{1^2 3^2}{\lfloor 2(8x)^2}-\dots\right\} \tag{138}$$

identically in terms of the ascending series for K_0?

To answer this question, we require a generalised formula for ϵ^{-x}. I first used a formula before referred to (§ 433), but it would not work, and on examination it was easily seen to be incorrect. To obtain the correct generalised formula, grope again. Suppose we alternate the signs of the terms in the generalised ϵ^x. Say

$$v=\dots-\frac{x^{r-1}}{\lfloor r-1}+\frac{x^r}{\lfloor r}-\frac{x^{r+1}}{\lfloor r+1}+\dots=\Sigma\pm\frac{x^r}{\lfloor r}. \tag{139}$$

Then v satisfies the differential equation of e^{-x}. It is e^{-x} itself when $r=0$ or any even integer; and is $-e^{-x}$ when r is an odd integer, so it must generally be Ae^{-x}, where A is a periodic function of r which is $+1$ when r is even, and -1 when r is odd. The simplest way of doing it is to say

$$v=\epsilon^{-x}\cos r\pi=\Sigma\pm\frac{x^r}{\lfloor r}. \tag{140}$$

Numerical tests of this formula are by no means so satisfying as those concerning the generalised ϵ^x. For instance, with $r=\frac{1}{2}$, we require $v=0$. This means that the generalised ϵ^x can then be divided into two equal parts, sum of even terms = sum of odd terms. This is satisfied as far as the initial convergence allows, but since v is the difference of the two series, we can only conclude that v is a small quantity. However, in this, and in the cases $r=\frac{1}{4}$ and $\frac{3}{4}$, the rule regarding the error limits is satisfied, so (140) may very well be true.

Now we have identified $H_0(x)$ with $U_0+U_{\frac{1}{2}}$, the suffixes meaning the values of r. Express $K_0(x)$ similarly in terms of two U's. In equation (138), put for ϵ^{-x} the ordinary ϵ^{-x} series; thus,

$$K_0(x) = \left(\frac{2\pi}{x}\right)^{\frac{1}{2}} \left\{ \frac{1}{(\underline{|-\frac{1}{2}})^2} - \frac{x^{-1}}{2(\underline{|-\frac{3}{2}})^2} + \frac{x^{-2}}{2^2\underline{|2}(\underline{|-\frac{5}{2}})^2} - \dots \right\}$$
$$\times \left(1 - x + \frac{x^2}{\underline{|2}} - \frac{x^3}{\underline{|3}} + \dots\right). \tag{141}$$

When arranged in powers of x, or preferably in powers of y, this makes a pair of series of the type U. To identify K_0, therefore, it is only necessary to size up the coefficients of the leading terms. We have

$$K_0(x) = \left(\frac{2}{\pi x}\right)^{\frac{1}{2}} \left[1 + \frac{(\frac{1}{2})^2}{2\underline{|1}} + \frac{\left(\frac{1.3}{2.2}\right)^2}{2^2\underline{|2}\underline{|2}} + \frac{\left(\frac{1.3.5}{2.2.2}\right)^2}{2^3\underline{|3}\underline{|3}} + \dots \right]$$
$$- \left(\frac{2x}{\pi}\right)^{\frac{1}{2}} \left[1 + \frac{(\frac{1}{2})^2}{2\underline{|2}\underline{|1}} + \frac{\left(\frac{1.3}{2.2}\right)^2}{2^2\underline{|3}\underline{|2}} + \frac{\left(\frac{1.3.5}{2.2.2}\right)^2}{2^3\underline{|4}\underline{|3}} + \dots \right]$$
$$+ \text{terms involving the other powers of } x. \tag{142}$$

The coefficient of $(2/\pi x)^{\frac{1}{2}}$ comes to 1·1795, and that of $-(2x/\pi)^{\frac{1}{2}}$ to 1·0782. Divide these by $\pi^{\frac{1}{2}}$. We get 0·665 and 1·217. The former is the value of $[g(-\frac{1}{4})]^2$, and the latter of $[g(\frac{1}{4})]^2$. We therefore prove that

$$K_0(x) = U_{-\frac{1}{4}} - U_{\frac{1}{4}}. \tag{143}$$

In the same way, by taking all signs positive in (141) and (142), we show that

$$H_0(x) = U_{-\frac{1}{4}} + U_{\frac{1}{4}}. \tag{144}$$

So far, then, we corroborate the formula (128), or

$$U = \tfrac{1}{2}(H_0 - \sin 2\pi r . K_0), \tag{145}$$

for it makes

$$U_r + U_{r+\frac{1}{2}} = H_0, \tag{146}$$

$$U_r - U_{r+\frac{1}{2}} = -K_0 . \sin 2\pi r, \tag{147}$$

for any r.

The Divergent $H_n(x)$ and $K_n(x)$ in Terms of Two Special Generalised Bessel Functions.

§ 440. We can now extend our results to Bessel functions of any order. We may verify (146), (147) with r left arbitrary by employing the generalised ϵ^x and the proposed ϵ^{-x} formula;

but this is unnecessary, because the work may be done upon the general formulæ for H_n and K_n. The characteristic is

$$(\Delta^2 + x^{-1}\Delta)\mathrm{U} = (1 + n^2/x^2)\mathrm{U}, \tag{148}$$

if Δ is d/dx, and a convergent solution is

$$\mathrm{I}_n(x) = y^{-\frac{1}{2}n}\left\{\frac{y^n}{\underline{|n}} + \frac{y^{n+1}}{\underline{|1}\,\underline{|n+1}} + \frac{y^{n+2}}{\underline{|2}\,\underline{|n+2}} + \dots\right\}. \tag{149}$$

The companion solution is $\mathrm{I}_{-n}(x)$, got by changing n to $-n$. Both these known functions are included in the generalised formula

$$\mathrm{U}_n = y^{-\frac{1}{2}n}\ \Sigma\ \frac{y^{n+r}}{\underline{|r}\,\underline{|n+r}}, \tag{150}$$

where, as usual, only the leading term is written. All the rest follow by changing the constant r to $r+1$, $r-1$, &c., the series to be made complete both ways.

The function U_n must be expressible in terms of I_n and I_{-n} because it satisfies the characteristic (148). The factors will be functions of r, though only one function is to be expected. We have to find out what U_n represents, and co-ordinate it with H_n and K_n, the divergent series also satisfying the characteristic, namely,

$$\mathrm{H}_n(x) = \left(\frac{2}{\pi x}\right)^{\frac{1}{2}}\epsilon^{x}\left\{1 + \frac{1^2 - 4n^2}{\underline{|1}(8x)} + \frac{(1^2-4n^2)(3^2-4n^2)}{\underline{|2}(8x)^2} + \dots\right\}, \tag{151}$$

$$\mathrm{K}_n(x) = \left(\frac{2}{\pi x}\right)^{\frac{1}{2}}\epsilon^{-x}\left\{1 - \frac{1^2 - 4n^2}{\underline{|1}(8x)} + \frac{(1^2-4n^2)(3^2-4n^2)}{\underline{|2}(8x)^2} - \dots\right\}. \tag{152}$$

First of all, see the effect of using the ordinary series for ϵ^x and ϵ^{-x} in these formulæ. Pick out the terms involving $x^{\frac{1}{2}}$ and $x^{-\frac{1}{2}}$, because we want only one term of each of the two series of the U_n type. Put $x = 2y^{\frac{1}{2}}$, then

$$\begin{aligned}\mathrm{H}_n = {} & \frac{y^{-\frac{1}{4}}}{\pi^{\frac{1}{2}}}\left\{1 + \frac{1^2-4n^2}{\underline{|1}(8)\underline{|1}} + \frac{(1^2-4n^2)(3^2-4n^2)}{\underline{|2}(8)^2\underline{|2}} + \dots\right\} \\ & + \frac{2y^{\frac{1}{4}}}{\pi^{\frac{1}{2}}}\left\{1 + \frac{1^2-4n^2}{\underline{|1}(8)\underline{|2}} + \frac{(1^2-4n^2)(3^2-4n^2)}{\underline{|2}(8)^2\underline{|3}} + \dots\right\} \\ & + \text{other terms.}\end{aligned} \tag{153}$$

We see that $r = \frac{1}{4}$ and $-\frac{1}{4}$ are involved; and since in the series

$$\mathrm{U}_{n,r} = \Sigma\frac{y^{r+\frac{1}{2}n}}{\underline{|r}\,\underline{|n+r}} \tag{154}$$

we have the exponent $r+\frac{1}{2}n$ in the leading term, we mus cancel the $\frac{1}{2}n$ by inclusion in the r. Thus

$$\left.\begin{aligned} \tfrac{1}{2}n+r &= -\tfrac{1}{4} \quad \text{makes} \quad r = -\tfrac{1}{2}n-\tfrac{1}{4}, \\ -\tfrac{1}{2}n+r &= \ \ \tfrac{1}{4} \quad \text{makes} \quad r = \ \ \tfrac{1}{4}-\tfrac{1}{2}n; \end{aligned}\right\} \qquad (155)$$

$$\left.\begin{aligned} &\text{therefore} \quad \frac{y^{r+\frac{1}{2}n}}{\underline{|r}\,\underline{|n+r}} = \frac{y^{-\frac{1}{4}}}{\underline{|-\frac{1}{4}-\frac{1}{2}n}\,\underline{|-\frac{1}{4}+\frac{1}{2}n}} \quad \text{in first case,} \\ &\text{and} \qquad\qquad\quad = \frac{y^{\frac{1}{4}}}{\underline{|\frac{1}{4}-\frac{1}{2}n}\,\underline{|\frac{1}{4}+\frac{1}{2}n}} \quad \text{in second case.} \end{aligned}\right\} (156)$$

So H_n is the sum of two U_n series with leading terms as just found, thus,

$$H_n = U_{n,\,(-\frac{1}{4}-\frac{1}{2}n)} + U_{n,\,(\frac{1}{4}-\frac{1}{2}n)}. \qquad (157)$$

Comparing with (153), we see that the coefficient of $y^{-\frac{1}{4}}$ in that equation is $g(-\frac{1}{4}-\frac{1}{2}n)g(-\frac{1}{4}+\frac{1}{2}n)$, and that of $y^{\frac{1}{4}}$ is $g(\frac{1}{4}-\frac{1}{2}n)g(\frac{1}{4}+\frac{1}{2}n)$.

In the same way, by alternating signs in H_n, we can show that

$$K_n = U_{n,\,(-\frac{1}{4}-\frac{1}{2}n)} - U_{n,\,(\frac{1}{4}+\frac{1}{2}n)}. \qquad (158)$$

The Divergent $H_n(x)$ and $K_n(x)$ in Terms of any Generalised Bessel Function of the same order.

§ **441.** Now, we found, subject to verification, that

$$U_{0,\,r} = \tfrac{1}{2}(H_0 - K_0 \sin 2\pi r). \qquad (159)$$

We also know, by (154), that the exponent r becomes $r+\frac{1}{2}n$ in passing from $U_{0,r}$ to $U_{n,r}$. So it is suggested that

$$U_{n,r} = \tfrac{1}{2}\{H_n - K_n \sin 2\pi(r+\tfrac{1}{2}n)\} \qquad (160)$$

is the general relation, true if (159) is true.

Changing n to $-n$, and, at the same time, increasing r by n, makes no difference on either side of the equation. Special cases of (160) are

$$U_{n,\,(-\frac{1}{4}-\frac{1}{2}n)} = \tfrac{1}{2}(H_n + K_n), \qquad (161)$$

$$U_{n\,(\frac{1}{4}-\ n)} \ \ = \tfrac{1}{2}(H_n - K_n). \qquad (162)$$

These are equivalent to (157) and (158).

Also, by negativing n in (160),

$$U_{-n,r} = \tfrac{1}{2}\{H_n - K_n \sin 2\pi(r-\tfrac{1}{2}n)\}\,; \qquad (163)$$

so, $$U_{n,r} + U_{-n,r} = H_n - K_n \cos \pi n \sin 2\pi r, \tag{164}$$

$$U_{n,r} - U_{-n,r} = -K_n \sin \pi n \cos 2\pi r; \tag{165}$$

and from these again,

$$H_n = U_{n,r} + U_{-n,r} + (U_{-n,r} - U_{n,r}) \cot \pi n \tan 2\pi r, \tag{166}$$

$$K_n = \frac{U_{-n,r} - U_{n,r}}{\sin \pi n \cos 2\pi r}. \tag{167}$$

We may also express H_n and K_n in terms of two U_n functions whose r's differ by $\frac{1}{2}$. Thus

$$H_n = U_{n,r} + U_{n,r+\frac{1}{2}}, \tag{168}$$

$$\sin \pi(n+2r) K_n = U_{n,r+\frac{1}{2}} - U_n. \tag{169}$$

The case $r=\frac{1}{4}$ is worth notice. It is easy to see by (160) that

$$U_{n,\frac{1}{4}} = U_{-n,\frac{1}{4}} = \tfrac{1}{2}(H_n - K_n \cos \pi n); \tag{170}$$

that is, the U function is the same for n as for $-n$ when the value of r is $\frac{1}{4}$. Similarly

$$U_{n,\frac{3}{4}} = U_{-n,\frac{3}{4}} = \tfrac{1}{2}(H_n + K_n \cos \pi n). \tag{171}$$

It looks unlikely at first that the expansion of U_n according to (154) should be valid here, because it is not the same for n as for $-n$, save when n is integral. When n is not integral there can only be equivalence. To examine this, take $n=\frac{1}{2}$. Comparing results by the two methods, we find that equivalence depends upon

$$\epsilon^x = 2\left\{ \dots + \frac{x^{-\frac{1}{2}}}{\lfloor -\frac{1}{2}} + \frac{x^{\frac{3}{2}}}{\lfloor \frac{3}{2}} + \frac{x^{\frac{7}{2}}}{\lfloor \frac{7}{2}} + \dots \right\}, \tag{172}$$

$$\epsilon^x = 2\left\{ \dots + \frac{x^{\frac{1}{2}}}{\lfloor \frac{1}{2}} + \frac{x^{\frac{5}{2}}}{\lfloor \frac{5}{2}} + \dots \right\}. \tag{173}$$

These agree with the $r=\frac{1}{2}$ case of the generalised formula (139) suggested above, but not yet verified algebraically.

To make $U_{n,r} = U_{-n,r}$ we require $4r = \text{odd}$, for any value of n, unless it be integral, when any r will do. Further, to reduce these equal series to $\frac{1}{2}H_0$ requires $2n$ to be integral. If odd, then $4r$ is odd, as before. If even, then n is an integer, and $2r$ is integral.

Going back to (167), the K_n formula. Or

$$K_n \sin \pi n \cos 2\pi r = U_{-n,r} - U_{n,r}. \tag{174}$$

Both members vanish when r is integral. Then differentiate to r to find K_n. The result is

$$-K_n\pi\cos\pi n\cos 2\pi r = (U_{n,r}+U_{-n,r})\log y^{\frac12}$$
$$+\Sigma\frac{y^{r+\frac12 n}}{|\underline{n+r}\,|\underline{r}}G(n+r)+\Sigma\frac{y^{r-\frac12 n}}{|\underline{-n+r}\,|\underline{r}}G(-n+r). \quad (175)$$

In the case $r=0$ we should reduce to known formulæ. Putting I_n for $\frac12 H_n$ we have

$$-\tfrac12\pi\cos\pi n\, K_n = I_n\log y^{\frac12}+\tfrac12\Sigma\frac{y^{\frac12 n}}{|\underline{n}\,|\underline{0}}G(n)$$
$$+\tfrac12\Sigma\frac{y^{-\frac12 n}}{|\underline{-n}\,|\underline{0}}G(-n). \quad (176)$$

In the summations only the leading terms are written. The step of r is 1 as usual. The two summations are stopping series now. The result may easily be expanded to the known formula

$$-\frac{\pi}{2}K_n\cos n\pi = I_n(\log\tfrac12 x+C)-\tfrac12\left(\frac{x}{2}\right)^n\sum_{r=0}^{r=\infty}\frac{S_r+S_{n+r}}{|\underline{r}\,|\underline{n+r}}\left(\frac{x}{2}\right)^{2r}$$
$$-\tfrac12\left(\frac{x}{2}\right)^n\sum_{r=1}^{r=n}\frac{\cos r\pi\,|\underline{r-1}}{|\underline{n-r}}\left(\frac{2}{x}\right)^{2r} \quad (177)$$

where $S_r = 1+\frac12+\frac13+\ldots+r^{-1}$.
To obtain the oscillating functions, put $x=zi$, and we have

$$K_n(x)=i^{-n}(G_n-iJ_n)(z). \quad (178)$$

To show the connections of H_n, K_n and U_n, U_{-n} briefly in a more general way, go back to (151) and (152). In the first use the generalised ϵ^x, and in the second the generalised ϵ^{-x}, or (140) above. Multiply up, and we obtain, by arrangement in powers of x or y,

$$H_n=\frac{1}{\pi^{\frac12}}\Sigma\frac{2^r}{|\underline{r}}y^{\frac12 r-\frac14}\phi(n,r), \quad (179)$$

$$K_n\cos n\pi=\frac{1}{\pi^{\frac12}}\Sigma\pm\frac{2^r}{|\underline{r}}y^{\frac12 r-\frac14}\phi(n,r), \quad (180)$$

where $\phi(n,r)$ is given by

$$\phi(n,r)=1+\frac{1^2-4n^2}{1.8.(r+1)}\left(1+\frac{3^2-4n^2}{2.8.(r+2)}\left(1+\ldots\right.\right. \quad (181)$$

Take half the sum and difference. Then

$$\tfrac12(H_n+K_n\cos r\pi)=\frac{1}{\pi^{\frac12}}\Sigma\frac{2^r}{|\underline{r}}y^{\frac12 r-\frac14}\phi(n,r), \quad (182)$$

$$\tfrac{1}{2}(\mathrm{H}_n - \mathrm{K}_n \cos r\pi) = \frac{1}{\pi^{\frac{1}{2}}} \Sigma \frac{2^{r+1}}{\underline{|r+1}} y^{\frac{1}{2}r+\frac{1}{4}} \phi(n, r+1). \qquad (183)$$

In these last summations the step of r is 2, not 1. But it is better to alter so that the step shall be 1, to harmonise with $\mathrm{U}_{n,r}$.

Put $\tfrac{1}{2}r - \tfrac{1}{4} = s + \tfrac{1}{2}n$, or $r = 2s + n + \tfrac{1}{2}$ in (182).

Also put $\tfrac{1}{2}r + \tfrac{1}{4} = s - \tfrac{1}{2}n$, or $r + 1 = 2s - n + \tfrac{1}{2}$ in (183).

These will make summations in which the step of s is 1. Then, after the change, put r for s. The results are

$$\tfrac{1}{2}\{\mathrm{H}_n - \mathrm{K}_n \sin \pi(2r+n)\} = \theta(n), \qquad (184)$$

$$\tfrac{1}{2}\{\mathrm{H}_n - \mathrm{K}_n \sin \pi(2r-n)\} = \theta(-n), \qquad (185)$$

where $$\theta(n) = \frac{1}{\pi^{\frac{1}{2}}} \Sigma \frac{2^{2r+n+\frac{1}{2}} y^{r+\frac{1}{2}n}}{\underline{|2r+n+\tfrac{1}{2}}} \left(1 + \frac{1^2 - 4n^2}{1.8(2r+n+\frac{3}{2})} + \dots\right) \qquad (186)$$

Comparing with the previous investigation, we see that

$$\theta(n) = \mathrm{U}_{n,r} = \Sigma \frac{y^{r+\frac{1}{2}n}}{\underline{|r}\,\underline{|n+r}}. \qquad (187)$$

This involves

$$\frac{1}{\underline{|r}\,\underline{|r+n}} = \frac{2^{2r+n+\frac{1}{2}}}{\pi^{\frac{1}{2}}\,\underline{|2r+n+\tfrac{1}{2}}} \left\{1 + \frac{1^2 - 4n^2}{1.8(2r+n+\frac{3}{2})}\left(1 + \frac{3^2 - 4n^2}{2.8(2r+n+\frac{5}{2})}\right. \right.$$
$$+ \dots, \qquad (188)$$

which is an identity.

Although the generalised ϵ^{-x} formula has so far only been used speculatively, yet the fact that its use, along with the generalised ϵ^{x} formula, enables us to connect algebraically the two sorts of expression of the two nth Bessel functions, the convergent and the divergent, makes it practically certain that it is the correct formula. An algebraical process leading to it will follow.

Product of the Series for ϵ^x and $\epsilon^{-x} \cos r\pi$. Possible Transition from ϵ^{-x} to ϵ^x.

§ **442**. The product of ϵ^x and ϵ^{-x} is 1. If, then, the product of the same when generalised came to 2 or to x, we could be sure there was something wrong. But we need not expect that the product should reduce to 1 in a plain manner. The product, say Q, of the generalised series for ϵ^x and $\epsilon^{-x} \cos r\pi$ makes the power series

$$Q = \ldots + x^{2r}\phi(r) - x^{2r+2}\phi(r+1) + x^{2r+4}\phi(r+2) - \ldots, \tag{189}$$

where $$\phi(r) = \frac{1}{(\lfloor r)^2}\left\{1 - \frac{2r}{r+1}\left(1 - \frac{r-1}{r+2}\left(1 - \frac{r-2}{r+3}\left(1 - \ldots\right\}\right.\right.\right., \tag{190}$$

which may be transformed to

$$\phi(r) = \frac{1}{(\lfloor r)^2}\frac{(1-r)(2-r)(3-r)\ldots}{(1+r)(2+r)(3+r)\ldots} = \frac{1}{(\lfloor r)^2}\frac{(1-1)^r}{(1+1)^r}. \tag{191}$$

So $\phi(r)$ is 0 when r is +, and is ∞ when r is −, unless integral. Q is 1 when r is 0, and is −1 when r is 1, and so on. But when r is fractional, Q assumes the indefinite form of a succession of infinities, not plainly reducible to $\cos r\pi$.*

Write (140) thus,

$$w = \Sigma \frac{x^r}{\lfloor r}\frac{1}{\cos r\pi}. \tag{192}$$

When x is +, w is ϵ^{-x}. Differentiate to r, then

$$w' = \Sigma \frac{x^r}{\lfloor r}\frac{1}{\cos r\pi}\left(\log x + G(r) + \pi \tan r\pi\right). \tag{193}$$

This is zero when x is +. Compare with u and u' before considered, representing ϵ^x and 0. The $r=0$ case of (193) is

$$w' = \epsilon^{-x}(\log x + C) - \left(\frac{1}{x} + \frac{1}{x^2} + \frac{\lfloor 2}{x^3} + \ldots\right) + \left(x - \frac{x^2}{\lfloor 2}(1 + \tfrac{1}{2}) + \ldots\right). \tag{194}$$

Compare with (93) above. That, by (97), makes a rapid approximation to the value of C. The present formula does not, for the same figures are differently arranged, which makes the l.c.t. large.

Since u satisfies $(\Delta - 1)u = 0$, it must be $A\epsilon^x$, where A is a function of r. This being 1 when x is +, if it is different when x is −, it is made a function of x. So there seems to be an irreconcilableness requiring some modification of ideas, perhaps. But u', u'', &c., also satisfy the characteristic, and so does

$$X = u + A_1 u' + A_2 u'' + A_3 u''' + \ldots. \tag{195}$$

Moreover, the value of X is simply u, or ϵ^x, when x is positive. Similarly, the characteristic $(\Delta + 1)w = 0$ is satisfied by

* In § 447 later, the improbability of Q coming to $\cos r\pi$ algebraically will appear, since the square of the expression for $\epsilon^{-x}\cos r\pi$ is the expression for $\epsilon^{-2x}\cos r\pi$.

$$Y = w + B_1 w' + B_2 w'' + B_3 w''' + \dots, \tag{196}$$

and the value of Y is w, or ϵ^{-x}, when x is positive. When x is negative, or complex, we do not know what they are. As the A's and B's may be various, we may have transitions from X to Y. Here is one. Say

$$\epsilon^{-x} = w + Bw'. \tag{197}$$

Put $x = -z = zi^2$. Then we get

$$\epsilon^z = \Sigma \frac{z^r}{\underline{|r}}(1 + i \tan r\pi)$$
$$+ B\Sigma \frac{z^r}{\underline{|r}}(1 + i \tan r\pi)(\log z + G(r) + i\pi + \pi \tan r\pi), \tag{198}$$

which splits into equations which may be written

$$\epsilon^z = u + Bu', \tag{199}$$

$$0 = \left(1 + B\frac{d}{dr}\right)\Sigma \frac{z^r}{\underline{|r}} \tan r\pi. \tag{200}$$

Whether this be a proper transformation or not, equations (197) and (199) are consistent. For with z positive and therefore x negative, both equations are true, if (197) is true for x negative as well as positive. But further investigation is reserved.*

Power Series for log x.

§ **443.** We may derive a power series for the logarithm of x from the binomial theorem

$$\frac{(1 + x)^n}{\underline{|n}} = \Sigma \frac{x^r}{\underline{|r}\,\underline{|n - r}}. \tag{201}$$

Differentiate to r. Then

$$-\log x = \Sigma \frac{x^r}{\underline{|r}\,\underline{|n - r}}\Big(G(r) - G(n - r)\Big). \tag{202}$$

In the simpler case of $n = 0$ this is

$$-\log x = \Sigma\, x^r \frac{d}{dr}\frac{\sin r\pi}{r\pi} = \Sigma \frac{x^r}{r}\left(\cos r\pi - \frac{\sin r\pi}{r\pi}\right). \tag{203}$$

---

* I do not think the above transformation is proper, though it is suggestive. There must be in existence, actual or potential, a theory of generalised functions of a complex.

When $r=0$ this makes

$$-\log x = \Sigma \frac{x^r}{r} \cos r\pi, \tag{204}$$

without the zeroth term for $r=0$. That is,

$$\log x = x - x^{-1} - \tfrac{1}{2}(x^2 - x^{-2}) + \tfrac{1}{3}(x^3 - x^{-3}) - \dots, \tag{205}$$

that is, the sum of the logarithms of $1+x$ and $(1+x^{-1})^{-1}$, according to the common formula.

In the case $r=\frac{1}{2}$,

$$\log x = \frac{1}{\pi} \Sigma \frac{x^r}{r^2} \sin r\pi; \tag{206}$$

or, putting $x=y^2$,

$$\frac{\pi}{2} \log y = y - y^{-1} - \tfrac{1}{9}(y^3 - y^{-3}) + \tfrac{1}{25}(y^5 - y^{-5}) - \dots. \tag{207}$$

The two formulæ (205) and (207) both lead to $\log i = \frac{1}{2}i\pi$.

Differentiating (204) to x, we get

$$-\frac{1}{x} = \Sigma x^{r-1}\left(\cos r\pi - \frac{\sin r\pi}{r\pi}\right). \tag{208}$$

Multiply by x and use the result (48); we then reduce to

$$0 = \Sigma x^r \cos r\pi. \tag{209}$$

The $r=0$ case is $\qquad 0 = (1+x)^{-1} - x^{-1}(1+x^{-1})^{-1}. \qquad (210)$

Also (208) seems to give a power series for x^{-1}. But it is only a particular form of (209).

Differentiating (209) to x gives

$$0 = \Sigma x^r \sin r\pi. \tag{211}$$

Equation (206) may be used numerically. I have also verified that the more general formula (203) goes when $r=\frac{1}{4}$, the error vanishing when $x=1$.

Put $y=xi$ in (207). We get the well-known

$$\frac{\pi}{4} z = \sin z - \tfrac{1}{9}\sin 3z + \tfrac{1}{25}\sin 5z - \dots. \tag{212}$$

Again, putting $x=\epsilon^{zi}$ in (203) we get

$$-z = \Sigma \frac{z^r}{\lfloor r}\left(\cos r\pi - \frac{\sin r\pi}{r\pi}\right), \tag{213}$$

which, when $r=0$, makes the well-known

$$\tfrac{1}{2}z = \sin z - \tfrac{1}{2}\sin 2z + \tfrac{1}{3}\sin 3z - \dots. \tag{214}$$

Although these substitutions of imaginary for real values are successful in the special cases chosen, they are only experimental, for the theory of generalised functions is *in nubibus* as yet.

Examination of some Apparent Equivalences, and Rectification.

§ 444. In § 433 this formula, equation (80), was arrived at from the binomial theorem *viâ* the logarithmic function :—

$$(1+x)^{-1} = \Sigma\, x^r g(r) g'(-r-1). \tag{215}$$

Put Δ^{-1} for x and algebrise with unit operand; then we get

$$\epsilon^{-x} \parallel \Sigma\, x^r [g(r)]^2 g'(-r-1). \tag{216}$$

Do the same again, and we get

$$J_0(x) \parallel \Sigma\, \frac{(\frac{1}{2}x)^{2r}}{(\underline{|r})^3} g'(-r-1). \tag{217}$$

Here $\parallel$ is used to indicate a possible or apparent equivalence; it may not be one, or it may. The method is a very speedy way of generating new formulæ, and sometimes gives true equivalences, at other times only partial ones.

Now (217) is easily seen to be incorrect. What about (216), then? And is not (215) to be suspected? It may be readily tested. Multiply by $1+x$. The result is

$$1 = \Sigma\, \frac{x^r}{\underline{|r}} \Big(g'(-r-1) + r g'(-r) \Big), \tag{218}$$

which is identically the same as

$$1 = \Sigma\, \frac{x^r}{\underline{|r}\ \underline{|-r}} = \Sigma\, x^r\, \frac{\sin r\pi}{r\pi}, \tag{219}$$

which formula was obtained before, equation (43). It asserts that $\epsilon^{\Delta} 1 = 1$, and often turns up as a connecting formula. It satisfies the characteristic

$$(1+x)\Delta u = 0, \tag{220}$$

i.e., $\Delta u = 0$, therefore u is constant as regards x. Numerically considered, however, it can only be used when x is near to 1. When $x=1$, we have such series as

$$(r=\tfrac{1}{2}) \qquad \frac{\pi}{4} = 1 - \tfrac{1}{3} + \tfrac{1}{5} - \tfrac{1}{7} + \dots, \tag{221}$$

$$(r=\tfrac{1}{4}) \qquad \frac{\pi}{4}\sqrt{2}=1+\tfrac{2}{15}-\tfrac{2}{63}+\tfrac{2}{143}-\tfrac{2}{255}+\ldots. \tag{222}$$

There is no reason, so far, to doubt (215).

Passing to (216), see on what the satisfaction of the characteristic of ϵ^{-x} by the right member depends. The result is

$$0 \parallel \frac{d}{dx}\Sigma\frac{x^r}{\lfloor r}\,\frac{\sin r\pi}{r\pi}. \tag{223}$$

If the $\parallel$ can be replaced by $=$, then (216) will go. Integrating to x,

$$1 \parallel \Sigma\frac{x^r}{\lfloor r}\,\frac{\sin r\pi}{r\pi}=v,\text{ say}, \tag{224}$$

because it is 1 for a special case, though the left side might be a function of r. On numerical testing, v seems to be 1 very closely. Thus take $r=\frac{1}{2}$; then $x=5$ makes $v=0{\cdot}9993$ by a long sum of about 20 terms. But this x may not be small enough to show an auxiliary satellitic function, if it exists. Now $x=2$ makes $v=1{\cdot}009$ with, and $1{\cdot}0009$ without the l.c.t. Again with $x=1$, I get $1{\cdot}063$ with and $1{\cdot}0093$ without the l.c.t. The error rule seems to fail here; for example, the last result is nearly 1 per cent. more than the size of the l.c.t. would allow. However, passing to $x=\frac{1}{2}$, the rule is satisfied, and also with $x=\frac{1}{4}$, although now the error itself has become large.

Testing (223) in the same way, with the same values of r and x, the value 0 comes out right (within the error limits) all through. Also with $x=3$. Unless, therefore, errors in calculation are involved, it would seem that the error rule, by which the l.c.t. finds the error, is not a complete rule.

Try with $r=\frac{1}{4}$. Here

$$g^2(\tfrac{1}{4})=1{\cdot}21690, \qquad g(\tfrac{1}{4})=1{\cdot}10313,$$

by the formula (188) above with $n=0$, and $r=\frac{1}{4}$. Applying these to (224), with $x=2$, 1, $\frac{1}{2}$, $\frac{1}{4}$, the results are a little less than 1 throughout, and within the error limits. We can say that v is certainly closely equal to 1 save when x is small, when the l.c.t. is too big for a certain conclusion.

But that (224) is not a real equivalence is to be seen by differentiating it to r. Thus

$$v'=v\log x+\Sigma\frac{x^r}{\lfloor r}\mathrm{G}(r)\,\frac{\sin r\pi}{r\pi}$$

$$+\Sigma\frac{x^r}{\lfloor r}\,\frac{\cos r\pi-(r\pi)^{-1}\sin r\pi}{r}. \tag{225}$$

When $r=0$, this reduces to

$$v' = \log x + \mathrm{C} - \left\{ x - \tfrac{1}{2}\frac{x^2}{\underline{|2}} + \tfrac{1}{3}\frac{x^3}{\underline{|3}} - \dots \right\} \tag{226}$$

$$= -\epsilon^{-x}\left\{ \frac{1}{x} - \frac{1}{x^2} + \frac{\underline{|2}}{x^3} - \frac{\underline{|3}}{x^4} + \dots \right\}. \tag{227}$$

Here v' is zero only when $x=\infty$. So $v \parallel 1$ is only an apparent equivalence. The characteristic of v is

$$(\Delta + 1)x\Delta v = 0, \tag{228}$$

$$\therefore v = a\int \frac{\epsilon^{-x}}{x}dx = \mathrm{A} + a\left[\log x + \mathrm{C} - \left(x - \tfrac{1}{2}\frac{x^2}{\underline{|2}} + \dots \right) \right], \tag{229}$$

where C, or G(0), is brought in to make $[\dots]=0$ when $x=\infty$. A and a are functions of r. From the preceding A is 1, and a is periodic. To harmonise with (226), we may write

$$v = 1 - \frac{\sin 2\pi r}{2\pi}\left[\left(x - \tfrac{1}{2}\frac{x^2}{\underline{|2}} + \dots \right) - (\log x + \mathrm{C}) \right]. \tag{230}$$

So $v=1$ when r is integral and midway between, and is nearly 1 for any r, if x is over 1.

The sine function thus reached may also be utilised thus. Equation (228), by (224), means the same as

$$\Sigma \frac{x^r}{\underline{|r}}\frac{\sin r\pi}{\pi} = a\epsilon^{-x}. \tag{231}$$

Divide by $\pi^{-1} \sin r\pi \cos r\pi$. Then

$$\Sigma \frac{x^r}{\underline{|r}}\frac{1}{\cos r\pi} = \frac{2\pi a}{\sin 2\pi r}\epsilon^{-x} = \epsilon^{-x}. \tag{232}$$

Comparing with (140), we see that the present investigation leads to the generalised ϵ^{-x} formula which harmonised the Bessel functions. Numerically tested, the formula (230) goes with $x=1, \frac{1}{2}, \frac{1}{4}$.

Determination of the Meaning of a Generalised Bessel Function in terms of H_0 and K_0.

§ 445. Passing now to the generalised formula

$$\mathrm{W} = \epsilon^x \pi^{\frac{1}{2}} \Sigma \frac{(2x)^r}{(\underline{|r})^2 \,\underline{|-\tfrac{1}{2} - r}}, \tag{233}$$

before obtained, equation (53). First test that W satisfies the characteristic of $I_0(x)$ identically. It therefore involves I_0 or K_0, or both. But can K_0 come in at all? W is the convergent I_0 when $r=0$, and is the equivalent divergent $\frac{1}{2}H_0$ when $r=\frac{1}{2}$. Now when r is arbitrary, we may, by the use of the corresponding generalised ϵ^x series, reduce W to an integral power series. But the solution of the characteristic equation of I_0 in an integral power series was shown to split into two series (even and odd terms), which satisfy the characteristic separately, and are, moreover, equivalent. This strongly suggests that W does not involve K_0 at all, but is $I_0(x) \times$ periodic function of r, which is 1 when r is 0 and $\frac{1}{2}$. Or, in terms of the even and odd series, $W = aA + bB$, where A is $I_0(x)$ and B is the equivalent series in odd powers as in (44), (45). I tried to test this numerically, to fix a and b. But the series to find a and b are divergent with wide error limits, so it was no good this way.

In order to detect the satellite, take $r=\frac{1}{4}$. The result (numerically) is that W is sensibly I_0 when x is big, but falls slightly below I_0 as x is reduced to 1, $\frac{3}{4}$, $\frac{1}{2}$. Then take $r=-\frac{1}{4}$, and test. As x is reduced W now *rises* slightly *above* I_0. This difference, + or − as the case may be, increases as x decreases.

The form of W is therefore probably $I_0 - cK_0$, as in the case of U before, and c has to be found. Finding the result of reducing W to an integral power series difficult to manage, try another way. Use the ordiuary ϵ^x series in (233). The result is then

$$W = \Sigma y^{\frac{1}{2}r}\phi(r), \tag{234}$$

where $y=\frac{1}{4}x^2$ and the step of r is 1, whilst

$$\phi(r) = 2^r\pi^{\frac{1}{2}}\left\{\frac{2^r}{(\lfloor r)^2\lfloor -r-\frac{1}{2}\lfloor 0} + \frac{2^{r-1}}{(\lfloor r-1)^2\lfloor -x+\frac{1}{2}\lfloor 1} + \dots\right\} \tag{235}$$

This is a stopping series, and is convergent, so can be closely calculated. With r integral, we get $I_0(x)$. With $r=$ integer $+\frac{1}{2}$, we get a result reducible to

$$W = \tfrac{1}{2}(U_{\frac{1}{4}} + U_{-\frac{1}{4}}) = \tfrac{1}{2}H_0(x). \tag{236}$$

In general, (234) makes

$$W_r = AU_{\frac{1}{2}r} + BU_{-\frac{1}{2}r}, \tag{237}$$

where A and B are functions of r to be found. Understand that U is the previous $U_{n,r}$ function, with $n=0$, as in § 439.

In the case of $r=\frac{1}{4}$ the result is

$$W_{\frac{1}{4}} = 0{\cdot}8525\, U_{\frac{1}{8}} + 0{\cdot}1460\, U_{-\frac{3}{8}}, \tag{238}$$

where $$U_r = \tfrac{1}{2}(H_0 - \sin 2\pi r . K_0)(x). \tag{239}$$

This makes $$W_{\frac{1}{4}} = \tfrac{1}{2}H_0 - \tfrac{1}{4}K_0, \tag{240}$$

very closely. Note that (238) is an identity, subject to the use of the ordinary ϵ^x series; whilst (240) is a conveniently substituted equivalence.

The general identical formula corresponding to (238) is

$$W_r = U_{\frac{1}{2}r} \cos^2 \tfrac{1}{2}\pi r - U_{-\frac{1}{2}r} \sin^2 \tfrac{1}{2}\pi r. \tag{241}$$

I have verified it numerically for eight values of r, namely, 0, $\frac{1}{8}$, $\frac{1}{4}$, &c., up to $\frac{7}{8}$. Using (239) in it, the result is

$$W_r = \tfrac{1}{2}(H_0 - \tfrac{1}{2} \sin 2\pi r K_0)(x). \tag{242}$$

The satellite for W_r has only half the mass of that for U_r. The value of $\phi(r)$ in (235) is

$$\phi(r) = [g(\tfrac{1}{2}r) \cos \tfrac{1}{2}\pi r]^2. \tag{243}$$

The reduction of W_r (involving K_0) to an integral power series, before referred to, implies that the series, which satisfies the characteristic as a whole, does not do so in two independent ways. The expansion (137) above illustrates this. Both the even and the odd powers have to be included to satisfy the characteristic. The characteristic is

$$(DyD - 1)K_0 = 0, \tag{244}$$

if $D = d/dy$, and this is satisfied because the odd power series when operated upon by $(DyD-1)$ gives $-D\Sigma y^r[g(r)]^2$, with $r=\frac{1}{2}$, whilst the even power series gives + (the same), with $r=0$. On addition, therefore, (244) is obeyed.

Of course, when W is expanded in the form (234), it does split into two series, as in (241), which separately satisfy the characteristic. But it seems that this is the exceptional case, and that there is no double satisfaction when the generalised ϵ^x is used.

The satellitic terms in both cases, U and W, become very important when x nears the origin. But then the series cannot be used for calculation, on account of the largeness of the possible error. So, practically, both U and W are apparently I_0; and the small correction may be eliminated by using two values of r; thus,

$$I_0(x) = \tfrac{1}{2}(U_r + U_{-r}) = \tfrac{1}{2}(W_r + W_{-r}). \qquad (245)$$

Some Apparent Equivalences.

§ 446. If the characteristic be of the third order, its three solutions will be all included in one generalised solution, and will be separable therefrom by giving special values to r. If $r=0$ makes the primary, the other two solutions are likely to enter satellitically in the general solution. Similar remarks may apply to higher orders, but there is some work to be done to investigate fully.

The apparent equivalence

$$I_0(x) \,||\, \Sigma \frac{y^r}{(\underline{|r})^2} \qquad (246)$$

arises from $\epsilon^{D^{-1}}1 = I_0(x)$; where $D = d/dy$, and $y = \frac{1}{4}x^2$. If we change y to Δ^{-1}, and algebrise again, with unit operand, we obtain

$$1 + y + \frac{y^2}{(\underline{|2})^3} + \frac{y^3}{(\underline{|3})^3} + \ldots \,||\, \Sigma \frac{y^r}{(\underline{|r})^3} = w_r. \qquad (247)$$

The characteristic is now

$$(DyDyD - 1)w = 0. \qquad (248)$$

It has three independent solutions. One of them, w_0, with $r=0$, is the left member of (247). All three are included in w_r. Nevertheless, $w_0 \,||\, w_r$ is an apparent equivalence.

Taking $r = \frac{1}{2}$, the original equation (246) is a true equivalence, but (247) is not. (Besides $r = 0$, $\frac{1}{3}$ and $\frac{2}{3}$ are probably the important values.) For w_0 is a little greater than $w_{\frac{1}{2}}$ when $y = 1$, the ratio being about 0·967. Increasing y improves the equivalence; $y = 3$ and above making $w_{\frac{1}{2}}$ a little greater than w_0, the ratio being 1·001 when $y = 8$.

Going a step further in the same way to the fourth order, say

$$w_r = \Sigma \frac{y^r}{(\underline{|r})^4}, \qquad (249)$$

and comparing w_0 with $w_{\frac{1}{2}}$, the numerical equivalence is so close as to make one think it may be an exact one here specially. Still $w_{\frac{1}{2}}$ is the smaller when $y=1$, though a trifle greater from $y=2$ up to 9.

Passing on to

$$w_r = \Sigma \frac{y^r}{(\underline{|r})^8}, \tag{250}$$

there is a falling off in agreement; for as y increases from 1 to 36, the ratio $w_{\frac{1}{2}}/w_0$ falls continuously from 1·37 to 0·907.

One more case to show further falling off. Let

$$w_r = \Sigma \frac{y^r}{(\underline{|r})^{16}}. \tag{251}$$

The peculiarity here (and in the last case less markedly) is the very large ratio, namely 2^{16}, of the coefficient of $y^{\frac{1}{2}}$ to that of $y^{-\frac{1}{2}}$ in the series $w_{\frac{1}{2}}$, whereas the succeeding ratios are quite small. This may affect the determination of the error in a complete rule. Anyhow the tendency shown in the last case is now more marked, for $w_{\frac{1}{2}}/w_0$ falls from 3·3 to 0·75 as y goes from 1 to 100. There is equality at about 49.

The above formulæ produce such good apparent equivalences at the beginning, though so bad later on, that it is suggested to examine the formulæ which arise from the equivalence $1 = \Sigma x^r(r\pi)^{-1} \sin r\pi$, equation (219), in a similar way. The first one, turning x to Δ and algebrising, is the apparent equivalence (224) above, shown to be excellent. But, others, of the types

$$w_r = \Sigma\, x^r \left(\frac{\sin r\pi}{r\pi}\right)^n, \qquad w_r = \Sigma \frac{x^r}{\underline{|r}}\left(\frac{\sin r\pi}{r\pi}\right)^n, \tag{252}$$

with n integral and greater than 1, are found by a cursory examination not to furnish apparent equivalences, at least as regards $r=0$ and $\frac{1}{2}$.

Cotangent Formula and Derived Formula for Logarithm. Various Properties of these and other Divergent Series.

§ **447.** So far O. P. M., Part 3. I regret that the condensation should tend to reduce it to the dry bones of mere formulæ. The original will not be published in Another Place before it is wanted. In the meantime I have only space left here for a few notes in connection with the preceding. They may be useful to investigators, as there are plenty of nuts to crack.

The formula (219) may be written

$$\frac{\pi}{\sin r\pi} = \Sigma \pm \frac{x^r}{r} = u. \tag{253}$$

The companion formula is

$$\pi \cot r\pi \pm \lambda = \Sigma \frac{x^r}{r} = v, \tag{254}$$

and possesses several points of interest. Just as u comes from the generalised ϵ^x, so does v come from the generalised ϵ^{-x}. Thus,

$$\epsilon^{\Delta} 1 = \Sigma \frac{\Delta^r}{\lfloor r} 1 = \Sigma \frac{x^{-r}}{\lfloor r \, \lfloor -r} = \Sigma \frac{x^r}{r} \frac{\sin r\pi}{\pi}, \tag{255}$$

$$\epsilon^{-\Delta} 1 \; || \; \Sigma \frac{\Delta^r}{\lfloor r} \frac{1}{\cos r\pi} \; || \; \Sigma \frac{x^r}{r} \frac{\tan r\pi}{\pi}. \tag{256}$$

Now (254) is exact when $x=1$, without λ. But when x is not 1, the functions u and v differ in behaviour. The first is continuous through $x=1$, the second is discontinuous. That is, there is a jump up through the amount λ when x is just above 1, and a jump down to the same extent when x is just below 1, λ being about 1·895, as found below. Why there should be a discontinuity in v, though not in u, at $x=1$, may be understood by considering that $\epsilon^{\Delta}\Delta 1$ is an impulse at $x=-1$, and $\epsilon^{-\Delta}\Delta 1$ an impulse at $x=+1$. So in the latter case integration produces a jump.

Or we may differentiate v to x. We get $0 = \Sigma x^r$. This means that Σx^r is an impulsive function. It is ∞ at $x=1$. Note also that in the theory of convergent functions the expression Σz^r, z being complex, plays the part of an impulsive function at the point $z=1$, although, of course, function-theorists would not, I think, admit that it had, by itself, any particular meaning.

By differentiation to r, the v formula makes

$$-\frac{\pi^2}{\sin^2 r\pi} = (\pi \cot r\pi \pm \lambda) \log x - \Sigma \frac{x^r}{r^2}, \tag{257}$$

and other formulæ may be derived. Perhaps the most interesting particular case of v is when $r=0$. Then there is an infinite term $x^0/0$ on the right side of (254). Deduct it from the left side. The result is

$$-\log x \pm \lambda = x + \tfrac{1}{2}x^2 + \tfrac{1}{3}x^3 + \ldots - x^{-1} - \tfrac{1}{2}x^{-2} - \ldots. \quad (258)$$

Remembering the potency of λ, this leads to

$$w = x + \tfrac{1}{2}x^2 + \tfrac{1}{3}x^3 + \ldots = \lambda - \log(x-1), \quad (x>1), \quad (259)$$

$$\text{or} = -\log(1-x), \quad (x<1). \quad (260)$$

This w function is infinite at $x=1$; beyond that it changes from $-\log(1-x)$ to $\log \mu(x-1)^{-1}$, where μ is a constant (about 6·653). It is then continuously divergent. When x is < -1 it is alternatingly divergent.

To show λ by itself, we have

$$\pm\lambda = 2(x + \tfrac{1}{3}x^3 + \tfrac{1}{5}x^5 + \ldots - x^{-1} - \tfrac{1}{3}x^{-3} - \tfrac{1}{5}x^{-5} - \ldots). \quad (261)$$

Numerical examination of various cases of the cot and log formulæ all lead to about the same value of λ when done roughly, usually from 1·8 to 2. But to find it more closely, special care must be taken. Some of the derived formulæ are much more convergent than the primitive; but this does not help as regards close estimation of λ, because the error becomes multiplied in another way. To find λ pretty closely, I used the w formula (259), with x between 1 and 2. The results are very regular when attention is paid to certain points. The error, if estimated by the size of the l.c.t., would be large unless a value of x so little greater than 1 were used that the labour became prohibitive. But there are certain considerations allowing of closer work with less labour. Thus, in the series $w = x + \tfrac{1}{2}x^2 + \ldots$, give x a value between 1 and 2, say 1·1, such that there are two equal bottom terms. Representing the series diagrammatically thus, we have to find the

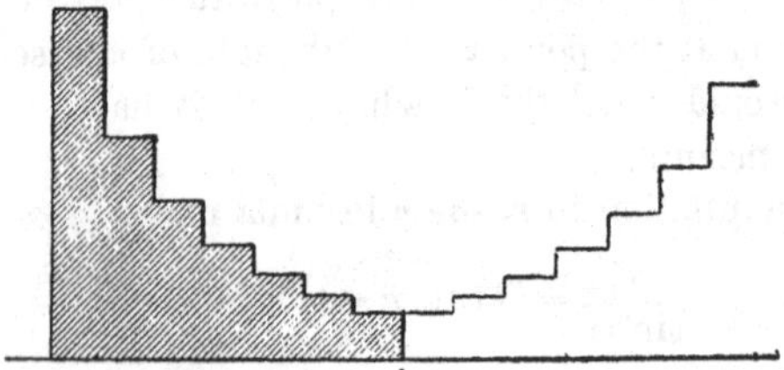

lowest point, when regarded as a continuous curve. If quite

symmetrical, the lowest point would be at A, and the shaded area would represent the value of the series. But there never is symmetry, so A is not quite, though nearly, right. In this way, choosing values of x producing two equal bottom terms, excellent results are obtainable by stopping at the end of the first of the two.

But if x is such that there is but one l.c.t., the results will be staggery if we always count the full l.c.t., or always one-half of it, &c. We should plainly count a portion only of the l.c.t. of such size as harmonises with the case of two equal bottom terms. Now to get the nth term to equal the $(n+1)^{\text{th}}$ when $x=1+m^{-1}$, requires $n=m$. Use this rule whether m be integral or fractional. For example, if $x=1{\cdot}3$, $n=3\frac{1}{3}$, count three terms and $\frac{1}{3}$ of the next. Then regular results arise, as may be seen by the following:—

x	n	λ	l.c.t.
2	1	2	
1·5	2	1·932	
1·4	2·5	1·921	0·916
1·3	3 33	1 9114	0·714
1·2	5	1·9026	0·4976
1·1	10	1 8974	0·2596
1·05	20	1·8953	0·1362
1·025	40	1·89520	0·0671
1·01	100	1 89518	0·0270

Observe that very large variation in the size of the l.c.t. makes little difference in the calculated value of λ, and that the last three results agree to four figures. It would seem that 1·895 is correct as far as written. But what it means (in terms of π or other constants) I do not know.

In the algebraical theory λ would not be a real constant, but an indeterminate imaginary. Thus

if $$y=x-\tfrac{1}{2}x^2+\tfrac{1}{3}x^3-\ldots,\quad\text{then}\quad \epsilon^y=1+x \tag{262}$$

is an identity, or $y=\log(1+x)$. This can be interpreted for any value of x, by using $\epsilon^{in\pi}=-1$, n being an odd number; or $\log(-1)=in\pi$. In accordance with this, $w=-\log(1-x)$ for any value of x, which makes the superinfinite w (when $x>1$) be complex to suit the algebra, whereas the divergent series w is real.

There is a similar conflict in the cot formula. By squaring the series u and v, it can be shown that

$$u^2_r = \pi \cot r\pi(v_r - v_{r+\frac{1}{2}}), \tag{263}$$

$$v^2_r = \pi \cot r\pi(v_r + v_{r+\frac{1}{2}}). \tag{264}$$

The formula for u is satisfied, but that for v, containing λ, will not do. If, however, we put $\lambda = \pm i\pi$, then we harmonise the algebra at the expense of making the series v be complex when x is real.

The generalised ϵ^x and ϵ^{-x} formulæ show a conflict also. Thus, by squaring,

$$\left(\Sigma \frac{x^r}{\lfloor r}\right)^2 = \Sigma\, x^{2r}\left(\frac{1}{\lfloor r\lfloor r} + \frac{2}{\lfloor r+1\lfloor r-1} + \frac{2}{\lfloor r+2\lfloor r-2} + \dots\right)$$

$$+ \Sigma\, x^{2r+1}\left(\frac{2}{\lfloor r\lfloor r+1} + \frac{2}{\lfloor r-1\lfloor r+2} + \dots\right)$$

$$= \Sigma \frac{x^{2r}2^{2r}}{\lfloor 2r} + \Sigma \frac{x^{2r+1}2^{2r+1}}{\lfloor 2r+1} = \Sigma \frac{(2x)^r}{\lfloor r}; \tag{265}$$

i.e., $(\epsilon^x)^2 = \epsilon^{2x}$, with the generalised series.* In a similar manner, we can show that

$$\left(\Sigma \pm \frac{x^r}{\lfloor r}\right)^2 = \Sigma \pm \frac{(2x)^r}{\lfloor r}. \tag{266}$$

This would be satisfied by $\epsilon^{-x} = \Sigma \pm x^r g(r)$. But it is not true, save when r is integral. The effective meaning of the series is not ϵ^{-x}, but $\epsilon^{-x}\cos r\pi$. I do not explain, but take things as I find them.

The w series above runs up to ∞ at $x = 1$, and then runs down again on the other side, with a constant multiplier introduced. But there is not necessarily a change of this kind in passing through infinity. For instance,

$$\mathrm{U} = 1 + \tfrac{1}{2}x(1 + \tfrac{3}{4}x(1 + \tfrac{5}{6}x(1 + \tfrac{7}{8}x(\dots \tag{267}$$

represents $(1-x)^{-\frac{1}{2}}$ when $x < 1$. It is convergent down to $x = -1$, and then alternatingly divergent and calculable. At $x = 1$ it is infinite. But it is directly divergent and calculable

* The generalised binomial theorem furnishes the expansions of $g(2r)$ and $g(2r+1)$ which appear above.

when $x > 1$, and seems to represent $(x-1)^{-\frac{1}{2}}$. Here there is conflict with the algebra of i.

But the series

$$V = 1 - \tfrac{1}{2}x(1 + \tfrac{1}{4}x(1 + \tfrac{3}{6}x(1 + \tfrac{5}{8}x(\ldots, \qquad (268)$$

which represents $(1-x)^{\frac{1}{2}}$, and is alternatingly divergent when $x < -1$, passes through zero when $x = 1$, and then becomes directly divergent. Considered as a function satisfying a differential equation, we should say it then represents $c(x-1)^{\frac{1}{2}}$ where c is a real constant. But c is not 1, as in the last case, but is $-\frac{1}{2}$, by rough calculation. The negativity can be easily explained, for $V = (1-x)U$ by algebra, and when U is $(x-1)^{-\frac{1}{2}}$, V is negative. A reason can also be given why the algebraical relation $V = (1-x)U$ is numerically violated. For though U and xU have the same point of convergence when $x > 1$, yet when they are united to make the V series, the point of convergence is shifted to quite another place. But I do not see why the factor should be $\frac{1}{2}$, if that be the true value. It is remarkable that the divergent Bessel functions should be more understandable, both algebraically and numerically, than elementary cases of Newton's binomial theorem.

Three Electrical Examples of Equivalent Convergent and Divergent Series.

§ **448.** The following electrical method of constructing formulæ for comparison is worthy of attention. If two electrical combinations whose resistance operators are Z_1 and Z_2 be put in sequence, and an impressed voltage e act at their junction, the current there is $C = e\,(Z_1 + Z_2)^{-1}$, and the voltage on the Z_2 side is

$$V = Z_2 C = \frac{e}{1 + Z_1/Z_2}. \qquad (269)$$

Taking various forms of Z_1 and Z_2 we may obtain interesting results by different ways of algebrisation. For instance, there are two ways of expansion of the operator $(1 + Z_1/Z_2)^{-1}$ by division. That either of the series of operations should give V correctly is remarkable. It will do so if the result is convergent. That the other way should give an equivalent result is much more remarkable. This occurs sometimes.

But it usually happens that the divergent result is not the full equivalent of the convergent result. Interesting generalised series connect them.

Say $Z_1/Z_2 = ap^{\frac{1}{2}}$, and $e=1$, starting at $t=0$. Denote the convergent result by u, and the divergent by v. Then

$$u = a^{-1}p^{-\frac{1}{2}} - a^{-2}p^{-1} + a^{-3}p^{-1\frac{1}{2}} - a^{-4}p^{-2} + \dots$$

$$= \left(\frac{t^{\frac{1}{2}}}{a\lfloor\underline{\tfrac{1}{2}}} + \frac{t^{\frac{3}{2}}}{a^3\lfloor\underline{\tfrac{3}{2}}} + \dots\right) - \left(\frac{t}{a^2\lfloor\underline{1}} + \frac{t^2}{a^4\lfloor\underline{2}} + \dots\right). \qquad (270)$$

This satisfies (269), and is right initially, and is convergent. It is not immediately evident what V rises to finally, though we know it should be 1. The other way makes

$$v = 1 - ap^{\frac{1}{2}} + a^2p - a^3p^{1\frac{1}{2}} + a^4p^2 - \dots$$

$$= 1 - \left(\frac{at^{-\frac{1}{2}}}{\lfloor\underline{-\tfrac{1}{2}}} + \frac{a^3t^{-1\frac{1}{2}}}{\lfloor\underline{-1\tfrac{1}{2}}} + \dots\right). \qquad (271)$$

This also satisfies (269); it is not evident what it means initially, but it is 1 when $t=\infty$. Here u and v are equivalent when a is +. Equating them, we obtain the generalised ϵ^{t/a^2}. But u and v are not equivalent when a is $-$ $V=u$ is the solution then, but not $V=v$. To see this put $a=-b$ and let b be +. Then use the generalised ϵ^{t/b^2} in the first part of u. Let u become w. It is

$$w = 1 - 2\epsilon^{t/b^2} + \left(\frac{bt^{-\frac{1}{2}}}{\lfloor\underline{-\tfrac{1}{2}}} + \frac{b^3t^{-1\frac{1}{2}}}{\lfloor\underline{-1\tfrac{1}{2}}} + \dots\right). \qquad (272)$$

This is the solution when a is negative. Comparing with v, we see that in passing from positive to negative a, the extra term $-2\epsilon^{t/a^2}$ is added.

The electrical interpretation is interesting. *See* § 242. Let a cable (having resistance and permittance only) of infinite length be earthed at $x=0$ through a resistance, and e be inserted there. Then u or v show the rise of V at the beginning of the cable from 0 to 1, when the terminal resistance is positive, and u or w show the rise of V from 0 to $-\infty$ when the resistance is negative. When a is reduced to zero, the rise from 0 to 1 takes place instantly. When b positive is reduced to zero, the rise from 0 to $-\infty$ takes place instantly. That is, if we had a negative resistance, even of infinitesimal amount, the finite voltage e would generate an infinite voltage instantly under the circumstances stated.

As a second case, let $Z_1/Z_2 = ap^{-\frac{1}{2}}$. Then

$$u = 1 - ap^{-\frac{1}{2}} + a^2p^{-1} - a^3p^{-1\frac{1}{2}} + \dots$$

$$= \epsilon^{a^2t} - \left(\frac{at^{\frac{1}{2}}}{\lfloor\underline{\tfrac{1}{2}}} + \frac{a^3t^{1\frac{1}{2}}}{\lfloor\underline{1\tfrac{1}{2}}} + \dots\right), \qquad (273)$$

$$v = a^{-1}p^{\frac{1}{2}} - a^{-2}p + a^{-3}p^{1\frac{1}{2}} - a^{-4}p^2 + \dots$$

$$= \frac{1}{at^{\frac{1}{2}}\lfloor\underline{-\tfrac{1}{2}}} + \frac{1}{a^3t^{1\frac{1}{2}}\lfloor\underline{-1\tfrac{1}{2}}} + \dots. \qquad (274)$$

Here u and v are equivalent again when a is $+$, as we see by equating them. V falls from 1 to 0. But when a is $-b$, and b is $+$, the result is

$$w = 2\epsilon^{b^2t} - \left(\frac{1}{bt^{\frac{1}{2}}\lfloor\underline{-\tfrac{1}{2}}} + \dots\right). \qquad (275)$$

This comes from u by the generalised exponential. V now rises from 1 to ∞. It is a terminal condenser, instead of a resistance, that is concerned. *See* § 243.

Thirdly, let $Z_1/Z_2 = ap^{1\frac{1}{2}}$. Then

$$u = a^{-1}p^{-1\frac{1}{2}} - a^{-2}p^{-3} + a^{-3}p^{-4\frac{1}{2}} - a^{-4}p^{-6} + \dots \qquad (276)$$

is the convergent solution, true whether a is $+$ or $-$. But the other result, dividing the other way, viz.,

$$v = 1 - ap^{1'} + a^2p^3 - a^3p^{4\frac{1}{2}} + a^4p^6 - \dots, \qquad (277)$$

is not equivalent to u whether a be $+$ or $-$. The reason is this. Let

$$X = \dots + 1 + bp^{1\frac{1}{2}} + b^2p^3 + b^3p^{4\frac{1}{2}} + \dots, \qquad (278)$$

the series being complete both ways. What is its value? Now

$$2\epsilon^t = \dots + 1 + p^{\frac{1}{2}} + p + p^{1\frac{1}{2}} + \dots, \qquad (279)$$

and I find that the sum of every third term beginning anywhere equals $\frac{1}{3}$ of the total. Therefore $X = \frac{2}{3}\epsilon^{t/c}$, if $c = b^{\frac{2}{3}}$, and b is $+$, and the equivalent solutions are

$$u = -(b^{-1}p^{-1\frac{1}{2}} + b^{-2}p^{-3} + b^{-3}p^{-4\frac{1}{2}} + \dots), \qquad (280)$$

$$w = -\tfrac{2}{3}\epsilon^{t/c} + 1 + bp^{1\frac{1}{2}} + b^2p^4 + \dots, \qquad (281)$$

when b is $+$. Here by p^{-n} is to be understood $t^n/\lfloor n$, to save useless work.

But when a is $+$,

$$u=(a^{-1}p^{-1\frac{1}{2}}+a^{-3}p^{-4\frac{1}{2}}+\dots)-(a^{-2}p^{-3}+a^{-4}p^{-6}+\dots) \tag{282}$$

$$=\tfrac{2}{3}\epsilon^{t/c}-(1+ap^{1\frac{1}{2}}+a^3p^{4\frac{1}{2}}+\dots)-2(a^{-2}p^{-3}+a^{-4}p^{-6}+\dots). \tag{283}$$

Here

$$1+a^{-2}p^{-3}+a^{-4}p^{-6}+\dots=\tfrac{1}{3}\{\epsilon^{t/c}+2\epsilon^{-t/2c}\cos{(t/c)}(\tfrac{3}{4})^{\frac{1}{2}}\}, \tag{284}$$

so u is transformed to

$$u=w=1-(ap^{1\frac{1}{2}}+a^3p^{4\frac{1}{2}}+\dots)-\tfrac{4}{3}\epsilon^{-t/2c}\cos{(t/c)}(\tfrac{3}{4})^{\frac{1}{2}}, \tag{285}$$

when a is $+$. This explains the result (36), § 244, relating to a terminal inductance.*

Sketch of Theory of Algebrisation of $(1-bp^r)^{-1}1$.

§ 449. The above three cases illustrate the general theory, which may be just sketched here. Let

$$\mathrm{U}=\frac{1}{1-bp^r}=-(b^{-1}p^{-r}+b^{-2}p^{-2r}+b^{-3}p^{-3r}+\dots). \tag{286}$$

This is the convergent solution when r is $+$. It is not necessary to consider r to be $-$, because the theory is quite similar. To transform U to the equivalent W when b is $+$, let

$$\mathrm{X}=\dots+1+bp^r+b^2p^{2r}+b^3p^{3r}+\dots, \tag{287}$$

the series to be complete. Then

$$\mathrm{W}=-\mathrm{X}+1+bp^r+b^2p^{2r}+b^3p^{3r}+\dots \tag{288}$$

is the required result.

When r is the reciprocal of any integer, say n^{-1},

$$\mathrm{X}=\frac{1}{r}\epsilon^{t/c}, \tag{289}$$

where $c=b^{1/r}$, because of the generalised ϵ^t formula. But

* In *Elec. Pa.*, Vol. I., pp. 153 to 169, are discussed problems equivalent to the above by the Fourier method with extensions. It is to be noted that the definite integral in that method corresponds to the divergent series in the present one. The extra term which comes in when a changes from positive to negative, is, in the Fourier method, accounted for by an extra root of the determinantal equation. The results by the two methods are in full harmony.

when b is negative, $X=0$. For when n is even, $\frac{1}{2}n$ generalised ϵ^t expressions are cancelled by the other $\frac{1}{2}n$. And when n is odd, the generalised ϵ^{-t} expressions occurring n times also annihilate the result. Thus, $n=5$, omitting the constant,

$$X=\epsilon^{-t}(1-\cos\tfrac{1}{5}\pi+\cos\tfrac{2}{5}\pi-\text{co }\tfrac{3}{5}\pi+\cos\tfrac{4}{5}\pi)=0. \quad (290)$$

It follows that in the physical case, when a is $+$, both the solutions, convergent and divergent, are obtained by division, as in the cases $r=\frac{1}{2}$ and $-\frac{1}{2}$ given above, whilst when a is $-$, there is an extra term, $-X$. In any case the matter is resolved into the evaluation of X. Putting $c=1$ for ease,

$$X=\ldots+1+p^r+p^{2r}+\ldots=r^{-1}\epsilon^t \quad (291)$$

is not only true for $r=1$, $\frac{1}{2}$, $\frac{1}{3}$, $\frac{1}{4}$, &c.; but also for their doubles, except when the doubling makes $r=2$. The generalised ϵ^{-t} formula shows the validity of this doubling. But when r is an integer, then

$$X=1+p^{-n}+p^{-2n}+\ldots=n^{-1}(\epsilon^t+\epsilon^{ct}+\epsilon^{c^2t}+\ldots), \quad (292)$$

where $1, c, c^2$, &c., are the n n^{th} roots of 1. [Prove by the Expansion Theorem, thus,

$$X=\frac{p^n}{p^n-1}=\Sigma\frac{\epsilon^{pt}}{n}.\Big]$$

So $X=r^{-1}\epsilon^t$ is not true for r integral, save 1, and there may be special peculiarities when $r>1$ and fractional, though it is true when $r=1\frac{1}{2}$. But continuity would seem to show that it is true from $r=0$ to $r=1$ generally. The following refers to $r=m/n$, m being a smaller integer than n. A special case will be easier to follow. Say $r=\frac{4}{5}$, then

$$\begin{aligned}X&=(1+p^{\frac{4}{5}}+p^{\frac{8}{5}}+p^{\frac{12}{5}}+p^{\frac{16}{5}})\;(\ldots.+1+p^4+p^8+\ldots)\\&=(1+p^{\frac{4}{5}}+p^{\frac{8}{5}}+p^{\frac{12}{5}}+p^{\frac{16}{5}})\tfrac{1}{4}(\epsilon^t+\epsilon^{ti}+\epsilon^{-t}+\epsilon^{-ti}).\end{aligned} \quad (293)$$

Here $\quad p^r\epsilon^t=\epsilon^t;\; p^r\epsilon^{-t}=\epsilon^{-t}\cos r\pi,\; p^r\cos t=\cos(t+\frac{1}{2}r\pi), \quad (294)$

so we get $X=$ sum of five correct real formulæ, and this sum comes to $\frac{5}{4}\epsilon^t$, as required.

If imaginary parts cancel one another at the end, the imaginary may be employed throughout. Thus, $r=\frac{3}{4}$.

$$\begin{aligned}X&=(1+p^{\frac{3}{4}}+p^{\frac{6}{4}}+p^{\frac{9}{4}})\,(\ldots+1+p^3+p^6+\ldots)\\&=(1+p^{\frac{3}{4}}+p^{\frac{6}{4}}+p^{\frac{9}{4}})\,\tfrac{1}{3}\,(\epsilon^t+\epsilon^{ct}+\epsilon^{c^2t}),\end{aligned} \quad (295)$$

where $c = i^{1\frac{1}{3}}$. Now take $p^r \epsilon^{ct} = c^r \epsilon^{ct}$. Then

$$X = \tfrac{4}{3}\epsilon^t + \tfrac{1}{3}(1 + i + i^2 + i^3)\epsilon^{ct} + \tfrac{1}{3}(1 + i^2 + i^4 + i^6)\epsilon^{c^2 t}, \qquad (296)$$

which is $\frac{4}{3}\epsilon^t$, as required. Numerical results are good. This process goes when $r = m/n$. But the individual results, when complex, are not true. *E.g.*, $p^r \epsilon^{i^2 t}$ is not $i^{2r} \epsilon^{-t}$, but only the real part thereof. There is more to be said on this subject, and I have no doubt a good deal more will be said when proper mathematicians will thoroughly explore divergent series for physical purposes. But this volume is now full up.

APPENDICES.

APPENDIX D.

ON COMPRESSIONAL ELECTRIC OR MAGNETIC WAVES.

There are no "longitudinal" waves in Maxwell's theory analogous to sound waves. Maxwell took good care that there should not be any. He knew what he was about, and having done a first-class piece of work in producing harmony between electrostatics and kinetics in a philosophical manner, by his invention of the electric current in non-conductors and his doctrine of the circuitality of the true current, he saw that it was good, and let it be. Moreover, the phenomena of light indicated the absence of longitudinal waves; to get rid of them was a difficulty in elastic solid theories; they could not even account satisfactorily for the elementary laws of reflection and refraction at the interface of transparent media. Now, Maxwell's theory went of itself in the directions required. Why, then, should he spoil his work by introducing longitudinal waves?

Although there does not, in my opinion, seem to exist at present any distinct evidence of longitudinal waves in reality, yet, if such should be superadded to Maxwell's theory, care should be taken that the modified or extended theory is constructed so as to harmonise with Maxwell. Now, there are compressional waves in Helmholtz's theory. But I am quite unable to see that that theory harmonises with Maxwell's, and I am not aware that anyone has shown that it does. It seems to me to be out of court.

There are many ways in which compressional waves can be introduced, some simple, others complicated. But in a primary theory we are naturally limited to the simplest ways possible. Let us, then, in the first place, see how the rotational ether will furnish compressional waves. The rotational ether is known to furnish a formal analogy which is useful so far as it can be followed with advantage. The moving force per unit volume arising from the stress in a medium which opposes finite elastic resistance to shear, compression and rotation is

$$\mathbf{F} = n(\nabla^2\mathbf{G} + \tfrac{1}{3}\nabla \operatorname{div}\mathbf{G}) + \lambda\nabla \operatorname{div}\mathbf{G} - \nu \operatorname{curl}^2\mathbf{G}, \tag{1}$$

if $\mathbf{G}$ is the spacial displacement, and n, λ, ν the elastic constants, connected with shear, compression and rotation ("Electromagnetic Theory," Vol. I., § 145, equation (313).) Here, for reasons explained (§ 153), put $n=0$, and keep ν finite instead; and, to have compressional waves, retain λ. Then

$$\mathbf{F} = \lambda\nabla \operatorname{div}\mathbf{G} - \nu \operatorname{curl}^2\mathbf{G} = c\ddot{\mathbf{G}} \tag{2}$$

is the equation of motion for small motions, without any impressed force, if c is the density of the ether.

The most convenient form of the analogy for present purposes is to compare $\mathbf{E}$, the electric force, with $\dot{\mathbf{G}}$, the velocity of the ether, as in § 159. Then

$$-\nu \operatorname{curl}\mathbf{G} = \mathbf{H}, \tag{3}$$

expressing that the rotation is elastically resisted, becomes by time differentiation, and putting $\nu^{-1} = \mu$,

$$-\operatorname{curl}\mathbf{E} = \mu p\mathbf{H}, \tag{4}$$

the second circuital law. Also $-\nu \operatorname{curl}^2\mathbf{G}$ is the same as $\operatorname{curl}\mathbf{H}$, so (2) above makes

$$\operatorname{curl}\mathbf{H} = cp\mathbf{E} - \lambda\nabla \operatorname{div} p^{-1}\mathbf{E}, \tag{5}$$

which is the modified first circuital law. The expansion is $\operatorname{div} p^{-1}\mathbf{E}$, and $\lambda\nabla \operatorname{div} p^{-1}\mathbf{E}$ means the moving force due to the space variation of the expansion (or compression), or of the pressure, if $\lambda \operatorname{conv} p^{-1}\mathbf{E}$ be considered to be the pressure, disregarding any constant pressure which is inoperative. The other moving force, curl $\mathbf{H}$, arises from the space variation of the torque.

Equations (4) and (5) being the working equations, it will be as well to see how λ affects the flux of energy. The mere convective flux of stored energy is disregarded. The stress on the $\mathbf{N}$ plane being

$$\mathbf{P}_{\mathrm{N}} = \mathbf{N}\lambda \operatorname{div}\mathbf{G} - \nu \mathrm{V}\mathbf{N} \operatorname{curl}\mathbf{G}, \tag{6}$$

(§ 145, equation (310),) the flux of energy representing its activity is the negative of the conjugate stress on the **E** plane (because **E** is the velocity), multiplied by the size of **E** (*see* § 145). This makes, after using (3),

$$\mathbf{W} = \mathrm{V}\mathbf{E}\mathbf{H} - \mathbf{E}\lambda\,\mathrm{div}\,p^{-1}\mathbf{E} \tag{7}$$

represent the flux required. Its convergence should represent the time rate of increase of energy per unit volume. Carrying this out, and using the circuital laws (4), (5) we get

$$\begin{aligned} -\mathrm{div}\,\mathbf{W} &= \mathbf{E}\,\mathrm{curl}\,\mathbf{H} - \mathbf{H}\,\mathrm{curl}\,\mathbf{E} + \mathrm{div}\,(\mathbf{E}\lambda p^{-1}\mathrm{div}\,\mathbf{E}) \\ &= \mathbf{E}cp\mathbf{E} + \mathbf{H}\mu p\mathbf{H} + \lambda\,\mathrm{div}\,p^{-1}\mathbf{E}\,\mathrm{div}\,\mathbf{E} \\ &= \frac{d}{dt}\left\{\tfrac{1}{2}c\mathbf{E}^2 + \tfrac{1}{2}\mu\mathbf{H}^2 + \tfrac{1}{2}\lambda(\mathrm{div}\,p^{-1}\mathbf{E})^2\right\}, \end{aligned} \tag{8}$$

which is correct, since $\frac{1}{2}c\mathbf{E}^2$ is the kinetic energy, $\frac{1}{2}\mu\mathbf{H}^2$ the energy of rotation, and $\frac{1}{2}\lambda(\nabla p^{-1}\mathbf{E})^2$ the energy of compression. (*See* also § 145).

To show the effect of electric conductivity, turn (5) to

$$\mathrm{curl}\,\mathbf{H} = (k + cp)\mathbf{E} - \lambda\nabla\,\mathrm{div}\,p^{-1}\mathbf{E}, \tag{9}$$

where $k\mathbf{E}$ is the conduction current. The translation of the ether is frictionally as well as inertially resisted (§ 159). **W** is unchanged in (7), but its convergence, by the introduction of k, produces an extra term $\mathbf{E}k\mathbf{E}$ in the right side of (8), expressing the waste of energy.

As regards the general interpretation of (4), (5) it is to be observed that circuital **E** and **H** are propagated identically as in Maxwell's theory. Further, if **H** is polar, or divergent without curl, it remains steady in time and place. But if **E** is polar and $\mathbf{H} = 0$ or is polar, then (5) makes

$$cp\mathbf{E} = \lambda\nabla\,\mathrm{div}\,p^{-1}\mathbf{E}, \tag{10}$$

which, if c is treated as a constant, means the same as

$$\nabla^2\mathbf{E} = c\lambda^{-1}p^2\mathbf{E}, \quad \text{or} \quad u^2\nabla^2\mathrm{P} = p^2\mathrm{P}, \tag{11}$$

if $\mathbf{E} = -\nabla\mathrm{P}$, and $u^2 = \lambda/c$. That is, polar **E**, and the divergence of **E**, and the associated potential, are propagated at speed $(\lambda/c)^{\frac{1}{2}}$, without magnetic force. This makes longitudinal electric waves.

The strict constancy of c is impossible in the fluid case, because it then stands for the density of the fluid (with some solid properties imitated by the rotational elasticity), and we have allowed it to be compressible. The equation of continuity is

$$\operatorname{div}(c\mathbf{E}) + \dot{c} = 0, \tag{12}$$

using which, equation (10), by a time differentiation, takes the form

$$-(\operatorname{div} c\mathbf{E})\dot{\mathbf{E}} + c\ddot{\mathbf{E}} = \lambda\nabla^2\mathbf{E}. \tag{13}$$

The first term must be negligible to reduce to the standard form (11).

But electrically there is no reason visible why c (the permittivity of the ether) should not be treated as usual, as an ether *constant*. Then we have simple longitudinal waves according to (11), or waves of compression and expansion which are propagated without change of type if they are plane waves. We are not bound to follow up the analogy when we find that it fails. That is the worst of analogies. Sooner or later they have to be given up.

A serious difficulty with all analogies which represent electrical phenomena by bodily motions of an ether is that the motions have to be continuous to represent certain steady states, and then the ether gets out of shape. So the analogy may be useful only for small oscillating motions. But there are many other difficulties. There are the mechanical forces to be explained, for instance. Even the existence of electrification isolated in a dielectric involves the stretching of a point. If, for instance, **E** is the velocity of the medium, then a charged body has to be imagined to be continuously emitting fluid in all directions. That is, we have to imagine an impressed source of fluid or something equivalent. On the other hand, if **H** is velocity, the case is far worse, for an impossibility is involved. The electric force becomes rotation or proportional thereto, and the impossibility is that we need to have **E** both circuital and polar at the same time roundabout an isolated charge! Dr. Larmor's determined attempt* to make the rotational ether go, with **H** as the velocity, labours under this

* J. Larmor, "A Dynamical Theory of the Luminiferous Medium," *Trans.* R.S., 1885-86.

apparently incurable defect. His electronic investigations have to be understood electromagnetically, but not in terms of the rotational ether. Of course Maxwell's theory is dynamical, without any specialised mechanical hypothesis about the ether.

Let us, then, have done with the analogy and its embarrassments, and let (4), (5) be the two circuital laws without specialised mechanical representation, or (4) and (9) in a conductor. We reduce to Maxwell by $\lambda = 0$ (or no resistance to compression in the fluid analogy). There are two kinds of electric energy shown in (8). If electrification is still to be measured by the divergence of the displacement, then we cannot have stationary electrification. Given any initial state with **H** circuital. What happens in an unbounded non-conducting uniform medium is that the circuital **E** and **H** make Maxwellian waves which go out to infinity, whilst the polar part of **E** makes longitudinal waves, which also go out to infinity. Nothing is left behind. In Maxwell's scheme, electrification persists in time and place. But according to (4), (5) if electrification has the same meaning, it only persists in the total. It does not keep its place, for one thing. Besides that, it may increase at some places and decrease at others equivalently. The persistence is merely one of total amount in the whole dielectric. To alter this amount a source is required.

The fact that div**D** does not persist at any place unless $\lambda = 0$, requires us to modify equation (5), in order to show electrification stationary, and still have longitudinal waves. But, whilst we are about it, we may as well remember that *two* fluxes are concerned in Maxwell's scheme, and that we can have longitudinal waves either of div**E** or of div**H**. Moreover, there is no present reason why we should not have both kinds at the same time, if we have any at all. Thus, let

$$\text{curl}\,\mathbf{H} = cp\mathbf{E} + \nabla P + \mathbf{g}, \tag{14}$$

$$-\text{curl}\,\mathbf{E} = \mu p\mathbf{H} + \nabla Q + \mathbf{f}, \tag{15}$$

be the circuital equations. Here **f** and **g** are introduced to represent sources; they are perfectly arbitrary vectors, and are impressed. We see at once that circuital **E** and **H** make Maxwellian waves, along with circuital **f** and **g**. Thus, if Γ

and $\mathbf{E}_2$ are the circuital and polar parts of **E**, and similarly for **H**, **f** and **g**, we have

$$\text{curl}\,\mathbf{H}_1 = cp\mathbf{E}_1 + \mathbf{g}_1, \tag{16}$$

$$-\text{curl}\,\mathbf{E}_1 = \mu p\mathbf{H}_1 + \mathbf{f}_1, \tag{17}$$

for the circuital parts, as in Maxwell's system, and

$$0 = cp\mathbf{E}_2 + \nabla\mathrm{P} + \mathbf{g}_2, \tag{18}$$

$$0 = \mu p\mathbf{H}_2 + \nabla\mathrm{Q} + \mathbf{f}_2, \tag{19}$$

for the polar parts. Now let

$$\mathrm{P} = -\lambda\,\text{div}\,p^{-1}\mathbf{E}_2, \qquad \mathrm{Q} = -\gamma\,\text{div}\,p^{-1}\mathbf{H}_2, \tag{20}$$

show the connection between P, Q and $\mathbf{E}_2$, $\mathbf{H}_2$. Then, using (20) in (18), (19), we obtain

$$(\lambda\nabla^2 - cp^2)\,\mathbf{E}_2 = p\mathbf{g}_2, \tag{21}$$

$$(\gamma\nabla^2 - \mu p^2)\,\mathbf{H}_2 = p\mathbf{f}_2, \tag{22}$$

showing that polar **E** when free is propagated at speed $(\lambda/c)^{\frac{1}{2}}$ by longitudinal waves, and that free polar **H** is similarly propagated at speed $(\gamma/\mu)^{\frac{1}{2}}$. Call these speeds u and w. These two kinds of longitudinal waves are quite independent of one another. The polar **E** waves have no **H** with them, and the polar **H** waves have no **E** with them.

Given any initial state of **E** and **H**, and no **f**, **g**. In an unbounded medium the result is automatic division into *three* sorts of waves, the circuital or Maxwellian at speed $v = (\mu c)^{-\frac{1}{2}}$, the polar **E** at speed u, the polar **H** at speed w. Nothing is left behind.

To have remanent **E** and **H**, we require **f** and **g** respectively. If $\mathbf{g}_1 = \text{curl}\,\mathbf{h}$, we see by (16) that $\mu\mathbf{h}$ may be regarded as the density of the intrinsic magnetisation. Its effect is known. It produces, when steady, a circuital state of magnetic induction without electric displacement. Intermediately, it produces both. Or we may regard $\mathbf{g}_1$ as impressed electric current. Similarly, if $\mathbf{f}_1 = -\text{curl}\,\mathbf{e}$, then **e** may be regarded as impressed electric force, or $c\mathbf{e}$ as intrinsic electrisation, or $\mathbf{f}_1$ as impressed magnetic current. It produces ultimately, when steady, a state of circuital electric displacement without magnetic induction, though both intermediately.

Lastly, we come to $\mathbf{f}_2$ and $\mathbf{g}_2$, and now we can represent stationary electrification and its analogue. To show this plainly, let

$$\mathbf{g}_2 = \lambda\nabla \operatorname{div} p^{-1}\mathbf{E}_0, \tag{23}$$

where $\mathbf{E}_0$ is polar, say $= -\nabla P_0$. Then (18) becomes

$$0 = cp\mathbf{E}_2 - \lambda\nabla \operatorname{div} p^{-1}(\mathbf{E}_2 - \mathbf{E}_0), \tag{24}$$

or, which is equivalent,

$$cp^2\mathbf{E}_2 = \lambda\nabla^2(\mathbf{E}_2 - \mathbf{E}_0). \tag{25}$$

Here it is $\mathbf{E}_0$ that is to be regarded as impressed arbitrarily. Now, if it is steady, or becomes steady after varying anyhow, (25) shows that $\mathbf{E}_2 = \mathbf{E}_0$ is the final state assumed. That is, P_0 is the electrostatic potential, so that

$$-\nabla^2 P_0 c = \rho, \tag{26}$$

if ρ is the electrification.

Similarly, if $$\mathbf{f}_2 = \gamma\nabla \operatorname{div} p^{-1}\mathbf{H}_0, \tag{27}$$
we have

$$\mu p^2\mathbf{H}_2 = \gamma\nabla^2(\mathbf{H}_2 - \mathbf{H}_0); \tag{28}$$

and if $\mathbf{H}_0$ is steady, the final result is $\mathbf{H}_2 = \mathbf{H}_0$, a polar state of $\mathbf{H}$ indicating stationary magnetification σ, according to

$$-\nabla^2 Q_0 \mu = \sigma, \qquad \text{if} \qquad \mathbf{H}_0 = -\nabla Q_0. \tag{29}$$

It is to be observed also that ρ and σ, or P_0 and Q_0 equivalently, need not be steady. This shows a striking difference from Maxwell's scheme, in which $\mathbf{E}$, $\mathbf{H}$, $\mathbf{f}_1$, $\mathbf{g}_1$ settle the state of induction and displacement. A knowledge of the electrification, and also of its analogue, is included, for that div $\mathbf{B}$ is zero in reality is a special experimental datum. But in our present system ρ and σ must be independently given, and may be varied.

As regards conductivity, turn cp to $k + cp$, and μp to $g + \mu p$ in (14) and (15), if k is the electric and g the magnetic conductivity, which is specialised to be zero in actual fact. So we may write

$$\operatorname{curl}(\mathbf{H} - \mathbf{h}) = (k + cp)\mathbf{E} - \lambda\nabla \operatorname{div} p^{-1}(\mathbf{E} - \mathbf{E}_0), \tag{30}$$

$$-\operatorname{curl}(\mathbf{E}-\mathbf{e})=(g+\mu p)\mathbf{H}-\gamma\nabla\operatorname{div}p^{-1}(\mathbf{H}-\mathbf{H}_0)\,; \qquad (31)$$

or, equivalently, in terms of ρ and σ,

$$\operatorname{curl}(\mathbf{H}-\mathbf{h})=(k+cp)\mathbf{E}-\lambda\nabla\operatorname{div}p^{-1}\mathbf{E}+\lambda\nabla p^{-1}\rho/c, \qquad (32)$$

$$-\operatorname{curl}(\mathbf{E}-\mathbf{e})=(g+\mu p)\mathbf{H}-\gamma\nabla\operatorname{div}p^{-1}\mathbf{H}+\gamma\nabla p^{-1}\sigma/\mu, \qquad (33)$$

where ρ and σ are arbitrary as well as **e** and **h**.

To illustrate, if it is given that there is no **E** or **H** initially, and no **e**, or **h**, or σ, at any time later, but merely ρ, then the solution will express the result of introducing ρ. If it is a steady point source, amount q, then

$$\mathrm{D}=\frac{q}{4\pi r^2} \qquad (34)$$

is the displacement at distance r from q finally. But at time t, although this is the solution still, it is only valid within the sphere of radius ut. Outside this sphere there is no disturbance. On the boundary is a state analogous to condensation, a longitudinal or normal wave, in fact. The total condensation in this wave increases uniformly with the time.

If q acts from time $t=0$ to t_1 and then ceases, the result is a shell of depth ut_1, which goes out to infinity. On the outer boundary is a condensational wave, and on its inner boundary a rarefactional wave. The total condensation and the total rarefaction increase uniformly with the time, but their difference remains constant, being proportional to qt_1. Between the waves is the steady state according to (34). If, temporarily, **E** be imagined to be velocity, then fluid is being steadily transferred from the inner to the outer boundary of the expanding shell. Similar remarks apply to a magnetic source σ.

But although we have a scheme which is really Maxwell's, with longitudinal waves of **E** and of **H** (or either alone) added, and have further added the means of exhibiting stationary and steady electrification if required, so that Maxwell's theory can also be imitated in this respect, yet I hasten to add that I have not the least faith in the physical possibility of the extensions. The price to be paid is too great.

To begin with, the representation of electrification by an impressed term may be regarded as objectionable, and open to suspicion. Its only justification would be that it worked out well, and did not lead to embarrassing difficulties. Now, though it is not immediately evident, examination shows that the representation of electrification by an impressed term in the above way cannot be admitted for energetic reasons. To see this, we must examine the activity products.

First put $\lambda=0$, $\gamma=0$ in (32), (33), reducing to Maxwell. It is then the same as (16), (17), if there is no conductivity. We may exhibit the equation of activity in two strikingly different ways. Let $(k+cp)\mathbf{E}$ be denoted by $\mathbf{J}$, and $(g+\mu p)\mathbf{H}$ by $\mathbf{G}$. They are the electric and magnetic currents respectively. Then one form of activity equation is

$$-(\mathbf{gE}+\mathbf{fH})=Q+pU+pT+\operatorname{div}\mathbf{W}_1. \tag{35}$$

On the left side is exhibited the sum of the activities of $\mathbf{f}$ and $\mathbf{g}$. On the assumption that these vectors indicate the sources of energy, $-\mathbf{f}$ and $-\mathbf{g}$ are impressed forces. The right side expresses the sum of the waste, the time rate of increase of stored energy, and the divergence of the flux of energy, according to

$$Q=k\mathbf{E}^2+g\mathbf{H}^2,\quad U=\tfrac{1}{2}c\mathbf{E}^2,\quad T=\tfrac{1}{2}\mu\mathbf{H}^2,\quad \mathbf{W}_1=\mathrm{V}\mathbf{EH}. \tag{36}$$

Now it is a consequence of the circuital equations that $\mathbf{f}$ and $\mathbf{g}$ are the sources of disturbances. But it cannot be proved from these equations that they are the sources of energy. The above form of activity equation suits the rotational ether. But in electromagnetics, though $\mathbf{f}$ and $\mathbf{g}$ are the sources of disturbances, it is $\mathbf{e}$ and $\mathbf{h}$ that are the sources of energy. This is concluded from experimental knowledge. The appropriate form of activity equation, instead of (35), is

$$\mathbf{eJ}+\mathbf{hG}=Q+pU+pT+\operatorname{div}\mathbf{W}, \tag{37}$$

where Q, U, and T are the same, but

$$\mathbf{W}=\mathrm{V}(\mathbf{E}-\mathbf{e})(\mathbf{H}-\mathbf{h}). \tag{38}$$

Equations (35) and (37) are mutually convertible, but differ entirely in the sources of energy and its flux, for the same varying disturbances $\mathbf{E}$ and $\mathbf{H}$. (Compare with § 70 and § 159, and observe the anomalies pointed out in the latter place.)

Now pass to the more general equations (32) (33). The sources of disturbance are $\mathbf{g}_1$, $\mathbf{f}_1$, ρ and σ. But we cannot conclude from the equations where the energy is taken in. We have the activity equation

$$-(\mathbf{gE}+\mathbf{fH}) = Q + pU + pT + pU_1 + pT_1 + \text{div}\,\mathbf{W}_2, \qquad (39)$$

where Q, U and T are as before, whilst there is fresh stored energy

$$U_1 = \tfrac{1}{2}\lambda(\text{div}\,p^{-1}\mathbf{E})^2, \qquad T_1 = \tfrac{1}{2}\gamma(\text{div}\,p^{-1}\mathbf{H})^2, \qquad (40)$$

and $\mathbf{W}_2$ is given by

$$\mathbf{W}_2 = \text{V}\mathbf{EH} - \mathbf{E}\lambda\,\text{div}\,p^{-1}\mathbf{E} - \mathbf{H}\gamma\,\text{div}\,p^{-1}\mathbf{H}. \qquad (41)$$

This form (41) suits the rotational ether, so far at least as it can be applied, for of course the equations are too general for it. For example, if $\mathbf{E}$ = velocity, then $cp\mathbf{E}$ is the rate of time increase of momentum (per unit vol.), and therefore $-\mathbf{g}$ is ordinary impressed Newtonian force (per unit vol.), and its activity is $-\mathbf{gE}$. This means the rate at which energy is being taken in on the spot.

The above is all very well for consistency, but it will not do in respect to stationary electrification. Thus, by (23) and (26),

$$-\mathbf{g}_2\mathbf{E} = -\mathbf{E}\lambda\nabla\,\text{div}\,p^{-1}\mathbf{E}_0 = -\mathbf{E}\lambda\nabla p^{-1}\rho/c. \qquad (42)$$

Apply this to the case of a sphere with constant charge. The activity increases uniformly with the time! This is accounted for by the longitudinal wave (here a spherical wave, of course) at the boundary of the region occupied by the $\mathbf{E}$ set up by ρ, that is, $\mathbf{E}_0$ ultimately, but practically only $\mathbf{E}_0$ in a sphere of finite radius, and which can never be anything but finite. Similar remarks apply to $\mathbf{f}_2$ and σ. We cannot possibly admit a scheme which requires a supply of energy to keep up an electrostatic field.

But perhaps the other way will work out better. Guided by Maxwell's electromagnetics, $\mathbf{e}$ and $\mathbf{E}_0$, $\mathbf{h}$ and $\mathbf{H}_0$ should be the sources of energy, although $\mathbf{g}_1$ and ρ, $\mathbf{f}_1$ and σ are the sources of disturbances. On this understanding, the equation of activity takes the form

$$(\mathbf{eJ}_1 + \mathbf{hG}_1) + (\mathbf{E}_0\mathbf{J}_2 + \mathbf{H}_0\mathbf{G}_2) = (Q + pU + pT) + pU_2 + pT_2 + \text{div}\,\mathbf{W}_3 \qquad (43)$$

where Q, U and T are as before, whilst the new quantities are thus defined. $\mathbf{J}_1$ and $\mathbf{G}_1$ are the circuital parts of the true currents $(k+cp)\mathbf{E}$ and $(g+\mu p)\mathbf{H}$. That is, $\mathbf{J}_1$ and $\mathbf{G}_1$ are the right members of (30) and (31). $\mathbf{J}_2$ and $\mathbf{G}_2$ are the polar parts of the true currents, that is

$$\mathbf{J}_2=\lambda\nabla\,\mathrm{div}\,p^{-1}(\mathbf{E}-\mathbf{E}_0),\qquad \mathbf{G}_2=\lambda\nabla\,\mathrm{div}\,p^{-1}(\mathbf{H}-\mathbf{H}_0). \qquad (44)$$

U_2 and T_2 are the new stored energies given by

$$\mathrm{U}_2=\tfrac{1}{2}\gamma\{\mathrm{div}\,p^{-1}(\mathbf{E}-\mathbf{E}_0)\}^2,\qquad \mathrm{T}_2=\tfrac{1}{2}\gamma\{\mathrm{div}\,p^{-1}(\mathbf{H}-\mathbf{H}_0)\}^2, \quad (45)$$

and, finally, $\mathbf{W}_3$ is given by

$$\mathbf{W}_3=\mathrm{V}(\mathbf{E}-\mathbf{e})(\mathbf{H}-\mathbf{h})-(\mathbf{E}-\mathbf{E}_0)\lambda\,\mathrm{div}\,p^{-1}(\mathbf{E}-\mathbf{E}_0)$$
$$-(\mathbf{H}-\mathbf{H}_0)\gamma\,\mathrm{div}\,p^{-1}(\mathbf{H}-\mathbf{H}_0). \qquad (46)$$

Superficially considered, we have now got rid of the energetic difficulty. If a stationary electric polar field exists, no work is apparently needed to keep it up. The activities of $\mathbf{E}_0$ and $\mathbf{H}_0$ are zero when their fields are established. But, looking closer, the former difficulty turns up in a new shape. Thus, considering a charged sphere. Formerly, the place of activity was the surface of the sphere, and the continuance and increase of the activity were obvious. At present, the seat of activity is the expanding spherical boundary of the radial **E**. But the activity is existent however far the region of **E** may extend.

We must therefore give up the representation of static states by impressed terms in the above way. That is, $\mathbf{E}_0=0$, $\mathbf{H}_0=0$, or $\rho=0$, $\sigma=0$. But when we do this we are no longer able to have stationary electrification in a non-conductor. Not only Maxwell's theory, but older theories, as W. Thomson's, are violated. The cure is to abolish the longitudinal waves, by $\lambda=0$, $\gamma=0$. Then we come back to Maxwell and his purely transverse waves. Good old Maxwell!

The way the transition takes place is worth notice. When λ and γ are reduced to zero the wave speeds u and w are also reduced to zero. Then, even if we do have the above ρ and σ in action, varying anyhow, they do nothing, for no disturbance leaves the sources. This agrees with Maxwell, inasmuch as his electrification cannot be introduced in a dielectric without conduction or convection. Also, if electrification exists, given by $\mathrm{div}\,\mathbf{D}$, it persists, unless there is conduction or convection.

Lastly, a few remarks about Helmholtz's theory must be added. I made acquaintance with it about 1886, and concluded that it would not do, being fundamentally in conflict with Maxwell's theory. Prof. J. J. Thomson seemed to be of the same opinion, in his Report on Electrical Theories, not being able to harmonise it with Maxwell. Dr. Larmor, too, had a go at it, without success. But Hertz appears to have believed in it, that is, as a possible extension of Maxwell's theory, and certainly Prof. Boltzmann and Dr. Curry have faith in it, as the recent work of the latter testifies.*

I think this state of dubiety is principally due to the extraordinary complication of Helmholtz's investigations. There is not only the usual cartesian complication, and the usual 4π anomalies, but an unusual display of constants, and worst of all, the exhibition of results in the form of equations of electric and magnetic force, which are very complicated, and the use of several potential functions. And yet the matter is essentially quite simple, if we look at it from another point of view, employing the simple methods which are coming into general use. Eliminate the potentials; rationalise the formulæ, and put in vectorial language and seek the circuital laws. Then we shall see what we shall see.

Transformed in the desired way, I make Helmholtz's equation of electric force, in my own notation, be

$$\mathbf{E} = -\nabla\phi - \mu_0 p \operatorname{pot}(k + c_1 p)\mathbf{E} + \mu_0 c_0 (1-\kappa)\nabla \operatorname{pot} p^2\phi \\ - p \operatorname{curl} \operatorname{pot} \mu_1 \mathbf{H}. \qquad (47)$$

Here understand that μ_0 and c_0 belong to the ether, and μ_1, c_1 are the extra parts due to matter, so that $\mu_0 + \mu_1 = \mu$, and $c_0 + c_1 = c$. Also ϕ is a potential, to be explained presently, and κ is a numeric, also to be interpreted. There is another potential ψ, in Helmholtz's equation, but I have put it in terms of ϕ, as in the third term, on the right. I also omit impressed electric force as unessential. There is no impressed magnetic force in the equation.

* Curry's "Theory of Electricity and Magnetism," 1887, may be consulted for Helmholtz's investigation. See also my review of that work in *The Electrician*, p. 643, Sept. 10, 1897, and Dr. Boltzmann's remarks thereon, p. 55, Nov. 5, 1897.

Operate on (47) by curl, and we get

$$-\operatorname{curl} \mathbf{E} = \mu p \mathbf{H}, \tag{48}$$

the second circuital law, simply. (Compare § 65, Vol. I.) Conversely, from (48), by vectorial work we may construct (47), except as regards the ϕ terms.

Similarly, from Helmholtz's equation of magnetic force, we obtain, by curling, another circuital law in the form

$$\operatorname{curl} \mathbf{H} = (k + c_1 p)\,\mathbf{E} - c_0 p \nabla \phi. \tag{49}$$

There are, initially, two remarks to be made. First, that the potential term, without further information, merely allows the electric current to have divergence. Next, notice that instead of $cp\mathbf{E}$, as in Maxwell, we have $c_1 p\mathbf{E}$. This is a fatal discrepancy. It is not an inadvertent mistake, however, but is intentional. Maxwell's current in a non-conductor is $pc\mathbf{E}$, or $p\mathbf{D}$, the time variation of the displacement. But Helmholtz's is $c_1 p\mathbf{E}$, the time variation of the polarisation only, $c_1\mathbf{E}$ being the electric polarisation. It is supposed that ϕ can be adjusted so as to give Maxwell's theory, or a wider theory, by varying a certain constant contained in ϕ, viz., κ in (47).

The next thing to do is to eliminate ϕ. Take the divergence of (49). Then

$$c_0 p \nabla^2 \phi = \operatorname{div}\,(k + c_1 p)\mathbf{E}. \tag{50}$$

Again, take the divergence of (47). We get

$$\operatorname{div} \mathbf{E} + \nabla^2 \phi = -\mu_0 p \operatorname{pot} \operatorname{div} (k + c_1 p)\mathbf{E} + \mu_0 c_0 (1 - \kappa) \operatorname{pot} \nabla^2 p^2 \phi, \tag{51}$$

which, by the use of (50) reduces to

$$\operatorname{div} \mathbf{E} + \nabla^2 \phi = \mu_0 c_0 \kappa p^2 \phi, \tag{52}$$

or

$$\frac{k + cp}{k + c_1 p} \nabla^2 \phi = \mu_0 c_0 \kappa p^2 \phi; \tag{53}$$

and therefore

$$c_0 \nabla p \phi = \frac{k + cp}{\mu_0 c_0 \kappa p^2} \nabla \operatorname{div} \mathbf{E}. \tag{54}$$

This gives $\nabla p \phi$ in terms of $\mathbf{E}$. The first circuital law is therefore, by (49),

$$\operatorname{curl} \mathbf{H} = (k + c_1 p)\mathbf{E} - \frac{c}{\mu_0 c_0 \kappa} \nabla \operatorname{div} p^{-1} \mathbf{E} - \frac{k}{\mu_0 c_0 \kappa} \nabla \operatorname{div} p^{-2} \mathbf{E}. \tag{55}$$

Now put $c/\mu_0 c_0 \kappa = \lambda$. Then

$$\operatorname{curl} \mathbf{H} = (k + c_1 p)\,\mathbf{E} - \lambda \nabla \operatorname{div} p^{-1} \mathbf{E} - \frac{k\lambda}{cp} \nabla \operatorname{div} p^{-1} \mathbf{E}. \tag{56}$$

This is perhaps the simplest form. Comparing with (9) we see the interpretation of Helmholtz's numeric κ. It is $\kappa = c/\mu_0 c_0 \lambda$, where λ is, in the rotational analogy, the elastic constant connected with compression.

But no possible legitimate manipulation of κ can reduce Helmholtz to Maxwell. The nearest approach is by $\kappa = \infty$, or $\lambda = 0$. Then we make

$$\text{curl}\,\mathbf{H} = (k + c_1 p)\mathbf{E}, \tag{57}$$

inconsistent for the reason before mentioned. The speed of propagation is $(\mu c_1)^{-\frac{1}{2}}$, and is only finite in polarisable media. It is infinite in the ether! This comes out of Helmholtz's conception of the current being the time variation of the polarisation instead of the displacement.

Take the case of κ finite, but no conductivity. Then

$$\text{curl}\,\mathbf{H} = (k + c_1 p)\mathbf{E} - \lambda \nabla\,\text{div}\,p^{-1}\mathbf{E}, \tag{58}$$

$$-\text{curl}\,\mathbf{E} = \mu p \mathbf{H}. \tag{59}$$

These indicate transverse waves at speed $(\mu c_1)^{-\frac{1}{2}}$ and normal waves at speed $(\lambda/c)^{\frac{1}{2}}$. *See* (9), (10), (11) above. When the conductivity is finite as in (56), the interpretation is more difficult, but of course still incapable of reduction to Maxwell.

It is important to note that c_0 cannot be equated to zero. If it could be, then $c_0 = 0$, making $c = c_1$ would reduce (49) to Maxwell in appearance. But this is nonsense. Note also that Helmholtz's theory, being reducible to (48) and (55), is not necessarily a distance action theory. ϕ has gone out. We do not need to have ϕ given to specify the state of the field.

Finally, remember that even if Helmholtz's theory could be reduced to Maxwell's, there would be, in its unreduced state, the electrification difficulties before described.

APPENDIX E.

DISPERSION.

A short time since Prof. J. J. Thomson in his Rede Lecture and then (*The Electrician*, July 17, 1896) Prof. Lodge directed attention to Helmholtz's (1893) electromagnetic theory of dispersion. This was followed next week by Dr. Howard's translation of that paper. The celerity was wonderful. I wish the paper itself could be as quickly understood as translated.

The subject is exceedingly important and exceedingly difficult. In one respect only is it easy. It is perfectly easy to make a mathematical theory of dispersion, or 20 theories in an hour, if desired. But it is not easy to make a mathematical theory which shall agree with the facts, which are rather complicated, and vary from one body to another. And as for a physical theory, the case is worse still. Our knowledge about atoms and molecules is quite nebulous, and an hypothesis concerning the mutual action of ether and matter, or of electromagnetic and material vibrations, must be highly uncertain, even if we have a fairly good mathematical theory. The subject demands and deserves study from several points of view. At present I only desire to direct attention to some obscurities and inconsistencies I find in Helmholtz's theory. The objections are made entirely in an enquiring spirit.

Let **D** be the displacement in Maxwell's theory, extended to formally include convection currents, ρ the density of electrification,* **u** the velocity of ρ. Then ρ**u** is the convection current, and **D** + ρ**u** the true current, the curl of the magnetic

* It is not implied that there is any volume electrification. In any ionic or electronic hypothesis, we must assume that the + and − charges balance one another in general, because there is no sign of volume electrification. (The footnotes to this paper, added December, 1897, did not appear in the original, which is represented by the text.)

force, in a dielectric. Let the time integral of $\rho\mathbf{u}$ be $\mathbf{d}$. Then the first circuital law is

$$\text{curl } \mathbf{H} = \dot{\mathbf{D}} + \rho\mathbf{u} = \dot{\mathbf{D}} + \dot{\mathbf{d}}, \qquad (1)$$

and the second is

$$-\text{curl } \mathbf{E} = \mu\dot{\mathbf{H}}, \qquad (2)$$

if $\mathbf{D} = \kappa\mathbf{E}$. If now $\mathbf{d}$ can be defined in terms of $\mathbf{D}$, we shall have the essentials of a complete system.

Now $\mathbf{D}$ is the (f, g, h) of Helmholtz's paper (translation), and $\mathbf{d}$ is the same in a thicker type. But I desire to use my usual vector notation (and rational units of course), and the reader will find that it produces a great simplification. $\mathbf{E}$ is (P, Q, R) the electric force, and it is given that $\mathbf{D} = \kappa\mathbf{E}$ and $\mathbf{d} = \theta\mathbf{E}$, and that κ and θ are constants. [Equation (2) and the one preceding (4)].* Here is the first inconsistency. If θ is constant, we cannot have a theory of dispersion. A homogeneous dielectric of the common kind is the result. θ must be a differential operator of some sort. And it is so, later on in the paper, although the supposed constant θ is retained, which makes a second inconsistency.

The density of the electric energy, say U_1 [equation (4)] is made to be

$$U_1 = \frac{\frac{1}{2}\mathbf{D}^2}{\kappa} + \frac{\frac{1}{2}\mathbf{d}^2}{\theta} - \frac{\mathbf{Dd}}{\kappa}, \qquad (3)$$

and if $\mathbf{D}$ and $\mathbf{d}$ are parallel, this reduces to

$$U_1 = \tfrac{1}{2}(\kappa - \theta)\mathbf{E}^2. \qquad (4)$$

I cannot clearly understand this, either when θ is regarded as a constant, or in the form (3).

The density of the magnetic energy is as usual [equation (5)]; say, if $\mathbf{B} = \mu\mathbf{H}$, where $\mathbf{B}$ is the induction, then $T = \frac{1}{2}\mu\mathbf{H}^2$ in my notation.

But Helmholtz introduces another sort of energy, called the electromagnetic, which is wholly incomprehensible to me. It is defined in terms of the vector potential $\mathbf{A}$, and the true electric current in the old-fashioned way [equation (8)], say in my notation, $\Sigma\,\mathbf{AC}$, if $\mathbf{C}$ is the true current. But by a well-known transformation, this is the same as $\Sigma\,\mathbf{HB}$, since $\mathbf{C}$ is

* These references in square brackets are to Dr. Howard's translation of Helmholtz's Paper in *The Electrician*, July 24, 1896.

the curl of **H**, and **B** the curl of **A**, as may be seen in the paper itself. [Equations (6), (7), (8), and the one between (12) and (13).] So the "electromagnetic energy" density is twice the magnetic energy density. What can it be?

There is also the kinetic energy of disturbed ions, which is represented by the product $\frac{1}{2}$ mass of ion × square of convection current. Obviously this makes the kinetic energy × ρ^2. That is a small matter. There is also the friction assumed to act on disturbed ions.

Finally, there is the Principle of Least Action. Now, Least Action has no more to do with the matter than the man in the moon, so far as I can see. It is quite unnecessary, to begin with. Next, it obscures and complicates the matter, so much so as sometimes to lead to serious error. I make this remark advisedly, remembering previous applications of the Principle of Least Action to electromagnetics, which is much clearer without it.

Lastly, we come to the circuital equations [equations (13), (14)]. One is equivalent to

$$\text{curl}\,\mathbf{H} = \dot{\mathbf{D}} + \dot{\mathbf{d}}, \tag{5}$$

in the sense before explained. The other is equivalent to

$$-\text{curl}\left(\frac{\mathbf{D}-\mathbf{d}}{\kappa}\right) = \mu\dot{\mathbf{H}}. \tag{6}$$

I understand (5). I cannot understand (6). Compare with (2) above. I do not see how $-\mathbf{d}$ gets in. The Principle of Least Action may do it, or else the reckoning of the electric energy as in (4) above, or the astonishing electromagnetic energy; but as no details are given, and there are the inconsistencies alluded to, the matter is hopeless at present.

Note that if the θ in $\mathbf{d} = \theta\mathbf{E}$ is really constant, then (5), (6) with **d** eliminated make a simple homogeneous dielectric, with no dispersion. But as a matter of fact, the relation between **D** and **d** is [equations (17)]

$$\mathbf{D} = (a^2 + hp + mp^2)\mathbf{d}, \tag{7}$$

(where I change k to h on account of another k), so that θ is an inverse operator, as is required in practical theories of dispersion.

Now (7) can be understood broadly without going through the previous obscurities. It says that the ordinary mechanical force $\mathbf{E}\rho$ on ρ produces motion of ρ, and that there is opposing elastic resistance, and frictional resistance, and inertial resistance. This is a reasonable elementary hypothesis, though when applied to atomic charges some process of averaging must be gone through.* Besides that, the real force on ρ is not $\mathbf{E}\rho$, but $\mathbf{E}\rho + \mathrm{V}\mathbf{CB}$. Pass this correction by.

If we take out the $\mathbf{d}$ from equation (6), reducing it to (2), then since (5) is the same as (1), I should be inclined to assert that the auxiliary hypothesis (7), without necessarily entering into details regarding the nature of the frictional force, &c., would give an intelligible dispersion theory. For (7) is simply an ordinary "equation of motion" introduced as an auxiliary to the circuital laws.

But I could not say the same if $\mathbf{d}$ is to be retained in equation (6), because of its unintelligibility. It might work, or it might not. Further examination, given below, raises doubts as to its physical possibility.

To construct electromagnetic theories of dispersion rapidly, proceed thus. Write the two circuital laws in this way:

$$\text{curl}\,\mathbf{H} = \mathrm{Y}\mathbf{E}, \qquad -\text{curl}\,\mathbf{E} = \mathrm{Z}\mathbf{H}, \tag{8}$$

where Y and Z are operators, functions of p, the time-differ-

* Perhaps the following transition from the continuity of the primary theory to the discontinuity of the secondary theory may help. In pure ether, we have curl $\mathbf{H} = \kappa\dot{\mathbf{E}} = \dot{\mathbf{D}}$. Now this equation will still hold good if we fill the space concerned with electrification, density $\frac{1}{2}\rho$ positive and the same amount negative, provided these electrifications cannot separate. But if they can separate, then $\mathbf{E}$ will do it, and $\rho\mathbf{u}$ is the additional current, making $\dot{\mathbf{D}} + \dot{\mathbf{d}}$ the true current as in (1) above. The etherial permittivity is κ. It will be increased by $\mathbf{d}$ to a variable amount in general, but to a constant amount if the two electrifications are merely elastically connected according to the simplest law. If the connection involves inertia and friction as well, then we have the auxiliary equation (7). So far, the theory remains a primary theory, inasmuch as there is a continuous structure assumed. But, if we concentrate the electrifications in numerous detached pairs separated by ether we come to a secondary theory, and the equations (1) and (7) can only be understood to result by some process of averaging. H. A. Lorentz's "Versuch einer Theorie der electrischen und optischen Erscheinungen in bewegten Körpern," and Dr. Larmor's work already referred to, should be consulted concerning ions and electrons in relation to electromagnetics.

entiator. In a common homogeneous dielectric, after Maxwell, $\mathrm{Y}=(k+\kappa p)$, $\mathrm{Z}=\mu p$. As is well known, this fails to account for the facts of dispersion, being only suitable for slow working. But Y and Z may be any operators we choose, provided they represent possible electrical arrangements of coils and condensers (elementary). They may be direct or inverse, rational or irrational. In any case, when we treat simply periodic states, we reduce the circuital laws to the form

$$\text{curl}\,\mathbf{H}=(\mathrm{K}+\mathrm{S}p)\mathbf{E}, \qquad -\text{curl}\,\mathbf{E}=(\mathrm{R}+\mathrm{L}p)\mathbf{H}, \tag{9}$$

by the assumption $p^2=-n^2$, where $n/2\pi$ is the frequency. If the four constants were independent of the frequency, the case would be that of a homogeneous dielectric of constant inductivity, permittivity, electric conductivity, and magnetic conductivity, whose theory I have described in Chap. IV., Vol. I, "Electromagnetic Theory." But in general they are functions of n^2. And note that even though there be no magnetic conductivity in reality, we are usually obliged in the above way to use an effective magnetic conductivity.

To make a large scale model, use the line integrals of **E** and **H**. Say, V and C. Then, for plane waves,

$$-\frac{d\mathrm{V}}{dx}=(\mathrm{R}'+\mathrm{L}'p)\mathrm{C}=\mathrm{Z}'\mathrm{C}, \qquad -\frac{d\mathrm{C}}{dx}=(\mathrm{K}'+\mathrm{S}'p)\mathrm{V}=\mathrm{Y}'\mathrm{V} \tag{10}$$

express the resultant forms. They may be interpreted in terms of a generalised telegraph circuit, with R′ the resistance of wires, K′ the conductance between them, L′ the inductance, S′ the permittance, all per unit length, and all being the effective values at the frequency concerned. We can concentrate the four quantities in lumps, and regard the circuit as made up of a number of Z′'s in series, with leaks represented by the Y′'s (*see* "Electromagnetic Theory," Vol. I., §§ 221, 222, for this generalisation of the circuital laws, and the resulting formulæ).

If we assume Y′ and Z′ to represent any real electrical arrangements of coils and condensers, we can, I think, be quite certain beforehand that the resulting theory of dispersion will be a consistent one in all parts, containing no electrical impossibilities. That is one thing. Another is, that if we translate a theory of dispersion to the telegraph circuit theory, or to the equivalent theory for plane waves in a doubly conducting medium, and do it properly, we can see whether it really represents a possible arrangement. If it does not, if the telegraph will not work properly, neither will the homogeneous medium; and, finally, it raises a very serious question whether the theory of dispersion is an admissible one.

Test Helmholtz's theory in the way described. Eliminate **d** from (5) and (6) by means of (7). For plane waves we get

$$-\frac{dH}{dx} = \kappa p\left(1 + \frac{1}{a^2 + hp + mp^2}\right)E, \qquad (11)$$

$$-\frac{dE}{dx} = \mu p\left(1 - \frac{1}{a^2 + hp + mp^2}\right)^{-1}H. \qquad (12)$$

Put $p = ni$, then

$$-\frac{dH}{dx} = \left\{\frac{h\kappa n^2}{X} + \kappa p\left(1 + \frac{a^2 - mn^2}{X}\right)\right\}E, \qquad (13)$$

$$-\frac{dE}{dx} = \left\{\frac{\mu hn^2 X + \mu p X(X - a^2 + mn^2)}{(X - a^2 + mn^2)^2 + h^2n^2}\right\}H, \qquad (14)$$

where

$$X = (a^2 - mn^2)^2 \times h^2n^2. \qquad (15)$$

Or, if we expand the reciprocal of Z, we have, more simply,

$$Z^{-1} = -\frac{p}{\mu n^2} + \frac{h\mu n^2 + \mu p(a^2 - mn^2)}{X\mu^2 n^2}, \qquad (14\text{A})$$

Comparing with

$$-\frac{dH}{dx} = (K + Sp)E, \qquad -\frac{dE}{dx} = (R + Lp)H, \qquad (16)$$

which are the special forms of (9) for plane waves, we see that the effective electric conductivity is positive, the effective permittivity may be positive or negative, likewise the effective inductivity, and the effective magnetic conductivity is positive. The rate of waste per unit volume is $KE^2 + RH^2$. (Divide by 2 if the amplitudes are in question.) The negativity of S and L is admissible at certain frequencies, but it is not allowable to have R negative under any real circumstances. It is impossible for R to be negative when we construct Y and Z out of really possible combinations of coils and condensers, either in the telegraph theory, or in its analogue in a medium. Effective resistances and conductances are always positive, because they arise out of the positive waste in Joules' law and its magnetic analogue, or from the positivity of the real ultimate electric conductivity and its analogue.

I make these remarks because in my first translation into the telegraph theory, a stupid mistake of my own made R come out negative. So far as the above goes, Helmholtz's system is not impossible. But let us see what arrangements the operators in Helmholtz's theory lead to. Translate to V and C. Say as in (10) above. Then, first we have

$$\mathrm{Y} = \kappa p + \frac{1}{\dfrac{a^2}{\kappa p} + \dfrac{h}{\kappa} + \dfrac{m}{\kappa}p}, \qquad \mathrm{Z}^{-1} = \frac{1}{\mu p} - \frac{1}{\mu p(a^2 + hp + mp^2)} \tag{17}$$

in (11), (12). Now imagine the proper changes made in the constants to suit V and C, instead of E, H. It is unnecessary to introduce fresh symbols. If we consider a short finite portion of the telegraph circuit, to which Y and Z belong, we see that the leak Y consists of a condenser without shunt, in parallel with another condenser without shunt which is itself in sequence with a resisting coil. *I.e.*, Y is a conductance operator; κp stands for the conductance operator of the first condenser, and the denominator of the big fraction in Y is the resistance operator of the other condenser $a^2/\kappa p$ in sequence with the coil $h/\kappa + mp/\kappa$. (Dimensions are of no importance here, since it is only the structure that is in question.) Thus Y makes a proper and intelligible arrangement.

But Z does not. It is the resistance operator of the short piece of the circuit, and Z^{-1} in (17) is the corresponding conductance operator. So we see that there are two arrangements in parallel, one of which is a coil of no resistance (resistance operator μp), and the other has the resistance operator $-\mu p\,(a^2 + hp + mp^2)$. Here, $-a^2\mu p$ represents a coil of no resistance and negative inductance, whilst the rest, $-\mu p^2(h + mp)$ is not electrically intelligible, although in simply periodic states, when $-\mu p^2$ becomes μn^2, it is equivalent to a coil of positive resistance and inductance. For this reason, then, in simply periodic states, we do get positive resistance, and if the effective inductance is negative, that does not matter. But in general, unlike Y, the operator Z^{-1} is electrically unintelligible.*

* Similar equations of Dr. Larmor, so far as I have examined them, are not open to this objection to Helmholtz's system made in this Paper.

This brings us back to Helmholtz's second circuital equation again, and the question what $-\mathbf{d}$ is doing there? If we omit it altogether, as before suggested, we do away with the unintelligibility. Y is as before in (17), whilst Z is μp. Then putting $Y = K + Sp$, $Z = R + Lp$, in the simply periodic case, we find

$$K = \frac{h\kappa n^2}{X}, \qquad S = \kappa\left(1 + \frac{a^2 - mn^2}{X}\right), \tag{18}$$

$$R = 0, \qquad L = \mu, \quad X = (a^2 - mn^2)^2 + h^2n^2, \tag{19}$$

The type of a wave is

$$E = E_0\epsilon^{-Px}\sin(nt - Qx), \tag{20}$$

when $\sqrt{YZ} = P + Qi$; and

$$P \text{ or } Q = (\tfrac{1}{2})^{\frac{1}{2}}\{(R^2 + L^2n^2)^{\frac{1}{2}}(K^2 + Sn^2)^{\frac{1}{2}} \pm (RK - LSn^2)\}^{\frac{1}{2}}, \tag{21}$$

as in "Electromagnetic Theory," Vol. I., § 221, equation (16). The same symbols are used here, only it is now E and H that are concerned, so R, L, K, S refer to unit volume.) Note that P and Q are always real and positive, for all values of the quantities R, K, L, S. For the square of $(RK - LSn^2)$ is less than $(R^2 + L^2n^2)(K^2 + Sn^2)$. When R is zero, take the positive value of $(L^2n^2)^{\frac{1}{2}}$, whether L is positive or negative.

So the wave speed n/Q may be zero or infinite, but cannot be negative or imaginary. Having the square of the wave speed negative at certain frequencies is not desirable.

It is true that S may be negative, and then $\sqrt{LS}$ is imaginary when L is positive, and the square of the wave speed seems to be negative. But that does not seem to me to be the right way to look at it. Say $R = 0$ first, then

$$P \text{ or } Q = (\tfrac{1}{2})^{\frac{1}{2}}\{Ln(K^2 + S^2n^2)^{\frac{1}{2}} \mp LSn^2\}^{\frac{1}{2}} \tag{22}$$

$$= (\tfrac{1}{2}Ln)^{\frac{1}{2}}\{(K^2 + Sn^2)^{\frac{1}{2}} \mp Sn\}^{\frac{1}{2}}. \tag{23}$$

Here L is + in the case under consideration, so the outside radical causes no trouble. Now if $K = 0$ also, we shall have dispersion but no waste; then $P = 0$, and $Q = (LSn^2)^{\frac{1}{2}}$ apparently; and therefore, apparently, Q is imaginary if S is negative, as it may be within certain frequencies. But this is wrong. We should take the + value of the radical; then, when S is negative, $= -S'$ say, it is P that is $(LS'n^2)^{\frac{1}{2}}$, and Q zero. In fact, the type of the wave is now $\epsilon^{-Px}\sin nt$; a stationary vibration.

It is easier to follow when K is not quite zero. If quite zero and $P=0$, then if by decrease of n, S becomes zero, we have Q zero, and v infinite; and when S goes on further the same way and becomes negative, Q remains zero, but P is finite. When, on further decrease of n, S become positive again, then Q becomes positive again, and P zero. If this is not the correct interpretation, I shall be glad to be corrected myself.

In the case (18), (19), L is constant, K is always + and is proportional to the square of the frequency, and by (13),

$$S = \kappa\left(1 + \frac{a^2 - mn^2}{(a^2 - mn^2)^2 + h^2n^2}\right). \tag{24}$$

Also, by (23), if V is the wave speed,

$$V^{-2} = \frac{Q^2}{n^2} = \frac{LK}{2n}\{(1+y^2)^{\frac{1}{2}} + y\}, \tag{25}$$

where $y = Sn/K$. Or

$$V^{-2} = \frac{Lh\kappa n}{2X}\{(1+y^2)^{\frac{1}{2}} + y\}. \tag{26}$$

Here X is always +, whilst y is generally +, but may vanish twice. When so, then $P=Q$; but in the region between the critical frequencies P and Q change places, as compared with their values when S was positive to the same extent as it is now negative. In the limit, if $K=0$, one of the vanishings of S on decrease of n takes place by the route $-\infty, 0, +\infty$. Then P rises to ∞ as n goes from the higher to the lower zero.

The two values of n^2 making $S=0$ are

$$n^2 = \frac{1}{2m^2}\{m(1+2a^2) - h^2 \pm \sqrt{(m-h^2)^2 - 4mh^2a^2}\}. \tag{27}$$

Between these frequencies, when possible, S is negative. But if

$$h^2 = m(1+2a^2) - 2ma(1+a^2)^{\frac{1}{2}}, \tag{28}$$

the two critical frequencies coincide, making

$$n^2 = \frac{a}{m}\sqrt{1+a^2}. \tag{29}$$

Therefore, when h^2 exceeds the value given in (28), the region of negative S does not exist. The general characteristics are much the same as in Helmholtz's theory.

If we imagine a plane electromagnetic sheet travelling through the ether to enter a material dielectric permeated by the ether, it is a matter of first principles that the very front of the disturbance goes on at the same speed as before, that is, as much or little as manages to get through at all at that speed. It may be infinitesimally little.* The practical speed of the sheet in the material dielectric is different, being less. The thin sheet becomes widened by retardation, and the bulk of the disturbance is left behind the first front. It is far more complicated in reality when we come to simply periodic waves (though so much simpler in the mathematics); the values of P and Q are determined by actions proceeding both ways, ultimately tending to establish a stably progressive condition under the influence of a continued simply periodic impressed wave.† To allow of this, frictional resistances causing waste should be positive.

---

* It seems to me very unlikely that the front of the wave should be unattenuated, for there is internal reflection and scattering to be considered, as well as true absorption by the interposed storers of energy. The reason why in Helmholtz's theory, or in the form I have substituted for it, the matter has no influence when the frequency is infinite, is simply the way inertia is associated with the ions or electrons. It takes time to move them, so they do not move when no time is allowed, and do not disturb the passage of a wave train of infinitely rapid oscillations.

† According to the above, we must, in the consideration of the passage of light through material bodies, always distinguish between the propagation of an impulsive wave and of a train of waves. They will in general behave quite differently, save when the action of the matter is such as to produce merely an increased permittivity of constant amount, which is, however, inconceivable in molecular theories. A simply periodic train of waves cannot be set up until all parts of a body are well under influence and reacting. We can also illustrate this effectively by the telegraph theory in its generalised form. Although an impulse will begin by travelling at the speed of light, it will be thrown back and redistributed in various ways, according to the structure of Y and Z, the generalised leakance and resistance operators. The finally resulting wave speed, when a simply periodic train of waves has been set up, comes about from the actions proceeding both ways. The distortionless circuit is the unique exception, for there is then no back action.

Another way of testing Helmholtz's theory is by forming the equation of activity, and seeing if it accounts for the work done. Take first the simple system consisting of equations (1), (2), and (7). Then the equation of activity is

$$\text{conv}\,\mathrm{V}\mathbf{EH} = \mathbf{E}\,\text{curl}\,\mathbf{H} - \mathbf{H}\,\text{curl}\,\mathbf{E}$$

$$= \mathbf{E}(\dot{\mathbf{D}} + \dot{\mathbf{d}}) + \mathbf{H}\dot{\mathbf{B}}. \tag{30}$$

Here $\mathbf{E}\dot{\mathbf{D}} = \dot{\mathrm{U}}$, the rate of increase of the electric energy $\mathrm{U} = \frac{1}{2}\mathbf{ED}$ according to my reckoning. Similarly, $\mathbf{H}\dot{\mathbf{B}} = \dot{\mathrm{T}}$, the rate of increase of the magnetic energy. The remaining term $\mathbf{E}\dot{\mathbf{d}}$ is $\mathbf{E}\rho\mathrm{u}$ or $\mathbf{F}\mathrm{u}$, if $\mathbf{F}$ is the moving force on ρ. That is, it is the activity of $\mathbf{F}$, and is, by (7) accounted for thus

$$\mathbf{F}u = u(a^2 + hp + mp^2)\frac{u\rho^2}{\kappa p}$$

$$= \frac{a^2\rho^2}{\kappa}u\frac{u}{p} + \frac{hu^2\rho^2}{\kappa} + \frac{m\rho^2}{\kappa}upu, \tag{31}$$

showing that the rate of waste is $hu^2\rho^2/\kappa$, the potential energy is $\frac{1}{2}(a^2\rho^2/\kappa)\,(p^{-1}u)^2$, and the kinetic energy is $\frac{1}{2}(m\rho^2/\kappa)u^2$.

The activity is therefore fully accounted for. But in Helmholtz's system equations (5), (6), and (7), we obtain

$$\text{conv}\,\mathrm{V}\mathbf{EH} = \dot{\mathrm{U}} + \dot{\mathrm{T}} + \mathbf{F}u - \mathbf{H}\,\text{curl}\,\mathbf{d}/\kappa, \tag{32}$$

and I am unable to see how to interpret the additional term. Of course, in (32) it is assumed that $\mathrm{V}\mathbf{EH}$ is the flux of energy, just as in (30). It may be objected that this is not the case, but that the flux of energy should be $\mathrm{V}(\mathbf{E} - \mathbf{d}/\kappa)\mathbf{H}$. But this does not help much. The convergence is

$$\frac{\mathbf{D} - \mathbf{d}}{\kappa}(\dot{\mathbf{D}} + \dot{\mathbf{d}}) + \mathbf{H}\dot{\mathbf{B}}. \tag{33}$$

The curl of $\mathbf{d}$ does not now appear, but (33) does not account for the increase of electrical energy as reckoned by Helmholtz, or for the strange "electromagnetic energy."

APPENDIX F.

ON THE TRANSFORMATION OF OPTICAL WAVE-SURFACES BY HOMOGENEOUS STRAIN.*

Simplex Eolotropy.

1. All explanations of double refraction (proximate, not ultimate) rest upon the hypothesis that the medium in which it occurs is so structured as to impart eolotropy to one of the two properties, associated with potential and kinetic energy, with which the ether is endowed in order to account for the transmission of waves through it in the simplest manner. It may be elastic eolotropy, or it may be something equivalent to eolotropy as regards the density. In Maxwell's electromagnetic theory the two properties are those connecting the electric force with the displacement, and the magnetic force with the induction, say the permittivity and the inductivity, or c and μ. These are, in the simplest case, constants corresponding to isotropy. The existence of eolotropy as regards either of them will cause double refraction. Then either c or μ is a symmetrical linear operator, or dyadic, as Willard Gibbs calls it. In either case the optical wave-surface is of the Fresnel type. In either case the fluxes displacement and induction are perpendicular to one another and in a wave-front, whilst the electric and magnetic forces are also perpendicular to one another. But it is the magnetic force that is in the wave-front, coincident with the induction, in case of magnetic isotropy and electric eolotropy, the electric force being then out of the wave-front, though in the plane of the normal and the displacement. And in the other extreme case of electric isotropy and magnetic eolotropy, the electric force is in the wave-front, coincident with the displacement,

* *Proceedings* R.S., Dec, 20, 1893.

whilst the magnetic force is out of the wave-front, though in the plane of the normal and the induction. Now, as a matter of fact, crystals may be strongly eolotropic electrically, whilst their magnetic eolotropy, if existent, is insignificant. This, of course, justifies Maxwell's ascription of double refraction to electric eolotropy.

Properties connected with Duplex Eolotropy.

2. When duplex eolotropy, electric and magnetic, is admitted, we obtain a more general kind of wave-surface, including the former two as extreme cases. It is almost a pity that magnetic eolotropy should be insensible, because the investigation of the conditions regulating plane waves in media possessing duplex eolotropy, and the wave-surface associated therewith, possesses many points of interest. The chief attraction lies in the perfectly symmetrical manner in which the subject may be displayed, as regards the two eolotropies. This brings out clearly properties which are not always easily visible in the case of simplex eolotropy, when there is a one-sided and imperfect development of the analysis concerned.

In general, the fluxes displacement and induction, although in the wave-front, are not coperpendicular. Corresponding to this, the two forces electric and magnetic, which are always in the plane perpendicular to the ray, or the flux of energy, are not coperpendicular. Nor are the positions of the fluxes in the wave-front conditioned by the effective components in that plane of the forces being made to coincide with the fluxes. There are two waves with a given normal, and it would be impossible to satisfy this requirement for both. But there is a sort of balance of skewness, inasmuch as the positions of the fluxes in the wave-front are such that the angle through which the plane containing the normal and the displacement (in either wave) must be turned, round the normal as axis, to reach the electric force, is equal (though in the opposite sense) to the angle through which the plane containing the normal and the induction must be turned to reach the magnetic force. These are merely rudimentary properties. I have investigated the wave-surface and associated matters in my paper "On the Electromagnetic Wave-surface" (*Phil. Mag.*, June, 1885; or "Electrical Papers," vol. 2, p. 1).

Effects of straining a Duplex Wave Surface.

3. The connection between the simplex and duplex types of wave-surface has been interestingly illustrated lately by Dr. J. Larmor in his paper "On the Singularities of the Optical Wave-surface" (*Proceedings* London Math. Soc., vol. 24, 1893). He points out, incidentally, that a simplex wave-surface, when subjected to a particular sort of homogeneous strain, becomes a duplex wave-surface of a special kind. To more precisely state the connection, let there be electric eolotropy, say c, with magnetic isotropy. Then, if the strainer, or strain operator, applied to the simplex wave-surface, be homologous with c, given by $c^{\frac{1}{2}} \times$ constant, the result is to turn it into a duplex wave-surface whose two eolotropies are also homologous with the original c; that is to say, the principal axes are parallel. This duplex wave-surface is, of course, of a specially simplified kind, though not the simplest. That occurs when the two eolotropies are not merely homologous, but are in constant ratio. The wave-surface then reduces to a single ellipsoid.

Conversely, therefore, if we start with the duplex wave-surface corresponding to homologous permittivity and inductivity, and homogeneously strain it, the strainer being proportional to $c^{-\frac{1}{2}}$, we convert it to a simplex wave-surface whose one eolotropy is homologous with the former two.

Remembering that the equation of the duplex wave-surface is symmetrical with respect to the two eolotropies, so that they may be interchanged without altering the surface, it struck me on reading Dr. Larmor's remarks that a similar reduction to a simplex wave-surface could be effected by a strainer proportional to $\mu^{-\frac{1}{2}}$. This was verified on examination, and some more general transformations presented themselves. The results are briefly these:—

Any duplex wave-surface (irrespective of homology of eolotropies), when subjected to homogeneous strain (not necessarily pure), usually remains a duplex wave-surface. That is, the transformed surface is of the same type, though with different inductivity and permittivity operators.

But in special cases it becomes a simplex wave-surface. In one way the strainer is $c^{-\frac{1}{2}}/[c^{-\frac{1}{2}}]$, where the square brackets indicate the determinant of the enclosed operator. In another the strainer is $\mu^{-\frac{1}{2}}/[\mu^{-\frac{1}{2}}]$. These indicate the strain operator to be applied to the vector of the old surface to produce that of the new one.

Now, these simplex wave-surfaces may be strained anew to their reciprocals with respect to the unit sphere, or the corresponding index-surfaces, which are surfaces of the same type. So we have at least four ways of straining any duplex wave-surface to a simplex one.

Furthermore, any duplex wave-surface may be homogeneously strained to its reciprocal, the corresponding index-surface, of the same duplex type. The strain is pure, but is complicated, as it involves both c and μ. The strainer is $c^{-1}(c\mu^{-1})^{\frac{1}{2}}$, divided by the determinant of the same. This transformation is practically the generalisation for the duplex wave-surface of Plücker's theorem relating to the Fresnel surface, for that also involves straining the wave-surface to its reciprocal.

Instead of the single strain above mentioned, we may employ three successive pure strains. Thus, first strain the duplex wave-surface to a simplex surface. Secondly, strain the latter to its reciprocal. Thirdly, strain the last to the reciprocal of the original duplex wave-surface. There are at least two sets of three successive strains which effect the desired transformation. The investigation follows.

Forms of the Index- and Wave-Surface Equations, and the Properties of Inversion and Interchangeability of Operators.

4. Let the electric and magnetic forces be **E** and **H**, and the corresponding fluxes, the displacement and induction, be **D** and **B**, then

$$\mathbf{D} = c\mathbf{E}, \qquad \mathbf{B} = \mu\mathbf{H}, \tag{1}$$

where c is the permittivity and μ the inductivity, to be symmetrical linear operators in general. We have also the circuital laws

$$\text{curl } \mathbf{H} = c\dot{\mathbf{E}}, \qquad -\text{curl } \mathbf{E} = \mu\dot{\mathbf{H}}. \tag{2}$$

Now, if we assume the existence of a plane wave, whose unit normal is **N**, propagated at speed v without change of type, and apply these equations, we find that **D** and **B** are in the wave-front, **E** and **H** are out of it, and that there are two waves possible. We are led directly to the velocity equation, a quadratic in v^2, giving the two values of v^2 belonging to a given **N**. Next, if we put $\mathbf{s}=\mathbf{N}/v$, then **s** is the vector of the index-surface, and its equation is

$$\mathbf{s}\frac{\mathbf{s}}{c^{-1}-\dfrac{\mu^{-1}}{[\mu^{-1}](\mathbf{s}\mu\mathbf{s})}}=0=\mathbf{s}\frac{\mathbf{s}}{\mu^{-1}-\dfrac{c^{-1}}{[c^{-1}](\mathbf{s}c\mathbf{s})}}, \tag{3}$$

which are, of course, equivalent to the velocity equation ("El. Papers," vol. 2, p. 11, equations (41)). Two forms are given for a reason that will appear later. I employ the vector algebra and notation of the paper referred to, and others. Sufficient to say here that c^{-1} and μ^{-1} are the reciprocals of c and μ; and that $\mathbf{s}c\mathbf{s}$ means the scalar product of **s** and $c\mathbf{s}$; for example, if referred to the principal axes of c,

$$\mathbf{s}c\mathbf{s}=c_1s_1^2+c_2s_2^2+c_3s_3^2, \tag{4}$$

if c_1, c_2, c_3 be the principal c's (positive scalars, to ensure positivity of the energy), and s_1, s_2, s_3 be the components of **s**. Also, $[c^{-1}]$ denotes the determinant* of c^{-1}, that is, $(c_1c_2c_3)^{-1}$.

The operators in the denominators of (3) may be treated, for our purpose, as linear operators themselves. But it is their reciprocals that occur. For example, the first form of (3) may be written

$$\mathbf{s}\left[c^{-1}-\frac{\mu^{-1}}{[\mu^{-1}](\mathbf{s}\mu\mathbf{s})}\right]^{-1}\mathbf{s}=0, \tag{5}$$

asserting that the vectors **s** and $[\dots]^{-1}\mathbf{s}$ are perpendicular. The expansion of (3) to Cartesian form may be done immediately if c and μ are homologous, for then we may take the reference axes **i**, **j**, **k** parallel to those of c and μ, and at once produce

$$\frac{s_1^2}{\dfrac{1}{c_1}-\dfrac{\mu_2\mu_3}{\mu\mathbf{s}}}+\frac{s_2^2}{\dfrac{1}{c_2}-\dfrac{\mu_3\mu_1}{\mathbf{s}\mu\mathbf{s}}}+\frac{s_3^2}{\dfrac{1}{c_3}-\dfrac{\mu_1\mu_2}{\mathbf{s}\mu\mathbf{s}}}=0, \tag{6}$$

* It occurs to me in reading the proof that the use of $[c]$ to denote the determinant of c, which is plainer to read in combination with other symbols than $|c|$, is in conflict with the ordinary use of square brackets, as in (5) and some equations near the end. But there will be no confusion on this account in the present paper.

where $s\mu s$ is as in (4), with μ written for c. Similarly as regards the second form of (3). When the operators are not homologous, the complication of the form of the constituents of the inverse operators makes the expansion less easy.

As regards the second form of (3), it is obtained from the first form by interchanging μ and c. It represents the same surface. The transformation from one form to the other, if done by ordinary algebra, without the use of vectors and linear operators, is very troublesome in the general case. But in the electromagnetic theory the equivalence can be seen to be true and predicted beforehand. For consider the circuital equations (2). If we eliminate **H**, we obtain

$$-\text{curl}\ \mu^{-1}\ \text{curl}\ \mathbf{E} = c\ddot{\mathbf{E}}, \tag{7}$$

whilst if we eliminate **E**, we obtain

$$-\text{curl}\ c^{-1}\ \text{curl}\ \mathbf{H} = \mu\ddot{\mathbf{H}}. \tag{8}$$

These are the characteristic equations of **E** and **H** respectively in a dielectric with duplex eolotropy, and we see that they only differ in the interchange of c and μ. When therefore we apply one of them, say that of **E**, to a plane wave to make the velocity equation, in which process **E** is eliminated, we can see that a precisely similar investigation applies to the **H** equation, provided μ and c be interchanged. So, if the **E** equation leads to the first form in (3), the **H** equation must lead to the second form. They therefore represent the same surface. The same property applies to any equation obtained from the circuital equations with the electrical variables eliminated, the equation of the wave-surface, for example. If we have obtained one special form, a second is got by interchanging the eolotropies.

The index equation being what we are naturally led to from the characteristic equation, it is merely a matter of mathematical work to derive the corresponding wave-surface. For **s** is the reciprocal of the perpendicular upon the tangent plane to the wave-surface, so that

$$\mathbf{rs} = 1, \tag{9}$$

if **r** is the vector of the wave-surface; and from the equation of **s** and its connection with **r**, we may derive the equation of

r itself. I have shown (*loc. cit.*, vol. 2, pp. 12 16) that the result is expressed by simply inverting the operators in the index equation. Thus, the equation of the wave-surface is

$$\mathbf{r}.\frac{\mathbf{r}}{c-\dfrac{\mu}{[\mu](\mathbf{r}\mu^{-1}\mathbf{r})}}=0=\mathbf{r}\frac{\mathbf{r}}{\mu-\dfrac{c}{[c]\,(\mathbf{r}c^{-1}\mathbf{r})}}, \tag{10}$$

where, as before, two forms are given. Now, the final equivalence of this transition from the index to wave-equation to mere inversion of the two eolotropic operators is such a simple result that one would think there should be a very simple way of exhibiting how the transition comes about. Nevertheless, I am not aware of any simple investigation, and, in fact, found the transition rather difficult, and by no means obvious at first. I effected the transformation by taking advantage of symmetrical relations between the forces and fluxes; in particular proving, first, that $\mathbf{rE}=0=\mathbf{rH}$, or that the ray is perpendicular to the electric and magnetic forces, comparing this with the analogous property $\mathbf{sD}=0=\mathbf{sB}$, and constructing a process for leading from the former to the wave-equation analogous to that leading from the latter to the index equation. It then goes easily. However, we are not concerned with these details here.

A caution is necessary regarding the interchangeability of μ and c. They should be fully operative as linear operators. If one of them be a constant initially, and therefore all through, we may not then interchange them in the simplified equations which result. For example, let μ be constant in (10). We have now

$$\mathbf{r}\frac{\mathbf{r}}{c-\dfrac{1}{\mu\mathbf{r}^2}}=0=\mathbf{r}\,\frac{\mathbf{r}}{\mu-\dfrac{c}{[c](\mathbf{r}c^{-1}\mathbf{r})}}. \tag{11}$$

The first form is what we are naturally led to by initial assumption of constancy of μ. Now observe that the interchange of μ and c in the second form gives us the first form, after a little reduction, remembering that $[\mu]$ is now μ^3. But the same interchange in the first form does not produce the second, because it is more general. So we have gained a relative simplicity of form at the cost of generality. The extra complication of the duplex wave-surface is accompanied by general analytical extensions which make the working operations more powerful. The equivalence of the two forms in (11) may be established by the use of Hamilton's general cubic equation of a linear operator, as done in Tait's work. Though not difficult to carry out, the operations are rather recondite. On the other hand, the much more general equivalence (10) is, as we saw for the reason following (7) and (8), obviously true. This suggests that some other transformations involving the general cubic may be made plainer by generalising it, employing a pair of linear operators.

General Transformation of Wave-Surface by Homogeneous Strain.

5. Now apply a homogeneous strain to the wave-surface. Let

$$\mathbf{q} = \frac{\phi}{[\phi]}\mathbf{r}. \qquad (12)$$

We need not suppose that the strain is pure. Use (12) in the first of (10). It becomes

$$\phi^{-1}\mathbf{q} \frac{\phi^{-1}\mathbf{q}}{c - \dfrac{\mu}{[\mu]\,[\phi]^2(\phi^{-1}\mathbf{q}\mu^{-1}\phi^{-1}\mathbf{q})}} = 0. \qquad (13)$$

Now the use of vectors and linear operators produces such a concise exhibition of the essentially significant properties, freed from the artificial elaboration of coordinates, that a practised worker may readily see his way to the following results by mere inspection of equation (13), or with little more. I give, however, much of the detailed work that would then be done silently, believing that the spread of vector analysis is not encouraged by the quaternionist's practice of leaving out too many of the steps.

In the first place, $\phi^{-1}\mathbf{q}$ is the same as $\mathbf{q}\phi'^{-1}$, if ϕ' is the conjugate of ϕ. So

$$\phi^{-1}\mathbf{q}\mu^{-1}\phi^{-1}\mathbf{q} = \mathbf{q}\phi'^{-1}\mu^{-1}\phi^{-1}\mathbf{q} \qquad (14)$$

in the denominator. Also, the first $\phi^{-1}\mathbf{q}$ in (13) may be written $\mathbf{q}\phi'^{-1}$, and the postfactor ϕ'^{-1} may then be transferred to the denominator. To do this, it must be inverted, of course, and then brought in as a postfactor. Similarly the ϕ^{-1} in the numerator may be merged in the denominator by inversion first, and then bringing it in as a prefactor. We may see why this is to be done by the elementary formula

$$a^{-1}b^{-1}c^{-1} = (cba)^{-1}, \tag{15}$$

where a, b, c are any linear operators. So (13) becomes

$$\mathbf{q}\frac{\mathbf{q}}{\phi c\phi' - \dfrac{\phi\mu\phi'}{[\mu][\phi]^2(\mathbf{q}\phi'^{-1}\mu^{-1}\phi^{-1}\mathbf{q})}} = 0. \tag{16}$$

Now introduce some simplifications of form. Let

$$\phi c\phi' = b, \qquad \phi\mu\phi' = \lambda. \tag{17}$$

It follows from the second, and by (15), that

$$\phi'^{-1}\mu^{-1}\phi^{-1} = (\phi\mu\phi')^{-1} = \lambda^{-1}. \tag{18}$$

We also have
$$[\lambda] = [\mu]\,[\phi]^2. \tag{19}$$

These three, (17) to (19), reduce (16) to

$$\mathbf{q}\frac{\mathbf{q}}{b - \dfrac{\lambda}{[\lambda]\,(\mathbf{q}\lambda^{-1}\mathbf{q})}} = 0 = \mathbf{q}\frac{\mathbf{q}}{\lambda - \dfrac{b}{[b]\,(\mathbf{q}b^{-1}\mathbf{q})}}, \tag{20}$$

where the second form is got from the first by interchanging λ and b, which is permissible on account of the interchangeability of μ and c.

Comparing (20) with (10), we see that there is identity of form. Consequently (20) represents a duplex wave-surface whose operators are b and λ, provided they are self-conjugate. They are, for, by the elementary formula

$$(abc)' = c'b'a', \tag{21}$$

it follows that

$$\phi c\phi' = (\phi c\phi')', \tag{22}$$

and similarly for the other one.

In case the strain is a pure rotation, we may take the form of ϕ (following Gibbs) as

$$\phi = \mathbf{I}.\mathbf{i} + \mathbf{J}.\mathbf{j} + \mathbf{K}.\mathbf{k}, \tag{23}$$

where **i**, **j**, **k** is one, and **I**, **J**, **K** another set of coperpendicular unit vectors. For, obviously, this makes

$$\phi\mathbf{r} = \mathbf{I}.\mathbf{ir} + \mathbf{J}.\mathbf{jr} + \mathbf{K}.\mathbf{kr} = \mathbf{I}x + \mathbf{J}y + \mathbf{K}z. \tag{24}$$

Special Cases of Reduction to a Simplex Wave-Surface.

6. Now take some special forms of ϕ. We see, by inspection of (17), that we can reduce either of b or λ to a constant. Thus, first,

$$\phi = \mu^{-\frac{1}{2}}, \qquad \lambda = 1, \qquad b = \mu^{-\frac{1}{2}}c\mu^{-\frac{1}{2}}. \tag{25}$$

Then (20) reduces to

$$\frac{\mathbf{q}}{b - \dfrac{1}{\mathbf{q}^2}} = 0 = \mathbf{q}\frac{\mathbf{q}}{1 - \dfrac{b}{[b]\,(\mathbf{q}b^{-1}\mathbf{q})}}, \tag{26}$$

showing that the original duplex wave-surface is reduced to a simplex one involving eolotropy b, given by (25).

Similarly, a second way is

$$\phi = c^{-\frac{1}{2}} \qquad b = 1, \qquad \lambda = c^{-\frac{1}{2}}\mu c^{-\frac{1}{2}}, \tag{27}$$

which reduces (20) to the simplex wave-surface

$$\mathbf{q}\frac{\mathbf{q}}{1 - \dfrac{\lambda}{[\lambda](\mathbf{q}\lambda^{-1}\mathbf{q})}} = 0 = \mathbf{q}\frac{\mathbf{q}}{\lambda - \dfrac{1}{\mathbf{q}^2}}, \tag{28}$$

involving the eolotropy λ.

The new surfaces (26), (28) may now be strained to their reciprocals. Thus, take the first of (26), and put

$$\mathbf{p} = \frac{b^{-\frac{1}{2}}}{[b^{-\frac{1}{2}}]}\mathbf{q}. \tag{29}$$

This makes

$$b^{\frac{1}{2}}\mathbf{p}\frac{b^{\frac{1}{2}}\mathbf{p}}{b - \dfrac{[b^{\frac{1}{2}}]^2}{(b^{\frac{1}{2}}\mathbf{p})^2}} = 0. \tag{30}$$

Here the initial and final $b^{\frac{1}{2}}$'s may be removed to the denominator, and, since we also have

$$(b^{\frac{1}{2}}\mathbf{p})^2 = b^{\frac{1}{2}}\mathbf{p}b^{\frac{1}{2}}\mathbf{p} = \mathbf{p}b\mathbf{p}, \tag{31}$$

we bring the first of (26) to

$$\frac{\mathbf{p}}{1 - \dfrac{b^{-1}}{[b^{-1}]\,(\mathbf{p}b\mathbf{p})}} = 0. \tag{32}$$

Now compare this with the second form of the same (26) They are identical, except that b is now inverted. Consequently (32) represents the index-surface corresponding to the wave-surface represented by the second of (26), and therefore by the first, since they are the same. In a similar manner the strain (29) applied to the second of (26) leads to the reciprocal of the first form.

In like manner the simplex surface (28) is strained to its reciprocal by

$$\mathbf{p} = \frac{\lambda^{-\frac{1}{2}}}{[\lambda^{-\frac{1}{2}}]}\mathbf{q}. \tag{33}$$

Applied to the first form of (28), we get the second form with λ inverted; and, applied to the second form, we get the first, with λ inverted. These inversions of simplex wave-surfaces by homogeneous strain are equivalent to Plücker's theorem showing that the Fresnel wave-surface is its own reciprocal with respect to a certain ellipsoid (Tait, "Quaternions," 3rd Ed., p. 342).

Transformation from Duplex Wave- to Index-Surface by a Pure Strain.

7. What is of greater interest here is the generalisation of this property for the duplex wave-surface itself. Take

$$\phi = c^{-1}(c\mu^{-1})^{\frac{1}{2}}. \tag{34}$$

Then we obtain

$$\phi c \phi = c^{-1}(c\mu^{-1})^{\frac{1}{2}} c c^{-1} (c\mu^{-1})^{\frac{1}{2}} = \mu^{-1}, \tag{35}$$

$$\phi \mu \phi = c^{-1}(c\mu^{-1})^{\frac{1}{2}} \mu c^{-1} (c\mu^{-1})^{\frac{1}{2}} = c^{-1}, \tag{36}$$

the first of which is obvious, whilst in the second we make use of

$$\mu c^{-1} = (c\mu^{-1})^{-1}. \tag{37}$$

There are other ways in which this ϕ may be expressed, viz.,

$$\phi = c^{-1}(c\mu^{-1})^{\frac{1}{2}} = \mu^{-1}(\mu c^{-1})^{\frac{1}{2}} = (\mu^{-1}c)^{\frac{1}{2}} c^{-1} = (c^{-1}\mu)^{\frac{1}{2}}\mu^{-1}, \tag{38}$$

all of which lead to $$\mu\phi c\phi = 1. \tag{39}$$

If this ϕ is self-conjugate, we see, by (17) and (35), that its use in (20) brings us to

$$\mathbf{q}\frac{\mathbf{q}}{\mu^{-1} - \dfrac{c^{-1}}{[c^{-1}](\mathbf{q}c\mathbf{q})}} = 0 = \mathbf{q}\frac{\mathbf{q}}{c^{-1} - \dfrac{\mu^{-1}}{[\mu^{-1}](\mathbf{q}\mu\mathbf{q})}}. \tag{40}$$

That is, the strain converts the first of (10) to the first of (40), and the second of (10) to the second of (40). But the first of (40) is the same as the second of (10) with μ and c inverted, and the second of (40) is the same as the first of (10) with the same inversions. In other words, the strain has converted the duplex wave-surface to its corresponding index-surface. Observe that the crossing over from first to second form is an essential part of the demonstration, which is the reason I have employed two forms.

In full, the strainer to be applied to **r** of the wave-surface to produce the vector **s** of the index-surface (or **q** in (40)) is

$$\frac{\phi}{[\phi]} = [c^{\frac{1}{2}}]\,[\mu^{\frac{1}{2}}]c^{-1}(c\mu^{-1})^{\frac{1}{2}}. \tag{41}$$

But to complete the demonstration it should be shown that this strain is pure, because we have just assumed $\phi = \phi'$ in equation (20) to obtain (40). Now the purity of this strain is not obvious in the form (41), nor in any of the similar forms in (38). But we may change the expression for ϕ to such a form as will explicitly show its purity. Thus, we have

$$c\mu^{-1} = c^{\frac{1}{2}}\,.\,c^{\frac{1}{2}}\mu^{-1}c^{\frac{1}{2}}\,.\,c^{-\frac{1}{2}},$$

identically, and this may be expanded to

$$c\mu^{-1} = c^{\frac{1}{2}}\,(c^{\frac{1}{2}}\mu^{-1}c^{\frac{1}{2}})^{\frac{1}{2}}c^{-\frac{1}{2}}c^{\frac{1}{2}}\,(c^{\frac{1}{2}}\mu^{-1}c^{\frac{1}{2}})^{\frac{1}{2}}c^{-\frac{1}{2}},$$

the right member reducing to the left by obvious cancellations. Therefore

$$(c\mu^{-1})^{\frac{1}{2}} = c^{\frac{1}{2}}(c^{\frac{1}{2}}\mu^{-1}c^{\frac{1}{2}})^{\frac{1}{2}}c^{-\frac{1}{2}},$$

by taking the square root. So, finally,

$$\phi = c^{-1}\,(c\mu^{-1})^{\frac{1}{2}} = c^{-\frac{1}{2}}\,(c^{\frac{1}{2}}\mu^{-1}c^{\frac{1}{2}})^{\frac{1}{2}}c^{-\frac{1}{2}}. \tag{42}$$

This is of the form $\phi_1\phi_2\phi_1$, where ϕ_1 is pure. Its conjugate is therefore $\phi_1\phi'_2\phi_1$. This reduces to ϕ itself if ϕ_2 is pure. But ϕ_2 is pure, because it is also of the form $\theta_1\theta_2\theta_1$, where θ_1 and θ_2 are both pure. So our single strain depending on ϕ is pure.

Substitution of three successive Pure Strains for one. Two ways.

8. This is dry mathematics. But it is at once endowed with interest if we consider the meaning of the expression of the strain ϕ as equivalent to the three successive strains ϕ_1, ϕ_2, and ϕ_1. First, the strain

$$\mathbf{q} = \frac{\phi_1}{[\phi_1]}\,\mathbf{r} = \frac{c^{-\frac{1}{2}}}{[c^{-\frac{1}{2}}]}\,\mathbf{r} \tag{43}$$

converts the duplex wave-surface to a simplex surface. This was done before, equation (28). Next, the strain

$$\mathbf{p} = \frac{\phi_2}{[\phi_2]}\,\mathbf{q} = \frac{(c^{\frac{1}{2}}\mu^{-1}c^{\frac{1}{2}})^{\frac{1}{2}}}{[c^{\frac{1}{2}}]\,[\mu^{-\frac{1}{2}}]}\mathbf{q} \tag{44}$$

converts the simplex surface $\mathbf{q}$ to another simplex surface whose vector is $\mathbf{p}$, and which is the index-surface corresponding to the wave-surface $\mathbf{q}$. This strain (44) is, in fact, the same as (33), and the result is

$$\mathbf{p}\frac{\mathbf{p}}{1 - \dfrac{\lambda^{-1}}{[\lambda^{-1}]\,(\mathbf{p}\lambda\mathbf{p})}} = 0 = \mathbf{p}\frac{\mathbf{p}}{\lambda^{-1} - \dfrac{1}{\mathbf{p}^2}}, \tag{45}$$

where $\lambda = c^{-\frac{1}{2}}\mu c^{-\frac{1}{2}}$. Finally, the strain

$$\mathbf{s} = \frac{\phi_1}{[\phi_1]}\mathbf{p} = \frac{c^{-\frac{1}{2}}}{[c^{-\frac{1}{2}}]}\mathbf{p} \tag{46}$$

converts the simplex surface $\mathbf{p}$ to a duplex surface $\mathbf{s}$, which is the reciprocal of the original duplex wave-surface, the result being (40).

The interchangeability of μ and c shows that we may also strain from $\mathbf{r}$ to $\mathbf{s}$ by a second set of three successive pure strains, thus,

$$\phi = \mu^{-\frac{1}{2}}(\mu^{\frac{1}{2}}c^{-1}\mu^{\frac{1}{2}})^{\frac{1}{2}}\mu^{-\frac{1}{2}}. \tag{47}$$

This is the same as first straining the surface $\mathbf{r}$ to the simplex surface (26); then inverting the latter, which brings us to the simplex surface (32); and finally straining the last to the duplex surface $\mathbf{s}$.

Transformation of Characteristic Equation by Strain.

9. In connection with the above transformations it may be worth while to show how they work out when applied to the characteristic equation itself of $\mathbf{E}$ or $\mathbf{H}$. Thus, take the form (7), or

$$-c\ddot{\mathbf{E}} = \mathrm{V}\nabla\mu^{-1}\mathrm{V}\nabla\mathbf{E}, \tag{48}$$

and let

$$\mathbf{r} = f\mathbf{r}', \qquad \nabla = f^{-1}\nabla', \qquad \mathbf{E} = f'^{-1}\mathbf{E}', \tag{49}$$

so that (48) becomes

$$-cf^{-1}\ddot{\mathbf{E}}' = \mathrm{V}f^{-1}\nabla'\mu^{-1}\mathrm{V}f^{-1}\nabla' f'^{-1}\mathbf{E}'. \tag{50}$$

Now employ Hamilton's formula

$$\mathrm{V}\mathbf{mn} = \frac{\phi\mathrm{V}\phi\mathbf{m}\phi\mathbf{n}}{[\phi]}, \tag{51}$$

ϕ being here any self-conjugate operator. Take $\phi = f^{-1}$, and we transform (50) to

$$-cf^{-1}\ddot{\mathbf{E}}' = \mathrm{V}f^{-1}\nabla'\mu^{-1}f\mathrm{V}\nabla'\mathbf{E}' \times [f^{-1}] \tag{52}$$

$$= \mathrm{V}f^{-}\nabla^{1\prime}f^{-1}(f\mu^{-1}f)\mathrm{V}\nabla'\mathbf{E}' \times [f^{-1}]. \tag{53}$$

In this use Hamilton's formula again, with $\phi = f^{-1}$, and we obtain

$$= f\mathrm{V}\nabla'(f\mu^{-1}f)\mathrm{V}\nabla'\mathbf{E}' \times [f^{-1}]^2. \tag{54}$$

Or, more conveniently written,

$$-\frac{(f^{-1}cf^{-1})}{[f^{-1}]}\ddot{\mathbf{E}}' = \mathrm{V}\nabla'\frac{(f\mu^{-1}f)}{[f]}\mathrm{V}\nabla'\mathbf{E}'. \tag{55}$$

So far, f is any pure strainer; we can now make various specialisations. For example, to get rid of μ^{-1} from the right side of (48), and substitute c. Take

$$\frac{f\mu^{-1}f}{[f]} = c, \quad \text{then} \quad \frac{f^{-1}cf^{-1}}{[f^{-1}]} = \mu^{-1}, \tag{56}$$

which brings (55) to the form

$$-\mu^{-1}\ddot{\mathbf{E}}' = \mathrm{V}\nabla' c\mathrm{V}\nabla'\mathbf{E}', \tag{57}$$

which should be compared with the other characteristic, that of **H**, which is (8), or

$$-\mu\ddot{\mathbf{H}} = \mathrm{V}\nabla c^{-1}\mathrm{V}\nabla\mathbf{H}. \tag{58}$$

The above process is analogous to our transformation from the duplex wave-surface to its reciprocal. As then, we have an inversion of operators and also a crossing over from one form to another.

Derivation of Index Equation from Characteristic.

10. We may also, in conclusion, exhibit how the index-surface arises from the characteristic, when done in terms of ∇ up to the last moment. Start from the last equation (58). Hamilton's formula (51) makes it become

$$-[c]\mu\ddot{\mathbf{H}} = \mathrm{V}\nabla\mathrm{V}c\nabla c\mathbf{H}. \tag{59}$$

The elementary formula in vector algebra,

$$\mathrm{V}\mathbf{a}\mathrm{V}\mathbf{bc} = \mathbf{b}(\mathbf{ca}) - \mathbf{c}(\mathbf{ab}), \tag{60}$$

transforms (59) to

$$-[c]\mu\ddot{\mathbf{H}} = c\nabla(\nabla c\mathbf{H}) - (\nabla c\nabla)c\mathbf{H}, \tag{61}$$

or $$\left[(\nabla c\nabla)c - [c]\mu\frac{d^2}{dt^2}\right]\mathbf{H} = c\nabla(\nabla c\mathbf{H}), \tag{62}$$

from which

$$\mu\mathbf{H} = \mu\left[(\nabla c\nabla)c - [c]\mu\frac{d^2}{dt^2}\right]^{-1} c\nabla(\nabla c\mathbf{H}). \tag{63}$$

So far we have merely a changed form of the characteristic But the induction $\mu\mathbf{H}$ is circuital. Therefore, taking the divergence of (63), we obtain

$$0 = \nabla\mu\left[(\nabla c\nabla)c - [c]\mu\frac{d^2}{dt^2}\right]^{-1} c\nabla(\nabla c\mathbf{H}), \tag{64}$$

or, which is the same,

$$0 = \nabla\left[(\nabla c\nabla)\mu^{-1} - [c]c^{-1}\frac{d^2}{dt^2}\right]^{-1} \nabla(\nabla c\mathbf{H}). \tag{65}$$

Here $\nabla c\mathbf{H}$ is the divergence of $c\mathbf{H}$. It is the same as $(c\nabla)\mathbf{H}$.

Now (65) only differs from the velocity equation (for plane waves) in containing ∇ instead of the unit normal $\mathbf{N}$ and d^2/dt^2 instead of v^2, v being the wave-velocity. Thus, let

$$\mathbf{H} = f(z - vt),$$

then we shall have $$v^2\nabla^2\mathbf{H} = \frac{d^2}{dt^2}\mathbf{H},$$

where, however, ∇^2 is specialised, being only ∇_3^2 or d^2/dz^2. We therefore put $v^2\nabla_3^2$ for d^2/dt^2 and $\mathbf{N}\nabla_3$ for ∇ in equation (65), thus making

$$0 = \mathbf{N}\nabla_3\left[(\mathbf{N}\nabla_3 c\mathbf{N}\nabla_3)\mu^{-1} - [c]\,c^{-1}v^2\nabla_3^2\right]^{-1}\mathbf{N}\nabla_3(\mathbf{N}\nabla_3 c\mathbf{H}), \tag{66}$$

We may now cancel out all the ∇_3's except the last, making

$$0 = \mathbf{N}\left[(\mathbf{N}c\mathbf{N})\mu^{-1} - [c]\,c^{-1}v^2\right]^{-1}\mathbf{N}(\mathbf{N}\nabla_3 c\mathbf{H}). \tag{67}$$

Now throw away the operand $\mathbf{N}\nabla_3 c\mathbf{H}$, and we get the velocity equation pure and simple, and the index equation (3) then comes by $\mathbf{s} = \mathbf{N}v^{-1}$.

But, although the above manipulation of the characteristic equation has some analytical interest, the process cannot be always recommended on the score of simplicity. It is, on the contrary, usually easier and simpler to work upon the component equations upon which the characteristic is founded.

APPENDIX G.

NOTE ON THE MOTION OF A CHARGED BODY AT A SPEED EQUAL TO OR GREATER THAN THAT OF LIGHT.

Mr. Searle remarks at the end of his paper* that it would seem to be impossible to make a charged body move at greater speed than that of light. Prof. J. J. Thomson† has made a similar remark in the same connection. Prof. G. F. Fitz-Gerald did so also, long ago, but in a different connection.‡

The argument implied in the cases of Thomson and Searle seems to be that since the calculated energy of a charged body is infinite when in steady rectilinear motion at the speed of light, and since this energy must be derived from an external source, an infinite amount of work must be done, that is, an infinite resistance will be experienced.

There is a fallacy here. One easy way of disproving the argument from infinity is to use not one, but two bodies, one positively and the other negatively charged to the same degree. Then the infinity disappears, and there you are, with finite energy when moving at the speed of light.

But I go much further than that, and assert that a single charged body may be moved at any speed, whether equal to or exceeding that of light, without any infiniteness of the energy or infinite resistance to motion—that, in fact, a charged body may be moved about anyhow. Remember that it is not a question of whether the mechanical construction of the ether will permit it—nothing is known about that—but merely one of electromagnetic laws. If they are valid at any

---

* Physical Society, 1897; *Phil. Mag.*, October, 1897.

† "Recent Researches in Electricity and Magnetism," Chap. I.

‡ Dr. Fleming has lately repeated Prof. J. J. Thomson's conclusion.

speed, then there is nothing to prevent speeds of motion greater than light.

To illustrate. Start with a stationary charged body; then move it anyhow, finally bringing it to steady rectilinear motion at speed u. Wait a little while, and the special distribution of displacement round the instantaneous position of the body will be assumed. But it will not extend to infinity. For beyond the distance from the first position of the body, travelled by light in the time occupied in shifting it to a second position, there is no disturbance of the original distribution of displacement. This holds for all time, so the energy is always finite, even when the speed u is that of light or more, unless there should be infinite collections of energy at a finite distance from the body, and there is no need for that.

Where does the energy come from? The body meets with resistance as its speed is increased. Not only that, but it requires force to maintain its speed steady until the steady distribution of displacement appropriate to the speed is assumed. This force tends to vanish when $u<v$. But it does not when $u=v$. (A single charged body is in question now.) There is a pull-back on the body exerted by the deformation of the tubes of force; their pull on the body is greater behind than in front until the regular steady distribution is fully assumed. This cannot be approximated to exactly when $u=v$. But we may get over this difficulty, as before, by means of two charges.

When $u>v$, the pull-back becomes very prominent, and in any case, whether there is one body or two, the constant exertion of force is required to maintain the motion. The displacement for a point charge is a conical sheet behind the charge, together with a supplementary distribution inside the cone. The pulling back is obvious, and the energy is being wasted at a steady rate by the constant growth of the cone at its apex, which is fully accounted for by the activity of the applied moving force. This is as I suspected in 1888* ("El. Papers," Vol. 2, p. 494), and I later corroborated it by

* But, on reference, I see that I described the electric *current* as being towards the charge inside the cone, and away from it on the outside. This should be the electric *displacement*.

mathematical investigation. The solution is described in my "El. Papers," Vol. 2, p. 516. It is not the same formula as when $u<v$. The infinities that are concerned in it are not essential. They arise from its being a *point* charge that is in question. Make it a surface distribution and it will all come plain, the cone (or other surface) behind the charge, and the constant pull-back exerted on the body by the displacement. [Fully worked out examples of this theory would be too lengthy here, and must be postponed.]

The tendency of the displacement to be left behind as the body moves is also the cause of the apparent increase of inertia of the body at slow speeds. We then calculate statically, as it were; in the approximate result it appears that when u varies anyhow (u/v being always very small), the resistance to motion is equivalent to an increase of inertia (J. J. Thomson). And yet there is no apparent pull-back by the electrical tension; but that is because of the erroneous assumption made, that the change of displacement is instantly assumed as the body moves. Allow for finite v, and this case is no exception.

The moral is—don't be afraid of infinity!

[The above arrived at the Physical Society too late for the discussion. The matter in question is quite visionary, superficially considered; but is of considerable theoretical importance.]

[Feb. 26, 1898.—It may also become of some practical importance in connection with "Cathode Rays" and "X-Rays," for J. J. Thomson and others have lately concluded from experiment that immense speeds of the charged particles, comparable with the speed of light, are concerned. If this be fully confirmed, we may well believe that increased voltage will produce speeds exceeding that of light, if they do not exist already, and so bring in the conical theory. Regarding the nature of the waves generated by starting or stopping charges, when the speeds are the same (or about the same) as that of light, *see* vol. 1, §§ 54 to 59.]

APPENDIX H.

NOTE ON ELECTRICAL WAVES IN SEA WATER.*

To find the attenuation suffered by electrical waves in sea water through its conductance, the first thing is to ascertain whether, at the frequency proposed, the conductance is paramount, or the permittance, or whether both must be counted. It is not necessary to investigate the problem for any particular form of circuit from which the waves proceed. The attenuating factor for plane waves, due to Maxwell, is sufficient. If its validity be questioned for circuits in general, then it is enough to take the case of a simply periodic point source in a conducting dielectric (Electrical Papers, Vol. II., p. 422, § 29). The attenuating constant is the same, viz. (equation (199) *loc. cit.*),

$$n_1=\frac{n}{v}\sqrt{\tfrac{1}{2}}\left[\left\{1+\left(\frac{4\pi k}{cn}\right)^2\right\}^{\frac{1}{2}}-1\right]^{\frac{1}{2}},$$

where $n/2\pi$ is the frequency, k the conductivity, c the permittivity, and $v=(\mu c)^{-\frac{1}{2}}$, where μ is the inductivity. The attenuator is then $\epsilon^{-n_1 r}$ at distance r from the source, as in plane waves, disregarding variations due to natural spreading. It is thus proved for any circuit of moderate size compared with the wave length from which simply periodic waves spread. This formula must be used in general, with the best values of k and c procurable. But with long waves it is pretty certain that the conductance is sufficient to make $4\pi k/cn$ large. Say with common salt solution $k=(30^{11})^{-1}$, then

$$\frac{4\pi k}{cn}=\frac{2k\mu v^2}{f},$$

* [Read at *Physical Society*, June 11, 1897; discussion on Mr. C. S. Whitehead's paper "The Effect of Sea Water on Induction Telegraphy." Mr. Whitehead worked out the case of diffusion of waves from a circular simply periodic source in a pure conductor.]

if f is the frequency. This is large unless f is large, whether we assume the specific c/c_0 to have the very large value of 80 or the smaller value effectively concerned with light waves. We then reduce n_1 to

$$n_1 = (2\pi\mu kn)^{\frac{1}{2}} = 2\pi(\mu kf)^{\frac{1}{2}}, \qquad \text{(A)}$$

as in a pure conductor. This is practically true, perhaps, even with Hertzian waves, of which the attenuation has been measured in common salt solution by P. Zeeman. If, then, $k^{-1} = 30^{11}$, and $f = 300$, we get

$$n_1 = \text{about } \frac{1}{5000}.$$

Therefore, 50 metres is the distance in which the attenuation due to conductivity is in the ratio 2·718 to 1, and there is no reason why the conductivity of sea water should interfere if its value is like that assumed above.

These formulæ and results were communicated by me to Prof. Ayrton at the beginning of last year, he having inquired regarding the matter, on behalf of Mr. Evershed, I believe. The doubtful point was the conductivity. I had no data, but took the above k from a paper which had just reached me from Mr. Zeeman. Now Mr. Whitehead uses $k^{-1} = 20^{10}$, which is no less than 15 times as great. I presume there is good authority for this datum. None is given. Using it, we obtain

$$n_1 = \frac{1}{1316}.$$

The 50 metres is reduced to 13·16 metres. But a considerably greater conductivity is required before it can be accepted that the statements which have appeared in the press, that the failure of the experiments endeavouring to establish telegraphic communication with a lightship from the sea bottom was due to the conductance of the sea, are correct. It seems unlikely theoretically, and Mr. Stevenson has contradicted it (in *Nature*) from the practical point of view. So far as I know no account has been published of these experiments, therefore there is no means of finding the cause of the failure.

APPENDIX I.

NOTE ON THE ATTENUATION OF HERTZIAN WAVES ALONG WIRES.*

The connection between the case investigated by Mr. Morton of a wave-train arising from a damped source and the standard case of an undamped source may be concisely exhibited thus. Using the notation of "Electromagnetic Theory," Vol. I., p. 452, we have

$$V = \epsilon^{-qx} V_0, \tag{1}$$

where

$$q = (R + Lp)^{\frac{1}{2}}(K + Sp)^{\frac{1}{2}}, \tag{2}$$

to express the wave-train V due to V_0 at $x = 0$. When V_0 is simply periodic, say $= e \sin nt$, then $p = ni$ reduces q to $P + Qi$, given by

$$\begin{aligned} P \text{ or } Q &= (\tfrac{1}{2})^{\frac{1}{2}}\{(R^2 + L^2n^2)^{\frac{1}{2}}(K^2 + S^2n^2)^{\frac{1}{2}} \pm (RK - LSn^2)\}^{\frac{1}{2}}, \\ &= (\tfrac{1}{2})^{\frac{1}{2}} \frac{n}{v}\left\{ \left(1 + \frac{R^2}{L^2n^2}\right)^{\frac{1}{2}}\left(1 + \frac{K^2}{S^2n^2}\right)^{\frac{1}{2}} \pm \left(\frac{RK}{LSn^2} - 1\right) \right\}^{\frac{1}{2}}, \end{aligned} \tag{3}$$

so that the solution is

$$V = e\,\epsilon^{-Px} \sin (nt - Qx). \tag{4}$$

Now if V_0 be damped, say $= e_0\epsilon^{-at} \sin nt$, the effect of shifting ϵ^{-at} to the left is to change p to $p - a$ in the operator ϵ^{-qx}, that is, in q. This is the same as changing R to R_0 and K to K_0, given by

$$R_0 = R - La, \qquad K_0 = K - La, \tag{5}$$

so that the wave-train is

$$V = e_0 \epsilon^{-at} \epsilon^{-P'x} \sin (nt - Q'x), \tag{6}$$

where P′ and Q′ are the same as P and Q with R_0 and K_0 instead of R and K. Of course R and K are positive, and q is in the first quadrant, but the new R_0 and K_0 may be posi-

* [Read at *Physical Society*, Nov. 11, 1898; discussion on Prof. Morton's paper "The Propagation of Damped Electrical Oscillations Along Parallel Wires." Prof. Morton showed that the fact that the circuit was not distortionless did not alter the results to an appreciable extent.]

tive or negative, and q may be in the second quadrant. [To be in the second quadrant $R_0/Ln + K_0/Sn$ must be negative.]

Practically, under the circumstances of the experiments

$$Q' = \frac{n}{v}, \quad \text{and} \quad P' = \frac{1}{2v}\left(\frac{R}{L} + \frac{K}{S}\right) - \frac{a}{v}, \tag{7}$$

where K is negligible, and

$$V = e_0 \epsilon^{-at} \epsilon^{ax/v} \epsilon^{-Rx/2Lv} \sin (nt - nx/v). \tag{8}$$

As regards the cause of the attenuating coefficient $R/2Lv$ coming out by Dr. Barton's calculations* from his experiments twice as great as when R is calculated by Lord Rayleigh's formula, I think it must be because the real circumstances do not correspond closely enough to those in the ideal theory. The external resistance of unknown amount is ignored, for one thing. Then, again, it is not to be certainly expected that the formula in question is true for millions of vibrations per second. We can conclude from the experiments, though, that it furnishes an approximation to the real resistance. But, even if it were rigorously true, the circumstances implied in it are not those in the experiments. The magnetic vibrations to which the wires are subjected are not long-continued and undamped, as assumed in the formula. When a wave-train passes any point on a wire, its surface is subjected to an impulsive vibration lasting only a very minute fraction of a second, a vibration, moreover, which is very rapidly damped. So there is no definite resistance, and the resistance is greater than according to Lord Rayleigh's formula.

Perhaps, also, the terminal reflections involved in Dr. Barton's calculations may introduce error.

Nov. 8, 1898.

Addition,† January 20, 1899.—Dr. Barton has calculated the change made in the formula for the resistance of the wires in his experiment, viz., $R' = (\frac{1}{2}R\mu n)^{\frac{1}{2}}$, R being the steady resistance, and μ the inductivity, by supposing the impressed

* [*See* Dr. Barton's paper "Attenuation of Electric Waves Along a Line of Negligible Leakage," *Physical Society*, June 10, 1898.]

† [Expansion of a short note read at *Physical Society*, Jan. 27, 1899; discussion on Dr. Barton's paper, "The Equivalent Resistance and Inductance of a Wire to an Oscillatory Discharge."]

vibrations to be damped, and finds that if R″ is the corrected value, then R″ = 1·054 R′. That is, there is only $5\frac{1}{2}$ per cent. increase. This R″ is to be used in the attenuator $\epsilon^{-Rx/2Lv}$ instead of R. It is, however, an under-estimate of the effect of the resistance, because the formula is not valid right up to the wave-front. The resistance is greater there, and decreases as the wave-train advances, tending towards R″ as the waves attenuate by the ϵ^{-at} damping. But it does not seem likely that the total attenuative effect of the variable resistance of the wires will be enough; and other causes, some of which have been suggested, are operative. In addition, it is not impossible that the conductivity of copper is less to vibrations 35 millions per second than to steady currents; and that the voltage at the wave-front is great enough to cause some leakage.

The following, down to (16), is equivalent to Dr. Barton's investigation brought into harmony with the above and simplified. The resistance operator of a straight round wire of radius b, inductivity μ, and steady resistance R per unit length, is

$$Z = R\tfrac{1}{2}hb\frac{I_0(hb)}{I_1(hb)}, \tag{9}$$

where $h = \sqrt{4\pi\mu kp}$, and k is the conductivity. It is such that in the case of a pair of parallel wires in which the current is C, and the tangential electric force at the boundary is E, the equation connecting them is

$$E = (L_0 p + 2Z)C, \tag{10}$$

where L_0 is the inductance of the dielectric (Elec. Pa., Vol. II., p. 63, or p 187). When $E = e_0\epsilon^{-at}\sin nt$, and C varies similarly, the potence of p or d/dt is $n(i-\kappa)$, if $\kappa = a/n$, and ni is understood not to be the complete time differentiator, but only as regards the simply periodic part of operands.

Now, when $n(i-\kappa)$ is big, the divergent series for I_0 and I_1 may be used, and $I_0 = I_1$. Then

$$\begin{aligned} Z &= R\tfrac{1}{2}b\sqrt{4\pi\mu kn(i-\kappa)} \\ &= (\tfrac{1}{2}R\mu n)^{\frac{1}{2}}(f+gi) \end{aligned} \tag{11}$$

where $$f \text{ or } g = \{(1+\kappa^2)^{\frac{1}{2}} \mp \kappa\}^{\frac{1}{2}}.$$

Here f and g are 1 when $\kappa = 0$, and when κ is big tend towards $f = (2\kappa)^{-\frac{1}{2}}$, $g = (2\kappa)^{\frac{1}{2}}$. So

$$Z = R'(f + gi), \tag{12}$$

where $$R' = L'n = (\tfrac{1}{2}R\mu n)^{\frac{1}{2}},$$

and $$R'' = (f + \kappa g)R', \quad L''n = R'g, \tag{13}$$

where R″ and L″ take the place of R′ and L′, which obtain when $a = 0$. This is for one wire. Passing to the complete pair with dielectric between, we have

$$E = (R'' - L''a + L''ni)C = (R'' + L''p)C \tag{14}$$

where $$R'' = 2(f + \kappa g)(\tfrac{1}{2}R\mu n)^{\frac{1}{2}} \tag{15}$$

$$L'' = L_0 + 2(\tfrac{1}{2}\mu R/n)^{\frac{1}{2}}g. \tag{16}$$

If $n/2\pi = 350^6$, and $\kappa =$ about 0·09545,

then $f = 0{\cdot}953$, $g = 1{\cdot}049$, $f + \kappa g = 1{\cdot}054$,

and the resultant effect is

$$\frac{R'}{R} = 31{\cdot}6, \quad \frac{R''}{R'} = 1{\cdot}054.$$

These express the multiplication of the resistance to periodic currents of constant amplitude, and the further multiplication when the currents subside according to ϵ^{-at}.

Both R″ and L″ go up to ∞ with infinite increase of κ. But practically there is little increase in the resistance under the circumstances supposed, and practically none in the inductance, which is sensibly L_0.

As regards the meaning of R″ and L″, they differ from R′ and L′. If $V = ZC$ is the equation of voltage of a combination, and we reduce Z to the form $R'' + L''p$, on the understanding that V and C are of the type $\epsilon^{-at} \sin nt$, we have the equation of activity

$$VC = R''C^2 + \frac{d}{dt}\tfrac{1}{2}L''C^2, \tag{17}$$

and also $$VC = Q + \dot{T}, \tag{18}$$

where Q is the waste and T the magnetic energy. But only when $a = 0$ can we say that the mean Q equals the mean $R''C^2$, and the mean T equals the mean $\frac{1}{2}L''C^2$. When a is not zero, the corresponding property is that the mean $Q \times \epsilon^{2at}$

equals the mean $R''C^2 \times \epsilon^{2at}$, and the mean $T \times \epsilon^{2at}$ equals the mean $\frac{1}{2}L''C^2 \times \epsilon^{2at}$.

Since the conduction is superficial, the penetration may be represented by a plane wave. Thus

$$E_y = \epsilon^{-hy}E \tag{19}$$

expresses E_y at distance y from the surface in terms of E at the surface. Here $h = s(f + gi)$, if $s = (2\pi\mu kn)^{\frac{1}{2}}$; so

$$E_y = e_0\epsilon^{-at}\epsilon^{-sfy} \sin(nt - sgy) \tag{20}$$

shows the penetration.

Printed in the United States
By Bookmasters